T0336062

ADVANCES IN PLASMA ASTROPHYSICS

IAU SYMPOSIUM No. 274

COVER ILLUSTRATION: VIEW OF GIARDINI NAXOS, ITALY

INTERNATIONAL ASTRONOMICAL UNION

UNION ASTRONOMIQUE INTERNATIONALE

ADVANCES IN PLASMA ASTROPHYSICS

PROCEEDINGS OF THE 274th SYMPOSIUM OF THE
INTERNATIONAL ASTRONOMICAL UNION
HELD IN GIARDINI NAXOS, ITALY
SEPTEMBER 6–10, 2010

Edited by

ALFIO BONANNO
INAF-Osservatorio Astrofisico di Catania, Italy

ELISABETE DE GOUVEIA DAL PINO
Instituto Astronômico Geofísico - Universidade de São Paulo, Brasil

and

ALEXANDER G. KOSOVICHEV
Stanford University, USA

CAMBRIDGE
UNIVERSITY PRESS

Shaftesbury Road, Cambridge CB2 8EA, United Kingdom

One Liberty Plaza, 20th Floor, New York, NY 10006, USA

477 Williamstown Road, Port Melbourne, VIC 3207, Australia

314–321, 3rd Floor, Plot 3, Splendor Forum, Jasola District Centre, New Delhi – 110025, India

103 Penang Road, #05–06/07, Visioncrest Commercial, Singapore 238467

Cambridge University Press is part of Cambridge University Press & Assessment, a department of the University of Cambridge.

We share the University's mission to contribute to society through the pursuit of education, learning and research at the highest international levels of excellence.

www.cambridge.org
Information on this title: www.cambridge.org/9780521197410

First published 2011

A catalogue record for this publication is available from the British Library

ISBN   978-0-521-19741-0   Hardback

Table of Contents

Plasma astrophysics in laboratory

Interstellar, space and planetary plasmas

Solar and stellar plasma

Plasma around compact objects

Observational and modelling programs for plasma astrophysics

Plasmas in galaxies and galaxy cluster

Plasma astrophysics in numerical simulations

Contents

Preface

The organization of this Symposium was first motivated by the fact that nowadays connecting astrophysical theory, observations, simulations and laboratory astrophysics is widely appreciated by the scientific community. In this respect this symposium was an important occasion to discuss recent observational, theoretical and experimental efforts in understanding the basic plasma processes in the Universe, with broad synergies among many areas of astrophysics, including the origin and dynamics of magnetic fields in astrophysical systems (the dynamo problem), the origin of x-ray emitting coronas and the role of magnetic reconnection, acceleration of charged particles, winds and jets from highly-evolved stars and supernova remnants, plasma radiation processes, turbulence of the magnetized plasma in astrophysical objects and in the interstellar and intergalactic media and the solar wind, quantum plasmas under extreme conditions in planetary interiors and in exotic stars, and other key problems in modern plasma astrophysics.

The most important goal of the symposium was therefore to bring together experts from plasma physics, MHD, laboratory experiments and numerical simulation communities. In fact, plasma astrophysicists have always been a fairly small group, often distinct from the main astrophysical community, holding their own workshops and special sessions at plasma physics conferences. Despite the identification of a rich class of physical problems of mutual interest, the plasma physics and astrophysics communities remain, for the most part, quite detached, with different societies and memberships, conferences and journals. This Symposium contributed to promote links and cooperation between these communities, to discuss the recent advances in understanding the fundamental plasma physics processes and their application to interpretation and understanding of phenomena observed in astrophysical plasmas at various scales. Despite the wide range of temporal and spatial scales and conditions the basic physics of these phenomena is often very similar. Therefore, it was a unique occasion to discuss these issues together.

Undoubtedly, such discussions and exchange of ideas from different fields have led to a better understanding of the basic mechanisms of many observational phenomena, their origin, structure and dynamics, and will guide future astrophysical observing programs, as well as theoretical and numerical modeling and laboratory experiments in plasma astrophysics. Such interdisciplinary and cross-discipline discussions become increasingly important as they provide a special opportunity to get a broader view of the field and new ideas about methodologies and approaches. This aspect is particularly crucial for younger researchers because the learning curves in various sub-disciplines become steeper and steeper. For this reason during the Symposyum, in addition to traditional review and contributed talks covering outstanding observational and theoretical problems of astrophysical plasmas, considerable time was devoted to exciting discussions at the end of each day session.

We would like to dedicate this meeting to Stirling Colgate, Gerhard Haerendel, Jumber Lominadze, Don Melrose, and Lucio Paternò, who made outstanding contributions to the field of plasma astrophysics.

It is also a great pleasure to acknowledge the financial support of our sponsors listed on page *xvi* of these Proceedings and the active support of the members of the LOC for performing so efficiently and enthusiastically the numerous tasks always associated with such a big meeting. In particular, our sincere thanks go to his competent and patient approach of Christian Napoli who helped the participants solving technical/computer problems, Gabriella Caniglia, Fatima Rubio da Costa and Enrico Corsaro who took care of the logistics of participants, Paolo Romano for his editorial work, to Elisabetta Palumbo, Luigia Santagati, Corrado Trigilio and Grazia Umana for assisting the participants in their numeous needs.

We also aknowledge the professional contribution made by Rainer Arlt who took the photos we published in this volume.

Finally, a special mention must be given to Daniela Recupero, whose professional skills and human gifts have been essential ingredients for the success of this meeting.

Unfortunately, a sad news arrived while we were finishing the editing of this volume, which we cannot help referring. Ilkka Tuominen, a close friend, a brilliant scientist and a mentor for many who attended the meeting, passed away in March 2011, leaving us astonished but at the same time grateful for the great heritage of human and scientific talents he left us. He attended this meeting with his usual enthusiasm and curiosity, providing the LOC with moral support and nice jokes.

We will all miss him. Ciao Ilkka.

Alfio Bonanno, Alexander Kosovichev and Elisabete de Gouveia Dal Pino, editors and co-chairs SOC

THE ORGANIZING COMMITTEE

Scientific

G.Belvedere (Italy)
A. Bonanno (co-chair Italy)
A. Brandenburg (Sweden)
E. de Gouveia Dal Pino (co-chair Brasil)
M. Goossens (Belgium)
G.Haerendel (Germany)
H. Ji (USA)
A. Kosovichev (co-chair USA)

K. Otmianowska-Mazur (Poland)
R. Rosner (USA)
M. Shats (Australia)
K. Shibata (Japan)
L. Vlahos (Greece)
D. Wu (China)
L. Zeleny (Russia)

Local

G. Belvedere
A. Bonanno (chair)
G. Caniglia
S. Gammino
M.E. Palumbo
D. Recupero

P. Romano
L. Santagati
C. Trigilio
S. Tudisco
G. Umana

Acknowledgements

The symposium is sponsored and supported by the IAU Divisions IV (Stars), VI (Interstellar Matter), VII (Galactic System), VIII (Galaxies) and XI (Space and High Energy Astrophysics); and by the IAU Commissions No. 26 (Binary and Multiple Stars), No. 28 (Galaxies), No. 29 (Stellar Spectra), No. 34 (Interstellar Matter), No. 35 (Stellar Constitution), No. 36 (Theory of Stellar Atmospheres), No. 37 (Star Clusters and Associations) and No. 44 (Space and High Energy Astrophysics).

Funding by the
International Astronomical Union,
Istituto Nazionale di Astrofisica,
Dipartimento di Fisica ed Astronomia dell'Universitá degli Studi di Catania,
Istituto nazionale di Fisica Nucleare-Sezione di Catania,
Laboratorio Nazionale del Sud-Sezione di Catania,
European Physical Society,
European Science Foundation

Stirling Colgate

In 2010 Stirling Colgate has turned 85. His career has spanned more than 60 years, starting as a PhD student in physics at Cornell University and working at Lawrence Livermore and New Mexico Institute of Mining and Technology. After the success of Bravo Test in 1950s, the first deliverable thermonuclear bomb, he was encouraged to begin research on thermonuclear fusion and plasma physics. Many of his scientific successes, however, have been realised at the Los Alamos National Laboratory where he arrived in 1976, joining the Theoretical Division.

He is recognized for negotiating the cessation of high-altitude and outer space nuclear tests. Colgate also has inspired the inertial fusion and astrophysics programs at Los Alamos and Lawrence Livermore and contributed basic science to fusion ignition and burn, plasma confinement and shock wave physics. In 2006 he has been awarded the Los Alamos Medal.

Stirling Colgate (right) and Alfio Bonanno

Jumber Georgievich Lominadze

September 20, 2010, was the 80th birthday of Jumber Georgivich Lominadze, one of the leading plasma astrophysicists, founder of the Plasma Astrophysics Center in Georgia, Head of the Center for Space Research, and Academician of the Georgian National Academy of Sciences. Jumber Lominadze was born in 1930 in Tbilisi. After the graduation from Moscow University in 1955 he worked at the Russian (Ural) Nuclear Center. In 1958 he returned to Tbilisi, and actively participated in the development of plasma physics and nuclear fusion research at the Georgian Institute of Physics. His studies were focused on the propagation and absorption of cyclotron waves in plasma, and were published in book 'Cyclotron Waves in Plasma' (Metsnierba, Tbilisi, 1975; Pergamon Press, Oxford, 1981). In 1976 he founded the Plasma Astrophysics Center, which under his leadership became one of the leading world-class research center. He actively developed international collaborations, and organized a series of legendary conferences, workshops, and schools on plasma astrophysics, which play very important role in the development of this field. He developed the electromagnetic theory of electron-positron plasma, which was used to explain mechanisms and properties of Crab pulsar radiation in different bands and other fundamental processes. More recently, he and his colleagues studied the physics of accretion disks, jets, resonance transformation of oscillations, excitation of waves by vortices, dynamical processes in shearing flows and instabilities in rotating plasma. For more than 40 years he has been teaching at Tbilisi University, and supervised the research of more than 20 PhD students. His former students now form a core of the Georgian Plasma Astrophysics school. On behalf of the IAUS 274 participants we sincerely congratulate Professor Jumber Lominadze on his 80th birthday and wish him all the best for the coming years.

Jumber Lominadze with his son Georgi

Gerhard Haerendel

Born in 1935, he graduated in Physics from the university of Munich in 1963. He is considered a pivotal figure in the European exploration of space, having more than 30 year of experience in space research, including the function of P.I. of several international rocket and satellite projects. His pioneering work opened a new view towards understanding of plasma in space and its interaction with the solar wind, small-scale magnetic reconnections events, high-beta plasma blobs in the magnetosphere and the in situ confirmation of reconnection, and fundamental theoretical works on basic plasma processes. He was recently awarded Jean Dominique Cassini Medal.

Gerhard Haerendel and Masaaki Yamada

Donald Melrose

Professor of Theoretical Physics since 1979, Donald Melrose made specific contributions to the theory of plasma emission and its application to solar radio burst, the theory of elector cyclotron maser emission and its application to planetary and the theory of pulsar radio emission. He is recognized as one of the leading experts of kinetic theory of plasmas, plasma instabilities and nonlinear processes with application in various fields of astrophysics.

Don Melrose (left) and Rainer Beck

Lucio Paternò

Professor at University of Catania since late 1960s, apart from a short parenthesis at Catania Astrophysical Observatory, Lucio Paternò is an outstanding figure in the Italian and international scene. His scientific activity encompasses astronomical photoelectric photometry, Solar site testing, Space physics and solar and stellar physics, in particular helioseismology and asteroseismology. He is a member of the French Academy of Sciences as well as member of the Accademia Gioenia of Natural Sciences of Catania.

Enrico Corsaro (left), Lucio Paternò (center), and Christian Napoli (right)

A table at the social dinner. From the right: Jim Drake, Klaus Strassmeier, Lucio Paternò, Alex Lazarian, Bob Rosner, Guenther Ruediger, Ilkka Tuominen, Jim Stone, Andrey Beresnyak

The LOC

Group picture

Participants

Antonello **Anzalone** INFN LSN-Catania, Italy, — anzalone@lns.infn.it
Rainer **Arlt** Astrophysikalisches Institut Potsdam,Germany — rarlt@aip.de
Rainer **Beck** Max-Planck-Institut für Radioastronomie, Germany — rbeck@mpifr-bonn.mpg.de
Gaetano **Belvedere** Dept. of Physics and Astronomy, University of Catania, Italy — gbelvedere@ct.astro.it
Svetlana **Berdyugina** Kiepenheuer Institut für Sonnenphysik, Germany — sveta@kis.uni-freiburg.de
Andrey **Beresnyak** University of Wisconsin-Madison, USA — andrey@astro.wisc.edu
Lapo **Bettarini** Centre for Plasma Astrophysics, Belgium — Lapo.Bettarini@wis.kuleuven.be
Matteo **Bocchi** Imperial College London, UK — m.bocchi@imperial.ac.uk
Alfio **Bonanno** INAF-Osservatorio Astrofisico di Catania, Italy — abonanno@oact.inaf.it
Axel **Brandenburg** NORDITA, Sweden — brandenb@nordita.org 5
Philippa **Browning** Jodrell Bank Centre for Astrophys., Univ. of Manchester, UK — p.browning@manchester.ac.uk
Jörg **Büchner** Max-Planck-Institut für Sonnensystemforschung, Germany — buechner@mps.mpg.de
Carlo **Burigana** INAF-IASF Bologna, Italy — burigana@iasfbo.inaf.it
Blakesley **Burkhart** University of Wisconsin Madison, USA — burkhart@astro.wisc.edu
Simon **Candelaresi** NORDITA, Sweden — iomsn@physto.se
Vincenzo **Capparelli** Dipartimento di Fisica UNICAL, Italy — vincenzocapparelli@hotmail.com
Monica **Cardaci** UAM & FCAGLP, Spain — monica.cardaci@uam.es
Giuseppe **Castro** Laboratorio Nazionale del Sud, Italy — giuseppe.castro@lns.infn.it
Stirling **Colgate** Los Alamos National Laboratory USA — colgate@lanl.gov
Enrico **Corsaro** Universitá di Catania, Italy — eco@oact.inaf.it
Neil **Cramer** University of Sidney, Australia — cramer@physics.usyd.edu.au
Claudio **Cremaschini** SISSA, Italy — cremasch@sissa.it
Serena **Dalena** Universitá della Calabria, Italy — serena.dalena@fis.unical.it
Gustavo Rocha **da Silva** Departamento de Astronomia - IAG/USP, Brasil — gustavords@astro.iag.usp.br
Garcia **De Andrade** University of Rio de Janeiro, Brasil — garciluiz@gmail.com
Elisabete **de Gouveia Dal Pino** Universidade de São Paulo - (IAG. -USP) — dalpino@astro.iag.usp.br
Fabio **Del Sordo** NORDITA, Sweden Sweden — fadiesis@gmail.com
Suzan **Doğan** University of Ege, Turkey — suzan.dogan@mail.ege.edu.tr
Sandro **Donato** UNICAL University of Calabria, Italy — kisspc@libero.it
James **Drake** University of Maryland, USA — drake@umd.edu
Anna **Dubinova** Institute of Applied Physics RAS, Nizhny Novgorod, Russia — anndub@gmail.com
Vincent **Duez** Argelander-Institut für Astronomie Bonn, Germany — vduez@astro.uni-bonn.de
Sergey **Dyadechkin** FMI, Helsinki, Finland — egopost@gmail.com
Natalia **Dzyurkevich** Max-Planck Institute for Astronomy, Germany — natalia@mpia.de
Rasha **Emara** German University in Cairo, Egypt — rasha.emara@guc.edu.eg
Adnan **Erkurt** Istanbul Univer., Depart. of Astr. and Space Sciences, Turkey — adnan.erkurt@ogr.iu.edu.tr
Luigina **Feretti** Inaf-IRA Bologna Italy — feretti@ira.inaf.it
Attilio **Ferrari** Universitá di Torino, Italy — ferrari@ph.unito.it
Markus **Flaig** Inst. for Computational Phys., Univer. of Tübingen, Germany — flaig@tat.physik.uni-tuebingen.de
Gregory **Fleishman** New Jersey Institute of Technology, USA — gfleishm@njit.edu
Kotaro **Fujisawa** The University of Tokyo, Japan — fujisawa@ea.c.u-tokyo.ac.jp 5
Nadia **Gambino** I.N.F.N. Laboratori Nazionali del Sud, Italy Italy — nadiagambino@lns.infn.it
Santo **Gammino** I.N.F.N. Laboratori Nazionali del Sud, Italy Italy — gammino@lns.infn.it
Urs **Ganse** Lehrstuhl für Astronomie, Universität Wuerzburg, Germany — uganse@astro.uni-wuerzburg.de
Ilknur **Gezer** Natural and applied science, Turkey — gezer.ilknur@gmail.com
Janusz **Gil** Kepler Institute of Astronomy, Zielonaga Gora, Poland — jag@astro.ia.uz.zgora.pl
Noémie **Globus** Observatoire de Paris, France — noemie.globus@obspm.fr
Daniel Osvaldo **Gómez** Department of Physics, University of Buenos Aires, Argentina — gomez@iafe.uba.ar
Antonella **Greco** Dipartimento di Fisica - Universitá della Calabria, Italy — greco@fis.unical.it
Oliver **Gressel** Queen Mary, University of London, UK — o.gressel@qmul.ac.uk
Salvatore **Guglielmino** Instituto de Astrofisica de Canarias, Spain — sgu@iac.es
Filippo **Guarnieri** University of Rome La Sapienza, Italy — guarnieri.filippo@gmail.com
Guillermo **Hagele** FCAGLP & UAM, Argentina — guille.hagele@uam.es
Gerhard **Haerendel** Max Planck Institute for Extraterrestrial Physics , Germany — hae@mpe.mpg.de
Michal **Hanasz** Centre for Astr., Nicolaus Copernicus University, Torun, Poland — mhanasz@astri.uni.torun.pl
Troels **Haugbølle** Niels Bohr Institute, Denmark — haugboel@nbi.dk
Mariko **Hirai** University of Tokyo, Japan — hirai@eps.s.u-tokyo.ac.jp
Subhon **Ibadov** Institute of Astrophysics, Tajik Academy of Sciences , Tajikistan — ibadovsu@yandex.ru
Stavro **Ivanovski** Universitá di Catania, Italy — stavro.ivanovski@gmail.com
Axel **Jessner** Max-Planck-Institute for Radio Astronomy, Germany — jessner@mpifr-bonn.mpg.de
Hantao **Ji** Princeton University, USA — hji@pppl.gov
Anders **Johansen** Lund Observatory, Sweden — anders@astro.lu.se
Marian **Karlický** Astronomical Institute, Ondrejov Observatory, Czech Republic — karlicky@asu.cas.cz
Subhash Chandra **Kaushik** School of Studies in Physics, Jiwaji Univ., India — subhash_kaushik@rediffmail.com
Koen **Kemel** NORDITA, Sweden — koen@nordita.org
Rony **Keppens** Centre for Plasma Astrophysics, K.U. Leuven, Belgium — Rony.Keppens@wis.kuleuven.be
Bernhard **Kliem** University of Potsdam, Germany — bkliem@uni-potsdam.de
Vladimir **Kocharovsky** Inst. of Applied Phys., Russian Academy of Scien., Russia — kochar@appl.sci-nnov.ru
Alexander **Kosovichev** Stanford University, USA — sasha@sun.stanford.edu
Manfred **Küker** Astrophysikalisches Institut Potsdam, Germany — mkueker@aip.de
Katarzyna **Kulpa-Dybel** Astronomical Observatory of the Jagiellonian University, Poland — kulpa@oa.uj.edu.pl
Alexey **Kuznetsov** Armagh Observatory, UK — aku@arm.ac.uk
Antonino Francesco **Lanza** INAF-Osservatorio Astrofisico di Catania, Italy — nlanza@oact.inaf.it
Alex **Lazarian** University of Wisconsin-Madison, USA — lazarian@astro.wisc.edu
Marcia Regina **Leão** Departamento de Astronomia - IAG/USP, Brasil — mrmleao@astro.iag.usp.br
Sergey **Lebedev** Imperial College, UK — s.lebedev@imperial.ac.uk
Martin **Lemoine** Institut d'Astrophysique de Paris, France — lemoine@iap.fr
Fabio **Lepreti** Universitá della Calabria, Italy — fabio.lepreti@fis.unical.it
Paolo **Leto** INAF - Osservatorio Astrofisico di Catania, Italy — pleto@oact.inaf.it
Harald **Lesch** University Observatory Munich ,Germany — lesch@usm.uni-muenchen.de
Jumber **Lominadze** Abastumani National Astrophysical Observatory, Georgia — contact@gsa.gov.ge
Richard **Lovelace** Cornell University, USA — lovelace@astro.cornell.edu
Nazzareno **Mandolesi** INAF-IASF, Italy — mandolesi@iasfbo.inaf.it
David **Mascali** INFN & Centro Sicil. di Fis. Nucl. e Strut. della Mat., Italy — davidmascali@lns.infn.it
Jin **Matsumoto** Kyoto University, Japan — jin@kusastro.kyoto-u.ac.jp
William **Matthaeus** University of Delaware, USA — whm@udel.edu
Andrew **McMurry** CMA, University of Oslo, Norway — andrew.mcmurry@astro.uio.no
Giorgi **Melikidze** Kepler Inst. of Astronomy, Univer. of Zielona Gora, Poland — gogi@astro.ia.uz.zgora.pl

Donald **Melrose** University of Sydney, Australia — melrose@physics.usyd.edu.au
Hana **Mészárosová** Astronomical Institute Ondrejov, Czech Republic — hana@asu.cas.cz
Natalia **Minkova** Tomsk State University, Russia Russia — nminkova@zmail.ru
Rosalba **Miracoli** INFN Laboratori Nazionali del Sud, Italy — rosalbamiracoli@lns.infn.it
Nishant **Mittal** Meerut College, India — nishantphysics@yahoo.com
Yosuke **Mizuno** UA Huntsville, USA — mizuno@cspar.uah.edu
Guillaume **Molodij** Observatoire de Meudon LESIA, France — guillaume.molodij@obspm.fr
Francesco **Musumeci** I.N.F.N. Laboratori Nazionali del Sud, Italy — fmusumeci@dmfci.unict.it
Cristian **Napoli** Universitá di Catania, Italy — chnapoli@gmail.com
Jacek **Niemiec** Institute of Nuclear Physics PAS, Poland — Jacek.Niemiec@ifj.edu.pl
Giuseppina **Nigro** Dipartimento di Fisica UNICAL, Italy — giusy.nigro@fis.unical.it
Åke **Nordlund** Niels Bohr Institute, Denmark — aake@nbi.dk
Martin **Obergaulinger** Max-Planck-Institut fuer Astrophysik, Germany — mobergau@mpa-garching.mpg.de
Andrea **Orlando** Catania Astrophysical Observatory, Italy — aorlando@oact.inaf.it
Viktor **Ostrovskiy** Karpov Institute of Physical Chemistry, Russia — kadyshevich@mail.ru
Katarzyna **Otmianowska-Mazur** Astronomical Obser. Jagiellonian Univer. Krakow, Poland — otmian@oa.uj.edu.pl
Lucio **Paternó** Dept. Physics & Astronomy, University of Catania, Italy — lpaterno@oact.inaf.it
Maria Elisabetta **Palumbo** INAF-Osservatorio Astrofisico di Catania, Italy — mepalumbo@oact.inaf.it
David **Pascoe** University of St Andrews, UK — dpascoe@mcs.st-and.ac.uk
Denise **Perrone** Dipartimento di Fisica UNICAL, Italy — denise.perrone@fis.unical.it
Martin **Pessah** Institute for Advanced Study, USA — mpessah@ias.edu
Gabriella **Piccinelli** Centro Tecnológico, FES Aragn, UNAM, Mexico — gabriela@astroscu.unam.mx
Arakel **Petrosyan** Space Research Inst. of the Russian Academy of Scien., Russia — apetrosy@rssi.ru
John **Podesta** Los Alamos National Laboratory, USA — jpodesta@solar.stanford.edu
Jens **Pomoell** University of Helsinki, Finland — jens.pomoell@helsinki.fi
Helen **Popova** Moscow State University, Russia — popovaelp@hotmail.com
Oliver **Porth** MPIA Heidelberg, Germany — porth@mpia.de
Pietro **Procopio** Istituto di Astrofisica Spaziale sez. Bologna, Italy — procopio@iasfbo.inaf.it
Tomasz **Rembiasz** Max Plank Institute for Astrophysics, Garching, Germany — rembiasz@mpa-garching.mpg.de
Maxim **Reshetnyak** Institute of the Physics of the Earth, Russia — m.reshetnyak@gmail.com
Brian **Reville** Max-Planck-Institut fuer Kernphysik,Germany — brian.reville@mpi-hd.mpg.de
Ronan **Rochford** National University of Ireland, Galway Ireland — ronan.rochford@nuigalway.ie
Paolo **Romano** INAF-Osservatorio Astrofisico di Catania, Italy — prom@oact.inaf.it
Marina **Romanova** Cornell University, USA — romanova@astro.cornell.edu
Robert **Rosner** University of Chicago, USA — r-rosner@uchicago.edu
Ilan **Roth** UC Berkeley, Space Sciences, USA — ilan@ssl.berkeley.edu
Fatima **Rubio da Costa** University of Catania, Italy — frdc@oact.inaf.it
Günther **Rüdiger** Astrophysikalisches Institut Potsdam , Germany — gruediger@aip.de
Arto **Sandroos** Finnish Meteorological Institute, Finland — arto.sandroos@fmi.fi
Reinaldo **Santos de Lima** Departamento de Astronomia - IAG/USP, Brasil — rlima@astro.iag.usp.br
Earl **Scime** West Virginia University, USA — escime@wvu.edu
Ildar **Shaikhislamov** Institute of Laser Physics SB RAS, Russia — ildars@ngs.ru
Kazunari **Shibata** Kyoto University, Japan — shibata@kwasan.kyoto-u.ac.jp
Hubert **Siejkowski** Astronomical Observatory of the Jagiellonian University, Poland — h.siejkowski@oa.uj.edu.pl
Mario **Scuderi** Dipartimento di Fisica ed Astronomia & INFN sez. Catania, Italy — mario.scuderi@ct.infn.it
Aimilia **Smyrli** University of Catania, Italy & University of St Andrews, UK — emilia@oact.inaf.it
Felix **Spanier** Lehrstuhl für Astronomie - Uni Wrzburg, Germany — fspanier@astro.uni-wuerzburg.de
Rodion **Stepanov** Institute of Continuous Media Mechanics, Russia — rodion@icmm.ru
James M. **Stone** Princeton University, USA — jmstone@Princeton.EDU
Klaus G. **Strassmeier** Astrophysical Institute Potsdam,Germany — kstrassmeier@aip.de
Giovanni **Strazzulla** INAF-Osservatorio Astrofisico di Catania, Italy — gstrazzulla@oact.inaf.it
Toshiki **Tajima** Ludwig-Maximilians-Universität,Germany — tajima.toshiki@gmail.com
Toshio **Terasawa** Institute for Cosmic Ray Research, Japan — terasawa@icrr.u-tokyo.ac.jp
Maurizio **Ternullo** INAF-Osservatorio Astrofisico di Catania, Italy — mternullo@oact.inaf.it
Corrado **Trigilio** INAF-Osservatorio Astrofisico di Catania, Italy — ctrigilio@oact.inaf.it
Enrico Maria **Trotta** Dipartimento di Fisica UNICAL, Italy — etrotta@thematica.it
Yuriy **Tsap** Crimean Astrophysical Observatory, Ukraine — yur@crao.crimea.ua
Salvatore **Tudisco** INFN-LNS, Italy — tudisco@lns.infn.it
Ilkka **Tuominen** Universiy of Helsinki, Finland — Ilkka.Tuominen@helsinki.fi
Grazia **Umana** INAF-Osservatorio Astrofisico di Catania, Italy — gumana@oact.inaf.it
Marek **Vandas** Astronomical Institute Ondrejov, Czech Republic — vandas@ig.cas.cz
Antonio **Vecchio** Dipartimento di Fisica Universitá della Calabria, Italy — antonio.vecchio@fis.unical.it
Loukas **Vlahos** Aristotle University of Thessaloniki, Greece — vlahos@astro.auth.gr
Miroslava **Vukácević** Military Academy, Belgrade University, Serbia — vuk.mira@gmail.com
Yörn **Warnecke** NORDITA, Sweden — Joern@nordita.org
Christopher **Watts** University of New Mexico, USA — cwatts@ece.unm.edu
Eli **Waxman** Weizmann Institute, Israel — eli.waxman@weizmann.it
Matthias **Weidinger** ITPA University of Wuerzburg,Germany — mweidinger@astro.uni-wuerzburg.de
Maasaki **Yamada** PPPL, Princeton University, USA — myamada@pppl.gov
Huirong **Yan** Kavli Institute of Astronomy and Astrophysics-PKU, China — IAU hryan@pku.edu.cn
Leonid **Yasnov** St.Petersburg State University , Russia — Yasnov@pobox.spbu.ru
Shinichiro **Yoshida** University of Tokyo, Japan — yoshida@ea.c.u-tokyo.ac.jp
Dario **Zappalá** INFN-Sezione di Catania — dario.zappala@ct.infn.it
Valentina **Zharkova** University of Bradford, UK — v.v.zharkova@brad.ac.uk
Francesca **Zuccarello** Department of Physics and Astronomy, Italy — fzu@oact.inaf.it

Address of the Director of Catania Astrophysical Observatory

I am happy to give you my warmest welcome to Giardini-Naxos to attend the meeting on Plasma Astrophysics. Giardini-Naxos is today a known touristic place with a beautiful shore and a very long history. Naxos was the most ancient of all the Greek colonies in Sicily, founded in 735 BC by a body of colonists from Chalcis in Eubea. The coins of Naxos, which are of fine workmanship, may almost all be referred to the period from 460 BC to 403 BC, which was probably the most flourishing in the history of the city. In 403 BC, Dionysius of Syracuse determined to turn his arms against the Chalcidic cities of Sicily. He sold all the inhabitants of Naxos as slaves and destroyed both the walls and buildings of the city.

As known, the Greek culture has been particularly relevant for the development of the Sicilian culture. And, last night thinking about what to say today, I had a dream: a meeting was held in the same place where we are now but in 450 BC. I saw Empedocles (a great Sicilian philosopher/scientist, ca. 490430 BC), a progenitor of the actual organizer. Empedocles philosophy is best known for being the originator of the cosmogenic theory of the four classical elements: air (the gaseous state), water (the liquid state), earth (the solid state) and fire (the fourth state that today we call Plasma). In the dream Empedocles/Alfio organized, exactly here, a meeting on the fire/plasma. I wish(and I am pretty sure that) we all are worthy of the great history that is at the root of this land. Have a great meeting and enjoy your stay in Sicily!

Gianni Strazzulla, Director of Catania Astrophysical Observatory
Giardini-Naxos, September 2010

Plasma Astrophysics in Laboratory

Hantao Ji

Advances in Plasma Astrophysics
Proceedings IAU Symposium No. 274, 2010
A. Bonanno, E. de Gouveia Dal Pino & A. G. Kosovichev, eds.

© International Astronomical Union 2011
doi:10.1017/S1743921311006491

The magnetized universe: its origin and dissipation through acceleration and leakage to the voids

Stirling A. Colgate[1], Hui Li[2], Philipp P. Kronberg[3]

[1]MS B227, Los Alamos Nat. Lab., Los Alamos, N.M., 87545, USA
email: colgate@lanl.gov
[2]MS B227, Los Alamos Nat. Lab., Los Alamos, N.M., 87545, USA
email: hli@lanl.gov
[3]MS T006, Los Alamos Nat. Lab., Los Alamos, N.M., 87545, USA
email: kronberg@lanl.gov

Abstract. *The consistency is awesome between over a dozen observations and the paradigm of radio lobes being immense sources of magnetic energy, flux, and relativistic electrons, – a magnetized universe.*

The greater the total energy of an astrophysical phenomenon, the more restricted are the possible explanations. Magnetic energy is the most challenging because its origin is still considered problematic. We suggest that it is evident that the universe is magnetized because of radio lobes, ultra relativistic electrons, Faraday rotation measures, the polarized emission of extra galactic radio structures, the x-rays from relativistic electrons Comptonized on the CMB, and possibly extra galactic cosmic rays. The implied energies are so large that only the formation of supermassive black hole, (SMBH) at the center of every galaxy is remotely energetic enough to supply this immense energy, $\sim(1/10)10^8 M_\odot c^2$ per galaxy. Only a galaxy cluster of 1000 galaxies has comparable energy, but it is inversely, (to the number of galaxies), rare per galaxy. Yet this energy appears to be shared between magnetic fields and accelerated relativistic particles, both electrons and ions. Only a large-scale coherent dynamo generating poloidal flux within the accretion disk forming the massive black hole makes a reasonable starting point. The subsequent winding of this dynamo-derived magnetic flux by conducting, angular momentum-dominated accreting matter, ($\sim10^{11}$ turns near the event horizon in 10^8 years) produces the immense, coherent magnetic jets or total flux of radio lobes and similarly in star formation. By extending this same physics to supernova-neutron star formation, we predict that similar differential winding of the flux that couples explosion ejecta and a newly formed, rapidly rotating neutron star will produce similar phenomena of a magnetic jet and lobes in the forming supernova nebula. In all cases the conversion of force-free magnetic energy into accelerated ions and electrons is a major challenge.

Keywords. Magnetic fields, MHD, galaxies: jets, acceleration of particles.

---

1. Introduction

This talk is concerned with the magnetic fields in the universe: AGN, radio lobes, jets, star formation jets, galactic fields, stellar dynamos, and galaxy cluster dynamos. We believe that all these phenomena depend upon the same physics and we believe have a physically related explanation.

1) *Observations*: Radio Lobes, magnetized Jets, (synchrotron emission and its polarization), CRs and UHECRs, Faraday rotation, and x-rays.

2) *Energy;* The magnetic energy: in radio lobes from luminosity and time is: $\sim 10^8 M_\odot c^2/10 \sim 10^{61}$ ergs. This energy is derived from mass accretion forming the SMBH. A minimum energy requires that the magnetic energy equals the relativistic electron energy in the lobes.

3) *The necesary large scale poloidal flux, requires a large scale magnetic dynamo in the accretion disk: (Despite good theoretical underpinning, three experiments have failed.)*

We are (still) trying to demonstrate by experiment that two naturally occurring, orthogonal coherent motions, α and Ω will lead to a coherent dynamo that transforms a very small fraction, $\sim 10^{-11}$, of the free energy of accretion within an AGN disk, into poloidal magnetic energy within the disk. In addition and comparably important a small, $\sim 10^{-2}\alpha$-viscosity that mediates a fraction of the angular momentum transport within the disk, presumably by MRI or Rossby instabilities, is mediated by a weak turbulence and hence, leads to a small, necessary and finite magnetic flux diffusion. Around AGN supermassive BH's this is a Keplerian disk with differential winding, the Ω motion, and star-disk-collision-driven plumes for the helicity motion. It is only the finite mass of the disk, when driven as a plume several scale heights above the disk that after expansion has enough moment of inertia so that it can twist toroidal to poloidal field. (Magnetic buoyancy leads to magnetic loops that are largely devoid of mass and so do not have enough inertia to create the necessary torque.) In stellar dynamos the same coherent dynamo most likely occurs where the entropy gradient stabilizes rotational shear at the base of the convective zone, the Ω motion, and semi-coherent scale height-size, convective plumes occur at the base of the convective zone above the stable shear, producing the α-helicity. We believe that the role of turbulence is not the conversion of kinetic to magnetic energy, but has the equally important function of allowing the magnetic flux to diffuse and self organize, between semi-coherent, otherwise perfectly conducting, orthogonal motions.

2) *The production of a still larger, $\sim \times 10^{11}$, magnetic energy and magnetic flux by winding or twisting of this poloidal flux, external to the disk in a conducting medium, by the angular momentum flux of the disk itself.*

Immense toroidal flux is produced by the differential winding of the AGN dynamo-generated poloidal flux by $N_{turns} \simeq 3 \times 10^{11}$ turns at the event horizon radius, r_g, in 10^8 years, (the AGN accretion time at the Eddington limit and the radio lobe life time). The foot-prints of poloidal flux are attached by conduction to different radii of the accretion disk and so the differential winding of the foot prints at different radii leads to the force-free helical jet.

At a much smaller scale, but still awesomely large, the collapse of a type II supernova should lead similarly to large differential winding of the flux trapped by conduction *between* the neutron star and the ejecta. The neutron star will make $\sim 5 \times 10^9$ turns at ~ 500Hz during the spin-down time of $\sim 1/3$ year with the flux trapped in a conducting medium. The result of this immense winding in a conducting medium, can be seen as a nested set of Force-Free cylindrical helical magnetic flux surfaces, a pinch in plasma physics or a jet when seen in radio emission and made visible by the emission or self-luminosity of the parallel current carriers supporting the Force-Free helical magnetic fields, (similar to laboratory pinch or tokamak experiments) $--$ and leading to the supernova remnant emissions observed in the radio, optical, x-rays, gamma rays, and cosmic rays.

3) *In galaxies, a galactic -scale dynamo may exist, and because of a small, but finite number of turns in a Hubble time, probably serves solely to maintain a previously trapped flux originating from the AGN's helical magnetic jet.*

The galaxy-scale field may be maintained, not generated, at just the back reaction limit where the galactic in-fall pressure from gas accretion (a galactic gas mass, $\sim 10\%$, in a Hubble time), from the galaxy halo onto the galaxy just balances the magnetic and cosmic ray pressure within the galaxy. This "seed" field flux is probably trapped during galaxy formation from the jet, the Force-Free helix, from the AGN dynamo and disk. It is a small, near trivial fraction, $\sim 10^{-3}$ of the flux producing the magnetic jets. The galactic dynamo then need only maintain, (not generate by many e-folds), its initial trapped seed field. Classically an $\alpha\Omega$-dynamo in the galactic disk can not make nearly enough revolutions (only < 100 turns in a Hubble time) needed to exponentiate by many e-folds to the final galactic field of $\sim 6 \times 10^{-6}$ G. The presumed starting, Biermann battery level seed fields are $\lesssim 10^{-20}$ G -- putatively from polarization currents from the dynamics at initial structure formation after de-coupling.

4) *Faraday Rotation measures and Synchrotron radiation maps in radio lobe sources confirm the existence of magnetic fields. These fields are then fundamental to the mechanism of the radio lobe emission, which, in addition, requires a comparable energy density in ultra-relativistic electrons.*

Faraday rotation measures the product of (the magnetic field aligned with the observer) x (the ambient electron density). Hence, when the field alters direction, the rotation measures may become confused. Fortunately the synchrotron polarization maps serve to establish a scale on which the field is coherent, namely, where the polarization vector is constant over a finite dimension. Furthermore the existence of polarized radio emission itself is almost uniquely indicative of synchrotron emission by energetic electrons in a magnetic field. The morphology of these emission vectors describe jets from the AGN where the direction oscillates back and forth ~ 45 degrees as a function of length as expected for a force-free helical field. Further in length the jet kinks and twists as expected for an unstable "pinch". In circumstances of AGN jets in galactic clusters where the electron density is large and measured independently by x-rays, the value and morphology of the Faraday rotation measures are consistent in magnitude and morphology (plus and minus change of flux direction across the symmetry axis) with a force-free helical jet from the AGN source and of strength consistent with the minimum energy model of synchrotron emission from the radio lobes. Theoretical models of these force-free helical jets predict much larger magnetic fields, $\sim \times 10^{8}$ at smaller radii, $\sim \times 10^{-8}$, but predicting emission is difficult because the ambient plasma is most likely excluded by the strong fields themselves.

5) *The equi-partition between accelerated non-thermal electron energy, and magnetic field energy implies the direct acceleration of electrons and ions without an intermediate step of converting the magnetic energy to heat through ηJ^2 through classical reconnection.*

This is because the magnitude of magnetic energy already challenges the upper limits of feasible mechanisms, and major heat losses before acceleration would create a highly visible plasma and a subsequent highly inefficient conversion of thermal plasma energy into a highly relativistic particle spectrum. This appears unlikely unless the thermal energy creates a strong shock wave in the IGM. There appears no evidence for such a shock in the ultra low density, $5 \times 10^{-6}/cm^3$, of the IGM, and to make such a relativistic shock with a relativistic fluid would require a magnetic field of $B_{shock} \simeq 2 \times 10^{-5}$ G, which also would require the radio lobes expanding to 10 Mpc in 10^8 y. Hence we must find an alternate, efficient mechanism to convert the free energy of Force-Free twisted magnetic fields into rapid, direct acceleration of the current carriers comprising the current, $J_{\parallel}$, that supports the helical field. Instead we have suggested a unique mechanism to preferentially accelerate the current carriers, which requires the effect of starving the helical field of current carriers so that the remaining current carriers must be accelerated to a limiting velocity c, in order to carry the current. Otherwise the current will be reduced and dB/dt will create a greater electric field, further accelerating the remaining current carriers. The result of current carrier starvation is the conversion of magnetic flux into electric field and consequential acceleration of the remaining current carrying particles.

The ultra strong gravity near the event horizon naturally creates current carrier starvation at the innermost boundary. An additional loss of accelerated particles occurs by random-walk diffusion, along the weakly tangled lines of force out of the helical field from weak instabilities on the ion plasma scale and furthermore "down" the field gradient. The return of ambient plasma or previously accelerated particles is similarly inhibited by the inverse, positive field gradient. The combination of diffusive instabilities and field gradient inhibition naturally maintains the current carrier starvation condition. These run-away, accelerated current carriers of the current carrier-starved, all $J_\parallel$ current, ($\sim 3 \times 10^{18}$ amperes) are then the accelerated electrons ($\Gamma \simeq 10^4$) of the radio lobe emission due to curvature radiation and truncated in energy by the electron radio emission itself. Similarly, as in our galaxy one expects the ions to be accelerated preferentially because of much smaller $\propto (m_e/m_p)^2$ radiation loss, leading to the cosmic rays of the IGM.

6) *The radiation spectrum from the accelerated electron current carriers of the Force-Free fields.*

These accelerated electrons, $\Gamma \simeq 10^4$, lead to the synchrotron radio spectrum from radio lobes. In the stronger magnetic fields of the jet at smaller radii, $B \propto 1/r$, these same accelerated electrons are a possible source of the polarized optical and possibly self-Compton x-rays from the synchrotron emission in the helical fields of the pinch or jets in the much stronger field at small radii. Finally x-rays from Compton collisions of these accelerated high energy electrons of the radio lobes with the CMB photons are expected, which have been interpreted from the morphology, spectrum, and x-ray luminosities.

7) *Galaxy cluster magnetic fields from the central SMBH*

The extreme motions within the cluster due to galaxy mergers with the cluster are predicted to amplify magnetic fields in the cluster by $\sim \times 10$ by the semi-coherent motions of ionized plasma. The higher density, $\sim \times 10^3$ of the cluster plasma compared to the IGM affords higher resolution Faraday rotation images of magnetic jets, intra-cluster radio lobe voids of mass density, and mergers. The PdV energies of the voids or bubbles within the cluster are comparable to the central AGN energy.

2. The Total Energy: Radio Lobes, Relativistic Electrons & Ions

The largest source of magnetic energy and magnetic flux in the universe is that of the giant radio lobes of the FRII radio galaxies. The evidence for this is the interpretation of the radio flux in a frequency band, $\sim 1 - 10 \times 10^9$ Hz and size interpreted from distance and angular size. It was pointed out over 50 years ago (Burbridge, 1956) that a single AGN, "quasar" must release a very large magnetic energy, up to $\sim 10^{61}$ ergs, and that gravitational energy is the only feasible source (Hoyle *et al.* 1964, Burbidge & Burbidge, 1965 and 1968). At that time the existence of a central massive black hole, (SMBH), of $10^8 M_\odot$ at the center, ($\sim 3 \times 10^{-4}$ of the exact center) of every flat, spiral galaxy was unknown. Now the confirmed existence of such awesome objects makes these early physical deductions seem prescient.

The interpretation is based upon the observed luminosity in the radio, the size, and the observation of polarization, coherent over large fractions of the physical extent of the emission, for example M87, (Owen *et al.*, 1989). Only synchrotron emission from relativistic electrons in a magnetic field seems feasible. The minimum total energy occurs when the magnetic and electron energies are comparable, $\sim 10^{61}$ ergs. Similarly when the size is estimated, a minimum total energy occurs for a given luminosity leading to magnetic fields of $\sim 10\,\mu G$ and a spectrum of relativistic electrons of mean energy of $\Gamma m_e c^2$, with $\Gamma \sim 10^4$, (50 Gev).

How could one accelerate so many electrons, $N_e \sim 10^{63}$, to such energies within an equal energy magnetic field? What other evidence exists for such an extreme result? Is it a ubiquitous phenomenon in the universe? Does the magnetic flux last long enough to "magnetize" the universe and affect the distribution of baryonic matter? or does the flux leak to the voids and reconnect in relatively short times, $\sim 10^8$ years?

Some years ago we, (Kronberg et al., 2001), asked this question and analyzed some 100 extra galactic radio lobes, (70 giants, > 0.6 Mpc, and ~ 30 Cluster-imbedded lobes) finding a distribution in sizes, luminosity, and energy, using the formulation of (Pacholczyk, 1970) for analyzing the minimum energy. The result is that there are two distributions of radio lobes, one in the IGM and the other within galaxy clusters. The highest energy lobes are in the IGM with energies up to $\sim 10^{61}$ ergs. However, the radio lobes within galaxy clusters are less in energy by $\sim 1/100$. This large energy difference is understood by the existence of the x-ray bubbles in the cluster medium (Fabian et al., 2003). Since the pressure in the inter cluster medium, ICM, is $\sim \times 10^6$ greater than the IGM, it is a wonder that there is similar emission at all. Nevertheless a simulation of these bubbles by the injection of large helical flux from a central source, (presumably by the AGN dynamo and accretion disk) produces just these voids or bubbles in the ICM and requires the same extreme energy, 10^{61} ergs to make the voids in the ICM. (Diehl et al., 2008).

If this magnetic energy is in near equi-partition with relativistic electron energy then these highly relativistic electrons, $\Gamma \sim 10^4$, should interact with the microwave background, scattering x-rays from the CMB. This is exactly what is seen in the x-rays, 1 to 5 Kev as mapped by Chandra, Croston et al. (2005). Thus if the size, luminosity, and relativistic electron spectrum fit the equi-partition model including the Comptonized x-rays, then the final convincing proof is evidence of the magnetic fields as seen first by polarization maps of the lobe emission itself, (Kronberg et al., 1986), and secondly by Faraday rotation of polarized background sources. Fortunately similar more distance sources supply just such a probe and numerous observations of the characteristic rotation of the plane of polarization vs. frequency have been made, indicative of fields up to the equi-partition value of $\sim 10^{-5}$ G, Kronberg 1986. The possibility of magnetic fields over much larger dimensions is demonstrated in the glow around the Coma cluster. (Kronberg et al., 2008).

Unfortunately both the model of the production mechanism of the fields and the plasma physics of the surrounding medium conspire to reduce the magnitude of the observed Faraday rotation signal. The winding of the dynamo flux by the AGN accretion disk leads to the averaging of many turns or reversals of the flux and hence reduces the Faraday rotation signal. In addition the likely background plasma density is reduced in the making the lobes as seen in the x-ray bubbles in clusters thereby reducing the expected Faraday rotation signal from within the lobes. Fortunately both these effects conspire to expect the largest rotation signal at largest radius, because the exclusion of plasma, ambient density, should be least at large radius just where the coherence of the rotation measure should be largest. In a few cases, such as seeing a helical structure edge on and within the higher density of the ICM, larger fields are confirmed, consistent with equi-partition of electron energy and magnetic field energy.

Combined this is awesome consistency of the model and therefore evidence for magnetic fields in the IGM on a scale of inter galactic structure, the spacing between galaxies and their organization into filaments. If 10^{61} ergs of magnetic energy is distributed within the volume defined by the mean galaxy spacing, ~ 1 Mpc in the filaments, then assuming every galaxy, had at the time of its formation, an engine producing such flux, then the mean field becomes $<B> \sim 1.6 \times 10^{-6}$ G. This field is larger by $\sim \times 10^2$ than the field as interpreted from Faraday rotation measurements of distant polarized sources. However, the possible multiple reversal averaging, the bubble reduction of ambient density, and finally the dissipation of magnetic flux and energy by reconnection or acceleration remain open questions. We believe the dissipation of magnetic flux actually occurs by acceleration rather than reconnection, because if the energy were converted first into heat by reconnection, then the IGM would be heated to $\simeq 10^4$ ev, and the radio emission would be missing from the kev electrons. We say heat, because current reconnection research leads to shocks and thus heat in the reconnection zone whereas we are suggesting a semi-coherent process where an $E_{\parallel}$ leads to the direct acceleration of the charged current carriers.

3. Extra-Galactic Cosmic Rays

Within the galaxy, the "milky way", cosmic rays of energy density $\sim$1eV/cm^3, 1.6×10^{-12}ergs/cm^3, are in equi-partition with the galactic magnetic field, $B_{gal} \simeq 6 \times 10^{-6}$ G. This equality suggests that the magnetic energy is the source of the particle energy or at least that the magnetic energy density determines the probability of the escape of accelerated cosmic rays from the galaxy. In either case the two are likely tightly dependent on each other. In our view extragalactic cosmic rays are likely to follow a similar acceleration mechanism and similar escape physics, that is escape from the filaments to the voids. Based on these assumptions, (Colgate & Li, 2004) analyzed the likely extra galactic cosmic ray spectrum and energy density starting with the flux and energy of UHEs and extrapolating to lower energy, $m_p c^2$, and concluded that the energy density of the extra galactic CR flux may be $\sim 1/200$ of the galactic flux. This ratio is the likely ratio of the infall pressure to the filaments from the voids that is necessary to double the filament mass in a Hubble time as compared to the infall pressure from the halo to the galaxy necessary to double the galaxy gas mass in a Hubble time. The magnetic field and particle pressure are then in equilibrium with the confining pressure of the in-falling, accreting matter. The total energy in extra galactic CRs per galaxy spacing volume is then $\sim 10^{60}$ ergs, somewhat less than what one might expect from the typical AGN-radio lobe. However, some leakage to the voids is expected at the highest energies in a Hubble time and furthermore the SMBH of our Galaxy is significantly smaller than the average implying a smaller initial AGN and radio lobe.

Thus the interpretation of immense extra galactic magnetic fields and equal immense energy in relativistic ions and electrons is supported by the polarization maps, radio luminosity, size, Faraday rotation, luminosity in x-rays, CRs, and CR energy. In turn this is consistent with the synchrotron emission of the equi-partition model of radio lobes.

4. The Origin of the Magnetic Flux and Energy

The magnetic flux of radio lobes is of order (area x field) or $\sim 10^{42}$ G-cm^2. This is $\times 10^3$ greater than the flux of the galactic magnetic field, yet to create a galactic dynamo that could make this much flux starting from the "Bierman battery" level of $\sim 10^{-20}$ G of the galaxy, seemed nearly impossible, (Kulsrud, R. M. & Zweibel, E. G., 2008). Hence we agree with their suggestion that the galactic flux was "seeded" from the AGN accretion disk dynamo before the galaxy formed and then only maintained by the galactic dynamo in pressure equilibrium with the in-fall pressure from the halo. Here the trapping of only

$\sim10^{-3}$ of the helical force-free jet flux within the galaxy is required to "seed" the galactic flux.

Nearly universally the $\alpha\Omega$ dynamo (Parker, 1955; Parker, 1979) has been the accepted mechanism for converting kinetic energy into magnetic energy, yet galactic motions alone are inadequate, because the galaxy makes only ~100 turns, the Ω-motion, in a Hubble time and the source of the helicity, the α-motion by supernova remnant bubbles occurs on a scale too large for the flux to merge in a Hubble time using laminar flow. Thus the generation of magnetic flux by the galaxy seems frustrated on both levels of the $\alpha\Omega$ dynamo. These problems were addressed in (Pariev, Colgate, & Finn, 2007; Pariev, Colgate, 2007) where it was recognized and calculated that the accretion disk forming the SMBH and AGN was a unique circumstance where the disk matter makes many turns, $\sim10^{11}$ during its life time. Consequently the dynamo gain can be so large that the question of seed field becomes moot. Equally important was the possibility that an expanding plume driven axially outwards from the disk in a rotating frame had the unique property of a $1/4$ turn of rotation before merging back with the disk and thus for creating the helicity necessary for an astrophysical dynamo. Such a dynamo was simulated and analyzed by axial vector simulation, mean field theory, and flux conversion, (Pariev, Colgate, & Finn, 2007). The formation of the large scale generation of magnetic flux in the accretion disk forming the SMBH of an AGN was a natural prediction. Here a small mass fraction of the disk, $\sim < 10^{-5}$, was assumed to be in stars in random orbits around the SMBH. This mass fraction of stars is less than the metalicity, usually assumed in pre-galactic stars, $\sim10^{-3}$. These stars collide with the disk creating buoyancy driven plumes, and thus the helicity necessary for the dynamo. Furthermore the α_T-viscosity turbulence gives rise to the necessary flux diffusion so that flux can merge, link, or multiply in a finite number of turns. We believe that this is a unique solution to this problem, of creating the immense, large scale magnetic flux and energy of the ISM. This is because:

1) Half the free energy of accretion, $\sim10^{62}$ ergs can potentially be converted to magnetic energy.

2) The total number of turns at $3r_g$ is $N_T \sim 10^{11}$ in the formation time, 10^8 years of the SMBH and thus the question of seed field moot.

3) The helicity produced poloidal flux is in the same direction and $1/4$ turn for every plume and thus the average poloidal flux is the arithmetic sum of the flux from each plume, not the statistical vector average as might be expected from turbulence alone.

4) When the plume falls back to the disk, the plume matter and imbedded flux mixes and diffuses with the turbulent diffusion coefficient of the plume-disk mixing turbulence, faster than the α_T-viscosity without angular momentum stabilization, thus allowing the small loop of poloidal flux to rapidly merge with others coherently. All loops are in the same direction, thus creating the large scale poloidal flux of the dynamo.

5) The mass density of the plumes is sufficient so that the moment of inertia of the plume will twist the fields up to the emission limit, or Eddington limit, $B_{rg} \simeq 3 \times 10^4$G.

6) The diffusive separation of circular poloidal flux foot prints by the α_T-viscosity in the disk allows a large differential winding of the Keplerian motion between the two cylindrical foot-prints, an inner one of flux leaving the disk and an outer one of flux returning to the disk. In the conducting medium of the IGM this leads to continuous differential winding, $\sim10^{11}$ turns of the flux surfaces, producing the near force-free helix of the jets and radio lobes. (Li, H. *et al.*, 2001)

5. Experimental Dynamos

There have been two liquid sodium dynamo experiments with positive results, both with flows that are turbulence-constrained by rigid walls. In three recent experiments using counter rotating converging flow, large turbulence in the unconstrained shearing mid-plane flow reduced the Ω-gain to $\simeq< \times 2$. In the New Mexico sodium experiment, where the flow is partially stabilized by angular momentum, Couette flow, the Ω-gain increased to $\simeq \times 8$.

6. Summary

The evidence for large scale and large energy magnetic fields with equi-partition with accelerated particles in the IGM is convincing. An $\alpha\Omega$ dynamo in the accretion disk of the SMBH creates poloidal field, (in stars as well). The winding of this poloidal flux by the accretion disk into a force-free helical jet is the likely explanation of the AGN jet, the radio lobes, and accelerated particles.

Acknowledgements

The experimental work has been supported by Howard Beckley, David Westpfahl, Jaihe Si, Joe Martinque, and a dozen undergraduate students at NMTechr. Vladimir Pariev first calculated the plume-driven $\alpha\Omega$ dynamo. The liquid sodium dynamo experiment has been supported by DOE through the MOU between Univ. of Calif. at Los Alamos Nation Lab and New Mexico Institute of Mining and Technology. The theory and experiment have both been supported though the LDRD funding by the DOE and LANS.

References

Burbidge, G. R. 1956, *Ap.J.*, 124, 416

Burbidge, G. R. 1958, *Ap.J.*, 129, 849

Burbidge, G. R. & Burbidge, E. M. 1965, *The Structure and Evolution of Galaxies*, Proc. 13th (Solvay) Conf. on Physics, Bruxelles, New York: Interscience, Wiley, 137

Chandrasekhar, S., 1960, *Proc. Natl. Acad. Sci.*, 46, 253

Colgate, S. A. & Li, Hui, 2004, *Reconnection of Force-Free Fields*, ed. Guenter Sigl and Murat Boratav in The French Academy of Sciences, Comptes rendus - Physique, C.R. Physique 5 431 (astro-ph/0509054)

Croston, J. H., Hardcastle, M. J., Harris, D. E., Belsole, E. Birkinshaw, M., & Worrall, D. M. 2005, *Ap.J.* 626, 733

Diehl, S., Li, Hui, Fryer, C., & Rafferty, D. 2008, *Ap.J.* 687, 173

Fabian, A. C., Sanders,J. S., Crawford, C. S., Conselice, C. J., Gallagher III J. S., & Wyse, R. F. G., 2003 *Mon. Not. R. Astron. Soc.* 344, L48 (2003)

Hoyle, F., Fowler, W. A., Burbidge, G. R., & Burbidge E. M. 1964, *Ap.J.*, 139, 909

Kronberg, P. P. Wielebinski, l. R., & Graham, D. A. 1986, *Astron. Astrophys.* 169, 63

Kronberg, P. P., Dufton, Q., Li, H., & Colgate, S. A. 2001, *Ap.J.*, 560, 178

Kronberg, P. P., Kothes, R., Salter, C., & Perillat, P. 2008, *Ap.J.*, 659.267

Kulsrud, R. M. & Zweibel, E. G. 2008, *Reports on Progress in Physics*, 71, 046901 (2008).

Li, H., Lovelace, R. V. E., Finn, J. M., & Colgate, S. A. *Ap.J.*, 561, 915

Mestel, L. 1999, Stellar Magnetism. (Oxford: Clarendon)

Owen, F. N., Hardee, P. E., & Cornwell, T. J. 1989, *Ap.J.*, 340 698

Pacholczyk, A. G. 1970, *Radio Astrophysics (San Francisco: Freeman)*

Pariev, I. V., Colgate, S. A., & Finn, J. M. 2007, *Ap.J.*, 658, 114.

Pariev, I. V., Colgate, S. A., & Finn, J. M. 2007, *Ap.J.*, 658, 129.

Parker, E. N. 1955, *Ap.J.* 121, 29

Parker, E. N. 1979, Cosmical Magnetic Fields, their Origin and their Activity. (Oxford: Claredon)

Advances in Plasma Astrophysics
Proceedings IAU Symposium No. 274, 2010
A. Bonanno, E. de Gouveia Dal Pino & A.G. Kosovichev, eds.
© International Astronomical Union 2011
doi:10.1017/S1743921311006508

Study of magnetic reconnection in collisional and collionless plasmas in Magnetic Reconnection Experiment (MRX)

Masaaki Yamada and Hantao Ji

Princeton Plasma Physics Laboratory,
Princeton University,
Princeton, New Jersey, U.S.A.
email: myamada@pppl.gov

1. Introduction

Magnetic reconnection (Parker, 1957; Sweet, 1958; Petschek, 1964; Yamada *et al.*, 2010; Biskamp, 2000; Tsuneta, 1996; Kivelson and Russell, 1995; Yamada, 2007; Birn *et al.*, 2001; Drake *et al.*, 2003) is considered important to many astrophysical phenomena including stellar flares, magnetospheric disruptions of magnetars, and dynamics of galactic lobes. Research on magnetic reconnection started with observations in solar coronae and in the Earths magnetosphere, and a classical theory was developed based on MHD. Recent progress has been made by understanding the two-fluid physics of reconnection, through space and astrophysical observations (Tsuneta, 1996; Kivelson and Russell, 1995), laboratory experiments (Yamada, 2007), and theory and numerical simulations (Birn *et al.*, 2001; Daughton *et al.*, 2006; Uzdensky and Kulsrud, 2006). Laboratory experiments dedicated to the study of the fundamental reconnection physics have tested the physics mechanisms and their required conditions, and have provided a much-needed bridge between observations and theory. For example, the Magnetic Reconnection Experiment (MRX) experiment (http://mrx.pppl.gov) has rigorously cross-checked the leading theories though quantitative comparisons of the numerical simulations and space astrophysical observations (Mozer *et al.*, 2002). Extensive data have been accumulated in a wide plasma parameter regime with Lundquist numbers of $S = 100 - 3000$, where S is a ratio of the magnetic diffusion time to the Alfven transit time.

In this article, we briefly review the recent major results on reconnection in MRX. The characteristics of the local reconnection layer have been studied both in the collisional (MHD) and collisionless (kinetic) regimes. An important scaling law has been obtained with respect to the ratio of the collisional mean free path to the size of reconnection layer. In the current MRX research, a special focus is put on magnetic energy dissipation due to reconnection, which has not been resolved in the past decades of research. Also we aim to address the important relationships between the local physics of the reconnection layer and the evolution of the global topology changes. One of our key questions is how large-scale systems generate local reconnection structures in realistic 3-D geometries, through formation of multiple current sheets or magnetic islands. For this purposes, we plan to upgrade our MRX facility substantially by extending our cross-discipline investigation of magnetic reconnection among theory, numerical simulation, laboratory experiments, and space and astrophysical observations.

2. Profiles of reconnection layer and reconnection rate

One of the most important questions has been why reconnection occurs so rapidly or impulsively with much faster speed than predicted by classical MHD theory. In recent numerical theory (Birn et al., 2001; Daughton et al., 2006) and experiments (Yamada, 2007; Mozer et al., 2002; Ren et al., 2005; Yamada et al., 2006; Brown et al., 2006; Ji et al., 1998; Ji et al., 2004) two-fluid effects have been utilized to explain the fast reconnection rate in the magnetosphere, laboratory plasmas, and even in stellar flares. The data from MRX show striking similarity to the magnetospheric measurements, in which both two-fluid Hall effects and magnetic fluctuations are detected together (Yamada et al., 2010; Mozer et al., 2002).

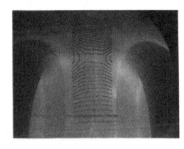

Figure 1. Reconnection layer in MRX with measured flux plots and time integrated photo.

The MRX device shown in Fig. 1, has been continuously generating fundamental data on the physics of magnetic reconnection (Yamada et al., 1997). The MRX plasma can be described globally by MHD, but the reconnection region has to be treated by two-fluid MHD model, since the width of reconnection layer becomes comparable to the ion skin depth.

In MRX, reconnection is driven in a controlled manner with toroidal shaped flux cores which contain two types of coil windings both in the toroidal and poloidal directions. By pulsing programmed currents in those coils, two annular plasmas are created by inductive formation around each flux core utilizing induced poloidal electric fields. After the plasmas are generated, the coil currents are programmed to drive magnetic reconnection producing a current layer or a reconnection region in the central region of Fig. 1 to study the dynamics of local reconnection. The evolution of the magnetic field lines can be seen by way of movies presented at the MRX Web site (http://mrx.pppl.gov/mrxmovies): this shows time evolutions of the measured flux contours of the reconnecting field. By monitoring the flux contours, the reconnection rate was measured as function of plasma parameters (Yamada, 2007; Ji et al., 1998).

The detailed profile of the current sheet was measured, demonstrating important two-fluid MHD features of magnetic reconnection. It was found that the thickness of the current sheet without a guide field was equal to a fraction of the ion skin depth (Ji et al., 2004; Ji et al., 2008) in the collissionless regime. In the collisional regime, a classical Sweet-Parker theory was verified based on the Spitzer resistivity. In the less collisional regime, a generalized Sweet-Parker model (Parker, 1957; Ji et al., 1998) explained the data with an enhancement of the resistivity over the classical value. The effects of plasma turbulence have been investigated. The cause of the enhanced reconnection rates in the collisionless regime has been studied intensively and significant breakthroughs have been recently made by an identification of a correlation of the enhanced resistivity with magnetic fluctuation level (Ji et al., 2004), as well as by an experimental verification of the Hall MHD effects (Yamada, 2007; Ren et al., 2005). The effect of the third component (guide field) of the magnetic field vector on reconnection was assessed, and it was found that the presence of the guide field would often slow down the reconnection rate (Yamada, 2007).

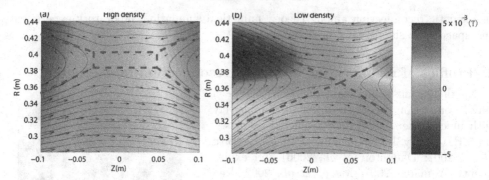

Figure 2. (left) Comparison of neutral sheet configuration described by measured magnetic field vectors and flux counters for high (collisional) and low density cases; (a) Collisional regime ($\lambda_{mfp} \sim 1mm \ll \delta$); (b) Nearly collisionless regime ($\lambda_{mfp} \sim 1cm \sim \delta$). Out-of plane fields are depicted by the color codes ranged -50 G < B_t <50 G.

3. Changes of the reconnection rate and neutral sheet profiles with respect to collisionality

In the study of the local two-fluid physics of the reconnection layer, an out-of-plane quadrupolar Hall field which was predicted by the recent two-fluid simulations has been verified in MRX (Yamada, 2007; Yamada et al., 2006). The measured profile of the neutral sheet changes drastically from high (collisional) to low density (nearly collsionless) cases. In the high plasma density case, shown in Fig. 2(a), where the mean free path is much shorter than the sheet thickness, a rectangular shape neutral sheet profile of the Sweet-Parker model is seen together with the observed classical reconnection rate. There is no recognizable out-of-plane Hall field in this case. In the case of low plasma density, shown in Fig. 2(b), where the electron mean free path is larger than the sheet thickness, the Hall effects become dominant as indicated by the out-of-plane field depicted by the color

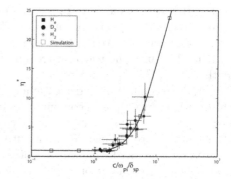

Figure 3. MRX scaling, Effective resistivity $\eta_{eff} = E/j$ normalized by the Spitzer value $\eta_{Spitzer}$ versus the ratio of the ion skin depth to the Sweet-Parker width is compared with numerical calculation of the contributions of Hall MHD effects to the reconnection electric field. The simulations were based on a 2-D 2-fluid code.

code. A double-wedge shape sheet profile of Petschek type, which is shown in the flux contours of reconnecting field in Fig. 2(b), is significantly different from that of the Sweet-Parker model [Fig. 2(a)], and a fast reconnection rate is measured. However, a slow shock, a signature of Petschek model, has not been identified even in this collisionless regime to date. This important observation supports the theoretical idea that the Hall effects originating from two-fluid dynamics contribute to the enhanced reconnection rate observed in collisionless reconnection.

It is important to know quantitatively under what conditions the two-fluid dynamics become important. The recent MRX data identified a criterion for the transition from the one-fluid MHD regime to the two-fluid Hall regime. Figure 3 presents an MRX scaling for effective resistivity $\eta^* = \eta_{eff}/\eta_{Spizer}$, ($\eta_{eff} = E/j$) normalized by the Spitzer value $\eta_{Spitzer}$ in the center of the reconnection region in comparison with a scaling from a Hall

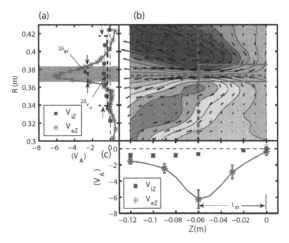

Figure 4. (a)The radial profiles of the electron outflow velocity, V_{eZ} (magenta asterisks), measured in ma Helium plasma. (b) The 2D profile of the out-of-plane field, B_T (color-coded contours), and the in-plane electron flow velocity, V_e (black arrows). (c) V_{eZ} as functions of Z. (a) and (b) represent the cuts at $Z = -6$cm and at $R = 37.5$cm, respectively, in (c). Taken from Ren *et al.* 2008.

MHD numerical simulation result. The classical rate of reconnection with the Spitzer resistivity is obtained (Yamada, 2007) in the collisional regime where $c/\omega_{pi} < \delta SP$. The horizontal axis represents the ratio of the ion skin depth to the classical Sweet-Parker width. This figure shows that the reconnection resistivity (or reconnection speed) increases rapidly from the Spitzer value as the ion skin depth (c/ω_{pi}) becomes large with respect to the Sweet-Parker width (δ_{SP}). The apparent agreement of the MRX scaling with two-fluid Hall MHD code indicates that the measured enhanced resistivity is primarily due to the laminar Hall effect, when the Spitzer resistivity is not large enough to balance the large reconnecting electric field in fast magnetic reconnection. We believe that this scaling observation should not exclude the effects of fluctuations particularly electromagnetic ones. Electrostatic and electromagnetic fluctuations have been observed in the neutral sheets of both laboratory and space plasmas, with notable similarities in their characteristics and theoretical interpretation. In MRX, a correlation was found between the reconnection rate and the amplitude of EM waves (Ji *et al.*, 2004), although the experimental operation range (a factor of 10) is rather narrow as shown in Fig. 3.

4. Identification of the electron diffusion layer

Utilizing high resolution magnetic probes, profiles of electron flow with finer scales were measured. In the neutral sheet of MRX, an electron diffusion region was identified in the reconnection layer for the first time in a laboratory plasma (Ren *et al.* 2008). The rate of reconnection can be controlled in part by dynamics in this small region, in which magnetic field lines tear and reconnect and energy dissipation occurs. The recent 2D numerical simulations by Daughton *et al.* (2006) and by Shay *et al.* (2007) predict a two-scale diffusion layer in which an electron diffusion layer resides in side of the larger ion diffusion layer of width the ion skin depth.

In MRX, the presence of an electron diffusion region was verified and it was found that de-magnetized electrons are accelerated in the outflow direction (Fig. 4). The width of the electron diffusion region was measured to scale with the electron skin depth (c/ω_{pe})

and the electron out-flow scales with the electron Alfvén velocity (0.11VA). The general features of both the electron and ion flow structures agree with simulations. But the thickness of the electron diffusion layer is much larger (5 times) than the values obtained by 2D simulations (Ji *et al.*, 2008). Careful checks of collisional effects have been made to determine how much of the enhanced diffusion layer thickness in MRX should be attributed to 3D effects and how much to collisions (Roytershteyn *et al.*, 2010).

Although the electron outflow seems to slow down by dissipation in the electron diffusion region, the total electron out flow flux remains independent of the width of the electron diffusion region. We note that even with the presence of the thin electron diffusion region, the reconnection rate is still primarily determined by the Hall electric field, as was concluded by the multi-code GEM project (Birn *et al.*, 2001). The ion outflow channel is shown to be much broader than the electron channel, which is also consistent with numerical simulations. Also this electron outflow often occurs impulsively as the collisionality of the plasma is reduced.

5. Outstanding issues on reconnection and future research plans in MRX

Our research plans cover major issues for both local reconnection physics and global reconnection dynamics and also address the interrelationship between the two regions as well. Regarding the local reconnection dynamics, the two fluid and kinetic physics are under intensive investigation in the reconnection research community: the effects of waves and turbulence, mechanisms of energy dissipation, and the dependence on the Lundquist number and system size. The latter covers multiple reconnections, impulsive reconnection and the effects of boundaries which includes line tying effects. These issues are often interrelated. For example, global reconnection often occurs impulsively which can be directly translated to a fast local reconnection. While line-tying effects have been studied based on MHD theory, new experiments are initiated to address this issue and to bridge between the space astrophysical observations and theoretical studies. At the moment, there is no consensus on one of the most important question, how magnetic energy is converted to particle energy during reconnection. We will explore the relationships between anomalous particle acceleration and heating and reconnection events in both laboratory and astrophysical plasmas. By comparing simulations and experimental data we hope to develop a theoretical model for particle acceleration and heating and apply it to explain solar/stellar flare energetic particle populations.

To extend the scope of our study, an upgrade of MRX has been considered. The main features are as follows; (1) To extend significantly the parameter range from fully collisional regime of ($\lambda_{mfp} \ll L$) to collisionless regime ($\lambda_{mfp} \gg L$). It is very important to obtain high T_e (> 30 eV) and large S ($S \sim 10^5$) for attaining fully collisionless regime ($\delta_{mfp} \gg L$), and for measuring temperature change. We plan to achieve effectively larger plasmas where the system size $\sim 10^3$ ion sound radii at high density ($> 10^{13}$ cm^{-3}). With new high-power flux cores, we plan to increase magnetic field by factor of 3 (to 1 kG) and increase T_e by at least 3 to 50 eV.

(2) We have been primarily using the steady pull mode of the local MRX reconnection operation. This mode has been very successful in simulating magnetosphere reconnection. In MRX-U we plan to move further by employing broader operation modes to address the above key issues including modes relevant to solar flare geometries, such as interaction of large plasma arcs or flares. For this we will use a merging plasma mode as well as line-tying solar flare plasma experiments. We will also address spontaneous reconnection in a large medium with new modes of operation.

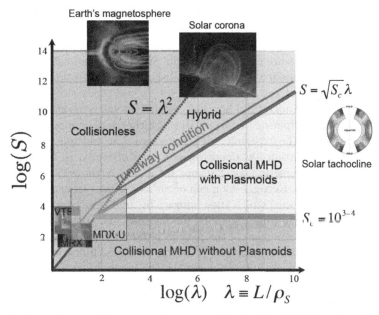

Figure 5. A phase diagram for magnetic reconnection in 2D. If either S or the normalized size, λ, is small, reconnection takes place in collisional MHD (without plasmoids) regime or in collisionless regime. When both S and λ are sufficiently large, two new regimes appear: a regime for collisional MHD with plasmoids and a regime for collisional MHD plasmoids with kinetic current sheets in between. The regimes for reconnection in Earths magnetosphere, solar corona, and solar tachocline are also shown. The existing experiments, such as MRX and VTF, do not have accesses to these new regimes. A new generation experiment, MRX-U based on MRX, is proposed to study reconnection in these new important regimes for direct relevance to reconnection in heliophysical plasmas.

(3) We plan to improve our diagnostics significantly so that we can address the above goals. Main additions are a Thomson scattering system, routine line density measurements by CO_2 laser, better spectroscopy extending into UV regions, and LIF (laser induced fluorescence) diagnostics for T_i measurement.

A new phase diagram) (Daughton and Roytershteyn, 2010; Ji and Daughton, 2011), was considered from recent large-scale numerical simulations as shown in Fig. 5 where four reconnection phases are illustrated: collisional MHD without plasmoids, collisional MHD with plasmoids (corresponding to reconnection in solar tachocline), collisional MHD plasmoids with kinetic current sheets in between (or hybrid, corresponding to solar corona) and collisionless (corresponding to Earths magnetosphere). In order to be directly relevant to reconnection phenomena observed in heliophysical plasmas, it is crucial for laboratory experiments to access all of these reconnection regimes. We note that reconnection in these new regimes is associated with multiple X-points, and thus can possibly provide solutions to the onset problem and the energy problem mentioned above.

In order to understand how magnetic energy is converted to particle energy we will explore the relationships between anomalous particle acceleration and heating and reconnection events in both laboratory and astrophysical plasmas. On the theoretical side, we will gain a new tool for probing how fluctuations are excited, and how they dissipate, through Particle-In-Cell (PIC) and hybrid (fluid electrons and kinetic ions) simulations of the reconnection layer. By comparing simulations and experimental data, we hope to develop a predictive theory of these fluctuations which can be applied to space and astrophysical plasmas and ultimately tested against models of solar/stellar flare energetic particle populations based on their x-ray, radio, and gamma ray emissions. In MHD reconnection theory, fluctuations are generally thought to enhance the resistivity in a position dependent manner; the same may happen in collisionless reconnection, but there may be other processes we have not yet identified.

It is very important to understand how large-scale systems generate local reconnection structures, through the formation of current sheets, either arising spontaneously or forced by boundary conditions. The experimental observation that multiple reconnections qualitatively alter self-organization has opened a new area of study. In upgraded MRX, we will study spontaneous triggering mechanisms for global reconnection phenomena, and magnetic self-organization. Reconnection can occur as a helical tearing instability. Multiple instabilities can also occur, which then interact by nonlinear coupling. With multiple reconnections, global momentum transport, ion heating, and magnetic self-organization should occur. Monitoring evolution of the plasma parameters, we will examine the effect of multiple reconnections on momentum transport in disks, flux conversion in jets and lobes (discussed below) and solar flares. A remarkably general feature of global reconnection phenomena is its impulsive nature. In the RFP, tokamak, magnetospheric substorms (Yamada *et al.*, 2010), and solar flares reconnection occurs suddenly in time. In most of these systems, theoretical ideas are evolving to explain the impulsive behavior, but none of the situations yet enjoys an established explanation. These questions will be addressed as a common issue of global reconnection phenomena.

References

Birn, J., Drake, J. F., Shay, M. A., Rogers, B. N., Denton, R. E., Hesse, M., Kuznetsova, M., Ma, Z. W., Bhattachargee, A., Otto, A., & Pritchett, P. L., *J. Geophys. Res.*, 106(A3): 3715, 2001.

Biskamp, D., Cambridge University Press, Cambridge, 2000.

Brown, M. R., Cothran, C. D., & Fung, J., *Phys. Plasmas*, 13(5):056503, 2006.

Daughton, W. & Roytershteyn, V., 2010, submitted to Proceedings of ISSI meeting (2010).

Daughton, W., Scudder, J., & Karimabadi, H., *Phys. Plasmas*, 13:072101, 2006.

Drake, J. F., Swisdak, M., Cattell, C., Shay, M. A., Rogers, B. N., & Zeiler, A., *Science*, 299: 873, 2003.

Ji, H. & Daughton, W., 2011, to be submitted to Phys. Plasmas.

Ji, H., Yamada, M., Hsu, S., & Kulsrud, R., *Phys. Rev. Lett.*, 80:3256, 1998.

Ji, H., Terry, S., Yamada, M., Kulsrud, R., Kuritsyn, A., & Ren, Y., *Phys. Rev. Lett.*, 92:115001, 2004.

Ji, H., Ren, Y., Yamada, M., Dorfman, S., Daughton, W., & Gerhardt, S. P., *Geophys. Res. Lett.*, 35:L13106, 2008.

Kivelson, M. & Russell, C. T., Cambridge University Press, London, UK, 1995.

Mozer, F. S., Bale, S. D., & Phan, T. D., *Phys. Rev. Lett.*, 89:015002, 2002.

Parker, E. N., *J. Geophys. Res.*, 62:509, 1957.

Petschek, H. E., *NASA Spec. Pub.*, 50:425, 1964.

Ren, Y., Yamada, M., Gerhardt, S., Ji, H., Kulsrud, R., & Kuritsyn, A., *Phys. Rev. Lett.*, 95 (5):055003, 2005.

Ren, Y., Yamada, M., Ji, H., Gerhardt, S., & Kulsrud, R., *Phys. Rev. Lett.*, 101:085003, 2008.

Roytershteyn, V., Daughton, W., Dorfman, S., Ren, Y., Ji, H., Yamada, M., Karimabadi, H., Yin, L., Albright, B. J., & Bowers, K. J., *Phys. Plasmas*, 17:055706, 2010.

Shay, M. A., Drake, J. F., & Swisdak, M., *Phys. Rev. Lett.*, 99(15):155002, 2007.

Sweet, P. A., In B. Lehnert, editor, *Electromagnetic Phenomena in Cosmical Physics*, page 123. Cambridge Univ. Press, New York, 1958.

Tsuneta, S., *Astrophys. J.*, 456:840, 1996.

Uzdensky, D. & Kulsrud, R., *Phys. Plasmas*, 13:062305, 2006.

Yamada, M., *Phys. Plasmas*, 14:058102, 2007.

Yamada, M., Ji, H., Hsu, S., Carter, T., Kulsrud, R., Bretz, N., Jobes, F., Ono, Y., & Perkins, F., *Phys. Plasmas*, 4:1936, 1997.

Yamada, M., Ren, Y., Ji, H., Breslau J., Gerhardt, S., Kulsrud R., & Kuritsyn, A., *Phys. Plasmas*, 13:052119, 2006.

Yamada, M., Kulsrud, R., & Ji, H., *Rev. Mod. Phys.*, 82:603, 2010.

Advances in Plasma Astrophysics
Proceedings IAU Symposium No. 274, 2010   © International Astronomical Union 2011
A. Bonanno, E. de Gouveia Dal Pino & A. G. Kosovichev, eds.   doi:10.1017/S174392131100651X

Current status and future prospects for laboratory study of angular momentum transport relevant to astrophysical disks

Hantao Ji

Princeton Plasma Physics Laboratory,
Princeton University,
Princeton, New Jersey, U.S.A.
email: hji@pppl.gov

Abstract. A concise review of the past and ongoing laboratory experiments on rotating flows and the associated angular momentum transport relevant to astrophysical disks is given in three categories: hydrodynamic, magnetohydrodynamic, gas and plasma experiments. Future prospects for these experiments, especially for those directly relevant to the magnetorotational instability (MRI), are discussed with an emphasis on a newly proposed swirling gas and plasma experiment.

Keywords. accretion disks, hydrodynamics, magnetohydrodynamics (MHD), instabilities, plasmas, turbulence, methods: laboratory

---

1. Introduction

Accretions disks consist of gas, dust, and plasma rotating around and gradually falling onto a central object. Many important astrophysical phenomena take place in accretion disks ranging from formation of stars and planets in proto-star systems, the mass transfer and energetic activity in binary star systems, to efficient energy releases in quasars and active galactic nuclei. A common puzzle across all these kinds of accretion disks is why accretion occurs fast compared to the predictions based on classical viscosity by many orders of magnitudes. This is equivalent to the question of how angular momentum is rapidly transported radially outwards. Cause of this rapid transport is generally attributed to the presence of rigorous turbulence in the accretion disk (Shakura and Sunyaev, 1973).

The difficulty of the problem, however, originates from the fact that there exist no known robust linear hydrodynamic instabilities in Keplerian disks where angular velocity, Ω, scales as radius, $R^{-3/2}$. The well known Rayleigh stability criterion (Rayleigh, 1916) against centrifugal instability $d(R^2\Omega)/dR > 0$ is satisfied. Therefore, there exist only two main mechanisms to generate turbulence in accretion disks: (1) nonlinear hydrodynamic instabilities and (2) linear magnetohydrodynamic instabilities, and more specifically Magnetorotational Instability or MRI. The possibility of nonlinear hydrodynamic instabilities was first pointed out by Zeldovich (1981) and followed up by Richard and Zahn (1999) based on experiments of Wendt (1933) and Taylor (1936) in Taylor-Couette flows (see Sec. 2). The idea is similar to what has been learned from pipe flows where subcritical transition was shown as the mechanism to generate turbulence despite linear stability (e.g., Eckhardt 2007). The argument is that in cool disks such as star-forming disks where ionization fraction is too low for the magnetic field to have any effects on disk dynamics, these non-magnetic, nonlinear hydrodynamic instabilities may be the only available mechanisms to generate the required turbulence. However, these nonlinear instabilities have not been demonstrated conclusively under astrophysically relevant conditions either theoretically, numerically, or experimentally.

In hot, thus sufficiently ionized accretion disks such as in binaries, the magnetic field effects are widely expected in generating turbulence (Shakura and Sunyaev, 1973). The well-accepted scenario there is that the turbulence is generated by the MRI, originally discovered by Velikhov (1959) and Chandrasekhar (1960) and re-discovered for astrophysical applications by Balbus and Hawley (1991), and (1998). The MRI is a linear ideal MHD instability which grows locally in Keplerian disks on the orbital timescale in a weak magnetic field. Because of the robustness of the MRI in numerical simulations there is little doubt on the existence of MRI, although it has not been positively confirmed either observationally or experimentally in the laboratory. The key problem currently under hot debate, however, becomes what turbulent viscosity the MRI can generate to transport angular momentum (Lesur and Longaretti, 2007; Fromang *et al.* 2007). Below we list the outstanding questions for angular momentum transport in astrophysical disks:

- Can nonlinear hydrodynamic turbulence efficiently transport angular momentum?
- How does the MRI saturate to generate angular momentum transport?
- How does the transport scale with Reynolds number?
- Are Hall effects and ambipolar diffusion important for the MRI and its saturation?
- Can kinetic effects significantly modify angular momentum transport?
- How do radiation and relativity affect the MRI and its saturation?

In principle, all of these issues (except the last one) can be studied in detail in the laboratory. We note that the role of laboratory experiments here is not to directly simulate astrophysical phenomena themselves, but to study their fundamental underlying physics. Therefore, numerical simulations combined with analytic theory play a crucial role in guiding laboratory experiments, in interpreting their results, and in making actual connections to astrophysical phenomena.

In the following Sections, three categories of experiments will be discussed in terms of their current status and future prospects in addressing the above issues.

2. Hydrodynamic Experiments

Current Status. Study of rotating flows between two concentric cylinders goes back to Couette (1890) who used it as a way to measure viscosity of water. Taylor (1923) significantly expanded the horizon of the Couette flow by successfully applying linear theory of Navier-Stokes equations using nonslip boundary conditions to accurately explain experimentally observed stability conditions for the centrifugal instability (Rayleigh, 1916). All subsequent work of Taylor-Couette flows focused on this centrifugal instability and its subsequent bifurcations eventually to full turbulence, except two specific work by Wendt (1933) and Taylor (1936) on linearly stable flows with $d\Omega/dR > 0$ which unfortunately are irrelevant to accretion disks, but are used in Richard and Zahn (1999). The relevant flows are quasi-Keplerian flows where $d\Omega/dR < 0$ and $d(R^2\Omega)/dR > 0$.

First modern hydrodynamic experiments on quasi-Keplerian were performed by Richard (2001) and Beckley (2002). Richard found signs of turbulence in quasi-Keplerian flows based on a visualization technique, but the end caps were either corotating with the outer cylinder (the "Ekman" configuration) or a half of the end caps corotating with the inner cylinder while the other half with the outer cylinder (the "Split" configuration). Both configurations are subject to Ekman circulation (Kageyama *et al.*, 2004) which can efficiently transport angular momentum and thus modify flow profile significantly away from the so-called Couette profiles (which are solutions with ideal axial boundary conditions). Thus, the observed turbulence is subject to the interpretation of Ekman

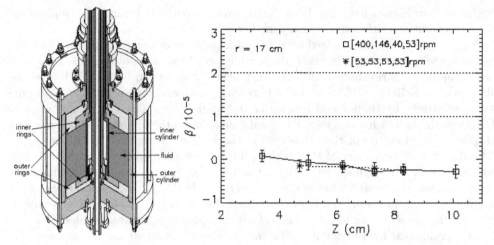

Figure 1. (left) End caps are divided into two rings in the Princeton MRI experiment; each of them are driven at the different speed to minimize the Ekman effect. (right) Measured angular momentum transport in quasi-Keplerian flows is indistinguishable from that in solid body flow with proper boundary controls, indicating no signs of turbulence.

circulation, rather conclusive evidence of nonlinear instability. Beckley (2002) measured torque in quasi-Keplerian flows in an Ekman configuration, and concluded that the measured torque was consistent with the expected angular momentum transport by Ekman circulation. Finally, higher transport levels inferred from torque measurements in quasi-Keplerian flows with end caps corotating with the outer cylinder in a more recent study Paoletti and Lathrop (2010) are also subject to the interpretation based on the Ekman circulation.

To minimize Ekman circulation, a Taylor-Couette device with active controls of axial boundary conditions was built, as shown in Fig.1(left) (Burin *et al.*, 2006; Schartman *et al.*, 2009). Each end cap is split into two rings, and each ring is driven separately at a selectable speed, different from those of the inner and outer cylinders. By choosing appropriate speeds for these rings, Couette profiles can be accurately restored, indicating minimization of Ekman circulation. Local Reynolds stress for angular momentum transport were determined (Ji *et al.*, 2006) through correlating radial and azimuthal velocity components measured by dual Laser Doppler Velocimetry (LDV). It was found that there are no signs of turbulence in quasi-Keplerian flows with boundary controls at Reynolds numbers up to 2×10^6 (Ji *et al.*, 2006; Schartman *et al.*, 2010), as shown in Fig.1(right). The measured transport efficiency is significantly below the requirements from the observed accretion rates. The conjectured nonlinear instability either has happened but at undetectable levels or does not happen until a Reynolds number higher than 2×10^6 is reached. Since the higher the critical Reynolds the less effective the turbulence will be (Lesur and Longaretti, 2005), it is highly unlikely that hydrodynamic turbulence is responsible for fast accretion even for cool disks.

Future Prospects. Despite the recent progress summarized above, an outstanding question remains as to why quasi-Keplerian flows can be so stable even at very high Reynolds numbers. This question may be related to the general observations of stable atmospheric vortices, such as hurricanes and tornados. Detailed studies of nonlinear stability of quasi-Keplerian flows are planned on a newly constructed, mechanically improved Taylor-Couette flow device at Princeton.

Another question regarding angular momentum transport in hydrodynamics concerns coherent wave activity, such as Rossby waves (Rossby *et al.*, 1939), in the context of accretion disks (e.g. Lovelace, 1999). Such waves exist due to the presence of radial variations of Coriolis force and can grow by flow shear into a large amplitude as seen in Earth's atmosphere and oceans. It is planned on the new Princeton hydrodynamic experiment to study Rossby wave excitation using radially varying axial height as in the earlier experiments (Sommeria *et al.*, 1988) but focusing on the associated angular momentum transport. Finally, compressible gas dynamics of quasi-Keplerian flows can be studied in the proposed Princeton plasma MRI experiment described in Sec.4.

3. Magnetohydrodynamic (MHD) Experiments

Current Status. Earlier MHD experiments using liquid metal in Taylor-Couette flow focused on stabilization of centrifugal instability by a magnetic field (Donnelly and Ozima, 1960). Modern MHD Taylor-Couette flow experiments were not proposed (Ji *et al.*, 2001; Rüdiger and Zhang, 2001 until a decade after the rediscovery of the MRI. The standard setup for such an experimental realization (Ji *et al.*, 2001; Goodman and Ji, 2002) is to apply a sufficiently strong axial magnetic field to destabilize an otherwise stable, quasi-Keplerian flow. To date, however, a definitely positive identification of the MRI in the standard setup has not been realized due to technical difficulties in achieving sufficiently fast speeds while maintaining mechanical integrity of the device, such as accurate controls of end rings (Schartman *et al.*, 2009). Currently, the highest speeds achieved in the Princeton MRI experiment are right on the edge of the required speeds to destabilize the MRI.

Nonetheless, there are two noteworthy milestones on the long journey towards demonstrating MRI in the laboratory. The first one is based on an experimental setup using spherical Couette flow (Sisan *et al.*, 2004). Before an axial magnetic field is imposed, the

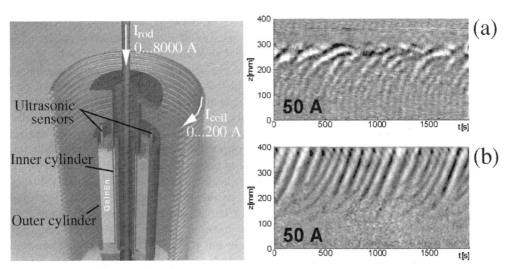

Figure 2. (left) PROMISE 2 device where end caps are split into two pieces with each attached to a cylinder, in contrast to PROMISE 1 (not shown) where the lower end cap rotates with the outer cylinder with the upper cap stationary. Both axial and azimuthal fields are imposed externally. (right) Measured axial velocity perturbations for PROMISE 1 (a) and PROMISE 2 (b) in quasi-Keplerian flows. Upward traveling waves were significantly disrupted by Ekman circulation in PROMISE 1 while much less so in 2, consistent with simulations. Figure courtesy of F. Stefani.

flow is hydrodynamically unstable since outer sphere is stationary. The flow exhibits instabilities with finite magnetic components of different nonaxisymmetric spherical modes at certain strengths of the imposed field, accompanied by increases in torque required to rotate the inner sphere. These instabilities were interpreted as the MRI which has similar expectations. Recently, a 3D MHD numerical study (Gissinger et al., 2010) has been performed in a similar setup but necessarily at lower Reynolds numbers. Most experimental results were successfully reproduced; the resultant instabilities are identified as instabilities of a shear layer, called the Schercliff layer (Gissinger et al., 2010), induced by a combination of sphere boundaries and the applied magnetic field. The Schercliff layer instabilities, however, are probably irrelevant to astrophysical disks due to the absence of these special boundaries there. They may be more relevant to planetary cores where similar boundaries and magnetic field can coexist. Interestingly, coherent oscillations in the external magnetic field (Nornberg et al., 2010) and in the internal flow (Roach et al., 2010) were also identified in the Princeton MRI experiment when a sufficiently strong magnetic field is imposed even at relatively low magnetic Reynolds numbers. These oscillations resemble the predicted Schercliff layer instabilities, but a positive identification is in order.

Hollerbach and Rüdiger (2005) reported a variation of the MRI, called the Helical MRI (HMRI), can be destabilized at much lower Reynolds numbers by imposing an azimuthal magnetic field, in addition to the standard axial magnetic field. The HMRI is an inductionless instability since it survives at the vanishing magnetic Reynolds number limit, in contrast to the standard MRI. It exhibits a form of traveling waves (Liu et al., 2007) which have been subsequently reproduced in the PROMISE experiments (Stefani et al., 2006; Stefani et al., 2009). Once again, the importance of axial boundary conditions was demonstrated: in the original device with significant Ekman circulation the traveling waves were disrupted by returning radial flow slightly above the mid height, while the traveling waves propagate smoothly throughout the flow when Ekman circulation is reduced by modified end caps, see Fig.2. The nature of the HMRI has been identified as a weakly destabilized inertial oscillation, which is unfortunately stable for Keplerian flows (Liu et al., 2006) although the role of some special radial boundary conditions is still subject to further clarifications (Rüdiger and Hollerbach, 2007; Liu, 2008).

Future Prospects. An experimental campaign is currently underway at Princeton to search for the standard MRI at higher Reynolds numbers considerably above the predicted threshold. Probably there is little doubt that some kind of magnetic instability will be found in the new regimes. Rather, the real challenge there is whether we can distinguish the MRI from the aforementioned boundary-induced instabilities which exhibit similar behaviors. This speaks for the crucial importance of numerical simulations and theoretical analyses as exemplified by the cases of Schercliff layer instability and HMRI.

A next challenge for laboratory experiments, also for numerical simulations, is extrapolating the results to astrophysically relevant parameters which are certainly beyond their reach. Therefore, an important step is for experiments to access a wide range of parameters so that meaningful scalings can be established, understood, and thus extrapolated with confidence to astrophysically relevant regimes. To this extent, Princeton liquid metal MRI experiment is planning an upgrade to broaden the accessible magnetic Reynolds numbers but still under well controlled boundary conditions. Separately, an ambitious facility, named "DRESDYN" (Stefani, 2010), is under planning in Dresden, Germany, for an even wider range of accessible parameters. Currently, other related instabilities, the so-called azimuthal MRI under a purely azimuthal magnetic field (Hollerbach et al., 2010) and current-driven Tayler Instability (Rüdiger and Schultz, 2010), are also under investigation in Dresden.

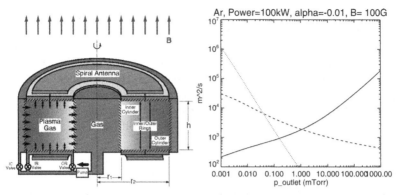

Figure 3. (left) Proposed Princeton Plasma MRI experiment where swirling gas flow is setup by an injection-pumping system in an annular geometry enclosed by two concentric cylinders and end caps. Gas is injected tangentially from the outer cylinder and is pumped out from the inner cylinder. Gas injection and pumping at the end caps are used to adjust rotation profile through adding or removing angular momentum. Spiral antennas are used to transmit RF power into the experiment to ionize the gas with a desirable degree of ionization. (right) Calculated Ohmic (solid line), Hall (dashed line), and ambipolar (dotted line) diffusivities as a function of gas pressure at the outlet. Different regimes can be accessed by selecting appropriate pressures.

Another potentially interesting possibility is resurrection of nonlinear transition to turbulence by a weak and diffusive magnetic field to break hydrodynamic constraints in otherwise stable hydrodynamic flows at large Reynolds number. In fact, resistive HMRI is a linear counterpart of such transitions. Numerical tests of such transitions could be helpful in determining the required initial perturbations and magnetic field strengths for such transitions.

4. Gas and Plasma Experiments

Current Status. There exist a number of physical effects absent in incompressible hydrodynamic or liquid metal experiments but they could be important in accretion disks. A subset of such effects can potentially be studied in gas and plasma experiments: MHD physics at larger magnetic Prandtl numbers than available in liquid metals, non-baroclinic effects due to non-coplanar temperature and pressure gradients, Hall effects due to different ion and electron masses, ambipolar diffusion due to ion-neutral collisions, and kinetic effects such as anisotropic viscosity, anisotropic thermal conduction, and micro-instabilities.

There existed quite a few rotating plasma experiments, but few of them were specifically targeted for MRI studies. Within the MHD framework, if a magnetic field is used to confine plasma from the first place, such as in tokamaks, the plasma β will be necessarily less than unity, and thus stable to the MRI. Instabilities found in these experiments are mostly electrostatic in nature similar to classical interchange instabilities. In two recent rotating plasma experiments (Wang *et al.*, 2008; Teodorescu *et al.*, 2010), magnetic fluctuations were detected but their sources are unidentified among many possible competing candidates other than the MRI.

Future Prospects. There are two active lines of approach to study the MRI in plasmas. The approach under development at Wisconsin, called the Plasma Couette Experiment (Collins *et al.*, 2010), uses the multipolar magnetic configuration of strong permanent magnets mounted on a cylinderical surface to confine high beta plasma inside. The plasma rotation is driven by biasing the plasma perpendicular to the local magnetic field.

The second approach being proposed at Princeton, called the Plasma MRI experiment (Ji, 2011), is based on a prototype experiment using helicon wave plasmas (Ji *et al.*, 2007). The idea is to set up a swirling gas flow with desired profiles by utilizing a gas injection-pumping system shown in Fig.3. Spiral antennas are used to transmit RF power into the experiment to ionize the gas with a desirable degree of ionization. A wide range of outstanding issues can be studied in such device, including: nonlinear hydrodynamic instability, baroclinic instability with axial or azimuthal temperature gradient, MRI in weakly ionized plasmas with Hall effect and ambipolar diffusion.

5. Conclusions

A concise review is given about three kinds of rotating flows in the laboratory: hydrodynamic, MHD, and gas/plasma experiments. Each of these can attack a unique set of outstanding questions in angular momentum transport relevant to astrophysical disks. Hydrodynamic experiments showed that nonlinear hydrodynamic turbulence is unlikely responsible for the turbulence in Keplerian disks while the mechanisms for stability of such a flow are still unclear. MHD experiments using liquid metal are on the verge of realizing the standard MRI while the related helical MRI has been already demonstrated. A challenge there is how to distinguish the MRI from boundary-induced instabilities, such as Schercliff layer instability. Gas and plasma MRI experiments are under development, and rich physics is expected be learned from these experiments such as Hall effect and ambipolar diffusion. In all of these cases, close interactions and collaborations with theory and simulation are crucial in understanding experimental results and in developing scalings applicable to astrophysical disks.

References

Balbus, S. A. & Hawley, J. F., *Astrophys. J.*, 376:214–222, 1991.
Balbus, S. A. & Hawley, J. F., *Rev. Mod. Phys.*, 70:1–53, 1998.
Beckley, H., PhD thesis, New Mexico Institute of Mining and Technology, 2002.
Burin, M. J., Schartman, E., Ji, H., R. Cutler, P. Heitzenroeder, W. Liu, L. Morris, & S. Raftopolous. *Experiments in Fluids*, 40:962–966, 2006.
Chandrasekhar, S., *Proc. Nat. Acad. Sci.*, 46:253–257, 1960.
Collins, C., Katz, N., Weisberg, W., Clark, M., Wallace, J., & Forest, C., *Bull. Am. Phys. Soc.*, 55:BAPS.2010.DPP.NP9.68, 2010.
Couette, M. M., *Ann. Chim. Phys.*, 6:433–510, 1890.
Donnelly, R. J. & Ozima, M., *Phys. Rev. Lett.*, 4:497–498, 1960.
Eckhardt, B., Schneider, T. M., Hof, B., & Westerweel, J., *Annual Review of Fluid Mechanics*, 39(1):447, 2007.
Fromang, S, Papaloizou, J., Lesur, G., & Heinemann T., *Astron. Astrophys.*, 476:1123–1132, December 2007. .
Gissinger, C., Ji, H., & Goodman, J., Instabilities in magnetized spherical Couette ow. to be submitted, 2010.
Goodman J. & Ji, H., *J. Fluid Mech.*, 462:365, 2002.
Hollerbach, R. & Rüdiger, G., *Phys. Rev. Lett.*, 95(12):124501–+, 2005. .
Hollerbach, R., Teeluck, V. & Rüdiger, G., *Phys. Rev. Lett.*, 104(4):044502, 2010.
Ji, H., Magnetorotational instability in a swirling gas and plasma annulus. to be submitted, 2011.
Ji, H., Goodman, J., & Kageyama, A., *Mon. Not. Astron. Soc.*, 325:L1–L5, 2001.
Ji, H., Burin, M., Schartman, E., & Goodman, J., *Nature.*, 444:343–346, 2006.
Ji, H., Foley, J., Levinton, F., Fetroe, B., Raitses, Y., Kefeli, J., Nornberg, M., Zweben, S., & Yamada, M., *Bull. Am. Phys. Soc.*, 52:BAPS.2007.DPP.BP8.86, 2007.

Kageyama, A., Ji, H., Goodman, J., Chen, F., & Shoshan, E., *J. Phys. Soc. Jpn.*, 73:2424–2437, 2004.

Lesur G. & Longaretti, P.-Y., *Astron. Astrophys.*, 444:25–44, 2005.

Lesur G. & Longaretti, P.-Y., *Mon. Not. Astron. Soc.*, 378:1471–1480, July 2007. .

Liu, W., *Phys. Rev. E*, 77:056314, 2008.

Liu, W., Goodman, J., Herron, I., & Ji, H., *Phys. Rev. E*, 74:056302, 2006.

Liu, W., Goodman, J., & Ji, H., *Phys. Rev. E*, 76(1):016310, 2007. .

Lovelace, R., Li, H., Colgate S., & Nelson, A., *The Astrophysical Journal*, 513:805, 1999.

Nornberg, M., Ji, H., Schartman, E., Roach, A., & Goodman, J., *Phys. Rev. Lett.*, 104:074501, 2010.

Paoletti, M. S. & Lathrop, D. P., *ArXiv e-prints 1011.3475*, 2010.

Rayleigh, L., *Proc. Roy. Soc. London A*, 93:148–154, 1916.

Richard, D., PhD thesis, Université Paris 7, 2001.

Richard, D. & Zahn, Z.-P., *Astron. Astrophys.*, 347:734–738, 1999.

Roach, A., Spence, E., Edlund, E., Sloboda, P. & Ji, H., *Bull. Am. Phys. Soc.*, 55: BAPS.2010.DPP.NP9.75, 2010.

C. G. Rossby *et al. J. Mar. Res*, 2(1):38–55, 1939.

Rüdiger, G. & Hollerbach, R., *Phys. Rev. E*, 76:068301, 2007.

Rüdiger, G. & Schultz, M., *Astronomische Nachrichten*, 331:121, 2010.

Rüdiger, G. & Zhang, Y., *Astron. Astrophys.*, 378:302–308, 2001.

Schartman, E., Ji, H. & Burin, M., *Rev. Sci, Instrum.*, 80:024501, 2009.

Schartman, E., Ji, H., Burin, M., & Goodman, J., Stability of quasi-keplerian shear ow in a laboratory experiment. submitted to Astron. Astrophys., 2010.

Shakura, N. I. & Sunyaev, R. A., *Astron. Astrophys.*, 24:337–355, 1973.

Sisan, D. R., Mujica, N., Tillotson, W. A., Huang, Y., Dorland, W., Hassam A.B., Antonsen, T. M. & Lathrop, D. T., *Phys. Rev. Lett.*, 93(11):114502, 2004.

Sommeria, J., Meyers, S. D. & Swinney, H. L., *Nature*, 331:689–693, February 1988.

Stefani, F. private communication, 2010.

Stefani, F., Gundrum, T., Gerbeth, G., Rüdiger, G., Schultz, M., Szklarski, J., & Hollerbach, R., *Phys. Rev. Lett.*, 97:184502, 2006.

Stefani, F., Gerbeth, G., Gundrum, T., Hollerbach, R., Priede, J., Rüdiger, R., & Szklarski, J., *Physical Review E*, 80(6):66303, 2009.

Taylor, G. I., *Philos. Trans. R. Soc. London, Ser. A*, 223:289–343, 1923.

Taylor, G. I., *Proc. Roy. Soc. London A*, 157:546–578, 1936.

Teodorescu, C., Young, WC., Swan, G. W., Ellis, R. F., Hassam, A. B., & Romero-Talamas, C. A., *Phys. Rev. Lett.*, 105:085003, 2010.

Velikhov, E. P., *Sov. Phys. JETP*, 36:995–998, 1959.

Wang, A., Si, J., Liu, W., & Li, H., *Phys. Plasmas*, 15:102109, 2008.

Wendt, F., *Ing. Arch.*, 4:577–595, 1933.

Zeldovich, Y. B., *Proc. Roy. Soc. London A*, 374:299–312, 1981.

Advances in Plasma Astrophysics
Proceedings IAU Symposium No. 274, 2010
A. Bonanno, E. de Gouveia Dal Pino & A. G. Kosovichev, eds.
© International Astronomical Union 2011
doi:10.1017/S1743921311006521

Laboratory simulations of astrophysical jets

**Sergey V. Lebedev[1], Francisco Suzuki-Vidal[1], Andrea Ciardi[2,3],
Matteo Bocchi[1], Simon N. Bland[1], Guy Burdiak[1],
Jerry P. Chittenden[1], Phil de Grouchy[1], Gareth N. Hall[1],
Adam Harvey-Thompson[1] Alberto Marocchino[1], George Swalding[1],
Adam Frank[4], Eric G. Blackman[4] and Max Camenzind[5]**

[1] The Blackett Laboratory, Imperial College London, SW7 2BW London, UK
email: s.lebedev@imperial.ac.uk

[2] LERMA, Université Pierre et Marie Curie, Observatoire de Paris, Meudon, France

[3] École Normale Supérieure, Paris, France. UMR 8112 CNRS

[4] University of Rochester, Department of Physics and Astronomy, Rochester, NY, U.S.A.

[5] University of Heidelberg, Centre for Astronomy Heidelberg (ZAH), Landessternwarte
Koenigstuhl D-69117, Heidelberg, Germany

Abstract. Collimated outflows (jets) are ubiquitous in the universe, appearing around sources as diverse as protostars and extragalactic supermassive black holes. Jets are thought to be magnetically collimated, and launched from a magnetized accretion disk surrounding a compact gravitating object. We have developed the first laboratory experiment to address time-dependent, episodic phenomena relevant to the poorly understood jet acceleration and collimation region (Ciardi *et al.*, 2009). The experiments were performed on the MAGPIE pulsed power facility (1.5 MA, 250 ns) at Imperial College. The experimental results show the periodic ejections of magnetic bubbles naturally evolving into a heterogeneous jet propagating inside a channel made of self-collimated magnetic cavities. The results provide a unique view of the possible transition from a relatively steady-state jet launching to the observed highly structured outflows.

Keywords. ISM: jets and outflows, hydrodynamics, plasmas, methods: laboratory

1. Introduction

The scaled study of astrophysical phenomena in laboratory experiments has seen major advances in recent years due to developments in the field of high-energy-density physics. The generation of extreme states of matter involving high temperatures (110 keV), densities ($10^{-5}10^1$ g/cm^3), and velocities (few x $100 km/s$) provides either a direct link to the physics encountered in a variety of astrophysical objects or a scaled representation of astrophysical dynamics, such as jets (see Ref.2 Remington *et al.*, 2006 for a review).

Although jets and outflows are associated with widely diverse astrophysical environments, they exhibit many common features independent of the central source (Livio, 2002). In all contexts, jets are believed to be driven by a combination of magnetic fields and rotation via, in most cases, an accretion disk (Lovelace, 1976, Blandford & Payne, 1982, Ferreira, 1997, Ouyed *et al.*, 1997). The standard magnetohydrodynamic jet models rely on rotation to twist a large-scale poloidal magnetic field B_P, producing a toroidal field B_Φ that drives and collimates the outflow. Our experiments are designed to model the acceleration and collimation of astrophysical jets taking place under the initial conditions $|B_\Phi| >> B_P$. The experiments also apply to magnetic tower models (Lynden-Bell, 1996, Nakamura *et al.*, 2007), where differential rotation of closed magnetic field lines creates a highly wound magnetic field which drives a magnetic cavity as well as collimating a jet on its axis. How jets survive such unstable field configurations is one of the

26

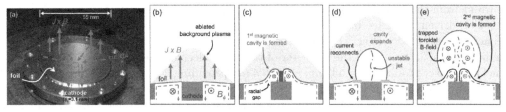

Figure 1. (a) A radial foil inside the discharge chamber of MAGPIE, with the cathode shown schematically (not to scale). (b) Schematic showing the mechanism of episodic magnetic cavity formation triggered by current reconnection at the base of the cathode. In both figures the current path (red-dashed arrows), toroidal magnetic field (blue arrows) and the resultant **JxB** force (green arrows) are shown

open astrophysical issues addressed by the experiments. Another aspect is the origin of the spatial and temporal variability that is observed on all scales in stellar jets (Hartigan *et al.*, 2005), and which is often interpreted as perturbations to a relatively steady flow.

By performing appropriately scaled, high-energy density plasma experiments, extreme laboratory astrophysics has recently emerged as a novel approach to complement our understanding of complex astrophysical phenomena (see Remington *et al.*, 2006 for a review). Jets have been the subject of several studies, which may be distinguished (Blackman, 2007) by whether they addressed problems related to the propagation (e.g. Foster *et al.*, 2005, Ciardi *et al.*, 2008), the launching (Hsu & Bellan, 2002), or both (Lebedev *et al.*, 2005, Ciardi *et al.*, 2007). Here, we present the first laboratory experiments exploring episodic, magnetohydrodynamic (MHD) jets.

Astrophysical jets and outflows are described to first approximation by ideal MHD and our experiments are designed to produce flows in that regime. Its applicability requires the dimensionless Reynolds (Re), magnetic Reynolds (Re_M), and Peclet (Pe) numbers to be much larger than unity; this implies that the transport of momentum, magnetic fields, and thermal energy, respectively, occurs predominantly through advection with the flow. It is important to stress that astrophysical jets have typical values $Re > 10^8$, $Re_M > 10^{15}$, and $Pe > 10^7$ that are many orders of magnitude greater than those obtained not only in the laboratory but also in numerical simulations, which have been so far the sole means of investigating time-dependent behavior of multi-dimensional MHD jets. In ideal MHD simulations, unphysical dissipation occurs at the grid level through numerical truncation errors (Ruy *et al.*, 1995, Lesaffre & Balbus, 2007). For global jet models, mostly performed assuming axisymmetry, the effective numerical Reynolds numbers are typically in the range of $Re_M \sim 10 - 10^3$ (see for example Goodson & Winglee, 1999), and we expect $Re \sim Pe \sim Re_M$. Severe limitations also exist in the range of plasma β, the ratio of thermal to magnetic pressure, which may be reliably modeled numerically (Miller & Stone, 2000). The (inherently) three-dimensional, scaled experiments discussed here extend the range of the dimensionless parameters obtained in the global modeling of jets. Typical values obtained are $Re \sim 5 \mathrm{x} 10^5 - 10^6$, $Re_M \sim 150 - 500$, $Pe \sim 20 - 50$, and $\beta \sim 10^{-3} - 10^3$; the plasma is highly collisional, and the fluid MHD approximation is a valid model. Finally, similar to astrophysical jets, the laboratory flows produced are radiatively cooled.

2. Experimantal Setup

The experimental configuration (Fig. 1) is similar to the radial wire array z-pinch used in our previous experiments (Lebedev *et al.*, 2005b). In the present experiments the current from the MAGPIE generator (peak current of 1.5MA in 250 ns) (Mitchell *et al.*, 1996) is driven into a $6-6.5$ μm thick aluminum foil, which is held radially between two concentric electrodes (Fig. 1). The central electrode (cathode) is a hollow cylinder with a diameter of 3.1mm, with the diameter of the outer electrode being 60 mm. Diagnostics included: laser probing ($\lambda = 532$ nm, $\delta t \sim 0.4$ ns) providing 2-frame interferometry, shadowgraphy and schlieren imaging; time resolved (~ 2 ns exposure) pinhole cameras which recorded emission in the XUV region (> 30eV) providing up to 8 frames per experiment; magnetic pick-up probes to measure any trapped magnetic field inside the outflows; an inductive probe connected to the cathode to measure voltage and thus Poynting flux driving the outflow. The imposed current path (Fig. 1) produces a toroidal magnetic field B_Φ below the foil which is directly proportional to the current and decreases with the radial distance from the cathode ($B_\Phi \propto I(t)/r$). For peak current the toroidal magnetic field can reach magnitudes of $B_\Phi \sim 100T$ (1MG) at the cathode radius.

A schematic diagram of the evolution of a typical experimental jet/outflow is shown in Figure 1. Two outflow components are generally present: a magnetic bubble (or cavity) accelerated by gradients of the magnetic pressure and surrounded by a shell of swept up ambient material, and a magnetically confined jet on the interior of the bubble. The confinement of the magnetic cavity itself relies on the presence of an external medium. The dynamics of the first magnetic bubble and jet are similar to those described in (Lebedev *et al.*, 2005b, Ciardi *et al.*, 2007). In the present work, however, we are able to produce and observe for the first time an episodic jet activity. The main difference with our previous experiments is an increased mass, as a function of radius, being available in the plasma source. The initial gap (see Figure 1) produced by the magnetic field is smaller and can be more easily refilled by the readily available plasma. Its closure allows the current to flow once again across the base of the magnetic cavity, thus re-establishing the initial configuration. When the magnetic pressure is large enough to break through this newly deposited mass, a new jet/bubble ejection cycle begins.

3. Experimental Results

3.1. *Distribution of the Ablated Plasma Above the Foil*

The distribution of plasma ablated from the foil and the initial axial motion of the foil were measured from side-on laser probing images. An example of a laser interferogram obtained at 172 ns after the start of the current pulse in Fig. 2a, where the bulk motion of the foil is seen as a dark non-transparent region above its initial position. The axial displacement increases at smaller radius, which is consistent with a larger magnetic pressure as radius decreases.

Above the boundary it is possible to measure the electronic density of the plasma by following the fringe shift of the interferogram. Fig. 2b shows a 2-D map of integrated electronic density across the plasma $\int n_e(r,z)dL \simeq n_e L$ obtained from the analysis of Fig. 2a, performed with the IDEA interferometric software (Hipp *et al.*, 2004). The contours in this figure represent regions of constant values of $n_e L$. Most of the plasma is concentrated in the region above the cathode, decreasing with distance from the axis. This agrees with a larger ablation rate of plasma at smaller radius, where the Lorentz $\mathbf{J} \times \mathbf{B}$ force is the strongest. The good degree of azimuthal symmetry in the ablated plasma allowed obtaining the radial electronic density $n_e(r)$ by using Abel inversion technique. Radial profiles at of $n_e L$ at different heights from the foil (z=1, 1.3, 2, 3 and 3.5 mm) are shown in Fig. 2c, with their respective Abel-inverted profiles in Fig. 2d. By integrating these profiles along the radius and assuming a value for the ionization of the plasma it is possible

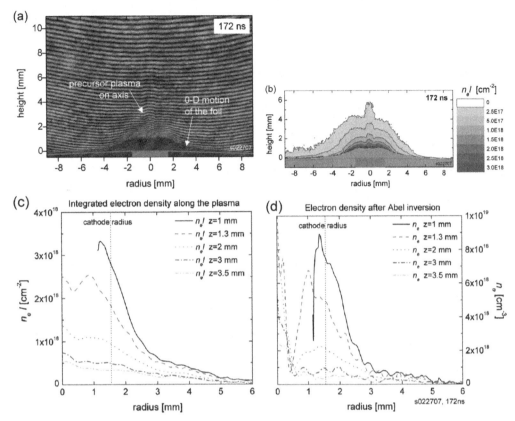

Figure 2. (a) Side-on laser interferogram at 172 ns showing the axial displacement of the foil (dark region) in respect to its initial position (red dotted lines) before the start of the current (the cathode is shown in red). Bending of the interference fringes is due to the plasma expanding from the foil and a hydrodynamic jet on axis. (b) 2-D map of electronic density, integrated along the line of sight $n_e L$ reconstructed from Fig. 2a. (c) Profiles of $n_e L$ as a function of radius at different heights above the foil and (d) the corresponding Abel-inverted electronic density profiles n_e.

to estimate the mass ablated from the foil at this time, which results in $< 1\%$ of the total mass from the foil inside a radius of 3 mm. At this time plasma is seen to expand to a maximum height of $z{\sim}6$ mm. Assuming that plasma is formed at the start of the current pulse this results in an axial expansion velocity of $V_Z \sim 35$ km/s. An interesting feature of the ablated plasma motion above the foil is the formation of a precursor plasma jet on the axis of the foil, which is seen in Fig. 2 as an increase of the electronic density in the region above the cathode. This jet is formed from the plasma which is redirected towards the axis by radial pressure gradients, as there is no ablation above the cathode, which initially leaves an empty region on the axis. The converging towards the axis plasma flow forms a standing shock which redirects the flow in the vertical direction, forming a plasma jet. The formation of this "hydrodynamic" plasma jet is similar to jet formation in our previous experiments with conical wire arrays (Lebedev *et al.*, 2002). The dynamics of the interaction of this hydrodynamic jet with an ambient gas introduced above the foil is relevant to the modeling of jet-ambient interaction, and first results of such experiments are presented in (Suzuki-Vidal *et al.*, 2009).

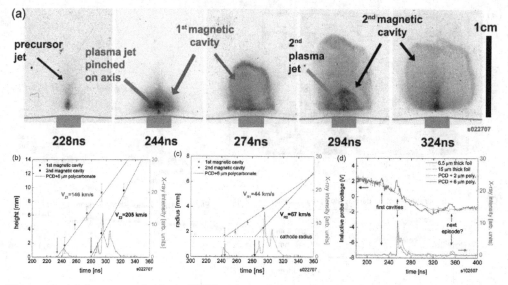

Figure 3. (a) Sequence of side-on XUV emission images showing the formation of two subsequent magnetic cavities during the same experiment. The initial positions of the foil and the cathode (with a diameter of 3.1 mm) are shown schematically. (b)-(c) Measured maximum height and radius of the two magnetic cavities from (a). The X-ray emission is also shown (in red). (d) Top: Voltage signals from an inductive probe from a 6.5 μm thick (solid black) and 15 μm thick (dashed blue) foils. Bottom: X-ray emission from a 6.5 μm thick foil, which is correlated to rapid changes in inductance

3.2. *Formation and Dynamics of Episodic Jets*

The formation of magnetically driven jets starts later in time, when the Lorentz $\mathbf{J}_R \times \mathbf{B}_\Phi$ force (which is strongest at the cathode radius) leads to ablation of all of the foil mass near the cathode and to the formation of a small radial gap between the cathode and the remainder of the foil. From this moment the Poynting flux can be injected through this gap into the region above the foil. The toroidal magnetic field pushes the ablated plasma axially and radially outwards and also pinches the plasma on axis, forming a magnetic tower jet configuration. At this stage the current flows along the jet on the axis of the magnetic cavity and along the walls of the cavity, in the same way as in our previous experiments (Lebedev *et al.*, 2005b, Ciardi *et al.*, 2007). The magnetic pressure from these rising toroidal loops inside the cavity inflates it both radially and axially, with measured velocities of $V_R \sim 50 - 60$ km/s and $V_Z \sim 130 - 200$ km/s respectively. Experimental results demonstrating such dynamics are shown in Fig. 3. The first cavity shows a high emitting region on the axis, which can be identified as pinching of plasma. It can be seen that the initial diameter of the bubble is given by the diameter of the cathode. The most prominent feature of this new experimental set-up is that we now observe several subsequent outflows formed in the same experiment. It is possible to follow the axial positions of the subsequent episodes of the outflows shown in Fig. 3, with Fig. 3b presenting the measurements that allowed the determination of their velocities along the axis. It is seen that each outflow is expanding with approximately constant velocity, and the extrapolation of the trajectories back in time allows determining the starting time for each episode. The second bubble expands with a faster velocity, $V_Z = 205$ km/s while the first with only 145 km/s. This increase in velocity is consistent with sweeping of the ambient plasma by the earlier episode, thus allowing the subsequent magnetic bubbles to propagate through a lower ambient density. Up to five subsequent magnetically driven cavities were observed (Ciardi *et al.*, 2009, Suzuki-Vidal *et al.*, 2009). Figure 3b also shows that the episodic outflows are accompanied by episodic outbursts of soft X-rays (detector sensitive to photon energy between 200-300 eV and above 800 eV), which can be well correlated with the formation of each new magnetic tower jet. This is an indication that each new episode starts from the pinching of plasma on the axis of the magnetic cavity and that pinched plasma is the source of the X-ray emission. Both the axial expansion dynamics and the periodicity of X-ray emission show a timescale of ∼30 *ns* for the formation of subsequent magnetic tower outflows. The fast rising (∼5 *ns*) part and peak of the emission are related to the maximum compression of the jet. We measured jet temperatures up to ∼300 eV using spatially resolved, time-integrated spectroscopic measurements of H-like to He-like line ratios (Ciardi *et al.*, 2009). Typical flow velocities observed are ∼100 − 400 *km/s*, the simulated sonic and the Alfvénic Mach numbers in the jet, defined as the ratios of the flow speed to the sound and Alfvén speed, respectively, are $M_S \sim M_A \sim 3 - 10$.

The voltage responsible for reconnection of current across the radial gap between the cathode and the leftover foil is induced by the dynamics of the magnetic cavity and was measured using an inductive voltage probe. The voltage measured by this probe is proportional to the time derivative of the magnetic flux produced by the current from the cathode and along the foil, connecting to ground. Thus the probe voltage V_{ind} is equal to $V_{ind} = -d(LI)/dt$, where the inductance L is related to the current path. An example of the inductive voltage probe signal obtained in a different experiment is shown in Fig.3d, where the voltage signal from a standard 6.5 *µm* thick foil is compared with a reference shot, in which a foil with a thickness of 15 *µm* was used. In the case of the reference shot the foil did not move on the time-scale of the experiment, the inductance was constant and measured voltage was equal to $V_{ind} = -LdI/dt$.

For the case when the episodic jets were produced, for a 6.5 *µm* foil, the inductive probe shows rapid changes in voltage in respect to the reference case, as seen at ∼220, 250 and 290 *ns*. The timing of these voltage spikes agrees with the X-ray emission measured in the experiment (shown at the bottom of Fig. 3d). The deviations of the voltage from the reference case come from an additional inductance due the change in the current path as the current now flows through the central jet returning along the wall of the cavity. This inductance is time dependent as the cavities expand and the jet is pinched. Fig. 3d also shows that the inductive probe voltage is correlated to the X-ray emission pulses, which are in turn correlated to the formation of two subsequent magnetic cavities seen from the imaging diagnostics. The use of an inductive probe allowed to measure the inductance associated with the formation of a magnetic cavity and thus to obtain estimates the energy balance during an episode, i.e. magnetic energy and Poynting flux. The electromagnetic energy delivered to a typical bubble via Poynting flux is ∼800 J, and we estimate the kinetic energy of the flow to be ∼100 − 400 J, the remainder is in the magnetic energy, internal energy of the plasma and partly lost to radiation. These results indicate that ∼25% of Poynting flux energy entering the magnetic cavity is converted into kinetic energy of the outflow. Details on these measurements will be presented in future publications.

The temporal evolution of the jets and bubbles on the longer time-scale is presented in Figure 4. A succession of multiple cavities and embedded jets are seen to propagate over length scales spanning more than an order of magnitude. The resulting flow is heterogeneous and clumpy, and it is injected into a long-lasting and well-collimated channel made of nested cavities. It is worth remarking that the bow-shaped envelope is driven by the magnetic field and not hydrodynamically by the jet.

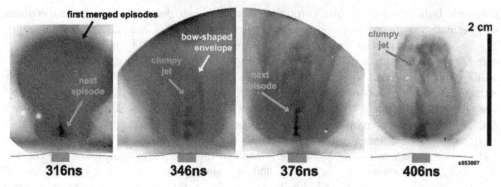

Figure 4. Time series of filtered XUV emission images showing late-time evolution of episodic jets

3.3. *Trapped magnetic field inside a magnetic cavity*

The estimates presented above show that the magnetic Reynolds number Re_M is much greater than unity and some magnetic flux should remain trapped inside the outflows. In particular, we can expect conservation of magnetic flux accumulated in the first cavity by the time the second cavity starts to form. The presence of toroidal magnetic field inside the expanding magnetic cavities was measured with a magnetic probe. The probe had five circular turns of 3mm diameter and it was placed 10 mm above the foil at a radial distance of ~13 mm from the axis, as shown in Fig. 5. The probe orientation was such that it measured the toroidal component of the magnetic field. To exclude capacitive coupling be between the probe and the cathode and also to prevent possible electron flow from the cathode region reaching the probe, a 1mm thick stainless steel diaphragm was installed at $z \sim 2\ mm$ above the foil. The magnetically driven jets were formed through a 10 mm diameter aperture in the diaphragm.

An XUV emission image in Fig. 5(b) shows two magnetic cavities expanding above the metallic diaphragm. The position of the cathode, the diaphragm, and the circular aperture on the axis are shown schematically in this figure. The position of the magnetic probe can also be seen from this image as a circular emitting boundary, due to the interaction of its outer shielding with the expanding cavity. This image shows that the addition of the probe and the diaphragm did not affect the overall dynamics of episodic jet formation.

The voltage measured by the magnetic probe is shown in Fig. 5(c). It can be seen that the voltage is zero until ~350 ns, which is consistent with the time when the magnetic cavity reaches the probe as seen from imaging diagnostics. The expected response of the magnetic probe assuming a sharp boundary for the toroidal magnetic field inside the expanding bubble can be calculated by taking into account the geometry of the probe and the dynamics of the radial expansion of the cavity, as shown in Fig. 5(b). For the relatively small size of the probe, we can assume that the wall of the magnetic cavity passing the probe at distance $R_B(t)$ from the axis is planar. When the wall of the cavity reaches the probe, the trapped toroidal magnetic field inside the expanding cavity B_{Φ_trap} will induce a voltage in the probe from the change in the magnetic flux Φ

$$V_{probe} = -\frac{d\Phi}{dt} = -\frac{d\left[B_{\Phi\_trap}(t) \cdot S(t)\right]}{dt} \tag{3.1}$$

where $S(t)$ is the cross-section area of the probe inside the cavity, related to the radial expansion velocity of the cavity V_R. The radial expansion velocity of the magnetic cavity

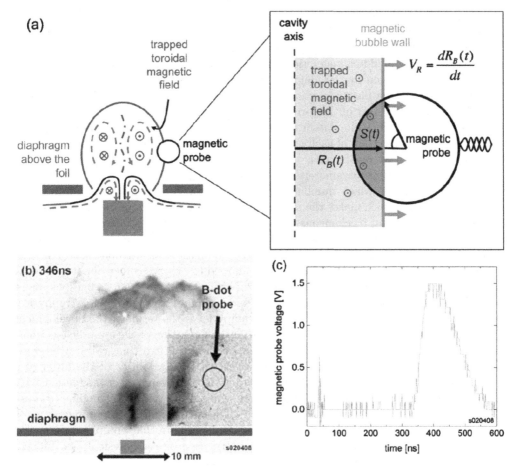

Figure 5. (a) Schematic setup of a radial foil with the addition of a magnetic probe and a metal diaphragm. (b) XUV emission at 346 *ns* and (c) signal from the magnetic probe obtained in the same experiment

for the particular shot presented in Fig. 5(b) was measured as $V_R \sim 90 \ km/s$, thus corresponding to a signal from the magnetic probe of $B_{\Phi\_trap} \sim 0.3$ T.

This estimate can be compared with the expected magnetic field in the expanded magnetic cavity, assuming that the toroidal magnetic flux present in the first cavity at the start of second cavity formation is conserved. The expected magnetic field ($B_\Phi \sim 1.5$ T) is of the right magnitude, though an accurate estimate of the expected field is not possible due to the uncertainty of the current loop inside the cavity at the time when it reaches the magnetic probe. To improve this comparison, spatially resolved measurements of the magnetic field inside the cavity are needed and three-dimensional MHD simulations (Ciardi *et al.*, 2009) indicate that the initially toroidal magnetic field becomes entangled due to the development of the kink mode of current-driven instability.

4. Summary

In this paper, we have presented results from high energy density plasma experiments designed to investigate the physics of magnetically driven, supersonic, radiatively cooled plasma jets. The most important new feature that appears in the radial foil configuration is the generation of several subsequent episodes of magnetically driven jets. The formation of the first outflow is similar to that previously observed in the radial wire array configuration. The outflow consists of a jet accelerated and confined by a toroidal magnetic field and embedded in a magnetic cavity. The cavity expansion into the surrounding ambient plasma is driven by the magnetic pressure. Reconnection of the current at the base of the cavity via a plasma expanding from the central electrode and the remnants of the foil leads to the start of the formation of the next outflow episode. The generation of several episodes is reproducible in the experiments and up to three to four eruptions are observed. The formation of each episode is correlated with a burst of x rays from the compressed/pinched jet on the axis.

The key dimensionless parameters Pe, Re, and Re_M are much greater than unity and, together with the Mach number and plasma beta, are all in the astrophysically appropriate regime, which makes the results of the experiments relevant to understanding the physics of launching mechanisms of astrophysical jets. The estimated values of the magnetic Reynolds number in these experiments, $Re_M \sim 200 - 1000$, are comparable or exceed those obtained in global numerical MHD simulations of astrophysical jets. The high values of the magnetic Reynolds number allowed to observe convection/trapping of the toroidal magnetic flux by the. The long term evolution of jets from radial foils might help addressing questions about the spatial and temporal variability of astrophysical jets. In our experiments, the formation of a clumpy jet is the result of the development of current-driven instabilities, which occur within the formation of the episodic ejections. The time variability in the experiments is characterized by two timescales of interest. The first is the growth time of the current-driven MHD instabilities of the order of a few nanoseconds in the experiments. The second is the relatively longer magnetic cavity ejection period of $\sim$30 ns. The ratio of both timescales are in the similar regime to those observed in protostellar jets (Ciardi *et al.*, 2009). The episodic formation of magnetically driven jets observed in the experiments allows us to speculate that an episodic scenario of jet formation could be also applicable to the formation of astrophysical jets. Indeed, steady jets confined by a toroidal magnetic field should be highly unstable to current-driven instabilities, unless stabilized by, e.g., a closely positioned rigid "wall" acting as a path for the return current. In the episodic jet formation scenario, the time for the growth of the current-driven modes is reduced to roughly the duration of one episode. The development of the instability could produce, as in the experiments, a clumpy outflow which still retains a high degree of collimation and propagates ballistically after the end of the episode. The resulting outflow in this scenario would have density and velocity variations due to both the current-driven instability and the episodicity of the ejection. We should note that episodic jet formation appeared in several numerical simulations of young stellar objects (YSO) jet launching (Goodson *et al.*, 1997, Goodson & Winglee, 1999, Goodson *et al.*, 1999b, Romanova *et al.*, 2006).

Acknowledgements

This work was supported by the EPSRC Grant No. EP/ G001324/1, by the NNSA under DOE Cooperative Agreement Nos. DE-F03-02NA00057 and DE-SC-0001063, and by a Marie Curie RTN fellowship, as part of the JETSET (Jet Simulations, Experiments and Theories) project.

References

Blackman, E. G., 2007, *A&SS*, 307, 7
Blandford, R. D. & Payne, D. G. 1982, *MNRAS*, 199, 883

Ciardi, A., Lebedev, S. V., Frank, A., Blackman, E. G., Chittenden, J. P., Jennings, C. J., Ampleford, D. J., Bland, S. N., Bott, S. C., Rapley, J., Hall, G. N., Suzuki-Vidal, F. A., Marocchino, A. Lery, T., & Stehle C. 2007, *Physics of Plasmas*, 14, 056501

Ciardi, A., Ampleford, D. J., Lebedev, S. V., & Stehle C., 2008, *The Astrophysical Journal*, 678, 968-973

Ciardi, A., Lebedev, S. V., Frank A., *et al.* 2009, *The Astrophysical Journal*, 691, L147-L150

Ferreira, J. 1997, *A&A*, 319, 340

Foster, J. M., Wilde, B. H., Rosen, P. A., Williams, R. J. R, Blue, B. E., Coker, R. F., Drake, R. P., Frank, A., Keiter, P. A., Khokhlov, A. M., Knauer, J. P., & Perry, T. S. 2005, *The Astrophysical Journal*, 634, L77-L80

Goodson, A. P., Winglee, R. M., & Böm, K. H. 1997, *The Astrophysical Journal*, 489, 199

Goodson, A. P. & Winglee, R. M. 1999, *The Astrophysical Journal*, 524, 159

Goodson, A. P., Böm, K. H., & Winglee, R. M., 1999, *The Astrophysical Journal*, 524, 142

Hartigan, P. *et al.* 2005, *Astronomical Journal*, 130, 2197

Hipp, M., Woisetschlaeger, J., Reiterer, P., & Neger, T., 2004, *Measurement*, 36, 53-66

Hsu S.C., Bellan, P. M., 2002, *MNRAS*, 334, 257

Lebedev, S. V., Chittenden, J. P., Beg, F. N., *et al.* 2002, *The Astrophysical Journal*, 564, 113-119

Lebedev, S. V., Ciardi, A., Ampleford, D. J., *et al.* 2005, *Plasma Physics and Controlled Fusion*, 47, B465-B479

Lebedev, S. V., Ciardi, A., Ampleford, D. J., *et al.* 2005, *MNRAS*, 361, 97-108

Lesaffre, P. & Balbus, S. A., 2007, *MNRAS*, 381, 319

Livio, M. 2002, *Nature*, 417, 125

Lovelace, R. V. E., 1976, *Nature*, 262, 649

Lynden-Bell, D., 1996, *MNRAS*, 279, 389

Miller, K. A. & Stone, J. M. 2000, *The Astrophysical Journal*, 534, 398

Mitchell, I. H., Bayley, J. M., Chittenden, J., *et al.* 1996, *Rev. Sci. Instr.*, 67, 1533-1541

Nakamura, M., *et al.* 2007, *The Astrophysical Journal*, 656, 721

Ouyed, R., *et al.* 1997, *Nature*, 385, 409

Remington, B. A., Drake, R. P., & Ryutov, D. D., 2006, *Review Modern Physics*, 78, 755

Romanova, M. M., Kulkarni, A., Long, M., *et al.* 2006, *Adv. Space Res.*, 38, 2887

Ryu, D., *et al.* 1995, *The Astrophysical Journal*, 452, 785

Suzuki-Vidal, F., Lebedev, S. V., Ciardi, A., *et al.* 2009, *Astrophysics and Space Science*, 322, 19-23

Advances in Plasma Astrophysics
Proceedings IAU Symposium No. 274, 2010
A. Bonanno, E. de Gouveia Dal Pino & A. G. Kosovichev, eds.
© International Astronomical Union 2011
doi:10.1017/S1743921311006533

Laboratory-generated Coronal Mass Ejections

Christopher Watts[1], Yue Zhang[1], Alan Lynn,[1] Ward Manchester[2] and C. Nick Arge [3]

[1]University of New Mexico, Albuquerque, NM 87131, USA;

[2]University of Michigan, Ann Arbor, MI 48109, USA;

[3]Air Force Research Laboratory, Kirtland AFB, Albuquerque, NM 87117, USA

Abstract. We have begun a series of laboratory experiments focused on understanding how coronal mass ejections (CME) interact and evolve in the solar wind. The experiments make use of the Helicon-Cathode (HelCat) plasma facility, and the Plasma Bubble eXperiment (PBeX). PBeX can generate CME-like structures (sphereomak geometry) that propagate into the high-density, magnetized background plasma of the HelCat device. The goal of the current research is to compare CME evolution under conditions where there is sheared flow in the background plasma, versus without flow; observations suggest that CME evolution is strongly influenced by such sheared flow regions. Results of these studies will be used to validate numerical simulations of CME evolution, in particular the 3D BATS-R-US MHD code of the University of Michigan. Initial studies have characterized the plasma bubble as it evolves into the background field with and without plasma (no shear).

Keywords. Plasmas, Coronal Mass Ejections(CMEs), Laboratory

1. Introduction

Coronal mass ejections (CMEs) are the most stunning activity of the solar corona in which typically$10^{12} - 10^{13}$ kg of plasma is hurled into interplanetary space with a kinetic energy of the order $10^{24} - 10^{25}$ J. They are of intense practical interest, as ultimately they are the cause of major geomagnetic storms here on Earth. Thus, it is important to understand the evolution of the CME as it propagates from the Sun to Earth through the solar wind. To this end, we have developed a laboratory experiment to "simulate" the dynamics of CME-like structures and their interactions with a magnetized background plasma.

There are three major goals to this project. First, we will attempt to gain a much better understanding of the dynamics of CMEs in the corona through direct and controlled laboratory experimentation that can, in many ways, realistically replicate the conditions found in the Sun's corona from 2-10 solar radii ($R_\odot$). Second, we will use the University of Michigan state-of-art 3D MHD numerical model to evaluate and validate how well it can simulate the results generated by the plasma physics experiment. And then finally, determine what new physics, or improvement in physics, may need to be added to the numerical model in order to more realistically simulate actual CMEs. The overall goal is to further develop and validate a sophisticated numerical code that can be applied with more confidence to a broad spectrum of interacting plasmas situations, including especially CME-solar wind interactions.

36

 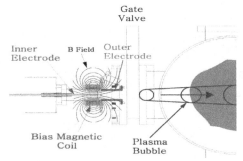

Figure 1. Left: Side view HelCat plasma device, with the PBeX electronics in the foreground. Right: Schematic of the PBeX gun system

2. Experimental Apparatus

The Plasma Bubble eXperiment (PBeX) magnetized coaxial gun is mounted on one side wall of the HelCat vacuum chamber (see Figure 1). HelCat is a large plasma facility (Christopher Watts 2005, Lynn *et al.*, 2009), 4 m long, 50 cm diameter with two plasma sources: an RF helicon and a thermionic cathode. The field strength and large diameter are sufficient to insure that the ions are magnetized. A large number of both large and small ports allows for easy diagnostic access. The two sources are able to generate background plasmas over an extensive regime of plasma operating parameters for a wide range of basic plasma studies, with relevance to fusion and astrophysical plasmas. These studies include turbulence dynamics, sheared flows, and Alfvénic waves.

Typically, the plasma is $\sim 100\,\%$ ionized. However, the sources can be adjusted to create a large neutral fraction, if desired. HelCat also makes use of several sets of electrodes/grids that can be biased to induced sheared plasma flows. Of particular interest in our studies is the propagation of CME-like structures through regions of sheared flow.

Also in Figure 1 is a schematic depiction of the PBeX gun. It injects a high density, magnetized, supersonic plasma jet into the magnetized background plasma generated independently in the HelCat device. This can, in effect, simulate the evolution of a CME into the solar wind. The central component of PBeX is coaxial plasma gun designed to launch a magnetized plasma bubble or jet with various magnetic configurations. Of particular interest for this project is spheromak geometry. The jet has an initial density of $\sim 10^{20}/\mathrm{m}^3$, a factor of 10 - 10^4 greater than the background plasma density; the magnetic field can be adjusted to up to 2 orders of magnitude larger than that of the background plasma. Thus, the bubble expands into the background at supersonic and super-alfvénic speeds, generating a shock front at the interface. Depending on the orientation of the background magnetic field, various Alfvénic (magnetic) shock configurations can be obtained, along with a range of magnetic reconnection geometries.

The laboratory is equipped with a wide array of diagnostic capabilities to allow us to study the PBeX dynamics. These include a fast imaging camera (1.5 ns gate time) for global visualization, a 4m spectrometer for detailed ion temperature measurements, 3D magnetic and electrostatic probe arrays for localized magnetic field, electron temperature and ion density characterization, two microwave interferometers for chord averaged density measurements, and a microwave reflectometer for localized density measurements. Thus we can characterize the expanding plasma bubble "CME" with unprecedented detail that would only be achieved with real CMEs in space using a large constellation of spacecraft.

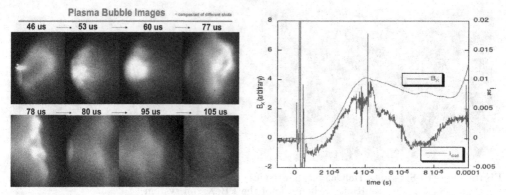

Figure 2. Left: Evolution of the PBeX plasma bubble in HelCat. Though the image sequence is from several different shots, the dynamics are very reproducible. **Right**: Plot of magnetic field and density (Isat) evolution during bubble expansion. At startup the two traces overlay, indicating the frozen flux in the expanding bubble shell

Solar prominence and flare formation have been modeled in the laboratory by several groups (Bellan, 1997, Hsu *et al.*, 2005) . These have used plasma generation mechanisms designed to create twisted flux ropes, and are focused on understanding formation and flux rope dynamics early in time. In contrast to this work, we are not interested in the formation dynamics of the CME. Rather, we specifically focus on the CME evolution long after the initial formation, as it is propagating and interacting with the solar wind. Here, the background plasma - not present in their experiments – plays a crucial role.

PBeX has been in operation only one year, and initial studies of the expanding plasma bubble have demonstrated the essential dynamics. Density, temperature and magnetic field measurements indicate the bubble has the expected parameters. Figure 2 shows a series of photographs taken with the fast imaging camera, depicting the evolution of the plasma bubble as it expands in the background HelCat plasma. Qualitatively, at least, the structure is similar to that of a CME as it expands into the interplanetary medium.

3. Modeling Connection

Our modeling effort will make use of the University of Michigan state-of-the-art 3-D MHD BATS-R-US code (Powell *et al.*, 1999, Groth *et al.*, 2000) to couple our PBeX laboratory experiments to solar wind observations. In the typical region of interest for CME study, 2-10 $(R_\odot)$, coronal particle densities range from 5×10^{11} to 10^{13} particles per cubic meter. Beyond 3 $(R_\odot)$ the plasma is essentially collisionless such that the ionization states remain frozen in as wind expands and cools. By contrast, the plasma experiments are performed with particle densities orders of magnitude higher, where particle interactions are dominated by collisions. The question naturally arises: Can a meaningful comparison be made between the experiments, coronal mass ejections and ideal MHD simulations?

Direct observations clearly indicate that the low-density plasma of the solar wind behaves in many ways like a much denser collision-dominated plasma. This fact is most obvious at the planetary bow shocks that form in the solar wind. The particle coupling and dissipation necessary for shock formation is attributed to wave-particle interaction, which makes the plasma behave like a much denser plasma in which coupling occurs directly by particle collisions. Shock-capturing MHD simulations performed with BATS-R-US accurately reproduce the behavior of CME-driven shocks in the solar wind. An example is found in Manchester *et al.* (Manchester *et al.*, 2008), where they accurately simulate the structure of a CME-driven shock observed with the LASCO C3 coronagraph onboard the SOHO observatory.

Conversely, the PBeX bubble evinces behavior of a much less collisional Ideal MHD plasma. Figure 2 also compares the evolution of the magnetic field and density (ion saturation current) measured at a single point as the bubble evolves past. Such measurements of density and magnetic field indicate that, at least initially (between 0 and 40 s), the field and plasma at the bubble's leading edge flow together, i.e., the flux is "frozen" into the expanding plasma. This, in spite of what the (low) magnetic Reynold's number might indicate. The magnetic Reynold's number R_s, and Lundquist number S, both characterize the coupling of the plasma and magnetic flux; a large value indicates the magnetic field is "frozen in" to the plasma. However, the highly collisional nature of the high-density laboratory plasma coupled with the short scale lengths artificially deflate Rm, which in HelCat/PBeX is about 100-1000 as opposed to $> 10^6$ in the solar wind, despite the fact that experiments confirm the flux is indeed convected with the plasma (or *vice versa*).

4. Perspective

Initial results of the PBeX experiment on the HelCat device, aimed at modeling CME propagation through the solar wind, have been presented. These measurements indicate that PBeX is highly relevant to understanding the less collisional, larger scale CME event, where Ideal (collisionless) MHD describes the plasma dynamics, and is also relevant for benchmarking the BATS-R-US code. Near term work will focus on the interaction of the bubble with the background plasma in terms of the evolution of magnetic helicity, magnetic and thermal energy, and azimuthal rotation. Also these experiments will investigate the dependence of flux amplification on gun parameters as well as background plasma parameters such as the background thermal and magnetic pressure. Longer term, we plan investigations with two major thrusts: 1) Understanding magnetized shock formation and evolution at the boundary between the bubble and background plasma, and 2) Characterizing the bubble evolution in the presence of sheared flows in the background plasma.

References

Bellan, P. M., 1997, Proceedings of the 4th IPELS Workshop, Maui, Hawaii

Groth, C. P. T., De Zeeuw, D. L., Gombosi, T. I., & Powell, K. G., 2000, *Journal of Geophysical Research*, 105, 25053

Hsu, S. C. & Bellan, P. M., 2005, *Phys. Plasmas*, 12, 1

Lynn, A., Gilmore, M., Watts, C., Herrea, J., Kelly, R., Will, S., Shuangwei, X., Lincan, Y., & Yue, Z., 2009, *Rev. Sci. Inst.*, 80, 103501

Manchester IV, W. B., Vourlidas, A., Toth, G., Lugaz, N., Roussev, I. I., Sokolov, T. V., Gombosi, T. I., De Zeeuw, D. L., & Opher, M., 2008, *Astrophysical Journal*, 684, 1448

Powell, K. G., Roe, P. L., Linde, T. J., Gombosi, T. I., & De Zeeuw, D. L., 1999, *Journal of Computational Physics*, 154, 284

Watts, C., 2005, *Rev. Sci. Inst.*, 75, 1975

Advances in Plasma Astrophysics
Proceedings IAU Symposium No. 274, 2010
A. Bonanno, E. de Gouveia Dal Pino & A. G. Kosovichev, eds.
© International Astronomical Union 2011
doi:10.1017/S1743921311006545

Region-1 field aligned currents in experiments on laser-produced plasma interacting with magnetic dipole

I. F. Shaikhislamov, Yu. P. Zakharov, V. G. Posukh, E. L. Boyarintsev, A. V. Melekhov, V. M. Antonov and A. G. Ponomarenko

Institute of Laser Physics SB RAS, pr.Lavrentyeva 13/3, 630090, Novosibirsk, Russia
e-mail: `ildars@ngs.ru`

Abstract. In previous experiments by the authors a generation of intense field aligned current (FAC) system on Terrella poles was observed. In the present report a question of these currents origin in a low latitude boundary layer of magnetosphere is investigated. Experimental evidence of such a link was obtained by measurements of magnetic field generated by tangential sheared drag. Results suggest that compressional and Alfven waves are responsible for FAC generation. The study is most relevant to FAC generation in the Earth and Hermean magnetospheres following pressure jumps in Solar Wind.

Keywords. Magnetosphere, field aligned current, low latitude boundary layer.

---

1. Introduction

Field-aligned currents play an important role in the magnetosphere-ionosphere coupling. It was proposed as early as in Eastman (1976) that a source of dayside Region-1 FAC is in the low-latitude boundary layer (LLBL) where transfer of plasma, momentum and energy from the magnetosheath to the magnetosphere takes place. Plasma in a thin layer on each flank moves antisunward and stretches frozen magnetic field lines. On the inner side of the layer tangential stress is mapped along closed field lines establishing convection pattern. The stress is loaded on ionosphere generating electric field and net cross-polar current. This dawn-dusk Region-1 current drags ions antisunward over poles and decelerates tangential plasma motion in LLBL far from the Earth.

There is a body of observational data on FACs at high latitudes, but so far nothing is known about related features in a driver region. While only *in sutu* observations can unequivocally link ionospheric FACs with LLBL, in this work we present exactly such evidence obtained in laboratory experiment. As described in the previous paper by Shaikhislamov *et al.* (2009), laser-produced plasma interacting with magnetic dipole, besides forming a well defined dayside magnetosphere, generates intense field aligned current system. Detailed measurements of total value and local current density, of magnetic field in equatorial and polar region revealed its similarity to the Region-1 system in the Earth. Such currents were found to exist only if they can closure via conductive cover of the dipole. Comparison of conductive and dielectric cases revealed specific magnetic features produced by FAC and their connection with electric potential generated in the LLBL. The next natural step is to look in the LLBL for a specific magnetic feature associated with FAC, namely for a tangential magnetic field produced by sheared plasma motion. Measurements presented in this paper reveal that such a field is indeed present.

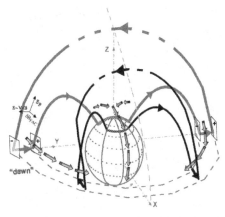

Figure 1. A scheme of magnetosfperic generator, On the flanks of LLBL plasma moves along the boundary across field and with shear along magnetic field. This generates electric field and azimuthal component of magnetic field δB_{FAC} such that the Pointing vector S_p is directed upward. Sequential arrows show that azimuthal field is positive at the dawnside and negative at the duskside. Magnetic field produced by FAC loops inside magnetosphere is sunward over the North pole and opposite to the dipole field at the equator.

It has specific spatial structure with maximum inside magnetosphere and immediately adjacent to LLBL.

2. Experimental results and conclusions

Fig. 1 demonstrates schematically magnetospheric generator driving Region-1 FAC in the Northern hemisphere. By thick lines two current loops are drawn, in the terminator plane and topologically similar at the dayside. Both loops closure through the ionosphere.

Experiment has been carried out at KI-1 space simulation Facility, which includes vacuum chamber $\varnothing 1.2 \times 5$ m. Two CO_2 laser beams of 70 ns duration and 150 J of energy each were focused and overlapped into a spot 1 cm in diameter on surface of a solid target. Plasma expanded inertially in a cone ≈ 1 sr with an average velocity $V \approx 1.5 \cdot 10^7$ cm/sec. A total kinetic energy and a number of ions in the plasma was ≈ 40 J and $\approx 5 \cdot 10^{17}$ respectively. At the axis of plasma expansion at a distance of 60 cm magnetic dipole with moment $\mu = 1.15 \cdot 10^6$ $G \cdot cm^3$ was placed. The dipole has a spherical stainless cover shield 8 cm in radius. The surface could be made un-conductive by placing over a thin dielectric film. At a time of about $t = 2$ μs after laser irradiation a well defined magnetopause is formed at a distance of about $R_m = 15$ cm. Due to specifically adjusted pulse and tail generation mode of laser oscillator, main plasma flow is followed by a second pressure jump. It slightly compresses magnetosphere and causes pronounced response in FAC. Such conditions resemble a pressure jump in the SW.

By measuring magnetic field above the equator plane ($Z = 4$ cm) and off the meridian plane at the afternoon and postnoon sections for North and South orientation of dipole moment it was found that the azimuthal component of magnetic field has quadruple structure that corresponds to Fig. 1. It changes sign at crossing either meridian or equator plane. Most importantly, the structure of the azimuthal field was different for conductive and dielectric dipole covers, that is in dependence on FAC system being present or not. The essence of the difference is revealed in spatial profiles given in Fig. 2 for a number of sequential times marked in μs. From the main δB_Θ component of magnetic field perturbation (upper panels) one can see how the plasma forms magnetosphere and

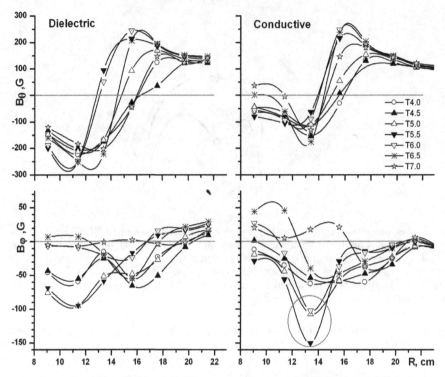

Figure 2. Spatial profiles of δB_Θ and δB_φ components at the duskside at different times for cases of dielectric and conductive dipole cover.

how the second pressure jump pushes it slightly closer to dipole. The boundary layer (from signal minimum to maximum) is about $3 \div 4$ *cm* wide. Field compression inside magnetosphere is substantially reduced for the conductive case. The second pressure jump makes the field even smaller than initial dipole one. This is interpreted as equatorial depression caused by Region-1 FAC as shown in Fig. 1. The azimuthal component δB_φ is presented in the lower panels of Fig. 2. For the dielectric case it shows oscillations and general increase by amplitude towards dipole. For the conductive case however, there is a much more pronounced structure – a compact minimum inside magnetosphere but immediately adjacent to the boundary layer (marked by the circle). It is formed solely by the second pressure jump.

Spatial structure of azimuthal field suggests current which flows downwards on the inner side of LLBL (for dusk sector). Total downward current can be estimated as $1.5 \div 2.5$ *kA* which is close to total cross-polar current measured independently by method described in Shaikhislamov *et al.* (2009). Thus, experiment provided three facts: The field corresponding to expected sheared stress in LLBL is indeed generated. Field-aligned current associated with it quantitatively agrees with independently measured total cross-polar current. Third, when „ionosphere" is non-conductive azimuthal field is also different and doesn't have well defined region of local maximums. Correlation analysis of magnetic probe signals shows that maximum of azimuthal component moves ahead and synchronously with magnetopause current. This indicates that compressional motion of boundary layer drives shear Alfven waves which carry FAC to the ionosphere. Such model aimed to explain observed transient geomagnetic variation was proposed by Glassmeier *et al.* (1992).

Acknowledgements: This work was supported by Russian Fund for Basic Research grant 09-02-00492 and OFN RAS Research Program 16.

References

Eastman, T. E. 1976, *Geophys. Res. Lett.*, 3(11), 685

Shaikhislamov, I. F., Antonov, V. M,. Zakharov, Yu. P., Boyarintsev, E. L., Melekhov, A. V., Posukh, V. G., & Ponomarenko A. G. 2009, *Plasma Phys. and Contr. Fusion*, 51(10), 105005

Glassmeier, K. H. & Heppner, C. 1992, *Geophys. Res.*, 97(A4), 3977

Advances in Plasma Astrophysics
Proceedings IAU Symposium No. 274, 2010
A. Bonanno, E. de Gouveia Dal Pino & A.G. Kosovichev, eds.
© International Astronomical Union 2011
doi:10.1017/S1743921311006557

Calculation of fusion rates at extremely low energies in laser plasmas

**D. Mascali[1,2], N. Gambino[1,2], S. Tudisco[1], A. Anzalone[1],
A. Bonanno[3], S. Gammino[1] and F. Musumeci[1,2]**

[1]INFN - Laboratori Nazionali del Sud,
via S. Sofia 62, 95123 Catania, Italy.
email: `davidmascali@lns.infn.it`

[2]Universitá di Catania, Dpt. di Fisica e Astronomia & CSFNSM,
Via S. Sofia 64, 95123 Catania, Italy

[3]INAF- Osservatorio Astrofisico di Catania
Via S. Sofia 78, 95123 Catania, Italy

Abstract. At temperatures and densities that are typical of plasmas produced by lasers pulses interacting with solid targets, at power intensities $I > 10^{12} \ W/cm^2$, the classical Debye screening factor in nuclear reactions becomes comparable with the one of the solar core. Preliminary calculations about the total number of fusion reactions have been performed following an hydrodynamical approach for the description of the plasma dynamics. This approach is propaedeutic for future measurements of D-D fusion reaction rates.

Keywords. plasmas, nuclear reactions, hydrodynamics

1. Introduction

At power intensities $I \geqslant 10^{12} \ W/cm^2$ laser generated plasmas supersonically expand in vacuum with ion temperatures of several hundreds of eV, and densities around $10^{20} - 10^{21} \ cm^{-3}$. Their importance in nuclear astrophysics is constantly growing, because some unexplored energy domains persist in calculation of fusion reaction rates, especially at low energy when clouds of cold electrons may play a role as electron screening, boosting the number of fusion events. By following a methodology largely employed in plasma physics, it is possible to rescale plasma parameters (e.g. temperature and density) in order to make more similar to the real world our laboratory conditions. In our laser generated plasmas, in fact, the classical Debye screening factor (it depends somehow by the temperature over density ratio) becomes comparable with the one of the solar core. Preliminary calculations about the total number of fusion reactions are reported in Mascali *et al.*, 2010, assuming a parametrization of the initial plasma temperature and density and following the evolution of such parameters by means of the 'one fluid' hydrodynamical model of Anisimov *et al.*, (1993). In this work we present an updated version of the numerical code, which easily evaluates the 3D fusion rate, and finally the number of expected fusions per laser shot, including the electron screening. The model predictions about the dependence of the plasma temperature and density on time, on the initial laser energy (fluence) and on type of ablated material are in good agreement with the experimental data collected in our laboratory during the last year.

2. Theory and models

Among the various treatments existing in literature about fusion reactions (Rolphs and Rodney, 1998), we used the Gamow theory, which includes the tunnel effect of the nuclear Coulomb barrier:

$$\langle \sigma v \rangle = \left(\frac{8}{\pi \mu} \right)^{1/2} \frac{1}{kT^{3/2}} \int_0^\infty S(E) \exp \left[-\frac{E}{kT} - \frac{b}{E^{1/2}} \right] dE \tag{2.1}$$

where $E = \frac{1}{2} m_r v^2$ is the energy of the particles pair with reduced mass m_r and relative velocity v, assuming a Maxwell Boltzmann distribution with temperature T. $\mu = \frac{m_r}{m_p}$ and $\sigma(E)$ can be obtained by the Gamow theory with an energy value $E_G = b^2 = (2\mu)\pi^2 e^4 \left(\frac{Z_1 Z_2}{\hbar} \right)^2$; here Z_1 and Z_2 are the atomic numbers of the pair and $S(E)$ is the astrophysical factor. For non resonant reactions $S(E)$ varies smoothly with energy and it can be developed in Mc-Laurin series, stopped at the quadratic term. For D-D reactions the series' coefficients are: $S(0) = 0.05 \ MeV \cdot b$, $S(0)' = 0.0183 \ b$, $S(0)'' = 4.24 \ MeV^{-1}b$ (NACRE website). Therefore, assuming that the plasma is made of identical particles, the total number of fusions will be:

$$f_{TOT} = \frac{1}{2} N \langle \sigma v \rangle \int_0^\infty \rho(t) dt \tag{2.2}$$

where N is the total number of particles in the plasma and $\rho(t)$ is the plasma density. Note that f_{TOT} depends on the density and the temperature variation in space and time (in laser produced plasmas they both strongly vary, spatially and temporally), other than on their absolute values. Hence eq. 2.2 must be discretized over small 3D cells in which density and temperature are evaluated step by step from the hydrodynamical code. Time intervals of 0.5 ns have been chosen, integrating over $300 \times 300 \times 300$ cells, each one having a volume of $10 \times 10 \times 10 \ \mu m^3$. ρ and T have been determined by the equations:

$$\rho(x, y, z, t) = \frac{M}{I_1 XYZ} \left[1 - \frac{x^2}{X^2} - \frac{y^2}{Y^2} - \frac{z^2}{Z^2} \right]^{\frac{1}{1-\gamma}} \tag{2.3}$$

$$T(x, y, z, t) = \beta \frac{\gamma - 1}{2\gamma} \frac{m_p}{k_B} \left[\frac{X_0 Y_0 Z_0}{XYZ} \right]^{\gamma-1} \left[1 - \frac{x^2}{X^2} - \frac{y^2}{Y^2} - \frac{z^2}{Z^2} \right]^{\frac{1}{1-\gamma}} \tag{2.4}$$

where $I_1, \beta, \gamma, m_p, k_B$ are parameters connected with the adiabatic expansion of the plume, and X, Y, Z are the coordinates of the plasma front at a given time t, which can be calculated by the following set of second order differential equations:

$$\xi \frac{d^2 \xi}{d\tau^2} = \eta \frac{d^2 \eta}{d\tau^2} = \left(\frac{\sigma}{\xi^2 \eta} \right)^{\gamma-1} \tag{2.5}$$

where $\xi(t) = X(t)/R_0$, $\eta = Z(t)/R_0$, $\sigma = Z(t)/R_0$, $\tau = t/t_0$ denote the spatial and temporal dimensionless coordinates, and Z_0, R_0 are the plasma dimensions at the end of the laser pulse. The model assumes a self-similar, isoentropic expansion of the plasma plume, whose shape is half-ellipsoidal at $t = 0$ (for more details see Anisimov *et al.*, (1993)). In our calculations we assumed $T_i \neq T_e$, with initial temperatures $T_{i0} \sim 450 \ eV$ and $T_{e0} \sim 45 \ eV$. This assumption is in agreement with empirical data; at power intensities comparable with the simulated ones, far from the target (i.e. in the free flight region) it is $T_i \leqslant 10^2 \ eV$ (these numbers are consistent with our initial condition if one considers the adiabatic cooling, which goes like $T \propto \rho^{\gamma-1}$); T_e may be determined by optical emission spectroscopy or electrostatic probes, and it is usually a factor 5 to 10 lower (Torrisi *et al.*, 2010). This discrepancy is mostly due to the plasma acceleration

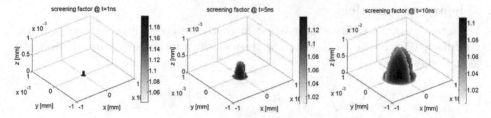

Figure 1. 3D calculation of the screening at different times after the laser shot.

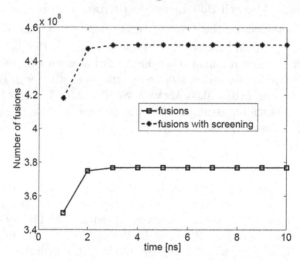

Figure 2. Cumulated number of fusions after 24h of operations in laser repetition rate (10 Hz) mode versus the time after each laser shot.

mechanism, partially driven by self-generated electric fields. The resulting velocity of the moving plume is too small to give any appreciable contribution to the electrons' kinetic energy (because of their small mass), but large enough to give a contribution to the ion energy content, that is even larger of the electrons thermal energy (collisional temperature equalization should give $T_e \sim T_i$). The ordered, forward peaked energy of the ions is then transformed by the ion-ion collisions in thermal motion, finally resulting in an effective ion temperature $T_i \gg T_e$.

The initial plasma density (that resulted to be around 10^{21} cm^{-3}, assuming the validity of the quasi-neutrality condition $\rho_e \simeq \rho_i$) was self-calculated by the code, assuming a laser pulse of duration $\tau_L = 6$ ns and spot diameter of 200 μm impinging on a virtual CD_2 thick target. Once known the temporal variation of ρ and T, the electron screening was calculated according to the well-known Debye-Huckel formula (Salpeter, 1954):

$$f_{scr} = exp\left(\frac{Z_1 Z_2 e^2}{kT_e \lambda_D}\right) \tag{2.6}$$

where λ_D is the Debye length ($\lambda_D \sim 743\sqrt{\frac{T_e\,[eV]}{n_e\,[cm^{-3}]}}\,[cm]$) and T_e the electron temperature.

3. Fusion rates calculations

Data coming from simulations are reassumed in figures 1 and 2. Figure 1 shows that the screening factor f_{scr} smoothly varies either in time and along the spatial coordinates; this occurs because the ratio ρ/T remains almost unvaried, although ρ and T rapidly drop during the plasma expansion. The above mentioned discrepancy between electron and ion temperatures confers to laser generated plasmas at medium-low power intensity the unique property to have high enough ion temperature to favorite a consistent number of fusion events, but low enough electron temperature to ensure a non negligible screening factor. About the absolute numbers, simulations results reported in figure 2 reveal that if the laser is operated in the repetition rate mode (10 Hz), after 24 hours the accumulated number of fusion is increased of a factor 1.2 by the electron screening. Note that almost the totality of the fusion events takes place in the first 2-3 ns after each laser pulse. The calculated screening factor is very similar to the stellar ones (Rolphs and Rodney, 1998).

The collected data put in evidence that an experiment with a deuterated target, and 4π detectors for fusion products like neutrons, is possible and physically meaningful; the strong temporal concentration of the significative events will help to adequately trigger the acquisition system, improving the experimental precision. The forecasted experiment will give the opportunity to calculate the electron screening in a stellar-like environment, to test the validity of the classical Debye-Huckel theory and finally to evaluate the astrophysical factor at energy domains of the order of hundreds of eV, never explored up to now by nuclear astrophysics.

References

Anisimov, S., Bauerle, D., & Lukyanchuk, B. S., 1993, *Phys. Rev. B*, 98.

Gambino, N., Mascali, D., Tudisco, S., Anzalone, A., Bonanno, A., Gammino, S., Grasso, R., Miracoli, R., Musumeci, F., Neri, L., Privitera, S., & Spitaleri A., 2010, *Proc. 37th EPS Conf. on Plasma Physics*, 21-26 June, Dublin, Ireland.

Mascali, D., Tudisco, S., Bonanno, A., Gambino, N., Ivanovski, I., Anzalone, A., Gammino, S., Miracoli, R., & Musumeci F., 2010, *Rad. Eff. Def. in Solids*, 165, 6.

NACRE – http://pntpm.ulb.ac.be/Nacre/nacre.htm

Rolphs, E. & Rodney, W. S., 1998, *Cauldrons in Cosmos*, Universty of Chicago Press.

Salpeter, E., 1954, *Aust. J. Phys.* 7 , 373.

Torrisi, L., Mascali, D., Miracoli, R., Gammino, S., Gambino, N., Giuffrida, L., & Margarone, D., 2010, *Journal of Applied Physics*, 107.

Interstellar, space and planetary plasmas

Advances in Plasma Astrophysics
Proceedings IAU Symposium No. 274, 2010
A. Bonanno, E. de Gouveia Dal Pino & A. G. Kosovichev, eds.
© International Astronomical Union 2011
doi:10.1017/S1743921311006569

A new viscous instability in weakly ionised protoplanetary discs

Anders Johansen[1], Mariko Kato[2] and Takayoshi Sano[3]

[1]Lund Observatory, Lund University
Box 43, 221 00 Lund, Sweden
email: anders@astro.lu.se

[2]Department of Earth and Planetary Science, Tokyo Institute of Technology
Ookayama 2-1-12-I2-10, Meguro-ku, Tokyo, Japan
email: marikok@geo.titech.ac.jp

[3]Institute of Laser Engineering, Osaka University
Suita, Osaka 565-0871, Japan
email: sano@ile.osaka-u.ac.jp

Abstract. Large regions of protoplanetary discs are believed to be too weakly ionised to support magnetorotational instabilities, because abundant tiny dust grains soak up free electrons and reduce the conductivity of the gas. At the outer edge of this "dead zone", the ionisation fraction increases gradually and the resistivity drops until the magnetorotational instability can develop turbulence. We identify a new viscous instability which operates in the semi-turbulent transition region between "dead" and "alive" zones. The strength of the saturated turbulence depends strongly on the local resistivity in this transition region. A slight increase (decrease) in dust density leads to a slight increase (decrease) in resistivity and a slight decrease (increase) in turbulent viscosity. Such spatial variation in the turbulence strength causes a mass pile-up where the turbulence is weak, leading to a run-away process where turbulence is weakened and mass continues to pile up. The final result is the appearance of high-amplitude pressure bumps and deep pressure valleys. Here we present a local linear stability analysis of weakly ionised accretion discs and identify the linear instability responsible for the pressure bumps. A paper in preparation concerns numerical results which confirm and expand the existence of the linear instability.

Keywords. accretion, accretion disks, (magnetohydrodynamics:) MHD, turbulence, (stars:) planetary systems: formation, (stars:) planetary systems: protoplanetary disks

1. Introduction

Consider a protoplanetary disc irradiated with a cosmic ray flux F (particles per area per unit time). The reduction of the flux is controlled by the equation

$$\frac{\mathrm{d}F}{\mathrm{d}z} = -\kappa \rho(z) F(z) \,, \tag{1.1}$$

where κ is the opacity and ρ is the z-dependent mass density. The solution is

$$F_{\downarrow}(z) = F_{\infty} \exp[-\kappa \Sigma_{\uparrow}(z)] \,, \tag{1.2}$$

$$F_{\uparrow}(z) = F_{\infty} \exp[-\kappa \Sigma_{\downarrow}(z)] \,. \tag{1.3}$$

Here $\Sigma_{\uparrow}(z) = \int_{z}^{\infty} \rho(z)\mathrm{d}z$ is the column density of gas above the given point, while $\Sigma_{\downarrow}(z) = \int_{-\infty}^{z} \rho(z)\mathrm{d}z$ is the column density below. We have $\Sigma = \Sigma_{\uparrow}(z) + \Sigma_{\downarrow}(z)$ at all z. Introducing the ionisation rate $\zeta(z)$ we get (Sano *et al.* 2000)

$$\zeta(z) = \frac{\zeta_{\mathrm{CR}}}{2} \left\{ \exp[-\Sigma_{\uparrow}(z)/\Sigma_{\mathrm{CR}}] + \exp[-\Sigma_{\downarrow}(z)/\Sigma_{\mathrm{CR}}] \right\} \,. \tag{1.4}$$

Here ζ_{CR} is the ionisation rate by cosmic rays in interstellar space and $\Sigma_{CR} = 1/\kappa$ is the penetration column density of cosmic rays. Free electrons are lost as they collide with dust grains. This yields the rate equation

$$\frac{\partial n_e}{\partial t} = \zeta n_n - n_e n_\bullet \langle \sigma v \rangle_{e,\bullet} \tag{1.5}$$

for electron number density n_e. Here $\zeta = \kappa m_n F$ is the ionisation rate. The equilibrium electron density fraction is

$$\frac{n_e}{n_n} = \frac{\zeta}{n_\bullet \langle \sigma v \rangle_{e,\bullet}} . \tag{1.6}$$

This expression is valid in the limit of negligible gas phase recombination and is equivalent to the *ion-dust plasma* limit of Okuzumi (2009). The resistivity of the electrons is given in c.g.s. units by

$$\eta - \frac{c^2}{4\pi\sigma_e} , \tag{1.7}$$

where c is the speed of light and σ_e is the electrical conductivity. In turn the conductivity is given by

$$\sigma_e = \frac{n_e e^2}{m_e \nu} . \tag{1.8}$$

Here n_e is the number density of electrons, e is the electron charge, m_e is the electron mass, and ν is the collision frequency of electrons with neutrals. The momentum rate coefficient $\langle \sigma v \rangle = \nu/n_n$ for transfer of momentum from electrons to neutrals is given by $\langle \sigma v \rangle = 8.3 \times 10^{-10} T^{1/2}$ cm^3 s^{-1}. This finally yields (Blaes & Balbus 1994)

$$\eta = 230 \left(\frac{n_n}{n_e} \right) T^{1/2} \text{ cm}^2 \text{ s}^{-1} . \tag{1.9}$$

Together with equation (1.6) and equation (1.4) this gives us a model for the resistivity in protoplanetary discs, provided that we know ζ_{CR} of the cosmic rays, $\rho(z)$ and Σ_{CR} for the gas, and number density n_d and collision cross section σ_d of the dust grains.

The fastest growing wavenumber for the MRI is

$$k_{BH} = \sqrt{\frac{15}{16} \frac{v_A}{\Omega}} , \tag{1.10}$$

where $v_A = B_0/\sqrt{\mu_0 \rho}$ is the vertical Alfvén speed and B_0 is the constant vertical magnetic field component. The MRI can grow when the Elsasser number Λ_{MRI} fulfills

$$\Lambda_{MRI} = \frac{v_A^2}{\eta\Omega} \gtrsim 1 . \tag{1.11}$$

For the Minimum Mass Solar Nebula the Elsasser number in the mid-plane scales with r^{-4}, assuming that $\Sigma \ll \Sigma_{CR}$ and constant $\beta = P_{gas}/P_{mag}$.

We set the rate coefficients for collisions between electrons and dust grains as

$$\langle \sigma v \rangle_{e,\bullet} = \pi a_\bullet^2 c_e , \tag{1.12}$$

where c_e is the thermal speed of the electrons.

$$c_e = \sqrt{\frac{8k_B T}{\pi m_e}} . \tag{1.13}$$

This simple approach allows us to calculate the resistivity of the gas anywhere in the disc at a relatively modest computational cost.

2. Stability analysis

We proceed now to analyse the stability of a turbulent accretion disc with turbulent stress S and turbulent diffusion D_t. We work in the shearing sheet formalism, representing a small corotating box at an arbitrary distance from the central star. The constant Keplerian angular frequency is Ω. The gas velocity relative to the main Keplerian flow is u and the velocity of the dust component is w. The gas density and the dust number are denoted ρ_g and n_d. We consider an isothermal equation of state with constant sound speed c_s. Dust is coupled to the gas via a drag force working on the time-scale τ_f. The dynamical equations are

$$\frac{Du_x}{Dt} = 2\Omega u_y - \frac{c_s^2}{\rho_g}\frac{\partial \rho_g}{\partial x}, \tag{2.1}$$

$$\frac{Du_y}{Dt} = -\frac{1}{2}\Omega u_x - \frac{1}{\rho_g}\frac{\partial S}{\partial x}, \tag{2.2}$$

$$\frac{D\rho_g}{Dt} = -\rho_g\frac{\partial u_x}{\partial x} + D_t\frac{\partial^2 \rho_g}{\partial x^2}, \tag{2.3}$$

$$\frac{Dw_x}{Dt} = 2\Omega w_y - \frac{1}{\tau_f}(w_x - u_x), \tag{2.4}$$

$$\frac{Dw_y}{Dt} = -\frac{1}{2}\Omega w_x - \frac{1}{\tau_f}(w_y - u_y), \tag{2.5}$$

$$\frac{Dn_d}{Dt} = -n_d\frac{\partial w_x}{\partial x} + D_t\frac{\partial^2 n_d}{\partial x^2}. \tag{2.6}$$

Here $D/Dt \equiv \partial/\partial t + (u \cdot \nabla) - (3/2)\Omega x\partial/\partial y$. Going in the limit of short friction times we get the simplification

$$w_x = u_x + \tau_f\frac{c_s^2}{\rho_g}\frac{\partial \rho_g}{\partial x}, \tag{2.7}$$

$$w_y = u_y + \tau_f\frac{1}{\rho_g}\frac{\partial S}{\partial x}. \tag{2.8}$$

The viscous instability arises from the dependence of S on n_d. We linearise the equation system around the state with gas density ρ_0 and dust number density n_0 and define the particle-stress coupling parameter of the background state $\chi = \partial \ln S/\partial \ln n_d$. When the Elsasser number is smaller than unity we expect that $\chi \sim 1$ (Pessah 2010). We ignore turbulent diffusion D_t in the linearisation. We consider axisymmetric perturbations with $f(x,t) = \hat{f}\exp[i(k_x x - \omega t)]$.

The resulting linearised equation system can be put on the matrix form $M\hat{f} = 0$, where $\hat{f} = (\hat{u}_x, \hat{u}_y, \hat{\rho}_g, \hat{n}_d)$ is a vector of complex amplitudes and M is

$$\begin{pmatrix} i\omega & 2\Omega & -c_s^2/\rho_0 ik_x & 0 \\ -(1/2)\Omega & i\omega & 0 & (\chi/n_0)(S_0/\rho_0)ik_x \\ -\rho_0 ik_x & 0 & i\omega & 0 \\ -n_0 ik_x & 0 & (n_0/\rho_0)\tau_f c_s^2 k_x^2 & i\omega \end{pmatrix} \tag{2.9}$$

We have only non-trivial solutions when the determinant of the matrix is zero. The dispersion relation is

$$0 = \omega^2(\omega^2 - c_s^2 k_x^2 - \Omega^2) - \frac{2k_x^2\Omega S_0\chi}{\rho_0}(c_s^2 k_x^2\tau_f - i\omega). \tag{2.10}$$

This equation can be solved numerically to find four complex frequencies for each

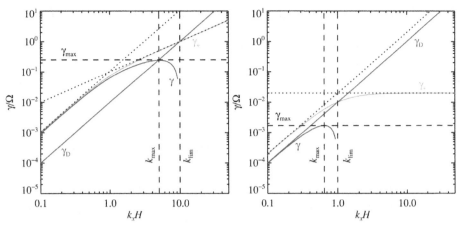

Figure 1. The growth rate γ_+ as a function of radial wavenumber k_x, based on a numerical solution to equation (2.10). Left plot: parameters are $\Omega = c_s = \rho_0 = \tau_f = 1$ and $S_0 = 0.01$. The maximum growth rate, $\gamma_{\max} = 0.25\Omega$, occurs at $k_{\max} = 5H^{-1}$, while the highest wavenumber for growth is $k_{\lim} = 10H^{-1}$. The two solution branches, following the limiting solutions found in equations (2.15) and (2.14), are indicated with dotted lines. Right plot: same as left plot but with passive particles ($\tau_f = 0$). The instability for passive particles occurs at longer wavelengths and at lower growth rates.

wavenumber. However, a simplification of the equation system allows us to find two analytical solutions instead.

Assuming geostrophic balance $0 = 2\Omega u_y - c_s^2(\partial \ln \rho_g/\partial x)$ instead of equation (2.1), the dispersion relation simplifies to the second order expression

$$0 = \omega^2(-c_s^2 k_x^2 - \Omega^2) - \frac{2k_x^2 \Omega S_0 \chi}{\rho_0}(c_s^2 k_x^2 \tau_f - i\omega). \tag{2.11}$$

The approximation that geostrophic balance is always maintained effectively filters away high frequency density waves that are of no importance for the viscous drift instability. The two solutions to equation (2.11) are

$$\omega_\pm = \frac{1}{1 + k_x^2 H^2} k_x^2 \frac{S_0 \chi}{\Omega \rho_0} \left[1 \pm \sqrt{1 + (1 + k_x^2 H^2)\frac{2c_s^2 \Omega \tau_f \rho_0}{S_0 \chi}}\right] i. \tag{2.12}$$

One solution is complex positive (instability), while the other is always complex negative (damped mode). In the two limits of $k_x H$ the growth rate $\gamma_+ = \mathrm{Im}(\omega_+)$ of the positive solution is

$$k_x H \gg 1 : \gamma_+ = \frac{1}{H^2}\frac{S_0 \chi}{\Omega \rho_0}\left(1 + \sqrt{1 + k_x^2 H^2 \frac{2c_s^2 \Omega \tau_f \rho_0}{S_0 \chi}}\right), \tag{2.13}$$

$$k_x H \ll 1 : \gamma_+ = k_x^2 \frac{S_0 \chi}{\Omega \rho_0}\left(1 + \sqrt{1 + \frac{2c_s^2 \Omega \tau_f \rho_0}{S_0 \chi}}\right). \tag{2.14}$$

The high wavenumber branch can further be expanded as

$$k_x H \gg \sqrt{S_0 \chi/(2c_s^2 \Omega \tau_f \rho_0)} : \gamma_+ = k_x \sqrt{\frac{2\Omega \tau_f S_0 \chi}{\rho_0}}. \tag{2.15}$$

This limit is only relevant if $S_0 \chi/(2c_s^2 \Omega \tau_f \rho_0) > 1$.

The growth rate of the viscous drift instability tends towards infinity for infinitely high wave numbers, according to equation (2.15). The linear scaling with k_x, however, implies that turbulent diffusion will stabilise the mode at high wavenumbers. Turbulent diffusion of the particles has a damping rate of

$$\gamma_{\mathrm{D}} = -k_x^2 D_{\mathrm{t}} = -k_x^2 \frac{S_0}{\Omega \rho_0}. \tag{2.16}$$

The limiting wavenumber for instability, where $\gamma = \gamma_{\mathrm{D}} + \gamma_+ = 0$, is

$$k_{\mathrm{lim}}^2 H^2 = 2\chi \left(1 + \frac{c_{\mathrm{s}}^2 \Omega \tau_{\mathrm{f}} \rho_0}{S_0} \right) - 1. \tag{2.17}$$

The most unstable wavenumber has no simple analytical form in the general case. However, in the case of mobile dust particles with $\Omega_{\mathrm{f}} \tau_{\mathrm{f}} > 0$ we find the most unstable wavenumber in the high wavenumber branch, because of the different wavenumber scaling of instability and turbulent damping. The most unstable wavenumber is

$$k_{\mathrm{max}} = \sqrt{\frac{\Omega^3 \tau_{\mathrm{f}} \chi \rho_0}{2 S_0}}, \tag{2.18}$$

which is two times the limiting wavenumber. Weaker (stronger) turbulence has maximum growth rate at shorter (longer) wavelengths. For typical parameters we find a wavelength for maximum growth around a few scale heights in the radial direction. The highest growth rate is

$$\gamma_{\mathrm{max}}/\Omega = \frac{1}{2} \chi \Omega \tau_{\mathrm{f}}, \tag{2.19}$$

which shows clearly the importance of freedom in the motion of the particles relative to the gas. The dependence of the growth rate on the wavenumber is shown in Figure 1 for typical values relevant to a protoplanetary disc.

2.1. Passive particles

For passive particles with $\Omega_{\mathrm{f}} \tau_{\mathrm{f}} = 0$ the most unstable wavenumber is

$$k_{\mathrm{max}}^2 H^2 = \sqrt{2\chi} - 1. \tag{2.20}$$

The wavenumber is real for $\chi \geqslant 0.5$. The maximum growth rate is

$$\gamma_{\mathrm{max}}/\Omega = (\sqrt{2\chi} - 1)^2 \frac{S_0}{c_{\mathrm{s}}^2 \rho_0}, \tag{2.21}$$

provided $\chi \geqslant 0.5$. For $\chi > 0.5$ there is growth even for zero friction time dust grains (which just trace the gas flow). In this case the increased gas density in the growing pressure bumps is enough to cause instability, from the passively advected dust grains. The growth rate with passive particles is also shown in Figure 1.

3. Outlook

To identify this new viscous instability in a numerical simulation we solve the resistive MHD equations in the standard shearing box approximation using the Pencil Code. We consider a box size of $L_x = 10.56H$, $L_y = 2.64H$, $L_z = 1.32H$ and a grid resolution of $256 \times 64 \times 32$. First we let the simulation run 20 orbits with only the constant hyper-resistivity needed to dissipate energy released by the turbulent stresses. After 20 orbits we turn on the density-dependent resistivity where regions of higher (lower) density have

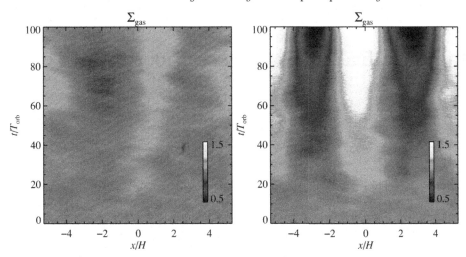

Figure 2. The evolution of the gas column density, averaged over the azimuthal direction, as a function of time. The left plot shows the results with constant hyperresistivity. Weak 5–10% level pressure bumps form on the largest scales of the turbulent flow (Johansen *et al.* 2009). Including dust-dependent resistivity, the right-hand plot shows the evolution of high-pressure regions with weak turbulence and low-pressure regions with high turbulence. This situation arises from the viscous instability driven by the dust-dependent resistivity of the gas.

higher (lower) resistivity. Figure 2 shows a space-time plot of the density. The emergence of a high-amplitude pressure bump is clear in the case of space-dependent resistivity.

The presence of pressure bumps in protoplanetary discs can have a positive effect on planet formation because the radial drift of particles is stopped in pressure bumps (particles seek the point of highest pressure, see e.g. Kato *et al.* 2009 and Johansen *et al.* 2009). Our new linear viscous instability can lead to the emergence of strong pressure bumps at the outer edge of the dead zone in protoplanetary discs. This makes the outer edge of the dead zone a prime site for planetesimal formation and thus for the rapid formation of the cores of gas giants. A paper in preparation details the non-linear evolution of the viscous instability (Kato *et al.* in preparation).

AJ and MK are grateful to Center for Planetary Sciences in Kobe for an extended visit that inspired this collaborative project. This work is supported in part by the Center for Planetary Science running under the auspices of the MEXT Global COE program entitled "Foundation of International Center for Planetary Science".

References

Blaes, O. M. & Balbus, S. A., 1994, *The Astrophysical Journal*, 421, 163

Johansen, A., Youdin, A., & Klahr, H., 2009, *The Astrophysical Journal*, 697, 1269

Kato, M. T., Nakamura, K., Tandokoro, R., Fujimoto, M., & Ida, S. 2009, *The Astrophysical Journal*, 691, 1697

Okuzumi, S., 2009, *The Astrophysical Journal*, 698, 1122

Pessah, M. E., 2010, *The Astrophysical Journal*, 716, 1012

Sano, T., Miyama, S. M., Umebayashi, T., & Nakano, T., 2000, *The Astrophysical Journal*, 543, 486

Advances in Plasma Astrophysics
Proceedings IAU Symposium No. 274, 2010
A.Bonanno, E. de Gouveia Dal Pino & A. G. Kosovichev, eds.
© International Astronomical Union 2011
doi:10.1017/S1743921311006570

Magnetic fractures or reconnection of type II

Gerhard Haerendel

Max Planck Institute for extraterrestrial Physics
85748 Garching, Germany
email: hae@mpe.mpg.de

Abstract. The importance of reconnection in astrophysics has been widely recognized. It is instrumental in storing and releasing magnetic energy, the latter often in a dramatic fashion. A closely related process, playing in very low beta plasmas, is much less known. It is behind the acceleration of auroral particles in the low-density environment several 1000 km above the Earth. It involves the appearance of field-parallel voltages in presence of intense field-aligned currents. The underlying physical process is the release of magnetic shear stresses and conversion of the liberated magnetic energy into kinetic energy of the particles creating auroral arcs. In this process, field lines disconnect from the field anchored in the ionosphere and reconnect to other field lines. Because of the stiffness of the magnetic field, the process resembles mechanical fractures. It is typically active in the low-density magnetosphere of planets. However, it can also lead to significant energy conversion with high-energy particle production and subsequent gamma ray emissions in stellar magnetic fields, in particular of compact objects.

Keywords. Auroral acceleration, magnetosphere, electric currents, neutron stars

---

1. Introduction

The auroral acceleration process is one of the best explored space plasma processes (Paschmann *et al.*, 2002). Accelerating electric fields with appreciable potential drops parallel to the magnetic field appear in concentrated currents and low densities above the ionosphere. A host of wave processes as well as double layers supporting the electric fields have been identified and related to specific particle distributions. Less well explored are the connections with the high-beta source regions in the outer magnetosphere and near-Earth magnetotail. The energy powering the auroral acceleration process is carried earthward in the Alfvénic wave mode and involves the build-up of magnetic shear stresses. The key issue is that the high-altitude plasma motions and related transverse electric field components are decoupled from those controlled by the collisional ionosphere due to the parallel potential drops, thereby releasing the shear stresses. The author has compared this process with mechanical fractures (Haerendel, 1980; Haerendel, 1988; Haerendel, 1980; Haerendel, 1994; Haerendel, 2007) There are several features shared by mechanical and 'magnetic' fractures: Shear stresses exerted on an elastic elongated medium, a rod or a flux tube, are concentrating in a region of structural weakness; the actual breaking occurs on the molecular or microphysical level; while the fracture propagates spontaneously, a stress release motion is initiated and elastic energy converted into kinetic energy of the elastic medium; subsequently, it is dissipated by some damping mechanism; in case of the aurora by acceleration and heating of particles. This paper deals with these aspects in the context of auroral arcs. A condensed formalism is presented allowing a simple evaluation of the relation between stress application, stress release, energy flux, and field-parallel potential drops, and is also easily adapted to other astrophysical systems.

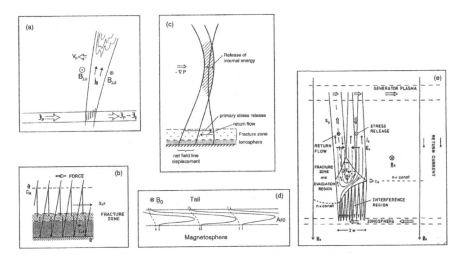

Figure 1. The fracture process or release of the differential shear stresses between two sides of a field-aligned current sheet (1a). Figures (1b) and (1c) show the stress release in the plane of the motion, and (1d) in ground projection in the frame of the propagating arc. The dashed lines in (1d) are the tracks of the ionospheric section. Figure 1e shows oblique Alfvén waves attached to the fracture zone and equipotential contours (dotted lines). (After Haerendel, 1994; Haerendel, 2007; Haerendel, 2009)

2. The Fracture Process

Figure 1 illustrates the stress release process in various projections. Figure 1a shows a sheet of upward directed field-aligned current extending normal to the plane of projections for the case of arcs embedded in a wider current system. The sheet current separates a region of highly stressed magnetic field from a less stressed one. It is the differential magnetic energy residing in the shear stresses that is being converted into kinetic energy of the auroral particles. Figure 1b shows that previously connected field lines (continuous inclined lines) are separated by the developing fracture zone into an upper magnetospheric and lower ionospheric section, while the attached plasma undergoes opposing stress relief motions. As the decoupling process is initiated somewhere above the ionosphere, the stress relief motions are propagated upward towards the source region of the shear stresses and affect the force balance. This is illustrated in Figure 1c. Since the shear stresses in the source region are being reduced, the plasma is accelerated in the direction of the primary force, here the pressure gradient force. After a while, the stress reduction fades and the plasma slows down. At the same time, the lower section of the magnetospheric part of the flux tube overshoots and is forced back, until the less strained configuration of the field at the rear of the current sheet is reached. This is shown in Figure 1d as a ground projection of the tracks of the magnetospheric as well as the ionospheric field lines in the frame of the arc. When the process is completed, new magnetic connections have been established. The auroral acceleration process is truly a reconnection process albeit in very low-β plasma.

Figure 1e contains an attempt to sketch the communication of the magnetic pertur-bations up and down the field lines during the fracture process. When the leading edge of the fracture zone hits so far unperturbed field lines, transverse motions are induced starting the stress release. The resulting perturbations are propagating in the Alfvén mode towards the source region while the 'fracture' progresses slowly into the current circuit. The large ratio of the two speeds causes the very slight obliqueness of the wave fronts. The field changes resulting from the dynamic reaction of the source plasma are again communicated towards the fracture zone. Meanwhile the low-altitude overshoot of the flux tubes is reversed and the field lines are being dragged into the lower shear configuration at the rear of the current sheet including the displacement of the source plasma. The detailed evaluation of Haerendel (1994) and 2007, shows that at least four transit times of an Alfvén wave between fracture zone and source plasma are required for this process. The key point is the near-incompressibility of the magnetic field near the planet (or star), which allows a representation of the electric fields as potential fields. Thus parallel fields are by necessity connected with transverse components and the field-aligned acceleration process with perpendicular stress relief motions. The dotted lines in Figure 1e depict these mainly U-shaped potential contours. The asymmetry of the transverse potentials between the leading and trailing edges corresponds to the asymmetry in the horizontal excursions in Figure 1d. The 'interference region' between fracture zone and lower ionosphere is probably dominated by Alfvénic turbulence and provides further decoupling (Haerendel, 2007).

There is rich experimental evidence for the U-shaped potentials at several 1000 km height from satellite crossings and electrostatic double probe measurements (Mozer et al., 1977, Paschmann et al., 2002). Another type of evidence comes from optical observations of auroral ray or fold motions when viewed along the magnetic field. By electron impact the counterflows of the stress release process are imaged on the atmosphere (Davis, 1978). Observations by Haerendel et al. (1996) clearly exhibit also the asymmetry between the leading and trailing edges. Simultaneous radar measurements confirm that the arcs are propagating, i.e. they are not frozen in the plasma frame.

3. Stress Concentration Mechanisms

Fractures of a strained elastic medium (e.g. a long rod) develop where stress concentration meets with structural weakness. In the magnetic case, this is achieved by concentrations of the field-aligned current in combination with a low density of the current carrying electrons (Kindel & Kennel, 1971). Various mechanisms exist creating field-parallel voltages and thus 'breaking' of the magnetic connections. Planetary magnetospheres with their low density and converging magnetic fields are particularly prone to develop such stress concentrations at relatively low altitudes (several 1000 km in the Earth's case). Knight (1973) and Fridman & Lemaire (1980) have developed a simple current-voltage relation, which has found rich application in the interpretation of observational data. It is due to the mirror effect creating a dearth of current carriers for a current imposed by the large-scale plasma dynamics. For large mirror ratios it can be simplified as:

$$\Phi_\parallel = K^{-1} \cdot j_\parallel, \tag{3.1}$$

where K is the conductance derived in the original papers for the inverse current-voltage relation.

While Equation 3.1 can be well applied to quasi-stationary arc models like the one shown in Figure 1e, propagating Alfvén waves have to be handled differently. According to Goertz & Boswell (1979), parallel electric fields exist in the inertial regime if the transverse scales are of the order of the electron inertial length, λ_e. In this case, the parallel and transverse potentials are of the same magnitude:

$$\Phi_\parallel \approx \Phi_\perp \quad \text{for} \quad k_\perp^2 \cdot \lambda_e^2 \geqslant 1 \tag{3.2}$$

In the much higher densities of stars or other astrophysical systems, the above conditions are hardly realized. However, substantial field-aligned voltages will arise if the current

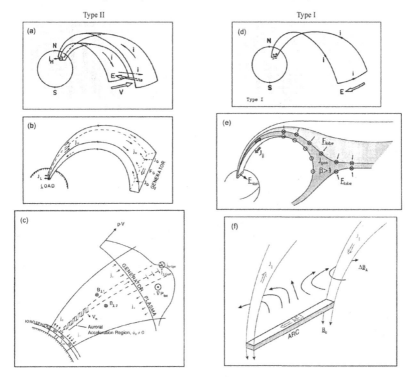

Figure 2. Two current systems after Boström (1964) and associated arcs: Type II with the generator forces acting parallel to the resulting arc, and Type I with the force acting perpendicular to it (Haerendel, 1988; Haerendel, 1994; Haerendel, 2007; Haerendel, 2009).

density exceeds a critical threshold, j_{crit}, which depends on the electron to ion temperature ratio (Papadopoulos, 1977). One can, for instance, scale the critical drift velocity, u_d, by the ion thermal speed and a scaling factor, f:

$$u_d \geqslant \frac{j_{crit}}{e\,n} = f \cdot v_{th,i} \qquad (3.3)$$

f is typically of the order of 10 for the most unstable ion cyclotron waves and $T_i \approx T_e$ (Treumann & Baumjohann, 1997).

There is a host of microphysical processes which actually sustain the parallel electric fields, such as pressure gradients, electron inertia, solitary waves associated with ion or electron phase space holes, large-amplitude ion acoustic or ion cyclotron waves, and double layers. Several processes relevant for auroral physics are discussed in detail in Paschmann *et al.*, 2002.

4. Stress Applications and Release

One of the pioneering contributions to auroral physics was the identification of the two basic current circuits underlying the interaction of the outer realms of the magnetosphere with the ionosphere and, for that matter, also with the auroral energy conversion regions (Boström, 1964). They are distinguished by the orientation of the driving forces with respect to the dissipation regions. Figure 2 shows the two principal current systems, referred to as Type II and Type I, the respective connections between source region and acting force, the current closure regions in the ionosphere, and the resulting arcs. The Type II system, displayed in Figure 2 a-c, are typical for the convective flows of the

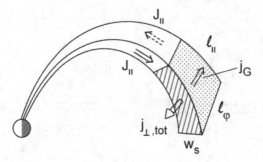

Figure 3. A high-β plasma layer as source of field-aligned currents, after Haerendel (2009).

magnetospheric and ionospheric plasmas, for instance along the auroral oval. They are driven by pressure gradient forces. Figure 2c shows the frequent situation of auroral arcs imbedded in the larger convection channel. These arcs carry only a fraction of the total current flowing between the generator region and the ionosphere in the normal direction to the arc. Figures 2d-f deal with the situation encountered at the boundary between the tail (polar cap) and the outer magnetosphere. The solar wind, by compressing the magnetotail, exerts a force on the outer more dipolar magnetospheric field. In contrast to the Type II case, the currents are flowing along the auroral arc, which is a region of enhanced conductivity. The downward transported shear stresses act in the normal direction to the arc.

5. Key Relations

The author has cast his concept of magnetic fractures into a small set of analytical relations, which also allow application to other than planetary environments (Haerendel, 2007; Haerendel, 2009). We assume a quasi-stationary situation, in which the arc system exists longer than the four Alfvénic transit times underlying the model of Figure 1. Quasi-stationarity implies a matching between the electromagnetic energy inflow from the generator, the Poynting flux, S, and the conversion rate into particle energy, $\dot{W}$, whereby any ohmic losses by the closure currents are being neglected. With the length of the field line, L_{par}, between generator and fracture zone and the respective Alfvénic transit time, τ_A, one can define an integral wave impedance

$$R_w = \frac{\mu_0 L_{\|}}{\tau_A} \tag{5.1}$$

and with the sheet current density, $J_{\|,arc}$, the Poynting flux becomes:

$$S_{arc} = R_w \cdot J_{\|,arc}^2 = \dot{W}_{arc} \tag{5.2}$$

In Figure 3, a current wedge of Type I is being displayed. The various dimensions characterizing the current system are the field-parallel and longitudinal extensions, $l_{\|,Gen}$ and $l_{\phi,Gen}$, and the projection of the latter into the fracture zone, l_{arc}. Following the derivation in Haerendel (2009), one gets:

$$J_{\|,arc} = \frac{l_{\varphi,Gen}}{l_{div}} \cdot \frac{l_{\|,Gen}}{l_{arc}} \cdot \frac{\beta_{Gen}}{2} \cdot \frac{B_{Gen}}{\mu_0} \tag{5.3}$$

The ratio, $l_{\varphi,Gen}/l_{div} \approx 0.1$ expresses that only a fraction of the current flowing in the generator layer is actually diverted towards the star. This means that most of the force acting on the generator plasma is balanced by magnetic normal stresses and only a small fraction by shear stresses transported towards low altitudes. The comparison of theory and data in the auroral context shows that typical auroral current densities and energy flows require $\beta_{Gen} \geqslant 1$. Furthermore, it is advantageous to decompose the second factor in Equation 5.3, since there is little a priori knowledge of the length of the arc:

$$\frac{l_{\parallel,Gen}}{l_{arc}} = \frac{l_{\parallel,Gen}}{l_{\varphi,Gen}} \cdot \frac{l_{\varphi,Gen}}{l_{arc}} \approx \frac{l_{\parallel,Gen}}{l_{\varphi,Gen}} \cdot \sqrt{\frac{B_{arc}}{B_{Gen}}} \qquad (5.4)$$

This way, current or Poynting flux near the fracture zone are entirely characterized by the plasma and field parameters and spatial dimensions of the generator and the magnetic field at the energy conversion level.

The accelerating parallel voltage, $\Phi_{\parallel}$, depends on the arc width, w_{arc}, which is either determined by the inverse conductance, K^{-1} (Equation 3.1), or the critical current density, j_{crit}:

$$\Phi_{\parallel} = \dot{W}_{arc} \cdot \frac{w_{arc}}{J_{\parallel,arc}} \qquad \text{and} \qquad \frac{J_{\parallel,arc}}{w_{arc}} \cong j_{crit} \qquad (5.5)$$

The last expression is better suited for the application to astrophysical systems.

6. Why Reconnection of Type II?

We have demonstrated that the auroral acceleration process involves reconnection, but the concept of magnetic fractures better describes what is happening. Since neither magnetic fractures nor their characterization as a reconnection process has yet found wide acceptance among auroral researchers, we will close this paper with a short list of features of this process justifying the designation 'reconnection of type II':

• In the decoupling process, the magnetospheric field undergoes new connections with the ionospheric field.

• There is a dissipation region, the fracture zone, corresponding to the so-called diffusion region in Type I reconnection.

• There is also a wave region, oblique quasi-stationary or fast propagating Alfvén waves, transferring momentum to the ambient plasma and channeling the energy flux.

• Type II, because momentum and energy inflow are widely separated from the dissipation region.

References

Boström, R. 1964, *J. Geophys. Res.*, 69, 4893, doi:10.1029/JZ69i023p04983
Davis, T. N. 1978, *Space Sci. Revs*, 22, 77
Fridman, M. & Lemaire, J. 1980, *J. Geophys. Res.*, 85, 664
Goertz, C. K. & Boswell, R. W. 1979, *J. Geophys. Res.*, 84, 7239
Haerendel, G. 1980, *ESA Journal*, 4, 197
Haerendel, G. 1988, *ESA-SP 285*, Vol.1, 37, Proc. ESA Symp. Varenna/Italy
Haerendel, G. 1994, *ApJS*, 90, 765
Haerendel, G. *et al.* 1996, *J. Atmos. Sol. Terr. Phys.*, 58, 71
Haerendel, G. 2007, *J. Geophys. Res.*, 112, A09214, doi:10.1029/2007JA012378
Haerendel, G. 2009, *J. Geophys. Res.*, 114, A06214, doi:10.1029/2009JA014139
Kindel, J. M. & Kennel, C. F. 1971, *J. Geophys. Res.*, 76, 3055
Knight, S. 1973, *Planet. Space Sci.*, 21, 741
Mozer, F. *et al.* 1977, *Phys. Rev. Lett.*, 38, 292
Papadopoulos, K. 1977, *Rev. Geophys. Spce Sci.*, 15, 113
Paschmann, G., Haaland, S., & Treumann, R. (eds.) 2002, *Space Sci. Revs*, 103, Nos. 1–4
Treumann, R. & Baumjohann, W. 1997, *Advanced Plasma Physics*, Imperial College Press

Advances in Plasma Astrophysics
Proceedings IAU Symposium No. 274, 2010
A. Bonanno, E. de Gouveia Dal Pino & A.G. Kosovichev, eds.
© International Astronomical Union 2011
doi:10.1017/S1743921311006582

Particle acceleration in fast magnetic reconnection

A. Lazarian[1], G. Kowal[2,3], E. de Gouveia Dal Pino[2] and E. Vishniac[4]

[1]Department of Astronomy, University of Wisconsin-Madison, 475 N. Charter St., Madison, WI, 53706, USA, email: lazarian@astro.wisc.edu

[2]Department of Astronomy of IAG, University of São Paulo, Rua do Matão, 1226, São Paulo, SP, 05508, Brazil,

[3]Astronomical Observatory, Jagiellonian University, Orla 171, 30-244 Kraków, Poland

[4]Department of Physics and Astronomy, McMaster University, 1280 Main St. W, Hamilton, ON, L8S 4M1, Canada

Abstract. Our numerical simulations show that the reconnection of magnetic field becomes fast in the presence of weak turbulence in the way consistent with the Lazarian & Vishniac (1999) model of fast reconnection. This process in not only important for understanding of the origin and evolution of the large-scale magnetic field, but is seen as a possibly efficient particle accelerator producing cosmic rays through the first order Fermi process. In this work we study the properties of particle acceleration in the reconnection zones in our numerical simulations and show that the particles can be efficiently accelerated via the first order Fermi acceleration.

Keywords. magnetic reconnection, cosmic rays, acceleration

1. Fast Magnetic Reconnection in the Presence of Weak Turbulence

A magnetic field embedded in a perfectly conducting fluid preserves its topology for all time (Parker 1979). Although ionized astrophysical objects, like stars and galactic disks, are almost perfectly conducting, they show indications of changes in topology, "magnetic reconnection", on dynamical time scales (Parker 1970, Lovelace 1976, Priest & Forbes 2002). Reconnection can be observed directly in the solar corona (Innes *et al.* 1997, Yokoyama & Shibata 1995, Masuda *et al.* 1994), but can also be inferred from the existence of large scale dynamo activity inside stellar interiors (Parker 1993, Ossendrijver 2003). Solar flares (Sturrock 1966) and γ-ray busts (Fox *et al.* 2005, Galama *et al.* 1998) are usually associated with magnetic reconnection. Previous work has concentrated on showing how reconnection can be rapid in plasmas with very small collisional rates (Shay *et al.* 1998, Drake 2001, Drake *et al.* 2006, Daughton *et al.* 2006), which substantially constrains astrophysical applications of the corresponding reconnection models.

We note that if magnetic reconnection is slow in some astrophysical environments, this automatically means that the results of present day numerical simulations in which the reconnection is inevitably fast due to numerical diffusivity do not correctly represent magnetic field dynamics in these environments. If, for instance, the reconnection were slow in collisional media this would entail the conclusion that the entire crop of interstellar, protostellar and stellar MHD calculations would be astrophysically irrelevant.

Here we present numerical evidence, based on three dimensional simulations, that reconnection in a turbulent fluid occurs at a speed comparable to the rms velocity of the turbulence, regardless of either the value of the resistivity or degree of collisionality. In particular, this is true for turbulent pressures much weaker than the magnetic field pressure so that the magnetic field lines are only slightly bent by the turbulence. These results are consistent with the proposal by Lazarian & Vishniac (1999, henceforth LV99) that reconnection is controlled by the stochastic diffusion of magnetic field lines, which produces a broad outflow of plasma from the reconnection zone. This work implies that reconnection in a turbulent fluid typically takes place in approximately a single eddy turnover time, with broad implications for dynamo activity (Parker 1970, 1993, Stix 2000) and particle acceleration throughout the universe (de Gouveia Dal Pino & Lazarian 2003, 2005, Lazarian 2005, Drake *et al.* 2006, de Gouveia Dal Pino, Piovezan & Kadowaki 2010, de Gouveia Dal Pino *et al.* 2010).

Astrophysical plasmas are often highly ionized and highly magnetized (Parker 1970). The evolution of the magnetic field in a highly conducting fluid can be described by a simple version of the induction equation

$$\frac{\partial \vec{B}}{\partial t} = \nabla \times (\vec{v} \times \vec{B} - \eta \nabla \times \vec{B}), \qquad (1.1)$$

where $\vec{B}$ is the magnetic field, $\vec{v}$ is the velocity field, and η is the resistivity coefficient. Under most circumstances this is adequate for discussing the evolution of magnetic field in an astrophysical plasma. When the dissipative term on the right hand side is small, as is implied by simple dimensional estimates, the magnetic flux through any fluid element is constant in time and the field topology is an invariant of motion. However, as mentioned earlier, the reconnection is observed in many astrophysical environments and quantitative general estimates for the speed of reconnection start with two adjacent volumes with different large scale magnetic field directions (Sweet 1958, Parker 1957).

The speed of reconnection, i.e. the speed at which inflowing magnetic field is annihilated by ohmic dissipation, is roughly η/Δ, where Δ is the width of the transition zone (see Figure 1). Since the entrained plasma follows the local field lines, and exits through the edges of the current sheet at roughly the Alfvén speed, V_A, the resulting reconnection speed is a tiny fraction of the Alfvén speed, $V_A \equiv B/(4\pi\rho)^{1/2}$ where L is the length of the current sheet. When the current sheet is long and the reconnection speed is slow this is referred to as Sweet-Parker reconnection. Observations require a speed close to V_A, so this expression implies that $L \sim \Delta$, i.e. that the magnetic field lines reconnect in an "X point".

The first model with a stable X point was proposed by Petschek (1964). In this case the reconnection speed may have little or no dependence on the resistivity. The X point configuration is known to be unstable to collapse into a sheet in the MHD regime (see Biskamp 1996), but in a collisionless plasma it can be maintained through coupling to a dispersive plasma mode (Sturrock 1966). This leads to fast reconnection, but with important limitations. This process has a limited astrophysical applicability as it cannot be important for most phases of the interstellar medium (see Draine & Lazarian 1998 for a list of the idealized phases), not to speak about dense plasmas, such as stellar interiors and the denser parts of accretion disks. In addition, it can only work if the magnetic fields are not wound around each other, producing a saddle shaped current sheet. In that case the energy required to open up an X point is prohibitive. The saddle current sheet is generic for not parallel flux tubes trying to pass through each other. If such a passage is seriously constrained, the magnetized highly conducting astrophysical fluids should behave more like Jello rather than normal fluids.

Finally, the traditional reconnection setup does not include ubiquitous astrophysical turbulence† (see Armstrong *et al.* 1994, Elmegreen & Scalo 2004, McKee & Ostriker 2007, Haverkorn *et al.* 2008, Lazarian & Opher 2009, Chepurnov & Lazarian 2010). Fortunately, this approach provides another way of accelerating reconnection. Indeed, an alternative approach is to consider ways to decouple the width of the plasma outflow region from Δ. The plasma is constrained to move along magnetic field lines, but not necessarily in the direction of the mean magnetic field. In a turbulent medium the two are decoupled, and fluid elements that have some small initial separation will be separated by a large eddy scale or more after moving the length of the current sheet. As long as this separation is larger than the width of the current sheet, the result will not depend on η.

LV99 we introduced a model that included the effects of magnetic field line wandering (see Figure 1). The model relies on the nature of three-dimensional magnetic field wandering in

---

† The set ups where instabilities play important role include Shimizu *et al.* (2009a,b). For sufficiently large resolution of simulations those set-ups are expected to demonstrate turbulence. Turbulence initiation is also expected in the presence of plasmoid ejection (Shibata & Tanuma 2001). Numerical viscosity constrains our ability to sustain turbulence via reconnection, however.

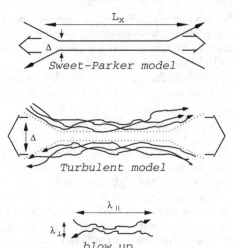

Figure 1. *Upper plot*: Sweet-Parker model of reconnection. The outflow is limited by a thin slot Δ, which is determined by Ohmic diffusivity. The other scale is an astrophysical scale $L \gg \Delta$. *Middle plot*: Reconnection of weakly stochastic magnetic field according to LV99. The model that accounts for the stochasticity of magnetic field lines. The outflow is limited by the diffusion of magnetic field lines, which depends on field line stochasticity. *Low plot*: An individual small scale reconnection region. The reconnection over small patches of magnetic field determines the local reconnection rate. The global reconnection rate is substantially larger as many independent patches come together. From Lazarian *et al.* 2004.

turbulence. This nature is different in three and two dimensions, which provides the major difference between the LV99 model and the earlier attempts to solve the problem of magnetic reconnection appealing to turbulence (Matthaeus & Lamkin 1985). The effects of compressibility and heating which were thought to be important in the earlier studies (Matthaeus & Lamkin 1985, 1986) do not play the role for the LV99 model either. The model is applicable to any weakly perturbed magnetized fluid, irrespectively, of the degree of plasma being collisional or collisionless (cf. Shay *et al.* 1998).

Two effects are the most important for understanding of the nature of reconnection in LV99. First of all, in three dimensions bundles of magnetic field lines can enter the reconnection region and reconnection there independently (see Figure 1), which is in contrast to two dimensional picture where in Sweet-Parker reconnection the process is artificially constrained. Then, the nature of magnetic field stochasticity and therefore magnetic field wandering (which determines the outflow thickness, as illustrated in Figure 1) is very different in 2D and the real 3D world (LV99). In other words, removing artificial constraints on the dimensionality of the reconnection region and the magnetic field being absolutely straight, LV99 explores the real-world astrophysical reconnection.

Our calculations in LV99 showed that the resulting reconnection rate is limited only by the width of the outflow region. This proposal, called "stochastic reconnection", leads to reconnection speeds close to the turbulent velocity in the fluid. More precisely, assuming isotropically driven turbulence characterized by an injection scale, l, smaller than the current sheet length, we find

$$V_{rec} \approx \frac{u_l^2}{V_A} \left(l/L\right)^{1/2} \approx u_{turb} \left(l/L\right)^{1/2}, \tag{1.2}$$

where u_l is the velocity at the driving scale and u_{turb} is the velocity of the largest eddies of the strong turbulent cascade. Note, that here "strong" means only that the eddies decay through nonlinear interactions in an eddy turn over time (see more discussion of the LV99). All the motions are weak in the sense that the magnetic field lines are only weakly perturbed.

It is useful to rewrite this in terms of the power injection rate P. As the perturbations on the injection scale of turbulence are assumed to have velocities $u_l < V_A$, the turbulence is weak at

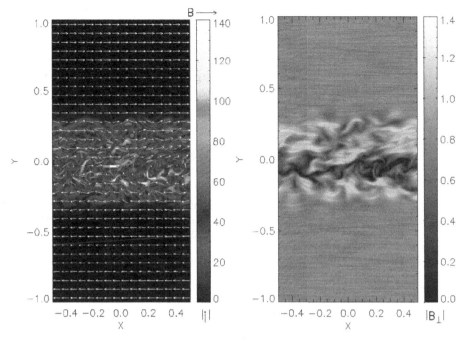

Figure 2. *Left panel:* Current intensity and magnetic field configuration during stochastic reconnection. We show a slice through the middle of the computational box in the xy plane after twelve dynamical times for a typical run. The shared component of the field is perpendicular to the page. The intensity and direction of the magnetic field is represented by the length and direction of the arrows. The color bar gives the intensity of the current. The reversal in B_x is confined to the vicinity of y=0 but the current sheet is strongly disordered with features that extend far from the zone of reversal. *Right panel:* Representation of the magnetic field in the reconnection zone with textures.

large scales. Therefore, the relation between the power and the injection velocities are different from the usual Kolmogorov estimate, namely, in the case of the weak turbulence $P \sim u_l^4/(lV_A)$ (LV99). Thus we get,

$$V_{rec} \approx \left(\frac{P}{LV_A}\right)^{1/2} l, \tag{1.3}$$

where l is the length of the turbulent eddies parallel to the large scale magnetic field lines as well as the injection scale. The reconnection velocity given by equation (1.3) does not depend on resistivity or plasma effects. Therefore for sufficiently high level of turbulence we expect both collisionless and collisional fluids to reconnect at the same rate.

Here we describe the results of a series of three dimensional numerical simulations aimed at adding turbulence to the simplest reconnection scenario and testing equation (1.3). We take two regions with strongly differing magnetic fields lying next to one another. The simulations are periodic in the direction of the shared field (the z axis) and are open in the reversed direction (the x axis). The external gas pressure is uniform and the magnetic fields at the top and bottom of the box are taken to be the specified external fields plus small perturbations to allow for outgoing waves. The grid size in the simulations varied from 256x512x256 to 512x1028x512 so that the top and bottom of the box are far away from the current sheet and the region of driven turbulence around it. At the sides of the box where outflow is expected the derivatives of the dynamical variables are set to zero. A complete description of the numerical methodology can be found in Kowal *et al.* (2009). All our simulations are allowed to evolve for seven Alfvén crossing times without turbulent forcing. During this time they develop the expected Sweet-Parker current sheet configuration with slow reconnection. Subsequently we turn on isotropic turbulent forcing inside a volume centered in the midplane (in the xz plane) of the simulation box and extending

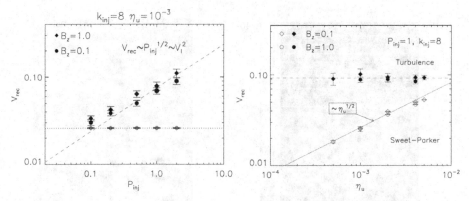

Figure 3. *Left panel*: Reconnection speed versus input power for the driven turbulence. We show the reconnection speed, defined by equation (4) plotted against the input power for an injection wavenumber equal to 8 (i.e. a wavelength equal to one eighth of the box size) and a resistivity ν_u. The dashed line is a fit to the predicted dependence of $P^{1/2}$ (see eq. (3)). The horizontal line shows the laminar reconnection rates for each of the simulations before the turbulent forcing started. Here the uncertainty in the time averages are indicated by the size of the symbols and the variances are shown by the error bars. *Right panel*: Reconnection speed versus resistivity. We show the reconnection speed plotted against the uniform resistivity of the simulation for an injection wavenumber of 8 and an injected power of one. We include both the laminar reconnection speeds, using the hollow symbols, fit to the expected dependence of η_u, and the stochastic reconnection speeds, using the filled symbols. As before the symbol sizes indicate the uncertainty in the average reconnection speeds and the error bars indicate the variance. We included simulations with large, $B_z = 1$, and small, $B_z = 0.1$, guide fields.

outwards by a quarter of the box size. The turbulence reaches its full amplitude around eight crossing times and is stationary thereafter.

In Figure 2 we see the current density on an xy slice of the computational box once the turbulence is well developed. As expected, we see that the narrow stationary current sheet characteristic of Sweet-Parker reconnection is replaced by a chaotic structure, with numerous narrow peaks in the current density. Clearly the presence of turbulence has a dramatic impact on the structure of the reconnection zone. In addition, we see numerous faint features indicating weak reconnection between adjacent turbulent eddies.

The speed of reconnection in three dimensions can be hard to define without explicit evaluation of the magnetic field topology. However, in this simple case we can define it as the rate at which the x component of the magnetic field disappears. More precisely, we consider a yz slice of the simulation, passing through the center of the box. The rate of change of the area integral of $-B_x$— is its flux across the boundaries of the box minus the rate at which flux is annihilated through reconnection (see more discussion in Kowal *et al.* 2009)

$$\partial_t \left(\int |B_x| dz dy \right) = \oint sign(B_x)\vec{E}d\vec{l} - 2V_{rec}B_{x,ext}L_z \qquad (1.4)$$

where electric field is $\vec{E} = \vec{v} \times \vec{B} - \eta\vec{j}$, $B_{x,ext}$ is the absolute value of B_x far from the current sheet and L_z is the width of the box in the z direction. This follows from the induction equation under the assumption that the turbulence is weak to lead to local field reversals and that the stresses at the boundaries are weak to produce significant field bending there. In other words, fields in the x direction are advected through the top and bottom of the box, and disappear only through reconnection. Since we have assumed periodic boundary conditions in the z direction the boundary integral on the right hand side is only taken over the top and bottom of the box. By design this definition includes contributions to the reconnection speed from contracting loops, where Ohmic reconnection has occurred elsewhere in the box and $|B_x|$ decreases as the end of a reconnected loop is pulled through the plane of integration. It is worth noting that this estimate is roughly consistent with simply measuring the average influx of magnetic field lines through the top and bottom of the computational box and equating the mean inflow velocity with the reconnection speed. Following equation (1.4) we can evaluate the reconnection speed for varying strengths and scales of turbulence and varying resistivity.

In Figure 3 (left panel) we see the results for varying amounts of input power, for fixed resistivity and injection scale as well as for the case of no turbulence at all. The line drawn through the simulation points is for the predicted scaling with the square root of the input power. The agreement between equation (1.3) and Figure 3 (left) is encouraging but does not address the most important aspect of stochastic reconnection, i.e. its insensitivity to η.

In Figure 3 (right panel) we plot the results for fixed input power and scale, while varying the background resistivity. In this case η is taken to be uniform, except near the edges of the computational grid where it falls to zero over five grid points. This was done to eliminate edge effects for large values of the resistivity. We see from the Figure 3 (right) that the points for laminar reconnection scale as $\sqrt{\eta}$, the expected scaling for Sweet-Parker reconnection. In contrast, the points for reconnection in a turbulent medium do not depend on the resistivity at all. In summary, we have tested the model of stochastic reconnection in a simple geometry meant to approximate the circumstances of generic magnetic reconnection in the universe. Our results are consistent with the mechanism described by LV99. The implication is that turbulent fluids in the universe including the interstellar medium, the convection zones of stars, and accretion disks, have reconnection speeds close to the local turbulent velocity, regardless of the local value of resistivity. Magnetic fields in turbulent fluids can change their topology on a dynamical time scale.

Finally, it is important to give a few words in relation to our turbulence driving. We drive our turbulence solenoidally to minimize the effects of compression, which does not play a role in LV99 model. The turbulence driven in the volume around the reconnection layer corresponds to the case of astrophysical turbulence, which is also volume-driven. On the contrary, the case of the turbulence driven at the box boundaries would produce spatially inhomogeneous imbalanced turbulence for which we do not have analytical predictions (see discussion of such turbulence in Beresnyak & Lazarian 2009). We stress, that it is not the shear size of our numerical simulations, but the correspondence of the observed scalings to those predicted in LV99 that allows us to claim that we proved that the 3D reconnection is fast in the presence of turbulence.

2. Acceleration of Cosmic Rays

In what follows we discuss the first order Fermi acceleration which arises from volume-filling reconnection†. The LV99 presented such a model of reconnection and observations of the Solar magnetic field reconnection support the volume-filled idea (Ciaravella & Raymond 2008).

Figure 4 exemplifies the simplest realization of the acceleration within the reconnection region expected within LV99 model. As a particle bounces back and forth between converging magnetic fluxes, it gains energy through the first order Fermi acceleration described in de Gouveia Dal Pino & Lazarian (2003, 2005, henceforth GL05) (see also Lazarian 2005, de Gouveia Dal Pino *et al.* 2010, Lazarian *et al.* 2010).

To derive the energy spectrum of particles one can use the routine way of dealing with the first order Fermi acceleration in shocks. Consider the process of acceleration of M_0 particles with the initial energy E_0. If a particle gets energy βE_0 after a collision, its energy after m collisions is $\beta^m E_0$. At the same time if the probability of a particle to remain within the accelerating region is P, after m collisions the number of particles gets $P^m M_0$. Thus $\ln(M/M_0)/\ln(E/E_0) = \ln P/\ln \beta$ and

$$\frac{M}{M_0} = \left(\frac{E}{E_0} \right)^{\ln P / \ln \beta} \tag{2.1}$$

For the stationary state of accelerated particles the number M is the number of particles having energy equal or larger than E, as some of these particles are not lost and are accelerated further.

---

† We would like to stress that Figure 1 exemplifies only the first moment of reconnection when the fluxes are just brought together. As the reconnection develops the volume of thickness Δ gets filled with the reconnected 3D flux ropes moving in the opposite directions.

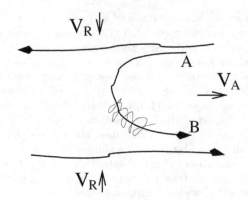

Figure 4. Cosmic rays spiral about a reconnected magnetic field line and bounce back at points A and B. The reconnected regions move towards each other with the reconnection velocity V_R. The advection of cosmic rays entrained on magnetic field lines happens at the outflow velocity, which is in most cases of the order of V_A. Bouncing at points A and B happens because either of streaming instability induced by energetic particles or magnetic turbulence in the reconnection region. In reality, the outflow region gets filled in by the oppositely moving tubes of reconnected flux which collide only to repeat on a smaller scale the pattern of the larger scale reconnection. From Lazarian (2005).

Therefore:

$$N(E)dE = const \times E^{-1+(\ln P/\ln \beta)} dE \qquad (2.2)$$

To determine P and β consider the following process. The particles from the upper reconnection region see the lower reconnection region moving toward them with the velocity $2V_R$ (see Figure 4). If a particle from the upper region enters at an angle θ into the lower region the expected energy gain of the particle is $\delta E/E = 2V_R \cos\theta/c$. For isotropic distribution of particles their probability function is $p(\theta) = 2\sin\theta\cos\theta d\theta$ and therefore the average energy gain per crossing of the reconnection region is

$$\langle \delta E/E \rangle = \frac{V_R}{c} \int_0^{\pi/2} 2\cos^2\theta\sin\theta d\theta = 4/3\frac{V_R}{c} \qquad (2.3)$$

An acceleration cycle is when the particles return back to the upper reconnection region. Being in the lower reconnection region the particles see the upper reconnection region moving with the speed V_R. As a result, the reconnection cycle provides the energy increase $\langle \delta E/E \rangle_{cycle} = 8/3(V_R/c)$ and

$$\beta = E/E_0 = 1 + 8/3(V_R/c) \qquad (2.4)$$

Consider the case of $V_{diff} \ll V_R$. The total number of particles crossing the boundaries of the upper and lower fluxes is $2 \times 1/4(nc)$, where n is the number density of particles. With our assumption that the particles are advected out of the reconnection region with the magnetized plasma outflow the loss of the energetic particles is $2 \times V_R n$. Therefore the fraction of energetic particles lost in a cycle is $V_R n/[1/4(nc)] = 4V_R/c$ and

$$P = 1 - 4V_R/c. \qquad (2.5)$$

Combining Eq. (2.2), (2.4), (2.5) one gets

$$N(E)dE = const_1 E^{-5/2} dE, \qquad (2.6)$$

which is the spectrum of accelerated energetic particles for the case when the back-reaction is negligible (see also GL05)†.

The first order acceleration of particles entrained on the contracting magnetic loop can be understood from the Liouville theorem. As in the process of the magnetic tubes are contracting,

† The obtained spectral index is similar to the one of Galactic cosmic rays.

the regular increase of the particle's energies is expected. The requirement for the process to proceed efficiently is to keep the accelerated particles within the contracting magnetic loop. This introduces limitations on the particle diffusivities perpendicular to magnetic field direction. The subtlety of the point above is related to the fact that while in the first order Fermi acceleration in shocks magnetic compression is important, the acceleration via LV99 reconnection process is applicable to incompressible fluids. Thus, unlike shocks, not the entire volume shrinks for the acceleration, but only the volume of the magnetic flux tube. Thus high perpendicular diffusion of particles may decouple them from the magnetic field. Indeed, it is easy to see that while the particles within a magnetic flux rope depicted in Figure 4 bounce back and forth between the converging mirrors and get accelerated, if these particles leave the flux rope fast, they may start bouncing between the magnetic fields of different flux ropes which may sometimes decrease their energy. Thus it is important that the particle diffusion parallel and perpendicular magnetic field stays different. Particle anisotropy which arises from particle preferentially getting acceleration in terms of the parallel momentum may also be important (Kowal *et al.* 2011, in preparation).

In the numerical studies of the cosmic ray acceleration we use data cubes obtained from the models of the weakly stochastic magnetic reconnection described in §2. For a given snapshot we obtain a full configuration of the plasma flow variables (density and velocity) and magnetic field. We inject test particles in such an environment and integrate their trajectories solving the motion equation for relativistic charged particles

$$\frac{d}{dt}(\gamma m \vec{u}) = q(\vec{E} + \vec{u} \times \vec{B}), \tag{2.7}$$

where $\vec{u}$ is the particle velocity, $\gamma \equiv (1 - u^2/c^2)^{-1}$ is the Lorentz factor, m and q are particle mass and electric charge, respectively, and c is the speed of light.

The study of the magnetic reconnection is done using the magnetohydrodynamic fluid approximation, thus we do not specify the electric field $\vec{E}$ explicitly. Nevertheless, the electric field is generated by either the flow of magnetized plasma or the resistivity effect and can be obtained from the Ohm's equation

$$\vec{E} = -\vec{v} \times \vec{B} + \eta \vec{j}, \tag{2.8}$$

where $\vec{v}$ is the plasma velocity and $\vec{j} \equiv \nabla \times \vec{B}$ is the current density.

In our studies we are not interested in the acceleration by the electric field resulting from the resistivity effects, thus we neglect the last term. After incorporating the Ohm's law, the motion equation can be rewritten as

$$\frac{d}{dt}(\gamma m \vec{u}) = q[(\vec{u} - \vec{v}) \times \vec{B}]. \tag{2.9}$$

In our simulation we do not include the particle energy losses, so particle can gain or loose through the interaction with moving magnetized plasma only. For the sake of simplicity, we assume the speed of light 20 times larger than the Alfvén speed V_A, which defines plasma in the nonrelativistic regime, and the mean density is assumed to be 1 atomic mass unit per cubic centimeter, which is motivated by the interstellar medium density. We integrate equation 2.9 for 10,000 particles with randomly chosen initial positions in the domain and direction of the motion.

In Figure 2 we show the particle energy evolution averaged over all integrated particles for two cases. In the left plot we used plasma fields topology obtained from the weakly stochastic magnetic reconnection models, in the right plot we use fields topology taken from the turbulence studies. Gray area shows the particle energy dispersion over the group of particles. In the case of reconnection model, the expected exponential acceleration is observed until time about 100 hours. Later on, the physical limitations of the computational domain result in a different growth rate corresponding to $E \sim t^{1.49}$. In the case of turbulence without large scale magnetic reconnection the growth of energy is slower $E \sim t^{0.73}$. This testifies that that the presence of reconnection makes the acceleration more efficient. The numerical confirmation of the first order acceleration in the reconnection regions is presented in our forthcoming paper.

In Figure 6 we show the evolution of particle energy distribution. Initially uniform distribution of particle energy ranging from 10^5 to 10^6 MeV evolves faster to higher energies if the reconnection is present. In this case, the final distribution is log-normal being more peaked over

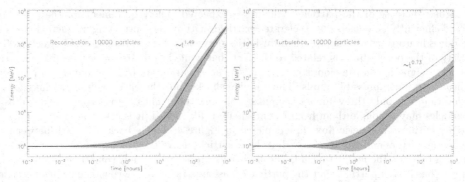

Figure 5. Particle energy evolution averaged over 10,000 particles with the initial energy $E_0 = 10^5$ MeV and random initial positions and directions. In the left panel we show results for the weakly stochastic turbulence environment and in the right panel results for turbulent environment without magnetic reconnection. In both cases we assume $c = 20V_A$ and $\langle \rho \rangle = 1$ u/cm^3.

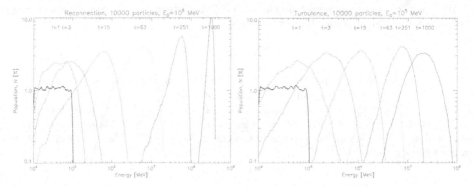

Figure 6. Particle spectrum evolution for 10,000 particles with the uniform initial energy distribution $E_0 = 10^5 - 10^6$ MeV and random initial positions and directions. In the left panel we show results for the weakly stochastic turbulence environment and in the right one results for turbulent environme nt without magnetic reconnection. In both cases we assume $c = 20V_A$ and $\langle \rho \rangle = 1$ u/cm^3.for the weakly stochastic turbulence environment.

the time and with decreasing dispersion of energies in logarithmic scale. On the contrary, in the case of pure turbulence, the energy distribution after evolving to the log-normal shape preserves its dispersion over the time.

As magnetic reconnection is ubiquitous process, the particle acceleration within reconnection process should be widely spread. Therefore accepting the preliminary character of these results above we are involved in more extensive studies of the acceleration-via-reconnection.

3. Summary

The successful testing of the LV99 model of fast reconnection opens avenues for the search of implications of that scheme. One of the implications of the model is the first order Fermi acceleration of energetic particles in the reconnection layer. As reconnection processes are expected to be ubiquitous in astrophysics, we expect the acceleration in reconnection layers to be also ubiquitous. Our simulations of the energetic particle acceleration in the reconnection layer provide results consistent with the first order Fermi acceleration. The origin of the anomalous cosmic rays may be related with the mechanism of particle acceleration via reconnection.

Acknowledgments. A.L. acknowledges NSF grants AST 0808118 and ATM 0648699, as well as the support of the NSF Center for Magnetic Self-Organization. GK acknowledges support from a grant of the Brazilian Agency FAPESP.

References

Armstrong, J. W., Rickett, B. J., & Spangler, S. R., 1995, *ApJ*, 443, 209

Beresnyak, A. & Lazarian, A. 2009, *ApJ*, 702, 460

Biskamp, D. 1986 *Physics of Fluids*, 29, 1520-1531

Ciaravella, A. & Raymond, J. C., 2008, *ApJ*, 686, 1372

Chepurnov, A. & Lazarian, A. 2010, *ApJ*, 710, 853

Daughton, W., Scudder, J., & Karimabadi, H., *Physics of Plasmas*, 072101-1-072101-15, 2006

de Gouveia Dal Pino, E. M., & Lazarian, A. 2003, arXiv:astro-ph/0307054 (preprint version of 2005, A&A, 441, 845-853)

de Gouveia Dal Pino, E. M. & Lazarian, A., 2005, *A&A*, 441, 845

de Gouveia Dal Pino, E. M., Kowal, G., Kadowaki, L. H. S., Piovezan, P., & Lazarian, A., 2010, *International Journal of Modern Physics D*, 19, 729

de Gouveia Dal Pino, E. M., Piovezan, P. P. & Kadowaki, L. H. S. 2010, *A&A*, 518, A5

Drake, J. F., 2001, *Nature*, 410, 525-526

Drake, J. F., Swisdak, M., Che, H., & Shay, M. A., 2006 *Nature*, 443, 553-556

Draine, B. T., & Lazarian, A. 1998, *ApJ*, 508, 157

Elmegreen, B. G. & Scalo, J. 2004, *ARA&A*, 42, 211

Fox, D. B. *et al.*, *Nature*, 437, 845-850 (2005)

Galama, T. J., *et al.*, 1998, *Nature*, 395, 670-672 (1998)

Haverkorn, M., Brown, J. C., Gaensler, B. M., & McClure-Griffiths, N. M. 2008, *ApJ*, 680, 362

Innes, D. E., Inhester, B., Axford, W. I., & Wilhelm, K., 1997, *Nature*, 386, 811-813

Kowal, G., Lazarian, A., Vishniac, E. T., & Otmianowska-Mazur, K. 2009, *ApJ*, 700, 63

Lazarian, A. 2005, Magnetic Fields in the Universe: From Laboratory and Stars to Primordial Structures., 784, 42

Lazarian, A., Kowal, G., Vishniac, E., & de Gouveia dal Pino, E., 2010, *Planet. Space Sci.*, doi:10.1016/j.pss.2010.07.020 (arXiv:astro-ph/1003.2637)

Lazarian, A., & Vishniac, E. T. 1999, *ApJ*, 517, 700-718

Lazarian, A., Vishniac, E. T., & Cho, J. 2004, *ApJ*, 603, 180

Lazarian, A., & Opher, M. 2009, *ApJ*, 703, 8

Lovelace, R. V. E., 1976, *Nature*, 262, 649-652

Masuda, S., Kosugi, T., Hara, H., Tsuneta, S., & Ogawara, Y., 1994, *Nature*, 371, 495-497

Matthaeus, W. H. & Lamkin, S. L. 1986, Physics of Fluids, 29, 2513

Matthaeus, W. H. & Lamkin, S. L. 1985, Physics of Fluids, 28, 303

McKee, C. F. & Ostriker, E. C. 2007, *ARA&A*, 45, 565

Ossendrijver M., 2003, A&A Rev., 11, 287-367

Parker, E., 1957, Journal Geophysical Research, 62, 509-520

Parker, E. N., 1970, *ApJ*, 162, 665-673

Parker, E. N., *Cosmical magnetic fields: Their origin and their activity*, Oxford, Clarendon Press (1979)

Parker, E. N., 1993, *ApJ*, 408, 707-719

Petschek, H. E. Magnetic field annihilation. The Physics of Solar Flares, AAS-NASA Symposium (NASA SP-50), ed. WH. Hess (Greenbelt, MD: NASA) 425

Priest, E. R. & Forbes, T. G., 2002, *A&A Rev.*, 10, 313-377

Shay, M. A. & Drake, J. F., 1998, *Geophys. Res. Letters Geophysical Research Letters*, 25, 3759-3762

Shibata, K. & Tanuma, S. 2001, *Earth, Planets, and Space*, 53, 473

Shimizu, T., Kondoh, K., Shibata, K. & Ugai, M. 2009a, *Physics of Plasmas*, 16, 000

Shimizu, T., Kondoh, K., Ugai, M., & Shibata, K. 2009b, *ApJ*, 707, 420-427

Stix, M. 2000, *A&A*, 42, 85-89

Sturrock, P. A., 1966, *Nature*, 211, 695-697

Sweet, P. A. 1958, in Electromagnetic Phenomena in Cosmical Physics, IAU Symposium no. 6, ed. Bo Lehnert, 123

Yokoyama, T. & Shibata, K., 1995, *Nature*, 375, 42-44

Advances in Plasma Astrophysics
Proceedings IAU Symposium No. 274, 2010
A. Bonanno, E. de Gouveia Dal Pino & A. G. Kosovichev, eds.
© International Astronomical Union 2011
doi:10.1017/S1743921311006594

Weakly imbalanced strong turbulence

Andrey Beresnyak

Los Alamos National Laboratory, Los Alamos, NM 87544

Abstract. The theory of strong MHD turbulence with cross-helicity has been a subject of many recent studies. In this paper we focused our attention on low-imbalance limit and performed high-resolution 3D simulations. The results suggest that in this limit both $\mathbf{w}^+ = \mathbf{v} + \mathbf{b}$ and $\mathbf{w}^- = \mathbf{v} - \mathbf{b}$ are cascaded strongly. The model for imbalance based on so-called "dynamic alignment" strongly contradicts numerical evidence.

Keywords. MHD – turbulence – ISM: kinematics and dynamics

1. Introduction

MHD turbulence has attracted attention of astronomers since mid 1960s. As most astrophysical media are ionized, plasmas are coupled to the magnetic fields (Biskamp, 2003). One-fluid MHD is applicable to most astrophysical environments on macroscopic scales. Turbulence covering huge range of scales has been observed in the ISM (Armstrong *et al.*, 1995).

As with hydrodynamics which has a "standard" phenomenological model of energy cascade (Kolmogorov, 1941), MHD turbulence has one too. This is the Goldreich-Sridhar model henceforth GS95 that uses a concept of critical balance, which maintains that turbulence will stay marginally strong down the cascade. The spectrum of GS95 is supposed to follow a $-5/3$ Kolmogorov scaling. However, a shallower slopes has been reported in numerical studies see, e.g., Müller 2005, which motivated modifications of GS95 (see, e.g., Boldyrev 2005, 2006 and Gogoberidze, 2007).

The other problem of GS95 is that it is incomplete, as it does not treat the most general imbalanced, or cross-helical case. As turbulence is a stochastic phenomenon, an average zero cross helicity does not preclude a fluctuations of this quantity in the turbulent volume. Also, most of astrophysical turbulence is naturally imbalanced, due to the fact that it is generated by a strong localized source of perturbations, such as the Sun in case of solar wind.

Several models of imbalanced turbulence appeared recently: Lithwick *et al.*, 2007 henceforth LGS07, Beresnyak & Lazarian (2008), Chandran (2008), Perez & Boldyrev (2009) henceforth PB09. The full self-consistent analytical model for strong turbulence, however, does not yet exist. In this situation observations and direct numerical simulations (DNS) of MHD turbulence will provide necessary feedback to theorists. We concentrated on two issues, namely that a) the power spectrum slopes of MHD turbulence can not be measured directly from available numerical simulations, supporting an earlier claim in BL09b, b) the ratio of Elsässer dissipation rates is a very robust quantity that can differentiate among many imbalanced models.

2. Numerical setup

We solved incompressible MHD or Navier-Stokes equations:

$$\partial_t \mathbf{w}^\pm + \hat{S}(\mathbf{w}^\mp \cdot \nabla)\mathbf{w}^\pm = -\nu_n(-\nabla^2)^{n/2}\mathbf{w}^\pm + \mathbf{f}^\pm, \qquad (2.1)$$

where $\hat{S}$ is a solenoidal projection and $\mathbf{w}^\pm$ (Elsasser variables) are $\mathbf{w}^+ = \mathbf{v} + \mathbf{b}$ and $\mathbf{w}^- = \mathbf{v} - \mathbf{b}$ where we use velocity $\mathbf{v}$ and magnetic field in velocity units $\mathbf{b} = \mathbf{B}/(4\pi\rho)^{1/2}$. The magnetic Prandtl number here is unity. The RHS of this equation includes a linear dissipation term which is called viscosity or diffusivity for $n = 2$ and hyper-viscosity or hyper-diffusivity for $n > 2$

Table 1. Three-dimensional simulations of balanced and imbalanced turbulence

Run	Resolution: $n_x \times n_y \times n_z$	Forcing	Dissipation	ϵ^+/ϵ^-	$(w^+)^2/(w^-)^2$
B1	1024×3072^2	$w^\pm$	$-3.3 \cdot 10^{-17} k^6$	~ 1	~ 1
B2	768×2048^2	v	$-3.1 \cdot 10^{-16} k^6$	~ 1	~ 1
B3	768×2048^2	$w^\pm$	$-3.1 \cdot 10^{-16} k^6$	~ 1	~ 1
B4	768×2048^2	$w^\pm$	$-6.7 \cdot 10^{-5} k^6$	~ 1	~ 1
I1	512×1024^2	$w^\pm$	$-1.9 \cdot 10^{-4} k^2$	1.187	1.35 ± 0.04
I2	768^3	$w^\pm$	$-6.8 \cdot 10^{-14} k^6$	1.187	1.42 ± 0.04
I3	512×1024^2	$w^\pm$	$-1.9 \cdot 10^{-4} k^2$	1.412	1.88 ± 0.04
I4	768^3	$w^\pm$	$-6.8 \cdot 10^{-14} k^6$	1.412	1.98 ± 0.03
I5	1024×1536^2	$w^\pm$	$-1.5 \cdot 10^{-15} k^6$	2	5.57 ± 0.08
I6	1024×1536^2	$w^\pm$	$-1.5 \cdot 10^{-15} k^6$	4.5	45.2 ± 1.5

and the driving force $\mathbf{f}^\pm$. The total dissipation per unit time of each Elsässer energy $(w^\pm)^2$ by the linear dissipation term are called Elsässer dissipation rates $\epsilon^\pm$. In most simulations from Table 1 we drove w^+ and w^- randomly and independently. Each $\mathbf{w}$ field can be represented as a mean field $\mathbf{v_A}$ plus perturbation $\mathbf{w}^\pm = \pm\mathbf{v_A} + \delta\mathbf{w}^\pm$. We will also use characteristic perturbation magnitudes on a particular scale l, $w_l^\pm$. We solved these equations with a pseudospectral code that was described in great detail in our earlier publications BL09a, BL09b. Table 1 enumerates latest high-resolution runs, which were performed in so-called reduced MHD approximation, where the $\mathbf{w}^\pm$ component parallel to the mean field (pseudo-Alfvén mode) is omitted and so are the parallel gradients in the nonlinear term $((\delta\mathbf{w}^\mp \cdot \nabla_\parallel)\delta\mathbf{w}^\pm$. The linear propagation term, i.e. $((\delta\mathbf{v_A} \cdot \nabla_\parallel)\delta\mathbf{w}^\pm$, however, is always finite. This is because we consider turbulence injected at such $k_\parallel$ that $v_A k_\parallel$ stays finite, in other words, we use computational box which is elongated in x direction by a factor which is proportional to v_A. In this situation the actual value of the mean field B_0 or v_A drops out of calculations. Physically this means that B_0 is "large enough" compared to perturbations. Under these assumptions one studies purely Alfvénic dynamics in a strong mean field, i.e., Alfvénic turbulence.

We started our high resolution simulations with earlier lower-resolution runs that were evolved for a long time, typically hundreds Alfvenic times and reached stationary state. The balanced runs were evolved, typically, for 6-10 Alfvenic times and the imbalanced runs were evolved for longer times, typically 10-40. The energy injection rates were kept constant in I1-6 and the fluctuating dissipation rate was within few percent of the former. The fluctuations in total energies $(w^+)^2$ and $(w^-)^2$ was a main source of uncertainty for I1-6 shown on Fig. 1 For more simulations and the analysis of anisotropy, see also BL09a, BL09b.

3. Nonlinear cascading and dissipation rate

One of the most robust quantities in numerical simulations of MHD turbulence is the energy cascading rate or dissipation rate. In high-Reynolds number turbulence energy has to cascade through many steps before dissipating and the dissipation is negligible on the outer (large) scale. Therefore, the nonlinear energy cascading rate and the dissipation rate are used interchangeably.

In hydrodynamic turbulence the dissipation rate and the spectrum of velocity are connected by the well-known Kolmogorov constant:

$$E(k) = C_K \epsilon^{2/3} k^{-5/3}. \tag{3.1}$$

The important fact that strong hydrodynamic turbulence dissipates in one dynamic timescale l/v is reflected by C_K being close to unity (~ 1.6). In MHD turbulence, however, there are two energy cascades (or "Elsasser cascades") and there are two dissipation rates, ϵ^+ and ϵ^-. The question of how these rates are related to velocity-like Elsasser amplitudes $\mathbf{w}^+$ and $\mathbf{w}^-$ is one of the central questions of imbalanced MHD turbulence. Each model of strong imbalanced

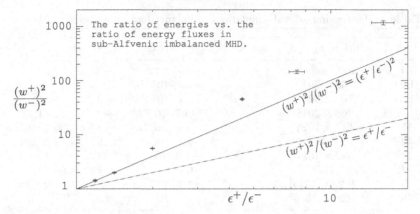

Figure 1. Energy imbalances versus dissipation rate imbalance. Predictions, solid: LGS07, dashed: PB09. Errorbars indicate fluctuation in time. On this plot I1 and I3 are omitted as they are close to I2 and I4. I1 and I3 are simulations with normal viscosity which have slightly lower energy imbalance than I2 and I4, see Table 1. This is an indication that in these simulations viscosity was affecting outer scales. Two high imbalance points are taken from BL09a. For a fixed dissipation ratio the energy imbalance has a tendency to only *increase* with resolution.

turbulence advocates a different physical picture of cascading and provides a different relation between the ratio of energies $(w^+)^2/(w^-)^2$ and the ratio of fluxes ϵ^+/ϵ^-.

Goldreich-Sridhar model (GS95) predicts that in the balanced case the cascading is strong and each wave is cascaded by the shear rate of the opposite wave, i.e.,

$$\epsilon^+ = \frac{(w_l^+)^2 w_l^-}{l}, \qquad \epsilon^- = \frac{(w_l^-)^2 w_l^+}{l}. \tag{3.2}$$

It is similar to Kolmogorov cascade with w's replacing v. Although this model does not make predictions for the imbalanced case, one could hope that in the case of small imbalance these formulae will still work. In this case we will obtain $(w^+)^2/(w^-)^2 = (\epsilon^+/\epsilon^-)^2$. LGS07 argued that this relation will hold even for large imbalances.

For the purpose of this short paper we mostly discuss the prediction of LGS07, $(w^+)^2/(w^-)^2 = (\epsilon^+/\epsilon^-)^2$ and the prediction of PB09 that nonlinear timescales are equal for both waves, which effectively lead to † $(w^+)^2/(w^-)^2 = \epsilon^+/\epsilon^-$. Note, that the last prediction is also true for highly viscous flows ($Re = Re_m \ll 1$). It could be rephrased that PB09 predicts turbulent viscosity which is equal for both components.

Compared to spectral slopes, dissipation rates are robust quantities that require much smaller dynamical range and resolution to converge. Fig. 1 shows energy imbalance $(w^+)^2/(w^-)^2$ versus dissipation rate imbalance ϵ^+/ϵ^- for simulations I2, I4, I5 and I6. We also use two data points from our earlier simulations with large imbalances, A7 and A5 from BL09a. I1 and I3 are simulations with normal viscosity similar to I2 and I4. They show slightly smaller energy imbalances than I2 and I4 (Table 1).

We see that most data points are above the line which is the prediction of LGS07. In other words, one can deduce that numerics strongly suggest that

$$\frac{(w^+)^2}{(w^-)^2} \geqslant \left(\frac{\epsilon^+}{\epsilon^-}\right)^2. \tag{3.3}$$

Although there is a tentative correspondence between LGS07 and the data for small degrees of imbalance, the deviations for large imbalances are significant. In the case of strong imbalances

---

† Both of these predictions are subject to intermittency corrections. We average $(w^+)^2$ and $(w^-)^2$ over volume and time. This averaging does not take into account possible fluctuations in ϵ^+ and ϵ^-. We believe, however, that these effects are small, as long as we use the second-order measures, such as energy.

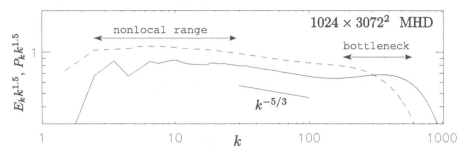

Figure 2. The spectrum from B1. Due to a higher resolution we are starting to see an asymptotic -5/3 slope of MHD turbulence. The Kolmogorov constant for this purely Alfvenic turbulence is $C_K \approx 3.2$, which is much higher than for hydro turbulence. This fact was missed in earlier lower resolution simulations (Biskamp, 2003). Also, higher C_K means less efficient energy transfer that is fully consistent with our picture of diffuse locality (BL10).

data deviates from GS95 and LGS07 and suggests that the strong component cascading rate is smaller than what is expected from strong cascade (BL08).

As to PB09 prediction, it is inconsistent with data for all degrees of imbalance including those with small imbalance and normal viscosity, i.e. I1 and I3.

4. Discussion

One of the most important measures not mentioned in this paper is the anisotropy of MHD turbulence. It had been considered in great detail in our earlier publication BL09a. In particular, we refer the reader to the result of BL08, BL09a that the anisotropy of strong component is smaller than the anisotropy of weak component. This fact is inconsistent with both the naive application of GS95 critical balance (which would have predicted the opposite), or the derivation in LGS07 that suggests that the both waves have the same anisotropy.

PB09 claims that the nonlinear timescales for both components are equal, i.e. there is a turbulent viscosity which is the same for both components, regardless of the degree of imbalance. This seems counter-intuitive for transition to freely-propagating Alfvenic waves. PB09 formula, $w^+/w^- = \sqrt{\epsilon^+/\epsilon^-}$ suggests that the asymptotic ($Re = Re_m \gg 1$) prediction for energy imbalance in this case will be the same as in highly viscous case ($Re = Re_m \ll 1$), i.e. $(w^+)^2/(w^-)^2 = \epsilon^+/\epsilon^-$. This is at odds with numerical evidence, which suggests $(w^+)^2/(w^-)^2 \geqslant (\epsilon^+/\epsilon^-)^2$.

References

Aluie, H. & Eyink, G. L., 2010, *Phys. Rev. Lett.* 104, 081101

Armstrong, J. W., Rickett, B. J. & Spangler, S. R. 1995, *ApJ*, 443, 209

Beresnyak, A. & Lazarian, A. 2008, *ApJ*, 678, 961 (BL08)

Beresnyak, A. & Lazarian, A. 2009a, *ApJ*, 702, 460 (BL09a)

Beresnyak, A. & Lazarian, A. 2009b, *ApJ*, 702, 1190 (BL09b)

Beresnyak, A. & Lazarian, A. 2010, *ApJ*, 722, L110 (BL10)

Biskamp, D. 2003, *Magnetohydrodynamic Turbulence.* (Cambridge: CUP)

Boldyrev, S. 2005, *ApJ*, 626, L37; Boldyrev, S. 2006, Phys. Rev. Lett., 96, 115002

Chandran, B. D. G. 2008, *ApJ*, 685, 646

Gogoberidze, G. 2007, *Phys. Plasmas*, 14, 022304

Goldreich, P. & Sridhar, S. 1995, *ApJ*, 438, 763

Kolmogorov, A. 1941, *Dokl. Akad. Nauk SSSR*, 31, 538

Lithwick, Y., Goldreich, P., & Sridhar, S. 2007, *ApJ*, 655, 269 (LGS07)

Müller W.-C. & Grappin, R 2005, *Phys. Rev. Lett.*, 95, 114502

Perez, J. & Boldyrev, S. 2009, *Phys. Rev. Lett.*, 102, 025003, (PB09)

Advances in Plasma Astrophysics
Proceedings IAU Symposium No. 274, 2010
A. Bonanno, E. de Gouveia Dal Pino & A. G. Kosovichev, eds.
© International Astronomical Union 2011
doi:10.1017/S1743921311006600

Plasma astrophysics implication in discovery and interpretation of X-ray radiation from comets

S. Ibadov

Sternberg Astronomical Institute, Moscow State University, Moscow, Russia
Institute of Astrophysics, Tajik Academy of Sciences, Dushanbe, Tajikistan
email: ibadovsu@yandex.ru

Abstract. Discovery of soft X-ray radiation from comet Hyakutake C/1996 B2 by space telescope ROSAT in March 1996 as well as establishing the regularity of the phenomenon for comets in general opened a new area of research for the plasma astrophysics. The first soft X-ray observations have been motivated by the results of a theoretical investigation on the efficiency of production of energetic photons, in the energy range 0.1-1 keV, by hot plasma clumps generated in dusty comets via high velocity collision with interplanetary dust at small heliocentric distances. Moreover, the soft X-ray luminosities measured significantly exceeded the value predicted. A short review of proposed theoretical models and mechanisms for explaining X-ray emission from comets as well as some prospects for the future X ray observations of comets are presented.

Keywords. plasma astrophysics general, comets, cosmic dust, high velocity collisions, high-temperature plasma, multicharge ions, X-ray radiation

---

1. Introduction

Development of the plasma astrophysics has accompanied formation of the physics of comets during more than hundred years. The origin and dynamics of comet plasma tails as well as mechanisms of ionization of cometary molecules in the coma were the basic subjects of plasma astrophysics and physics of comets. The study of comets up to 1970 years carried out mainly using ground based observations, i.e. in the optical range of the spectrum (see, e.g., Dobrovolsky 1966; Brandt & Hodge 1967).

Space UV observations of comets, including HI Lyman alpha 1216 A, discovered the presence of giant atomic hydrogen coma (Donn *et al.* (1976) and references therein). Space missions to comet Halley 1986 III and other comets presented new data on the structure of comets, including comet nucleus, gas, dust and plasma (Reinhard 1986; Sagdeev *et al.* 1986, and references therein).

Comet X-ray radiation in the 0.09 to 2.0 keV band was discovered by space telescope ROSAT during observations of comet Hyakutake C/1996 B2 in 27 March 1996 (Lisse *et al.* 1996). These soft X-ray observations have been motivated by results of a theoretical investigation on the efficiency of generation of energetic photons, within the energy range 0.1-1 keV, by hot plasma blobs/clumps produced in dusty comets via high velocity collision, V > 70 km/s, with interplanetary dust at small heliocentric distances, R < 1 AU (Ibadov 1990).

At the same time the cometary soft X-ray luminosity measured, of the order of 10^{15} erg/s with considerable variations, significantly exceeded the value predicted.

In the present paper a short review of proposed theoretical models and basic mechanisms for explaining X-ray emission from comets as well as some prospects in the field are presented.

2. Generation of X-rays in comets

Classical mechanism of comet radiation due to scattering of solar radiation by cometary dust and resonance fluorescence of solar photons on cometary atmosphere molecules is unable to produce detectable comet X-rays, at least this mechanism is weaker than that connected with collision between cometary and interplanetary dust at R < 1 AU (Ibadov 1985).

High-temperature intensively radiating plasma clumps in comets may be generated via high-velocity collisions between cometary and interplanetary dust particles (Ibadov 1980, 1983, 1987, 1990, 1996a). Expanding plasma clumps, "compound particles", produced in such process consist of heavy multicharge ions of elements like Fe, Si, O etc. (Ibadov 1986). For this reason the intensity of recombination radiation from clumps, i.e. due to bound-bound transitions of electrons in the plasma clump, tens times greater than that for bremsstrahlung, i.e. free-free transitions. Moreover, plasma clumps may radiate as a black body. In this case the efficiency of X-ray radiation k_x reaches 0.1, i.e. around tenth time larger than that for hot expanding plasma produced by ultrashort/picosecond laser pulses on light/ fusion elements (Basov *et al.* 1971).

The comet X-ray luminosity due to high-velocity collisions between cometary and interplanetary dust particles is determined as a sum of emissions from collisionally thick and collisionally thin volumes of the comet dust coma, namely

$$L_x(dd) = k_x \frac{\rho V^3}{4} \left[(\pi r_x^2) + 4r_c r_x \ln(r_c/r_x) \right]. \tag{2.1}$$

Here $\rho = \rho(R) = 5 x 10^{-22} R_*^{-1.3}$ g/cm^3 is the mean spatial mass density of interplanetary dust (has mainly prograde orbit near the ecliptic, see, e.g., Grun *et al.* 1985), and this value can rise up to thousands times within interplanetary dust clouds/trails (e.g., Lebedinets 1984), V=V(R) is the relative velocity of colliding comet and IPD particles, r_x is the radius of the collisionally thick zone of the comet dust coma relative IPD, r_c is the effective radius for comet X-rays: $r_c = \min(r_a, r_c), r_a$ is the aperture radius of the photon detector/telescope, r_c is the radius of a comet dust coma, depends on the dust production rate of the nucleus and size distribution of dust particles (Reinhard 1986, Sagdeev *et al.* 1986, Utterback & Kissel 1990, Lisse *et al.* 1996), R_* is the heliocentric distance in AU (Ibadov 1990, 1996a, 1998, 2000, 2001; Ibadov & Dennerl 2000a, b).

The most probable energy of photons from plasma clumps is

$$h\nu_m = 2.8kT = \frac{0.7 A m_p V^2}{3(1 + z + 2x_1/3)} = T_* \left(\frac{V}{V_*} \right)^2 \approx \frac{h\nu_{m*}}{R_*} \text{ eV}, \tag{2.2}$$

where T is the initial temperature of the expanding plasma clump, A is the mean atomic number for colliding dust particles, m_p is the proton's mass, z is the mean multiplicity of charge of ions, x_1 is the mean relative ionization potential, $T_* = 10^6 K, V_* = 70$ km/s, $h\nu_{m*} = 70$ eV for comets like 1P/Halley with retrograde orbits and $T_* = 7 x 10^5 K, V_* = 50$ km/s, $h\nu_{m*} = 50$ eV for Hyakutake C/1996 B2 type comets with polar/quasipolar orbits (Ibadov 1990, 1996b).

Using (2.1) and (2.2) for $R = 1$ AU with realistic values of parameters of the radiation process $k_x = 0.1, \rho_* = 5 \times 10^{-22}$ g/cm^3, $V = 50$ km/s, $r_x = 10^2$ km, $r_c = r_a = 10^5$ km we have $L_x(dd) = 3 \times 10^{15}$ erg/s, $h\nu_m = 50$ eV.

Gas and dust flows from comet nuclei are ejected mainly to the Sun direction, according to results of space missions to comet 1P/Halley in March 1986 as well as ground-based observations of comet Mrkos 1957d with 5-meter Palomar telescope at high angular resolution (Greenstein 1958, Greenstein & Arpigny 1962, Reinhard 1986, Sagdeev et al. 1986, and references therein). In addition, the cometocentric velocities of ejected dust particles are negligibly small than comet orbital one. Therefore maximum X-ray brightness due to dust-dust collisions should be shifted in the Sun direction, independently of the position/direction of a comet relative to its perihelion. The ROSAT observations in March 1996 showed similar picture. So, rapid variability of the detected flux of X-rays on a time scale around 1 hour may be caused by considerable IPD nonhomogeneity.

The further searches, followed the discovery of comet X-rays, revealed other possible plasma processes resulting in generation of soft X-rays in comets (Brandt et al. 1996, Ip & Chow 1997, Uchida et al. 1998, and references therein). In particular, soft X-rays may be generated due to recombination, charge exchange, of the solar wind multicharge ions like C^{6+}, O^{6+}, N^{6+} etc., escaping from the solar corona plasma with temperature around 100 eV, on cometary coma molecules. The corresponding X-ray luminosity is determined as

$$L_x(SW) = \sum A_i E_i n_i V_i S_{i*}, \qquad (2.3)$$

where A_i, E_i, n_i, V_i are the abundance of multicharge ions of kind i in the solar wind relative to protons, recombination energy of these ions, protons number density and velocity in the solar wind, respectively, S_{i*} is the cross section of the cometapause in the comet coma where charge exchange process between solar wind multicharge ions and cometary molecules intensively occurs (Cravens 1997).

From (2.3) with $A_i = 10^{-3}$, $E_i = 300$ eV, $n_i = 5$ cm^{-3}, $V_i = 400$ km/s, $S_{i*} = 10^{19}$ cm^2 we have $L_x(SW) = 10^{15}$ erg/s. This value is four times less than the mean value obtained from ROSAT observations of comet Hyakutake in March 1996. Hence, contribution of other possible mechanisms for comet X-ray emission should be taken into account.

3. Conclusions

High velocity collisions between cometary and interplanetary dust particles at small heliocentric distances, $R \leqslant 1$ AU, are possible generators of high-temperature plasma and soft X-rays as well as multicharge ions in dusty comets, especially in comets like 1P/Halley (retrograde orbits) and Hyakutake C/1996 B2 (polar and quasipolar orbits).

Searches for soft, 0.09–2 keV, and ultrasoft, 30–90 eV, X-ray radiation from dusty comets with retrograde and polar/quasi-polar orbits around R ⩽1 AU by space telescopes like ROSAT and XMM are of interest for plasma astrophysics, physics of comets as well as for studying cosmic dust in the inner heliosphere.

Acknowledgements

The author is grateful to the IAU Symposium 274 SOC/LOC for a travel grant. The hospitality of SAI MSU, DIC MSU as well as assistance of Dr. Georgij M. Rudnitskij and Firuz S. Ibodov are acknowledged.

References

Basov, N. G., Zakharov, S. D., Krokhin, O. N. et al. 1971, Quantum Electronics USSR, 1, 4
Brandt, J. C., & Hodge P. W. 1967, Solar System Astrophysics, Mir Publ. Comp., Moscow, 488

Brandt, J. C., Lisse, C. M., & Yi, Y. 1996, *BAAS*, 189, 19

Christian, D. J., Lisse, C. M., Dennerl, K., Marshal, F., Mushotsky, R. F., Petre, R.,Snowden, S., Weaver, H., Stoozas, B., & Wolk, S. 2000, *HEAD*, 42.22, *http://chandra.harvard.edu/press/00_releases/press_072700.html*

Cravens, T. E. 1997, *Geophys. Res. Lett.*, 24, 105

Dennerl, K., Englhauser, J., & Trumper, J. 1997, *Science*, 277, 1625

Dobrovolsky, O. V. 1966, *Comets, Nauka, Moscow*, 288

Donn, B., Mumma, M., Jackson, W., A'Hearn, M., & Harrington, R. (eds.) 1976, *The Study of Comets, NASA SP-393, Proc. IAU Colloquium No. 25*, 1083

Greenstein, J. L. 1958, *ApJ*, 128, 106

Greenstein, J. L. & Arpigny, C. 1962, *ApJ*, 135, 892

Grun, E., Zook, H. A., Fechtig, H., & Giese, R. H. 1985, *Icarus*, 62, 244

Ibadov, S. 1980, *Comet. Circ. USSR*, 266, 3

Ibadov, S. 1983, in: *Cometary Exploration, Proc. Int. Conf., Budapest, Hungarian Acad. Sci. Publ.*, 227

Ibadov, S. 1985, in: *Properties and Interactions of Interplanetary Dust, Proc. IAU Colloquium No. 85, Reidel Publishing Company, Dordrecht*, 365

Ibadov, S. 1986, in: *ESA SP-250, Proc. 20th ESLAB Symposium on the Exploration of Halley's Comet*, 377

Ibadov, S. 1987, in: *ESA SP-278, Proc. Symposium on the Diversity and Similarity of Comets*, 655

Ibadov, S. 1989, in: *NASA CP-3036, Proc. IAU Symposium No. 135 on Interstellar Dust*, 49

Ibadov, S. 1990, *Icarus*, 86, 283

Ibadov, S. 1996a, *Adv. Sp. Res.*, 17:12, 93

Ibadov, S. 1996b, *Physical Processes in Comets and Related Objects, Cosmosinform Publ. Comp., Moscow*, 181

Ibadov, S. 1998, *Proc. First XMM Workshop on Science with XMM*, *http://astro.estec.esa.nl/XMM/news/ws1.top.html*, *http://xmm.esac.esa.int/external/xmm_science/workshops/1st_workshop*

Ibadov, S. 2000, *Astron. Astrophys. Transac.*, 18:6, 799

Ibadov, S. 2001, in: *Astrophysical Sources of High Energy Particles and Radiation, Kluwer Academic Publisher, Dordrecht*, 245

Ibadov, S. & Dennerl, K. 2000a, *IAU Colloquium 181/COSPAR Colloquium 11 Abstracts*, 71

Ibadov, S. & Dennerl, K. 2000b, *JENAM-2000: 9th European and 5th Euro-Asian Astron. Soc. Conf. Abstracts*, 206

Ip, W. H. & Chow, V. W. 1997, *Icarus*, 130, 217

Lebedinets, V. N. 1984, *Astron. Vestnik*, 18, 35

Lisse, C. M., Dennerl, K., Englhauser, J., Harden, M., Marshall, F. E., Mumma, M. J., Petre, R., Pye, J. P., Ricketts, M. J., Trumper, J. & West, R.G. 1996, *Science*, 274, 205

Reinhard, R. 1986, *Nature*, 321, 313

Sagdeev, R. Z., Blamont, J., Galeev, A. A., Moroz, V. I., Shapiro, V. D., Shevchenko, V. I., & Szego, K. 1986, *Nature*, 321, 259

Uchida, M., Morikawa, M., Kubotani, H., & Mouri, H. 1998, *ApJ*, 498, 863

Utterback, N. G. & Kissel, J. 1990, *AJ*, 100, 1315

Advances in Plasma Astrophysics
Proceedings IAU Symposium No. 274, 2010
A. Bonanno, E. de Gouveia Dal Pino & A. G. Kosovichev, eds.

© International Astronomical Union 2011
doi:10.1017/S1743921311006612

Large eddy simulations in plasma astrophysics. Weakly compressible turbulence in local interstellar medium

A. A. Chernyshov, K. V. Karelsky and A. S. Petrosyan

Theoretical section, Space Research Institute of the Russian Academy of Sciences,
Profsoyuznaya 84/32, 117997, Moscow, Russia
email: `apetrosy@iki.rssi.ru`

Abstract. We apply large eddy simulation technique to carry out three-dimensional numerical simulation of compressible magnetohydrodynamic turbulence in conditions relevant local interstellar medium. According to large eddy simulation method, the large-scale part of the flow is computed directly and only small-scale structures of turbulence are modeled. The small-scale motion is eliminated from the initial system of equations of motion by filtering procedures and their effect is taken into account by special closures referred to as the subgrid-scale models. Establishment of weakly compressible limit with Kolmogorov-like density fluctuations spectrum is shown in present work. We use our computations results to study dynamics of the turbulent plasma beta and anisotropic properties of the magnetoplasma fluctuations in the local interstellar medium.

Keywords. turbulence, ISM: kinematics and dynamics, methods: numerical

1. Introduction

Turbulence represents one of the most important phenomena, both in astrophysical and in laboratory plasmas. There is increasing evidence of the key role played by turbulence within different physical processes taking place in magnetofluids, like transport phenomena or the nonlinear dynamics of such complex systems. The presence of velocity and magnetic field fluctuations in a wide range of space and time scales has been directly detected in the interplanetary medium, while there are strong indications of their presence also in the solar corona. In contrast, large-eddy simulation (LES) is a multiscale computational modeling approach that offers a more comprehensive capturing of unsteady turbulent flow. So far, LES have been mainly associated with flows of modest complexity and primarily of academic interest. However, LES hold much promise for becoming the future research and development strategy for astrophysical applications in which turbulent flow is of pivotal importance. Some first important developments in this direction in which LES is being applied to astrophysical applications have recently emerged by development LES model for compressible MHD. Turbulent flows are inherently unsteady in this model and LES capture their major properties. The emergence of LES follows the rapidly growing computer capabilities that now allows for describing, not only the average behavior, but also most of the time evolution of the solar flow dynamics.

2. Formulation of Large Eddy Simulation

For study of compressible MHD turbulence in interstellar, medium we use large eddy simulation (LES) method (Chernyshov *et al.* 2007, Chernyshov *et al.* 2008). The filtering procedure is applied to the initial equations in the LES method. Each physical parameter is expanded into large- and small-scale components. The effects on large scales are calculated directly and those on small scales are modeled. The filtered part $\bar{f}(x_i)$ is defined as follows:

$$\bar{f}(x_i) = \int_{\Theta} f(\acute{x}_i)\xi(x_i, \acute{x}_i; \bar{\triangle})d\acute{x}_i, \tag{2.1}$$

where ξ is the filter function satisfying the normalization property, Θ is the domain, $\bar{\triangle}$ is the filter-width and $x_j = (x, y, z)$ are axes of Cartesian coordinate system.

In order to simplify the resulting equations describing turbulent MHD flow with variable density, it is convenient to use the Favre filtering (it is also called mass-weighted filtering) to avoid additional subgrid-scale (SGS) terms. Therefore, Favre filtering will be used further. The mass-weighted filtration is used for all parameters of charged fluid flow except for the pressure and the magnetic field. Mass-weighted filtering is determined as follows: $\tilde{f} = \overline{\rho f}/\bar{\rho}$. To denote the filtration in this relation we use two symbols, viz. the overbar denotes the ordinary filtration and the tilde specifies the mass-weighted filtration. Using the mass-weighted filtration operation, we rewrite the compressible MHD equations as:

$$\frac{\partial \bar{\rho}}{\partial t} + \frac{\partial \bar{\rho}\tilde{u}_j}{\partial x_j} = 0; \tag{2.2}$$

$$\frac{\partial \bar{\rho}\tilde{u}_i}{\partial t} + \frac{\partial}{\partial x_j}\left(\bar{\rho}\tilde{u}_i\tilde{u}_j + \bar{p}\delta_{ij} - \frac{1}{Re}\,\tilde{\sigma}_{ij} + \frac{\bar{B}^2}{2M_a^2}\delta_{ij} - \frac{1}{2M_a^2}\,\bar{B}_j\bar{B}_i\right) = -\frac{\partial \tau_{ji}^u}{\partial x_j}; \tag{2.3}$$

$$\frac{\partial \bar{B}_i}{\partial t} + \frac{\partial}{\partial x_j}\left(\tilde{u}_j\bar{B}_i - \tilde{u}_i\bar{B}_j\right) - \frac{1}{Re_m}\frac{\partial^2 \bar{B}_i}{\partial x_j^2} = -\frac{\partial \tau_{ji}^b}{\partial x_j}, \tag{2.4}$$

The filtered nondivergent (solenoidal) property of magnetic field is $\partial \bar{B}_j/\partial x_j = 0$.

Here, ρ is the density; p is the pressure; u_j is the velocity in the direction x_j; B_j is the magnetic field in the direction x_j; $\sigma_{ij} = 2\mu S_{ij} - \frac{2}{3}\mu S_{kk}\delta_{ij}$ is the viscous stress tensor; $S_{ij} = 1/2\left(\partial u_i/\partial x_j + \partial u_j/\partial x_i\right)$ is the strain rate tensor; μ is the coefficient of molecular viscosity; η is the coefficient of magnetic diffusivity; δ_{ij} is the the Kronecker delta.

The right-hand sides terms in equations (2.3) - (2.4) designate influence of subgrid terms on the filtered part: $\tau_{ij}^u = \bar{\rho}\left(\widetilde{u_i u_j} - \tilde{u}_i\tilde{u}_j\right) - \frac{1}{M_a^2}\left(\overline{B_i B_j} - \bar{B}_i\bar{B}_j\right)$; and $\tau_{ij}^b = \left(\overline{u_i B_j} - \tilde{u}_i\bar{B}_j\right) - \left(\overline{B_i u_j} - \bar{B}_i\tilde{u}_j\right)$.

To close the system of MHD equations, we assume that the relation between density and pressure is polytropic and has the following form: $p = \rho^\gamma$, γ is a polytropic index.

The effect of the subgrid terms which appear in the right-hand side of the magnetohydrodynamics equations (2.3) - (2.4) on the filtered part is modelled by the SGS terms. We use Smagorinsky model for compressible MHD case for subgrid-scale parameterization. The Smagorinsky model for compressible MHD turbulence showed accurate results under various range of similarity numbers (Chernyshov *et al.* 2007):

$$\tau_{ij}^u - \frac{1}{3}\tau_{kk}^u\delta_{ij} = -2C_1\bar{\rho}\bar{\triangle}^2|\tilde{S}^u|\left(\tilde{S}_{ij} - \frac{1}{3}\tilde{S}_{kk}\delta_{ij}\right), \tag{2.5}$$

$$\tau_{ij}^b = -2D_1\bar{\triangle}^2|\bar{j}|\bar{J}_{ij}, \tag{2.6}$$

$$\tau_{kk}^u = 2Y_1\bar{\rho}\bar{\triangle}^2|\tilde{S}^u|^2 \tag{2.7}$$

The parameters C_1, Y_1 and D_1 in equations (2.5) - (2.7) are model constants, their

A. A. Chernyshov, K. V. Karelsky & A. S. Petrosyan

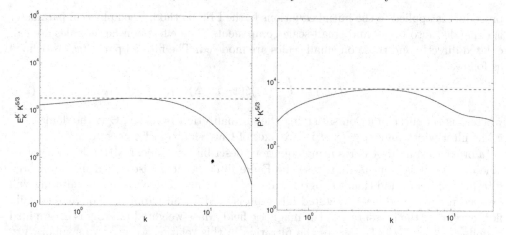

Figure 1. Normalized and smoothed spectrum of kinetic energy E_k^k (left) and ensity fluctuations P^k (right), multiplied by $k^{5/3}$. Notice that the spectrum is close to $\sim k^{-3}$ in a forward cascade regime of decaying turbulence. However, there is well-defined inertial Kolmogorov-like range of $k^{-5/3}$ that confirms observation data.

values being self-consistently computed during run time with the help of the dynamic procedure.

We perform three-dimensional numerical simulation of decaying compressible MHD turbulence for study of interstellar turbulence. The numerical code of the fourth order accuracy for MHD equations in the conservative form is used in our work. The skew-symmetric form for nonlinear terms is applied to reduce error of discretization when finite difference scheme is employed for modeling of turbulent flow. The third order low-storage Runge-Kutta method is applied for time integration. The explicit LES method is used in this work. To separate the turbulent flow into large and small eddy components, Gaussian filter of the fourth order of accuracy is applied. Periodic boundary conditions for all the three dimensions are applied. The simulation domain is a cube with dimensions of $\pi \times \pi \times \pi$. Initial hydrodynamic turbulent Reynolds number is chosen $Re \approx 2000$ and magnetic Reynolds number is chosen $Re_m \approx 200$. The initial isotropic turbulent spectrum was chosen for kinetic and magnetic energies in Fourier space to be close to k^{-2} with random amplitudes and phases in all three directions. The choice of such spectrum as initial conditions is due to velocity perturbations with an initial power spectrum in Fourier space similar to that of developed turbulence.

3. Results

Compressible MHD turbulence evolves under the effect of nonlinear interactions in which larger eddies transfer energy to smaller ones through forward turbulent energy cascade. Notwithstanding the fact that supersonic flows with high value of large-scale Mach numbers are characterized in interstellar medium, nevertheless, there are subsonic fluctuations of weakly compressible components of interstellar medium. These weakly compressible subsonic fluctuations are responsible for emergence of a Kolmogorov-type spectrum in interstellar turbulence which is observed from experimental data. It is shown that density fluctuations are a passive scalar in a velocity field in weakly compressible magnetohydrodynamic turbulence and demonstrate Kolmogorov-like spectrum in a dissipative range of the energy cascade (Fig. 1). The spectral indexes of density fluctuations

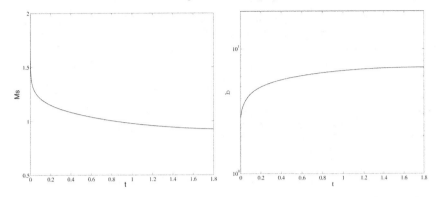

Figure 2. Decay of turbulent small-scale Mach number $\breve{M}_s$ with time(left). A transition from a supersonic $\breve{M}_s > 1$ to a subsonic $\breve{M}_s < 1$ can be observed. Time dynamics of the turbulent plasma beta $\breve{\beta}$ in compressible MHD turbulence (right). The MHD plasma is strongly magnetized initially and then, as the turbulence evolves, the plasma becomes less magnetized.

and kinetic energy are almost coincident. Notice that the range with Kolmogorov-like spectrum exists the same as kinetic energy spectrum, with the same wave numbers $2 \leqslant k \leqslant 5$. On the whole, the density fluctuation spectrum demonstrates similar behaviour in Fourier space, as kinetic energy spectrum. Consequently, we infer that density fluctuations are passive scalar in weakly compressible subsonic turbulent flow. Furthermore, theoretical models of turbulence support that any physical characteristic of flow, that propagates passively in large-scale or ambient velocity component of the background turbulence, demonstrates similar spectrum.

It is shown in Fig. 2(left), that the turbulent sonic Mach number decreases significantly from a supersonic turbulent regime ($\breve{M}_s > 1$), where the medium is strongly compressible, to a subsonic value of Mach number ($\breve{M}_s < 1$), describing weakly compressible flow. This fact indicates that turbulent cascades associated with the nonlinear interactions in combination with the dissipative effects at the small scales predominantly cause the supersonic MHD plasma fluctuations to damp strongly leaving primarily subsonic fluctuations in the MHD fluid.

In the interstellar medium, the transition of MHD turbulent flow from a strongly compressible to a weakly compressible state not only transforms the characteristic supersonic motion into subsonic motion, but also attenuates plasma magnetization, which is shown in Fig. 2(right) because plasma beta $\breve{\beta}$ increases with time, thus, role of magnetic energy decreases in comparison with plasma pressure. Fig. 2(right) demonstrates that the thermal pressure do not exceeds the magnetic energy (that is $\breve{\beta} \leqslant 1$) in initial time interval in fully compressible magnetohydrodynamic flow. Plasma particles coupled to the magnetic field lines are expelled from their gyro orbits due to increase of plasma pressure role in comparison with magnetic energy. This fact leads ultimately to a reduced plasma magnetization and hence plasma fluctuations, and transit into $\breve{\beta} > 1$ regime and subsonic weakly compressible flow. Besides, the gradual increase of the turbulent plasma beta $\breve{\beta}$ leads to change of speed of turbulent cascade in subsonic regime of the compressible MHD flow. The high plasma beta $\breve{\beta}$ state implies that the shear Alfvenic modes propagate more slowly than sound waves. When magnetized compressible plasma decreases and the turbulent plasma pressure evolves to exceed the turbulent magnetic energy, the perturbations are essentially non-magnetized, that is, the situation is hydrodynamic-like.

A. A. Chernyshov, K. V. Karelsky & A. S. Petrosyan

Mixing properties of compressible turbulence in the local interstellar medium predicted by LES method are important to understand effects of radio waves propagation and their scattering in observations data.

References

Chernyshov A., Karelsky K., & Petrosyan A. 2007, *Phys. Fluids*, 19, 5, 055106
Chernyshov A., Karelsky K., & Petrosyan A. 2008, *Phys. Fluids*, 20, 8, 085106

Advances in Plasma Astrophysics
Proceedings IAU Symposium No. 274, 2010   © International Astronomical Union 2011
A. Bonanno, E. de Gouveia Dal Pino & A. G. Kosovichev, eds.   doi:10.1017/S1743921311006624

Similarity of Jupiter and RRATs

Ilknur Gezer and E. Rennan Pekünlü

Department of Astronomy and Space Sciences, Faculty of Science, University of Ege, Bornova, 35100, Izmir, Turkey
email: gezer.ilknur@gmail.com
email: rennan.pekunlu@ege.edu.tr

Abstract. In the present investigation, radial diffusion of equatorially trapped electrons in the magnetospheres of Jupiter and Rotating Radio Transients (RRATs) are examined and compared. It is assumed that electrons lose energy through synchrotron radiation and the wave-particle interaction. The phase space density of the electrons, which go through gradB drift in Jupiter's and RRATs magnetospheres and thus resonate with the plasma waves, changes and this change predicted by the model seems to be consistent with the Pioneer 10 and Pioneer 11 data for Jupiter's case and a similar result obtained for RRATs.

Keywords. RRATs, Jupiter, radio bursts, wave-particle interaction

---

1. Introduction

Parker Multi beam Survey began in August 1997 and was completed in March 2002 and over 700 new pulsars were discovered. 11 of these pulsars are previously unknown type of sources. These sources have been described as a Rotating Radio Transients. They show single, dispersed burst having durations within the range of 2-30 ms. Maximum flux density of radio bursts are $\sim 0.1 - 4$ Jy and the average time intervals between bursts range from 4 minutes to 3 hours. Periodicities 10 of the 11 sources are in the range 0.4-7 s. Period variations were measured for 3 of these 11 sources, through the measurements of magnetic fields (10^{12} - 10^{14} G) and ages (0.1 3 My) (M.A. McLaughlin, 2006). Nature of these sources is not known yet, but it is assumed that the sources which show different radio properties in comparison with known radio sources, are field neutron stars of our galaxy.

Luo and Melrose (2007) have proposed that radiation belts that are similar to the planetary magnetosphere may exist for a pulsar which have relatively long period and a strong magnetic field. Polar cap model does not explain observed intense radio burst of these sources because most of RRATs have long periods and pair production is not effective. Similarity of Jupiter and pulsars magnetospheres are known subject for a long time (e.g. de Pater, 2004, Hill, 1995). Luo and Melrose (2007) have suggested that pulsars may have radiation belts in the closed field lines region where relativistic plasma are trapped. The trapped region is believed to be so similar to the Earth's radiation belts. The same authors suggest that low-frequency waves occurring in pulsar magnetosphere can disrupt the trapped plasma and it can lead to intense precipitation toward the neutron star surface. Intense radio bursts are thought to be produced by these precipitating particles toward the star's surface. Equatorially trapped particles while radially diffusing towards the surface of RRATs at the same time may resonate with the plasma waves and be raised to relativistic energies. In situ measurements showed that in van Allen radiation belts electrons are raised by this process to energies as high as 10 MeV (Home, 2005). These energetic particles then precipitate towards the stellar surface and may initiate the intense radio bursts as proposed by Lou and Melrose (2007).

Radial diffusion of equatorially trapped particles in Jupiter's magnetosphere is examined by Pekünlü (1992, 1995). The author assumed that the first and the second adiabatic invariants are conserved and the third one is violated. Pekünlü (1992, 1995) also assumed that particles lose energy with synchrotron radiation and wave particle interactions. In this respect, the mechanism proposed by Luo and Melrose (2007) to explain unusual radiation properties of RRATs is similar to that occurring during radial diffusion in Jupiter's magnetosphere. In this investigation we try to apply the method Pekünlü(1992, 1995) used in his early study to RRATs.

2. Radial Diffusion Mechanism

Generally there are four processes in the radiation belts of planets magnetosphere.These are:(i) injection of charged particles into the trapping region of the magnetosphere from the solar wind through the day side neutral points, (ii) acceleration, (iii)diffusion,(iv) loss (Schulz and Lanzerotti, 1974).Observation by Simpson *et al.* (1974) and Simpson (1974) with Pioneer 10 have shown that radial diffusion is the main process in radiation belts. The type of radial diffusion that conserves both magnetic moment μ (the first adiabatic invariant) and flux density J (the second adiabatic invariant).Convenient conditions for both μ and J is conserved case are magnetic sudden impulses and other magnetic disturbances operating on the same time scale. Actually there are three motion in the magnetosphere; longitudinal drift, bounce motion and cyclotron motion. Their time scales are given by, $\tau_D \geqslant \Delta t \geqslant \tau_b \geqslant \tau_c$ and Δt where drift period,bounce period,cyclotron period and period of disturbance.Radial diffusion mechanisms which μ and/or J is not conserved processes are less important than sudden impulses in radiation belt dynamics since particles can not be energized efficiently in the process.(Pekünlü,1992).Pekünlü has indicated in the same paper that radial diffusion requires that at least the third invariant should change.Since the period of the drift in longitude is the longest of the three periods mentioned above.Random variations of potential electric fields can cause violation of the third invariant and thus lead to radial diffusion.(Roederer, 1970).Radial diffusion formula is given by(Pekünlü,1992)

$$D_o L^n \frac{\partial^2 f}{\partial L^2} + (n+2)D_o L^{n-1}\frac{\partial f}{\partial L} + \left\{ \frac{2(n-1)D_o}{L^{2-n}} + \frac{\beta B_0^2}{m_0 c^2 L^6}\left[3(1 + \frac{2\mu B_0}{m_0 c^2 L^3})^{1/2} \right] \right\}f$$

$$+ \frac{\beta B_0}{L^3}(1 + \frac{2\mu B_0}{m_0 c^2 L^3})\left[(1 + \frac{2\mu B_0}{m_0 c^2 L^3})^{1/2} - 2\right]\frac{\partial f}{\partial \mu} = 0 \quad (2.1)$$

where $f(\mu, J, L, t)$ represents the particle distribution function, L is McIlwain's magnetic shell parameter D_{LL} is the diffusion coefficient. $L = R/R_j$, $B = B_0 L^{-3}$, B_0 is the equatorial surface field strength, $D_{LL} = D_0 L^n$ is the radial diffusion coefficient where D_0 is the non-radial dependence and is calculated from the Pioneer data as having a value $D_0 \sim 2.5 \times 10^{-6} R_j^2 s^{-1}$ and n indicating the power dependence on radial position. n=2 value have been used since it provides a very good fit to Pioneer 10 data. Finally the steady state assumption,$\partial f/\partial t= 0$, have been used (Pekünlü,1992). In the same paper Pekünlü have calculated that the numerical integration of radial diffusion equation for Jupiter over the interval L=1 to L=5, using Gaussian elimination method.

In the present investigation,radiation processes in the magnetosphere of RRATs', as described by Rotating Radio Transients and of Jupiter are compared. Radial diffusion of electrons trapped in the magnetic equator of Jupiter and RRATs is explored. A similar analysis, which has been done for Jupiter, was carried out for RRATs. As a result of the exploration we have obtained the above result.

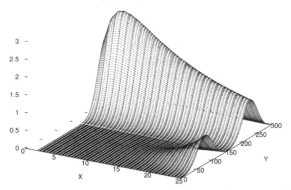

Figure 1. Jupiter's case; If the electrons lose energy only synchrotron radiation, f phase space distribution is in range $0 \leqslant Y \leqslant 150$. In the fig.1 x- direction is L and y- direction is μ. If the electrons lose energy both synchrotron radiation and the wave-particle interaction, f phase space distribution is in range $150 \leqslant Y \leqslant 300$. Fig.1 has been drawn for $212 MeV/G$.

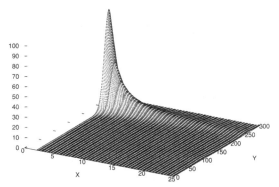

Figure 2. RRATs's case; Fig.2 has been drawn for calculated three RRATs'average magnetic moment value which is $330 MeV/G$.

3. Results

In this examination we considered only the radial diffusion. Fig.2 shows that the phase-space density of electrons diffusing radially inward increases toward the surface of a RRATs. This enhancement in the number of electrons may be the reason for the coherent emission.

In the present study we could not work out the time scale for bursts, neither the flux density of the radiation to see if the radial diffusion could explain the observed properties of the radio bursts. Second shortcoming of the present study is the uncertainity in the diffusion coefficient, D_{LL} for RRATs.

If the burst we receive from RRATs are generated by streaming instability caused by downward moving particles along the magnetic field lines, as proposed by Lou and Melrose (2007) then both the processes of radial diffusion and the pitch-angle diffusion should be taken into account.

References

Becker, W. 2009 Springer-Verlag Berlin Heidelberg, *Neutron Stars and Pulsars*, p. 41

de Pater, I., Kassim, N. & ve Rucker, H. O. 2004, *Planetary and Space Science*,15, 1339-1341

Hill, W. T. ve Dessler, A. J 1995, *Earth in Space*, 8, No.2, 6.

Horne, R. B. 2005, *Nature*, 437, 227

Luo, Q. & Melrose, D. 2007, *MNRAS*,378, 1481L
Mc Laughlin 2006, *Nature*, 439, 817M
Pekunlu, E. R. 1992, *Earth, Moon and Planets*, 59, 201
Pekunlu, E. R. 1995, *Earth, Moon and Planets*
Popov, S. B., Turolla, R., & Possenti, A. 2006, *MNRAS*, 369, L23-L26
Shannon, R. & Cordes, J. M. 2006, *AAS*, 20915910S

Advances in Plasma Astrophysics
Proceedings IAU Symposium No. 274, 2010
A. Bonanno, E. de Gouveia Dal Pino & A.G. Kosovichev, eds.

© International Astronomical Union 2011
doi:10.1017/S1743921311006636

On the development of a spherical hybrid model - Lessons and applications

S. Dyadechkin[1,2], E. Kallio[1], R. Jarvinen[1], P. Janhunen[1], V. S. Semenov [2] and H. K. Biernat[3]

[1]Finnish Meteorological Institute, Helsinki, Finland
P.O. BOX 503 FI-00101 Helsinki Finland
email: egopost@gmail.com

[2]Saint-Petersburg State University, Russia
198504, 1 Ulyanovskaya, Petrodvorets, St.-Petersburg, Russia
[3]Space Research Institute Austrian Academy of Sciences
Schmiedlstrasse 6, 8042 Graz, Austria

Abstract. We are developing a spherical hybrid model to study how the solar wind interacts with the solar system bodies. In this brief status report we introduce some lessons from the spherical grid development and illustrate the usage of the new model by showing a preliminary test run.

Keywords. Solar wind, methods:numerical, stellar dynamics

1. Introduction

The solar wind plasma interacts with the different Solar System bodies in several ways. The different celestial objects reveal specific properties: The Moon does not have either an atmosphere or a global intrinsic magnetic field. Mercury does not have an atmosphere but it has a weak intrinsic magnetic field that forms a "pocket magnetosphere". Venus and Mars do not have a strong global intrinsic magnetic field but they have atmospheres dense enough to produce shields against the solar wind by their ionospheres. Martian magnetic anomalies can also be considered as "mini magnetospheres". Titan's dense atmosphere, in turn, can meet subsonic plasma flow.

A hybrid approach provides an efficient way to model how the cosmic plasma interacts with non-magnetized and magnetized planetary objects. In a hybrid model ions are considered as particles, while electrons form a massless, charge neutralizing fluid. The basic properties of the HYB hybrid model are described in Kallio & Janhunen (2003). The HYB hybrid model family is being developed at the Finnish Meteorological Institute (FMI) during the last decade. The model has been used successfully to describe how the flowing plasma interacts with various solar system bodies such as Mercury, Venus, the Moon, Mars, Saturnian moon Titan and asteroids.

One geometrical limitation of the HYB model is, however, that it assumes cube shaped grid cells. In order to expand the usage of the HYB model we have initialized a project aiming to develop a spherical coordinate version of the model. A spherical grid would give us important advantages compared with a Cartesian grid, such as: 1) A good grid resolution, because the grid size decreases automatically near the obstacle (the planetary surface) and 2) Natural boundary conditions for the obstacle, because the planetary surface covers r-constant surface of the grid. However, the implementation of a spherical grid into the hybrid code also entails some challenges: first, a complexity of the interpolation on the spherical grid in comparison with the Cartesian, and second, pole problems.

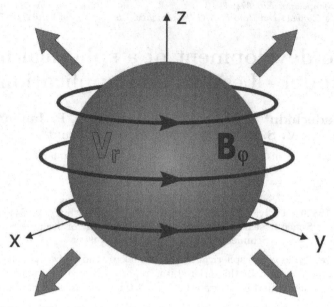

Figure 1. Initial conditions for the spherical expanding test. Initial magnetic field is posed on the spherical surface and it is purely toroidal (blue circles). Initial velocities of protons are perpendicular to the surface (red arrows).

2. Spherical grid in a hybrid model

In this section we briefly describe main advantages and challenges of spherical grid when it is used in a hybrid model:

2.1. *Advantages*

1 *Spherical worlds.* Planetary worlds are spherical and to simulate them it seems to be more natural to use a spherical grid instead of a Cartesian grid.

2 *Grid resolution.* In spherical coordinates there is "natural" grid refinement: the closer the cells are to the obstacle the smaller grid cells become. This property could be used for the introducing of a self-consistent ionosphere into the hybrid model, where we need to decrease the cell size at low altitudes.

3 *Obstacle boundary condition.* Another "natural" property of spherical coordinates is the geometrical interpretation of the obstacle. Planetary surface is covered by the $r = constant$ surface of the spherical grid, which simplifies the implementation of the boundary condition.

2.2. *Challenges*

1 *Interpolations.* Hybrid approach imply a number of vector and scalar value interpolations between different grid elements: interpolation from cell center to cell nodes (CN - interpolation), interpolation from cell faces to cell center (FC - interpolation) etc. As the spherical grid is not homogeneous, the realization of interpolation methods is not as straightforward as in the Cartesian coordinates.

2 *Pole Problems.* Spherical grid includes two singular points, poles, where the numerical values are not defined. Moreover, in circumpolar region numerical values "feel" singularities, because the spherical cell elements, edge lengths, face areas and cell volume become smaller when the cell is closer to the pole. There are several techniques to exclude the poles from the grid, most of them based on the decomposition of spherical grid into several grids (Usmanov, 1995; Kageyama & Tetsuya, 2004; Ronchi *et. al.*, 1996).

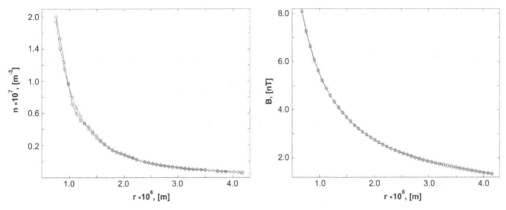

Figure 2. The figure shows analytical and experimental profiles of the number density and the magnetic field for a steady-state solution of a super-Alfvenic plasma flow from a spherical surface at $r = 0.6 \cdot 10^7 m$. Analytical curve is shown in red and experimental in blue. For the super-Alfvenic regime we can find an analytical approximation for the dependency of the number density and magnetic field on the radial distance. In the equatorial plane ($\theta = 0$) the number density is proportional to $1/r^2$, and the magnetic field to $1/r$.

3. Spherical expanding test

To illustrate preliminary results of the spherical hybrid model let's consider the super-Alfvenic plasma flow from the spherical surface. The initial conditions of the spherical expanding test are depicted on Fig. 1.

Initial state: A spherical empty box and a magnetized wind from the spherical surface. Initially there is no plasma in the simulation box; only internal (obstacle) boundary conditions are defined.

Initial magnetic function: Initially there is only a toroidal component of the magnetic field $B = (0, 0, B_\phi)$, which is defined as: $B_\phi = B_0 sin\theta$, where $B_0 = 1nT$. To exclude pole problems from our simulation we used $sin\theta$ dependence in the initial magnetic configutation.

Particle population: We consider only one H^+ population with number density $n = 14 \cdot 10^6 m^{-3}$ and temperature $T = 10^5 K$. The initial velocity has only a radial component $V = (V_r, 0, 0)$, where $V_r = 4 \cdot 10^5 m/s$. The average number of macroparticles per a cell is 30.

4. Summary

A spherical hybrid model is anticipated to provide a powerful tool for the modeling of self-consistent cosmic plasma interactions with various solar system bodies and exoplanets. Developing such a model is, however, a challenging task bacause of the complexity of the interpolation and the polar region of the grid.

References

Kallio, E. & Janhunen P. 2003, *Annales Geophysicae*, 21, 2133
Usmanov, A. V. 1993, *Saint Petersburg State Univ., International Solar Wind 8 Conference*, 65
Kageyama, A. & Tetsuya, S. 2004, *Geochemistry Geophysics Geosystems*, 5, 9
Ronchi, C., Iacono, R., & Paolucci, P. S. 1996, *Journal of Computational Physics*, 124, 1, 93

Advances in Plasma Astrophysics
Proceedings IAU Symposium No. 274, 2010
A. Bonanno, E. de Gouveia Dal Pino & A. G. Kosovichev, eds.
© International Astronomical Union 2011
doi:10.1017/S1743921311006648

Solar plasma generated by sungrazing comets

F. S. Ibodov[1] and S. Ibadov[1,2]

[1]Sternberg Astronomical Institute, Moscow State University, Moscow
email: mshtf@sai.msu.ru, firuz@pochta.ru

[2]Institute of Astrophysics, Tajik Academy of Sciences, Dushanbe
email: ibadovsu@yandex.ru

Abstract. It is analytically shown that passages of comets near the Sun's surface with velocities more than 600 km/s is accompanied by aerodynamic crushing of their nuclei within the solar chromosphere and transversal expansion of the crushed matter. The deceleration of the flattened hypervelocity body within the solar photosphere has sharply impulsive and strongly explosive character. The specific energy release in the explosion zone near the solar surface 10-100 thousand times exceeds the evaporation heat of the nucleus material, so that the process is accompanied by generation of high-temperature plasma and non-stationary explosive phenomena around the photosphere. Spectral observations of these phenomena by SOHO and SDO type space observatories with high spatial and temporal resolutions are of interest for the plasma astrophysics as well as the physics of solar flares.

Keywords. Sun: general, sungrazing comets, solar impact plasma, solar photospheric flares

1. Introduction

Progress in discoveries of sungrazing comets (COSPAR 1998, ESA 2008, NASA 2008) achieved owing to coronagraphic observations of circumsolar region by such space missions as SOLWIND, SMM (Solar Maximum Mission) and SOHO (Solar and Heliospheric Observatory) as well as results of theoretical investigations on the origin and dynamics of sungrazing comets, on the basis of many-body celestial mechanics (Bailey *et al.* 1992), made actual the study of physical evolution of comet nuclei within the solar dense atmosphere, i.e. taking into account aerodynamic effects.

The physical evolution of comet nuclei in the field of solar photon radiation is sufficiently studied. In particular, it is established that the decrease of a radius of comet nuclei passing near the photosphere of the Sun is limited by a value no more than 15-20 meters, while aerodynamic effects are not considered at all, i.e. vacuum approximation is used (e.g. Weissman 1983, MacQueen & St. Cyr 1991).

Meantime, parabolic velocity of comets near the solar surface is equal to 617 km/s, while the density of the solar atmosphere within the chromosphere considerably exceeds $10^{-15} g/cm^3$ being of the order of $10^{-7} g/cm^3$ in the photosphere. In such situation aerodynamic pressure on the comet nuclei considerably exceeds tidal disruption of the nuclei (Grigoryan *et al.* 1997).

Besides, the specific kinetic energy of comet nuclei in the inner heliosphere, especially near the surface of the Sun, essentially exceeds the evaporation/sublimation energy of comet nuclei material $10^{10} - 10^{11}$ erg/g, so that plasma phenomena in the region should be taken into account (Ibodov *et al.* 2006).

In the present paper some properties of plasma produced by sungrazing comet nuclei near the solar surface are considered and connection of the process with solar photospheric flares is marked.

2. Solar plasma generated by sungrazing comets

Calculations show that comet nuclei moving through the chromosphere with orbital velocities more than 600 km/s will be aerodynamically destructed. Besides, the intensity of energy flux related to the flow of bombarding atmospheric particles to the comet nucleus surface, J_a, essentially exceeds the intensity of energy flux to the comet nucleus due to thermal photon radiation from the solar photosphere, J_λ. It means that near the photosphere comet nuclei with practically all radii will be destroyed/fragmented and evaporated, in principle, due to aerodynamic effect (Ibadov *et al.* 2007).

Moreover, the aerodynamic pressure gradient on the frontal surface of the comet nucleus leads to transversal expansion of the fragmented comet nucleus matter. Due to this process an aerodynamic deceleration of hypervelocity matter acquires impulse character. Namely, calculations show that the length of the sharp deceleration of flattened comet nucleus matter will be around 0.7 H, where H=200 km is the characteristic altitude scale of the solar photosphere (Ibadov *et al.* 2008).

The atmospheric mass within sharp deceleration layer is of the order of comet nucleus mass. The situation in this region is similar to that for high velocity collision of two dust particles (Ibadov 1992, Ibadov 1996), so that the initial temperature of the impact region in the sub-photosphere layer can be estimated as

$$T = \frac{A m_p V^2}{12k(1 + z + 2x_1/3)}. \tag{2.1}$$

Here A is the mean atomic number for the falling comet nucleus material and matter of solar photosphere, m_p is the proton mass, k is the Boltzmann constant, z is the mean multiplicity of charge of plasma ions produced during thermalization of the kinetic energy of comet nucleus in the sub-photosphere layer, x_1 is the mean relative ionization potential.

Using (2.1) with realistic values of $A = 20$, $z=5$, $x_1 =3$ we obtain for the initial temperature of impact produced solar plasma $T = 10^7$ K. This value is considerably more than the temperature of the solar corona plasma, so that the impact generated sub-photosphere explosive process may be observed as a solar photospheric flare.

3. Conclusions

Passages of nuclei of sungrazing comets through the solar chromosphere are accompanied by aerodynamic destruction and transversal expansion of the fragmented hypervelocity mass. A sharp aerodynamic deceleration of the nucleus having flattened form in the sub-photosphere layer leads to impulse thermalization of the kinetic energy of the nucleus and generation of solar impact plasma with initial temperature more than the temperature of the solar corona plasma.

Space spectral observations of solar impact plasma / solar photospheric flare, generated by sungrazing comets, with high spatial and temporal resolutions using space observatories like SDO (Solar Dynamics Observatory) are of interest for the physics of solar flares and plasma astrophysics.

Acknowledgements

The authors are grateful to the IAU Symposium No. 274 SOC/LOC for a travel grant. The hospitality of SAI MSU and DIC MSU as well as the assistance of Dr. Konstantin V. Bychkov and Oleg V. Egorov are acknowledged.

References

Bailey, M. E., Chambers, J. E., & Hahn, G. 1992, *A&A*, 257, 315
COSPAR Inform. Bull. 1998, 142, 22
ESA 2008, *ESA-Space Sci.*, "SOHO Discovers its 1500th Comet",
    http://www.esa.int/esaSC/,
    http://sohowww.nascom.nasa.gov/hotshots/2008 06 23/
Grigoryan, S. S., Ibodov, F. S., & Ibadov, S. 1997, *Dokl. Akad. Nauk*, 354, 187,
    Engl. Transl.: *Phys.-Dokl.*, 42, 262
Grigoryan, S. S., Ibadov, S., & Ibodov, F. S. 1998, *Cometary Nuclei in Space and Time, IAU
    Colloquium No. 168 Abstracts, Nanjing, China*, 13
Grigoryan, S. S., Ibadov, S., & Ibodov, F. S. 2000, *Dokl. Akad. Nauk*, 374, 40
    Engl. Transl.:*Dokl.-Phys.*, 45, 463
Ibadov, S. 1992, *AZh*, 69, 737
Ibadov S. 1996, *Physical Processes in Comets and Related Objects*, Moscow, Cosmosinform Pub-
    lishing Company
Ibadov, S., Ibodov, F. S., & Grigoryan, S. S. 2007, *Star-Disk Interaction in Young Stars, Proc.
    IAU Symp. No. 243*, Grenoble, France,
    www.iaus243.org
Ibadov, S., Ibodov, F. S., & Grigorian, S. S. 2008, *Universal Heliophysical Processes, Proc. IAU
    Symp. No. 257*, N. Gopalswamy & D. F. Webb, eds., Cambridge University Press, 341
    http://iau257.uoi.gr/

Ibodov, F. S., Grigoryan, S. S., & Ibadov, S. 2006, 36th *COSPAR Scientific Assembly Abstracts*,
    Beijing, China, B0.4-0068-06
    www.cospar-assembly.org
MacQueen, R. M. & St. Cyr, O. C. 1991, *Icarus*, 91, 96
NASA 2008, http://sohowww.nascom.nasa.gov/hotshots/2008_06_23/
Weissman, P. R. 1983, *Icarus*, 55, 448

Advances in Plasma Astrophysics
Proceedings IAU Symposium No. 274, 2010   © International Astronomical Union 2011
A. Bonanno, E. de Gouveia Dal Pino, & A. G. Kosovichev, eds.   doi:10.1017/S174392131100665X

Development of the PFO–CFO hypothesis of solar system formation: Why do the celestial objects have different isotopic ratios for some chemical elements?

Elena A. Kadyshevich[1] and Victor E. Ostrovskii[2]

[1] Obukhov Institute of Atmospheric Physics RAS
Pyzhevskii str. 3, Moscow, 119017 Russia
email: **kadyshevich@mail.ru**

[1] Karpov Institute of Physical Chemistry,
Vorontsovo Pole str. 10, Moscow, 105064 Russia
email: **vostrov@cc.nifhi.ac.ru**

Abstract. The Solar System formation PFO–CFO hypothesis is developed in the direction of creation of a phenomenological model focused on solution of a number of paradoxes and answering to a number of mysterious questions under the same cover. For explanation of the events and processes that occurred over the period from the middle ages of the pre-solar star to the Solar System formation, original approaches are applied.

Keywords. Sun: evolution, solar system: formation, planets and satellites: formation

---

1. Introduction. Statement of the problem

The results of the Solar System (SS) observations lead to some paradoxes and require answers to a number of principle questions.

The most important paradoxes are as follows. (1) Any isolated star early in its life is electrically neutral, and its electron and proton amounts are equal. As neutronization of a star proceeds, its electron and proton amounts decrease to the same degrees. Thus, at the stage of full neutronization, the collapsed neutron stars should have no electrons and should have zero magnetic moment. Meanwhile, the measured magnetic moments of neutron stars are extremely high. Why is it so? (2) If the SS is the product of explosion of the pre-solar star, what is the mechanism of the transfer of the major portion of the star angular momentum to the planets, and, if the angular momentum was received from any other source, what is its nature? (3) If the SS is the product of explosion of the pre-solar star, why is the total mass of all SS planets less than the Sun mass by a factor of almost 1000? (4) The fourth paradox was recently formulated by the US National Research Council: "If only one nebula is the progenitrix for the SS, why are the planets principally different?" Solution of this paradox was qualified as the most important astrophysical problem (http://books.nap.edu/openbook.php?record_id=12161&page=9).

Some of the questions are as follows. (1) What is the nature of the 11-year solar-activity cycle, and is there a causal relationship between the variations in the solar activity and solar magnetic moment? (2) What is the mechanism of formation of chemical elements, including the heavy ones? (3) Why do the SS celestial objects contain chemical elements in different isotopic compositions? (4) Why is the corona temperature much higher than the photosphere temperature? (5) Why are most of the biggest celestial bodies located within a space belt along the ecliptic plane? (6) Can the Earth's localizations of minerals result from any space processes? (7) Were the terrestrial planets melted at the initial step of their origin?

Some of these paradoxes and questions are quite mysterious. Most of them have hypothetical solutions (e.g., Basu *et al.* 2009; Guerrero & de Gouveia Dal Pino 2008; Bonanno *et al.* 2008; Erdelyi & Ballai 2007). However, there is no hypothesis that could consider all them under one cover. In astronomy, there are a number of brilliant physico-mathematical works that give numerical values of different parameters, lead to definite limitations, and put lids on different phenomena. Meanwhile, some processes, phenomena, and states of matter, for which these parameters are calculated, are hypothetical and the values of the parameters can not be verified. When discussing the problems of solar interior, we should take into account that there are no knowledge on the properties of the matter and energy in very dense and hot media when the translational terms are degenerated and the entire giant energy of the matter is concentrated in the electron energy and in the oscillatory and rotational energy of nucleons, interactions between which in similar media can not be realized and studied under real conditions. On frequent occasions, calculations are based on some model notions, which are applicable until observational data indescribable by the conclusions from these notions are collected.

This stimulates new competitive phenomenological models. Below, we consider an original model explanation for a wide circle of the phenomena of the SS formation era and of the present epoch. Many of its components are available for subsequent testing.

2. The advanced Solar System formation PFO–CFO hypothesis

General provisions. The PFO–CFO hypothesis of the SS formation is considered in a general way in the papers by Kadyshevich & Ostrovskii (2010a,b); Kadyshevich (2009a,b); Ostrovskii & Kadyshevich (2009a,b); Ostrovskii & Kadyshevich (2008); Ostrovskii & Kadyshevich (2007). The main goal of the previous version of this hypothesis consisted in solution of the paradox (4) in its wording given in Section 1. The title "PFO–CFO hypothesis" reflects its central idea that the cold celestial objects formed from light elements through physical processes, such as condensation, physical adsorption and absorption, aggregation, occlusion, etc., and are the Physically Formed Objects (PFO) and the warm celestial objects formed mainly from medium-weight elements through chemical syntheses and are the Chemically Formed Objects (CFO). This paper is aimed at more extensive consideration of the processes that proceeded in the pre-solar star and initiated the SS formation phenomenon. The pre-solar star history is considered as the typical history of transformations of the stars similar to this star in the age and size. We proceed from the following starting positions. The SS represents an electrically neutral system that originated on the basis of pre-solar star, which was formed, apparently, from an object similar to a medium-size red giant. The middle-aged mother star was similar in its principal characteristics to the present Sun: it consisted of a core, a radiation zone, a convection zone, and outer zones. (Therefore, below, we alternate the consideration of the middle-aged pre-solar star with the consideration of the present Sun.) The star was compressing under the gravity forces; therewith, its body and surface temperatures increased steadily. The degree of compression heightened inward along the radius. The pre-solar star core and radiation zone consisted (similarly to the Sun) of a non-atomized unstructured p–n–e dense-plasma matter which is unknown under the Earth's conditions. The principal physical specificity of this matter consists in the fact that each its mass unit possesses an enormous kinetic energy realized in rotational and vibrational degrees of freedom with no translational component; similar analogs are unknown under the Earth's conditions, and the physical laws accessible for them can be only objects of guesses. The degree of matter compression in the core was higher than that in the radiation zone. In the core and in the radiation zone, processes of neutronization and of ionization of the p–n–e matter proceeded. We assume that the notion (see the book by Bisnovatyi-Kogan (2001)) on the necessity of extremely high degrees of compression for neutronization is inapplicable to the hot p–n–e systems with low neutron and electron concentrations for the following reasons. (1) The number of neutrons is greater than that of protons in the atomic nucleus of all stable and radioactive chemical isotopes (except protium); therewith, the bigger are the atomic nuclei, the higher is the n/p ratio (e.g., Greiner & Zagrebaev 2006, Fig. 1). This means that such a situation is thermodynamically preferential. (2) Neutronization is the exothermic process, and the entropy change is insignificant; this fact confirms that neutronization can proceed with no external energy. (3) Statements of some authors about the necessity of an external energy for this process relates to the energy barrier, i.e., to the rate of neutronization but not to its possibility; in addition, they relate to isolated atoms but not to the p–n–e matter where the potential barriers can be much lower. Note also that this process can proceed by tunneling. Bearing on this conclusions, we take that neutronization proceeded in the core and in the radiation zone. We don't understand the authors who take that neutronization begins deep in stars at a density of about 10^6 g/cm^3 and at a temperature of about 10^6 K; the point is that, in the process of compressing and heating, this matter should lose (as a result of thermal ionization) the major portion of its electrons necessary for neutronization.

Similarly to the authors of the widely distributed hypotheses (e.g., Caroll & Ostlie 2006), we assume formation of the degenerate electron gas, but we see its origin in thermal ionization of the p–n–e matter. Namely, the p–n–e matter is characterized by a potential of thermal ionization and, thus, it is capable of emitting electrons at any definite temperature. Compression of the p–n–e matter is accompanied by the thermal ionization starting with a temperature level, which can be dependent on the p–n–e composition. This ionization process has its analog in the process of ionization of weakly-ionized plasma. The degree of ionization of such plasma depends on the temperature, density, and ionization energies of the atoms (Saha 1921). We believe that ionization of the n–p–e matter proceeds in dense layers and leads to separation of the electrons from the stratum with formation of an electron-gas "pillow" between the star core and the radiation zone. In the Saha process, the Debye length is large. However, in the star core, the outflow of the electrons from the reaction zone and consumption of a portion of the electrons for neutronization should stimulate ionization. We assume that, in rather hot and dense media, neutronization is attendant with ionization and, apparently, some of the acts of ionization are conjugated with acts of neutronization, the conjugation proceeding by tunneling. The deeper is located a layer of the p–n–e matter, the higher is its temperature. The electron-gas layer is located over the field of the ionization temperature and compensates the pressure of the overlying layers of the radiation zone. Thus, the field of the ionization temperature demarcates the core from the radiation zone and the degenerate electron gas accumulates in this field and forms a "pillow" between the zones. Just this pillow is the cause of a significant difference between the Sun-core rotational speed and the Sun radiation zone rotational speed and of the mysterious 11-year cycles of solar activity. The difference between the rotational speeds is discovered recently by the SOHO mission (Garcia *et al.* 2007). The 11-year cycles are explained by us as follows. In the course of any era, the depth of electron-gas burial and the rate of its formation as a result of thermal ionization of the p–n–e matter are approximately constant. The 11-year cycles are of hydraulic nature and are determined by the time period in which the increment of the electron-gas pressure becomes sufficient for overcoming the pressure of the radiation-zone superstratum. When passing through the radiation zone, the electron-gas flows enrich themselves with the p–n–e matter, and the rather powerful ones cross the photosphere–chromosphere boundary. Some narrow channels are passable for electrons always or almost always, but their traffic capacity in the present era is less than the inflow of the core electrons to the electron layer. Between the events of maximum solar activity, the quantity of "working" narrow and medium channels for the electron outflow depends of the electron-gas pressure. The spots are caused by these flows, and the opposite polarities of the observable local magnetic fields in the photosphere are caused by the quantitative relations between electrons and positive ions in the powerful flows. The Sun's global magnetic field variations are, apparently, caused by the periodic alternation in the nature, amounts, and polarity of the charge-carrier particles (e.g., electrons and metal ions) in the global flows of electric current. The compression and warming-up increased the rotational speed of the star and its kinetic energy; the last tendency revealed itself to the utmost along the equatorial belt of the star.

The mechanism of formation of chemical elements. We assume that atoms of chemical elements originated and originate now not within the star core but over the space in the outer vicinity of the radiation-zone boundary as a result of tangential sputtering of radioactive unstructured nanodrops (NDs) of the star matter from the radiation-zone–convective-zone boundary. Outside this boundary, the energy of the NDs is mainly realized in the translational movement. In the beginning, these NDs were minimum in their sizes, composed predominantly of 1, 2, 3 nucleons, and increased in size as the star rotational speed increased. Continuous increasing in the rotational speed was initially caused by the gravity densification of the star matter and then was also caused by its neutronization. The ND flows occurred predominantly in the vicinity of the equatorial belt, i.e., in the field, where the rotational speed was maximal. The size of the NDs steadily increased with the rotational speed. The unstructured NDs, after their isolation from the unstructured star matter, quickly formed radioactive atoms and then steadily transformed to the stable atoms of corresponding sizes. As a result of radioactive decays, the atoms got additional impulses and formed, depending on their energy, convection zone, photosphere, chromosphere, or corona. The convection zone temperature is rather low because a significant portion of atoms passed it before their decays, and the corona atoms are very hot because they underwent several decays or collided in the convection zone with hot atoms and captured a portion of their energy. Therefore, the corona is populated with the most energetic atoms. Thus, the angular momentum of all SS atoms or of the planetary system as a whole is not limited by the momentum taken from the star but represents its sum with a contribution obtained from radioactive decays.

The major components of the mechanism of the initial period of planet formation. According to the PFO–CFO hypothesis, the SS originated as a result of the loss of some mass by the pre-solar star with no effect of other objects or phenomena. Apparently, only the phenomenon of the so-called isotopic anomalies was proposed by advocates of the effect of outside events on this process. We will show that the isotopic anomalies are explainable on the basis of our concept.

With time, the star core and radiation zone compressed and their rotational speed rose. Therewith, the ND flows over the equatorial region intensified and the ND sizes increased. At each rotational speed, a definite size of the NDs prevailed. The flows formed clouds, which flowed away from the star when their energy obtained from the star and from the radioactive decays became sufficient. The lighter were the atoms, the farther flowed the clouds into the cold space regions. The difference in the rotational speeds of the star core and the star radiation zone increased progressively, and the radiation-zone pressure on the electron-gas layer heightened. Under some critical conditions, the radiation zone was destroyed as a result of a restricted explosion. Namely, the radiation zone was separated from the star core, divided into clouds of radioactive NDs, and the blockade of the electron-gas layer was called off. These processes liberated a huge amount of energy as a result of depression of the system and formation of atoms and their radioactive decays and led to formation of a magnetic field around the core on the basis of liberated electron gas. Apparently, these processes could not lead to dramatic transformations of the core. After this event, almost all newly formed radioactive heavy elements returned to the core before long and only a small portion (about 0.11%) of the pre-solar star mass transformed to the SS celestial bodies. Their retrieval led to formation of a rather loose radioactive stratum that initiated chain fission reactions. The stratum transformed progressively to the p–n–e mass under the action of the heat of these reactions and gravity, its luminosity increased, and we obtained our Sun in all its glory.

The solar nebula composed of the pre- and post-explosion clouds that flowed away from the star core near the ecliptic plain, progressively cooled, and could come apart or mix. Therewith, the earlier they were born and the lighter atoms they contained, the farther from the core they advanced; therefore, the clouds composed of light atoms could not mix with the clouds composed of heavy atoms.

PFO formation started, when the nebula began to collapse after its outer H_2 and He clouds cooled to the H_2 condensation temperature. Hydrogen droplets absorbed light Li, Be, B, LiH, and BeH atoms and molecules, which formed the agglomerate cores. Steadily, the agglomerates increased in size as a result of their competition with each others for the mass and gravitational attraction. Obeying the law of conservation of the angular momentum, the nebula collapsed as a whole because its objects were bound by gravity forces. Heavy atoms and hydrides remained in that nebula section where the temperature was too high for physical absorption processes and the temperature and concentrations of atoms and molecules were too low for chemical reactions. The system progressively decreased in its volume as a result of the condensation processes and under the action of the gravitational attraction. Meanwhile, the activity of the pre-solar star core began to increase. The less was the distance from the Sun, the greater was the degree of compression and the higher was the temperature. These trends initiated chemical reactions over the warm region of the nebula. As the gravity compressing of the space matter increased, chemical combination reactions exponentially accelerated. These reactions stimulated localizations of the substances and reaction heat and initiated compressible vortexes, within which hot cores originated, and triggered the process of CFO formation. The reaction heat was capable of melting the cores. In the vicinities of the giant vortexes, low-pressure and gravitational attraction zones arose. The occurrence of such zones stimulated flows of light cold vaporous and gaseous substances and asteroid-like agglomerates from the outer space and flows of asteroid-like agglomerates of not so light substances from the intermediate regions of the space to the hot cores. The flows precipitated over the hot core surfaces of the CFO and cooled these surfaces. The "spherical thermoses" obtained as a result of these precipitation processes steadily became the young terrestrial planets and their satellites. The following period of the planet formation is detailed in the above-cited preceding version of the PFO–CFO hypothesis.

Why do the celestial objects have different isotopic ratios for chemical elements? We return to the period of the SS formation just before the explosive disruption of the pre-solar radiation zone. The n/p ratio of the radiation-zone matter progressively increased, matter consolidated, rotational speed increased, and atomized clouds composed of more and more heavy radioactive atomic nuclei floated away from the star. The ND nucleon number gradually increased. The NDs of any definite composition left the star when the centrifugal force came up with the force of bond between these NDs and the star body. The before-explosion emission, which consisted of low- and moderate-sized atoms, was

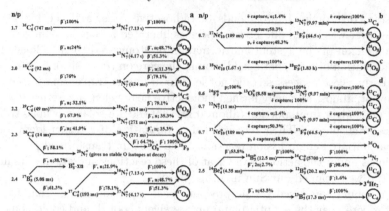

Figure 1. Possible ways of formation of stable O-isotopes (a, b, c) and C-isotopes (d) from nanodrops; stable isotopes are encircled, radioactive isotopes are marked by asterisks, arrows show the directions of the reactions; the n/p values for the source nanodrops, the half-life, and the types and percentages of the radioactive decays are indicated.

multi-cloud. The isolated NDs transformed quickly to the corresponding radioactive atoms and then step by step transformed to the stable atoms; this process began just after isolation of NDs from the radiation-zone p–n–e mass.

Figure 1 gives the available data (Tuli 2005; Audi *et al.* 2003) on all possible chains of radioactive decays that can lead to ^{16}O, ^{17}O, ^{18}O (a, b, c) and ^{12}C, ^{13}C (d) formation. The ways of formation of these elements are considered as examples. The side isotopes that form in these chains are also given. In each cloud, O-isotopes are of a specific origin and of a specific isotopic ratio, both depending on the ND size and nuclide composition in the period of ND formation. Later, the O-containing clouds could mix with other clouds and with each other and form minerals with different ^{16}O/^{17}O/^{18}O ratios. The most active mixing proceeded during vortex processes of formation of the hot planets.

An analysis of the data of the figure in the context of the PFO–CFO hypothesis led us to the following general conclusions. Each of the elements was presented in several clouds formed at different n/p ratios characteristic for the outer surface of the radiation zone. Each element in any one cloud after termination of radioactive decays could be presented by one isotope or by several isotopes. The relative amounts of the stable isotopes formed from any one source radioactive isotope can be calculated but no quantitative data on the relative powers of different flows from the star are available. Each of the clouds and the secondary clouds resulted from full or partial mixing of the primary clouds participated in further chemical reactions and in agglomerations. In each such a process, the isotopic compositions of the products were determined by the isotopic compositions of the clouds that were initial ones for this process.

Relative to carbon, the following conclusions can be made. Carbon could be obtained from four sources: ^{14}F*, ^{17}Ne*, ^{12}N*, and ^{14}Be*. ^{13}C is a small admixture to ^{12}C, we conclude that the ^{12}N* flow was most powerful. Apparently, three carbon-containing clouds participated in formation of the SS. They are: α-cloud (100% ^{13}C) formed at n/p = 0.6 from the ^{14}F* flow; β-cloud (a mixture of ^{12}C, ^{13}C, ^{16}O, and ^{17}O, where (^{17}O/^{13}C) = 36.0) formed at n/p = 0.7 from the low-power ^{17}Ne* flow and high-power ^{12}N* flow (apparently, β-cloud is the main contributor to the Earth's carbon and ^{17}O); and γ-cloud formed at n/p = 2.5 and composed of the mixture of (^{14}N/^{13}C/^{12}C/^{4}He) = 1/0.81/0.048/0.0055 and of (^{12}C/^{13}C) = 1/0.059 (apparently, γ-cloud along with β-cloud played an important role in formation of some cold celestial objects such as Saturn and

Titan). It is impossible to exclude the occurrence of SS objects in which the entire C is obtained from the flow of ^{14}F* and carbon is all-^{13}C.

Oxygen could be obtained from the flows of ^{17}Ne* (n/p = 0.7), ^{18}Ne* (n/p = 0.8), ^{16}C* (n/p = 1.7), ^{18}C* (n/p = 2.0), ^{19}C* (n/p = 2.2), ^{20}C* (n/p = 2.3), and ^{17}B* (n/p = 2.4). It is impossible to exclude the existence of the celestial objects that contain oxygen only in the form of ^{16}O or in the form of ^{18}O. The maximum content of ^{17}O in the mixture with ^{16}O can be 51%. Different ^{16}O/^{17}O/^{18}O compositions are possible as a result of mixing of several flows. For our consideration, the cloud formed by the ^{17}Ne* and ^{18}Ne* flows is of principal interest. Such a cloud should have ^{18}O twice as much as ^{17}O if the powers of these two flows are equal. Apparently, this condition is satisfied because the n/p ratios and the sizes of the source NDs in these flows are rather close. This means that additions of portions of this cloud to the cloud, on the basis of which SMOW was formed, contributed to ^{18}O twice as much as to ^{17}O in the resulted clouds and in the products of their subsequent chemical reactions. These reactions led to formation of different mineral localizations, each of which contained different additions of ^{18}O and ^{17}O relative to the SMOW isotopic content, but, in all localizations, the ^{18}O addition was twice as large as the ^{17}O addition. Just such a relation is observed for different minerals over the Earth and Moon (Clayton 1993).

3. Conclusion

The dense p–n–e matter is scarcely realizable at the Earth. Only specialized analyzing of astronomic data and new phys.-math. studies are capable of testing our hypothesis.

References

Audi, G., Bersillon, O., Blachot, G., & Wapstra, A. H. 2003, *Nuclear Physics A*, 729, 3

Basu, S., Chaplin, W. J., Elsworth, Y., New, R., & Serenelli, A. M. 2009, *ApJ*, 699, 1403

Bisnovatyi-Kogan, G. S. 2001, *Stellar Physics* (Springer)

Bohr, N. & Wheeler, J. 1939, *Phys. Rev.*, 56, 426

Bonanno, A., Schlattl, H., & Paterno, L. 2008, *A&A*, 390, 1115

Burbidge, E. M., Burbidge, G. R., Fowler, W. A., & Hoyle, F. 1957, *Rev. Modern. Phys.*, 29, 547

Caroll, B. W. & Ostlie, D. A. 2006, *An Introduction to Modern Astrophysics, 2nd. ed.* (Addison-Wesley Publ. Co.)

Clayton, R. N. 1993, *Annu. Rev. Earth Planet. Sci.*, 21, 115

Erdelyi, R. & Ballai, I. 2007, *Astron. Nachr.*, 328, 726

Garcia, R. A., Turck-Chieze, S., Jimenez-Reyes, S. J., Ballot, J., Palle, P. L., Eff-Darwich, A., Mathur, S., & Provost, J. 2007, *Science*, 316, 1591

Greiner, W. & Zagrebaev, V. 2006, *J. Nucl. Radiochem. Sci.*, 7, R1

Guerrero, G. & de Gouveia Dal Pino, E. 2008, *A&A*, 485, 267

Kadyshevich, E. A. & Ostrovskii, V. E. 2010a, *EPSC Abstracts*, 5, EPSC2010-3

Kadyshevich, E. A. & Ostrovskii, V. E. 2010b, *274 IAU Symp. Abstracts Booklet*, 65

Kadyshevich, E. A. 2009a, *Meteorit. Planet. Sci.*, 44, A105

Kadyshevich, E. A. 2009b, *EPSC Abstracts*, 4, EPSC2009-1

Myers, W. D. & Swiatecki, W. J. 1966, *Nucl. Phys.*, 81, 1

Ostrovskii, V. E. & Kadyshevich, E. A. 2009a, *Orig. Life Evol. Biosph.*, 39, 217

Ostrovskii, V. E. & Kadyshevich, E. A. 2009b, *Geochim. Cosmochim. Acta*, 73, A979

Ostrovskii, V. E. & Kadyshevich, E. A. 2008, in A. N. Dmitrievskii & B. M. Valyaev (eds.) *Degassing of the Earth* (Moscow: GEOS), p. 374

Ostrovskii, V. E. & Kadyshevich, E. A. 2007, *Physics-Uspekhi*, 50, 175

Saha, M. N. 1921, *Pros. Roy. Soc. London*, Ser.A., 99, 135

Tuli, J. K. 2005, *Nucl. Wallet Cards* (N. Y.: Nucl. Data Center, Brookhaven Nat. Lab., Upton)

Advances in Plasma Astrophysics
Proceedings IAU Symposium No. 274, 2010
A. Bonanno, E. de Gouveia Dal Pino & A. G. Kosovichev, eds.
© International Astronomical Union 2011
doi:10.1017/S1743921311006661

A note on using thermally driven solar wind models in MHD space weather simulations

Jens Pomoell[1] and Rami Vainio[1]

[1]Department of Physics, University of Helsinki
P.O. Box 64, 00014 University of Helsinki, Finland
email: `jens.pomoell@helsinki.fi`, `rami.vainio@helsinki.fi`

Abstract. One of the challenges in constructing global magnetohydrodynamic (MHD) models of the inner heliosphere for, e.g., space weather forecasting purposes, is to correctly capture the acceleration and expansion of the solar wind. In many current models, the solar wind is driven by varying the polytropic index so that a desired heating is obtained. While such schemes can yield solar wind properties consistent with observations, they are not problem-free. In this work, we demonstrate by performing MHD simulations that altering the polytropic index affects the properties of propagating shocks significantly, which in turn affect the predicted space weather conditions. Thus, driving the solar wind with such a mechanism should be used with care in simulations where correctly capturing the shock physics is essential. As a remedy, we present a simple heating function formulation by which the polytropic wind can be used while still modeling the shock physics correctly.

Keywords. shock waves, methods: numerical, solar wind, solar-terrestrial relations, Sun: coronal mass ejections

1. Introduction

Although more than 50 years has passed since Parker theorized the existence of a supersonic solar wind (Parker 1958), the physical processes responsible for accelerating the fast solar wind and heating the solar corona remain unknown (see, e.g., Cranmer 2010 for a review). In spite of this, a number of magnetohydrodynamic (MHD) models have successfully been developed that are capable of reproducing reasonably accurately the physical conditions in the inner heliosphere under the steady state assumption (see, e.g., Cohen *et al.* 2008 for a short overview). Common to these models is that they all use some form of (more or less) ad-hoc heating mechanism to drive the solar wind. For instance, a popular choice used in many studies investigating coronal mass ejections (CMEs) is to use a polytropic index γ smaller than $5/3$, the value expected for a monoatomic plasma such as the solar corona.

Altering the polytropic index is not problem-free. For instance, the maximal compression ratio r of a MHD shock is given by $r = (\gamma + 1)/(\gamma - 1)$, which is equal to 4 for a monoatomic gas, and increases for smaller values of γ. However, the efficiency of shocks as particle accelerators is very sensitive to the magnitude of the compression ratio of the shock (e.g., Reames 1999). Thus, for applications where capturing the shock physics is essential, it is not clear if models using an altered polytropic index can be used. In this paper, we demonstrate the effects that altering the polytropic index has on the evolution of shocks in MHD simulations. We also show how it is possible to retain the correct shock physics and still use a solar wind solution generated by a model with a polytropic index not equal to five thirds.

2. Model

We solve numerically the equations of ideal MHD, given by

$$\partial_t \rho = -\nabla \cdot (\rho \mathbf{v}) \qquad\qquad \partial_t \mathbf{B} = \nabla \times (\mathbf{v} \times \mathbf{B})$$

$$\rho \mathcal{D} \mathbf{v} = -\nabla P + \frac{1}{\mu_0}(\nabla \times \mathbf{B}) \times \mathbf{B} + \rho \mathbf{g} \qquad\qquad 0 = \nabla \cdot \mathbf{B}$$

$$S = \mathcal{D}(P/\rho^\gamma) \qquad\qquad \mathcal{D} \equiv \partial_t + \mathbf{v} \cdot \nabla,$$

using spherical coordinates in a two-dimensional azimuthally symmetric setting. Here ρ is the mass density, $\mathbf{v}$ is the velocity field, $\mathbf{B}$ is the magnetic field, P is the thermal pressure, $\mathbf{g} = \frac{GM_\odot}{r^2}\hat{\mathbf{r}}$ is the gravitational acceleration, γ is the polytropic index and S is an energy source term. Note that we solve the equations not in the primitive form given above, but in conservative form using the variables $\rho, \rho\mathbf{v}, \mathbf{B}$ and energy density $u = \frac{P}{\gamma - 1} + \frac{1}{2}\rho\mathbf{v}^2 + \frac{1}{2\mu_0}\mathbf{B}^2$.

2.1. *Solar wind model 1: $S = 0$, $\gamma = 1.05$*

To obtain a steady state solar wind solution, we set $S = 0$ and choose the adiabatic index to be $\gamma = 1.05 \equiv \Gamma_1$, a value commonly used in the literature. The initial magnetic field is set to a dipole field, and the density and radial velocity is initialized according to Parker's hydrodynamical solar wind solution. We then integrate the MHD equations in time until a converged solution is reached. For the upcoming discussion, we denote the obtained steady state solution by $\{\rho_1, P_1, \mathbf{v}_1, \mathbf{B}_1\}$.

2.2. *Solar wind model 2: $S \neq 0$, $\gamma = 5/3$*

As the next step, we wish to obtain an identical solar wind solution as obtained in the previous case, but instead use $\gamma = 5/3 \equiv \Gamma_2$. To achieve this, an energy source S driving the wind is necessary. We derive S by requiring that in the steady state the two solutions are identical, i.e. $\{\rho_1, P_1, \mathbf{v}_1, \mathbf{B}_1\} = \{\rho_2, P_2, \mathbf{v}_2, \mathbf{B}_2\}$ when $\partial_t = 0$. Plugging into the MHD equations gives

$$S = \mathbf{v}_1 \cdot \nabla \left(P_1 \rho_1^{-\Gamma_2} \right) \qquad\qquad (2.1)$$

Applying this energy source term, we can retain $\gamma = 5/3$ and still have the identical solar wind solution as obtained by using a lower polytropic index.

2.3. *CME model*

The final step in our simulation is to generate a CME. In this work, we simply superimpose on the solar wind solution a circular region with a higher density and an initial radial velocity. Additionally, we launch the CME $30°$ to the North from the equatorial plane.

3. Results and Discussion

The erupting CME launches a coronal shock wave evolving initially in a quasi-circular manner with the strongest regions near the nose of the shock. While the early evolution of the shock is morphologically similar for both models, there are important differences to note. Fig. 1 shows the compression ratio for the two simulation runs 48 minutes after the start of the eruption, with the left (right) figure corresponding to the simulation using solar wind model 1 (2). As can be seen, the compression ratio of the shock in the simulation using $\gamma = 1.05$ exceeds 4 for the entire shock front except for the flank. On the other hand, with $\gamma = 5/3$, the compression ratio $r > 3$ for a large part of the shock, but

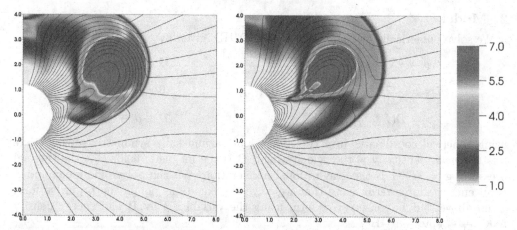

Figure 1. Snapshots from the MHD simulations showing the compression ratio and magnetic field lines (black curves) 48 minutes after the start of the eruption. The image to the left corresponds to the case where $\gamma = 1.05$ was used, and the image to the right to the case where $\gamma = 5/3$ and a energy source term S was used. Note that except for the mechanism driving the background solar wind the two simulation setups were identical, using identical initial and boundary conditions as well as the same computational grid and numerical method. See on-line version for color figures.

does not exceed 4 anywhere. This indicates that our model including the energy source treats the shock physics consistently in terms of compression, which is important, e.g., for particle acceleration applications.

It is not only the compression ratio that is different for the two simulations. In fig. 1, it is evident that the shock with $\gamma = 5/3$ has reached further out. Moreover, the shock structure at the southern flank is significantly different; not only has it reached further, but also the morphology near the skirt of the shock is different. This might be of importance to studies using MHD simulations that discuss the origin of so called EIT waves commonly observed in conjunction with CMEs.

4. Summary and Conclusions

We have used MHD simulations to obtain two identical steady state solar wind solutions, but using different methods to drive the wind. In the first case a polytropic index $\gamma = 1.05$ was used, while in the second case the polytropic index was set to $\gamma = 5/3$, but an appropriate extra source term in the energy equation was included. On top of these solutions, we launched a CME, and studied the shock launched by the eruption.

From the simulations we can conclude that using $\gamma = 5/3$ is vital in order for the model to treat the shock physics consistently in terms of compression and dynamical evolution. This is especially important for studies where the shock obtained by the MHD simulation is used as input to particle acceleration simulations.

References

Parker, E. N. 1958, *ApJ*, 128, 664
Cranmer, S. R. 2010, *SSRv*, doi:10.1007/s11214-010-9674-7
Cohen, O., Sokolov, I. V., Roussev, I. I., & Gombosi, T. I. 2008, *JGR* 113, A03104
Reames, D. V. 1999, *SSRv* 90, 413

Solar and Stellar Plasma

I. Kitiashvili, S. Berdyugina and G. Melikidze

Advances in Plasma Astrophysics
Proceedings IAU Symposium No. 274, 2010
A. Bonanno, E. de Gouveia Dal Pino & A.G. Kosovichev, eds.
© International Astronomical Union 2011
doi:10.1017/S1743921311006673

Anomalous momentum transport in astrophysical return-current beam plasmas - the two-dimensional electromagnetic case

Kuang Wu Lee[1] and Jörg Büchner[2]

Max-Planck-Institut für Sonnensystemforschung, 37191 Katlenburg-Lindau, Germany
[1] email: lee@mps.mpg.de [2] email: buechner@mps.mpg.de

Abstract. Anomalous momentum transport in a typical astrophysical return-current beam plasma system is studied by means of two-dimensional PIC code simulations. A forward going hot electron beam compensated by a cold return beam is considered. A linear dispersion analysis predicts the linearly unstable wave modes. Our simulation reveals that the nonlinerly generated waves and the consequent wave-particle interactions cause the electron heating and the relaxation of the electron drifts. Both, the developments of electrostatic and electromagnetic waves are analyzed as well as the roles they play in energy conversion. In particular it is found that the relaxation of electron drifts is stronger if the electromagnetic turbulence is taken into account.

1. Return-current in astrophysical plasma

In various astrophysical plasma environments return-current beams have to be assumed which compensated the intense electron beams generated in acceleration sites like reconnecting current sheets. Return current beam instabilities have to be taken into account also in the production of thermal X-rays by the bombardment of the neutron star surface by return-current electrons (Cheng & Zhang, 1999) and as cosmic rays streaming instabilities behind supernovae shock waves. Induced return-currents can compensate the cosmic ray currents (Amato & Blasi, 2009, Niemiec *et al.*, 2008). The spectrum of X-rays generated by energetic electrons precipitating in the solar atmosphere is likely to be influenced by return currents (Zharkova & Gordovskyy, 2006, Lee *et al.*, 2008). The localized large-amplitude electrostatic and electromagnetic structures excited by beam and return current electrons are of significant importance for astrophysical plasmas. One of the impacts of this non-linear wave-particle interaction is the anomalous transport caused by the current driven instabilities. The latter has been studied, e.g., by means of electrostatic Vlasov code simulations with open boundary conditions (Büchner & Elkina, 2006). The instability of a return-current beam system has been investigated by means of a three-dimensional electromagnetic particle-in- cell (PIC) simulation code (Karlický & Bárta, 2009). In their study the authors considered the injection of a cold electron beam into a background of electron and ions. They analyzed the parallel electric field perturbations, and found a strong background magnetic field in the electron drift direction suppresses the Weibel- (filamentation-) instability. The electrons and ions are heated preferentially in the parallel direction while the background magnetic field is weak. The anomalous transport in a return-current beam plasma was compared by means of electrostatic Vlasov-code and multifluid simulations (Lee & Büchner, 2010). It is known, however, that electromagnetic waves might accelerate and heat electrons more effectively than electrostatic waves (Tsironis & Vlahos, 2005), i.e the associated anomalous transport could be stronger. For the common case of magnetic reconnection

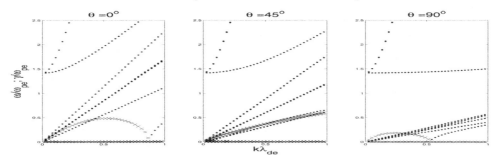

Figure 1. Wave dispersion for different propagation angles: $\theta = 0^o$; $\theta = 45^o$; $\theta = 90^o$. The dots depict the normalized real frequencies $\omega(k)$ of waves in the system while the crosses depict the corresponding growth rates $\gamma(k)$.

outflow, we consider in this paper a hot electron beam injection into the cold background plasma with the compensating the return current flow.

2. Linear dispersion analysis of warm magnetized plasma

Our multifluid linear dispersion considers thermal plasma effects, electromagnetic waves (full set of Maxwell equations), different electron drifts and oblique wave propagation as well as the real electron-to-ion mass ratio $(m_i/m_e = 1836)$. We concentrate on a situation where the forward going electron beam is twice as hot as the background ions and return-current electrons as cold as the ions $(T_{e,beam} = 2T_{ion} = 2T_{e,RC})$. The densities of beam and return-current electron beams are equal $n_{e,beam} = n_{e,RC} = n_{ion}/2$ and the net current vanishes with their drift velocities $(V_{de,beam} = -V_{de,RC} = -v_{te})$.

Fig. 1 shows the dispersion relation for different prapagation angles, where θ is the angle between $\vec{k}$ and $\vec{V}_{de,beam}$. Without a background magnetic field, the instability (growth rates are depicted by crosses and the wave frequencies by the dots) is due to the interaction of two electron acoustic waves. The electron-electron acoustic wave grows fastest at an oblique propagation angle (cf. the second panel of Fig.1 for $\theta = 45^o$). Note that due to the limits of a fluid description this result is correct only for wavelength longer than the Debye length. At the nonlinear stage of the instability the inhomogeneity of the electron drift distribution results in localized currents which cause the growth of localized magnetic field structures.

3. Self-generated anomalous transport

The nonlinear stage of the return-current beam relaxation is studied via (2D EM PIC) simulation, for which the periodic boundary conditions are implemented. In the past mainly anomalous transport caused by electrostatic waves has been investigated, and it was concluded that the relaxation of electron drifts corresponds closely to the development of electric field structures. In the first panel of Fig. 2 the temporal evolution of total field energy in the simulation domain is shown. Indeed, at the nonlinear stage first the electric field energy strongly grows and after some time it decreases and saturates at a lower level, as shown previously by electrostatic simulations (cf. also Lee & Büchner, 2010). Howeverm as one can see further in the Figure, the total (electrostatic plus electromagnetic) field energy continues to increase even after the electrostatic waves did saturate (blue dashed line in first panel of Fig.2).

The second and third panels of Fig.2 depict the temporal evolution of electron drifts and plasma temperature. The energy conversion from the electron drift decreases at the

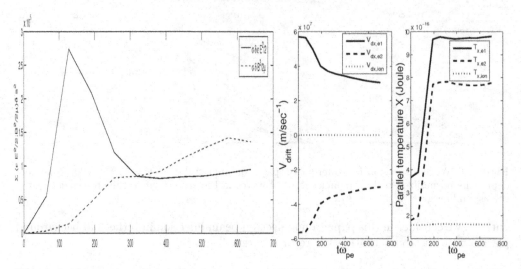

Figure 2. Temporal evolution of the electromagnetic field energy shown in the first panel. The solid line depicts the electrostatic and the dashed line the magnetic field energy. The second and third panels are the temporal evolutions of electron drifts and temperatures.

nonlinear stage of the instability evolution $T \approx 70\omega_{pe}^{-1}$. At the same time the electron temperature increases and the energy of the electric field energy rapidly decreases. At the late stage of electric field evolution $T \approx 200\omega_{pe}^{-1}$ the electron heating stops. Drift relaxation and electron heating effects correspond to those obtained in electrostatic approaches. However, the relaxation of electron drifts it is not completely stopped, as the case in electrostatic simulation. Hence, the drift relaxation continues as well as the anomalous transport. The reason is that if the excitation of electromagnetic waves are taken into account the electron drift energy is converted, in addition to electric field fluctuations, also to magnetic field fluctuations, as indicated in Fig. 2.

To conclude, the interaction of the two beams causes anomalous transport and electron heating. These processes are kinetic in nature and might be due to the interaction of the beam particles with unstably generated waves. At the nonlinear evolution stage, however, the inhomogeneity of the electron drift distribution creates local current and associated magnetic field structures. Contrary to the widely studied case of relaxation by electron interaction with the self-generated electrostatic waves, the magnetic structures do intensify the drift relaxation by extracting kinetic energy from the electron drift continuously in the late stage of evolution. We argue that electromagnetic effects play a most significant role in the anomalous transport and plasma heating at the late stage of the drift evolution.

4. Acknowledgement

The authors thank the Max-Planck Society for funding this work by the Inter-Institutional Research Initiative, Project No. MIF-IF-A-AERO8047.

References

Amato, E. & Blasi, P. 2009, *Mon. Not. R. Astron. Soc.*, 392, 1591
Büchner, J., Elkina, N. 2006, *Phys. Plasmas*, 13, 082304
Cheng, K. S. & Zhang, L. 1999, *ApJ*, 515, Issue 1, 337

Karlický, Bárta, M. 2009, *Nonlin. Processes Geophys.*, 16, 525

Lee, K. W., Büchner, J., & Elkina, N. 2008, *A&A*, 478, 889

Lee, K. W. & Büchner, J. 2010, *Phys. Plasmas*, 17, 042308

Niemiec, J., Pohl, M., Stroman, T., & Nishikawa, K.-I. 2008, *ApJ*, 684, 1174

Tsironis, C. & Vlahos, L. 2005, *Plasma Phys. Control. Fusion*, 47,131

Zharkova, V. V. & Gordovskyy, M. 2006, *ApJ*, 651, 553

Advances in Plasma Astrophysics
Proceedings IAU Symposium No. 274, 2010
A. Bonanno, E. de Gouveia Dal Pino & A. Kosovichev, eds.

© International Astronomical Union 2011
doi:10.1017/S1743921311006685

On radiation-zone dynamos

Günther Rüdiger, Marcus Gellert and Rainer Arlt

Astrophysikalisches Institut Potsdam,
An der Sternwarte 16, D-14482 Potsdam, Germany
email: gruediger@aip.de, mgellert@aip.de, rarlt@aip.de

Abstract. It is shown that the magnetic current-driven ('kink-type') instability produces flow and field patterns with helicity and even with α-effect but only if the magnetic background field possesses non-vanishing current helicity $\bar{\boldsymbol{B}} \cdot \mathrm{curl}\bar{\boldsymbol{B}}$ by itself. Fields with positive large-scale current helicity lead to negative small-scale kinetic helicity. The resulting α-effect is positive. These results are very strict for cylindric setups without z-dependence of the background fields. The sign rules also hold for the more complicated cases in spheres where the toroidal fields are the result of the action of differential rotation (induced from fossil poloidal fields) at least for the case that the global rotation is switched off after the onset of the instability.

Keywords. magnetic instability, helicity, radiation zone, dynamo

1. Introduction

Open questions in stellar physics led to the idea that a dynamo operates in the radiative cores of early-type stars (Spruit 2002). Even the helioseismologic observation of rigid rotation of the solar interior shows in this direction. The angular momentum transport by the large-scale magnetic field pattern (fossil field plus toroidal field induced by differential rotation) does *not* lead to a solid-body rotation unless the viscosity of the plasma exceeds the molecular value by a few orders of magnitude (Rüdiger & Kitchatinov 1996, Eggenberger *et al.* 2005). Other examples are given by the evolution of the fast rotating early-type stars which can only be understood if i) there is a basic transport of angular momentum outwards and ii) the radial mixing of chemicals remains weak (Yoon *et al.* 2006, Brott *et al.* 2008). Hence, if a (magnetic-induced) instability existed in the radiative stellar cores then the corresponding Schmidt number $\mathrm{Sc} = \nu/D$ must be rather large. We have shown that the kink-type instability (or Tayler instability, TI) of toroidal magnetic fields forms a much-promising candidate for the instability. A Schmidt number larger than ten results as the ratio of the effective viscosity and the diffusion coefficient (Rüdiger *et al.* 2009).

The unstable modes of the TI are basically nonaxisymmetric driven by the energy of the electrical current which produces the toroidal field. Interestingly enough, there exists even an instability of a toroidal magnetic field which in the fluid is current-free. In this case the energy comes from a differential rotation which itself is stable but which is unstable under the influence of the (current-free) toroidal field. We have called this instability as Azimuthal MagnetoRotational Instability (AMRI) as – like for the standard MRI (with axial fields) – the field itself is current-free and does not exert forces. In opposition to the standard MRI the AMRI is always nonaxisymmetric and it is, therefore, much more interesting for the dynamo theory. For complicated radial profiles of the toroidal field we shall always have a mixture of TI and AMRI. Generally, the latter is more important for fast rotation ($\Omega > \Omega_{\mathrm{A}}$) and v.v. Here the Alfvén frequency Ω_{A} for the toroidal field is used which derives from the Alfvén velocity $v_{\mathrm{A}} = B_\phi/\sqrt{\mu_0 \rho}$ as the related frequency. Between two cylinders with different radii the toroidal field profile with $B_\phi = AR + B/R$

(R radius) is free of dissipation. The 'perfect' AMRI appears for $A = 0$ while the 'perfect' TI results for $B = 0$.

One can compute the necessary electrical currents to excite both sorts of instabilities in a columnar Taylor-Couette experiment with gallium as fluid conductor. As we have shown the critical Hartmann numbers for self-excitation of axi- and nonaxisymmetric perturbation modes do not depend on the magnetic Prandtl number of the fluid which is as small as 10^{-3} for stellar plasma and 10^{-5} for liquid sodium (Rüdiger & Schultz 2010).

It is typical for the nonaxisymmetric TI and AMRI that always the two modes with $m = \pm 1$ are excited for the same critical Hartmann number and also – if supercritical – with the same growth rates (Fig. 1). Despite their simultaneous existence they can be excited as singles with different initial conditions. However, if the initial conditions are as neutral as possible with respect to a preferred helicity, in the majority of the cases one of the modes dominates after our experiences.

This behavior may have dramatic consequences with respect to the dynamo theory. Both the modes with $m = \pm 1$ have opposite helicity with the same total amount. The mode with $m = -1$ is identical to the mode with $m = 1$ but in a left-hand system. The helicity of $m = 1$ in the right-hand system equals the helicity of $m = -1$ in the left-hand system. So it is obvious that in one and the same coordinate system the sum of the helicity of $m = -1$ and $m = 1$ is zero. As a consequence the instability of a toroidal field can only develop helicity if by some reasons one of the modes with $m = \pm 1$ dominates the other. There are the two possibilities that i) one mode dominates the other by chance (so as the matter dominates the antimatter) or ii) the existence of a poloidal field prefers one of the modes. We have shown that indeed in stellar radiation zones – if the background field has a positive current helicity $\bar{\boldsymbol{B}} \cdot \mathrm{curl}\bar{\boldsymbol{B}}$ – the resulting kinetic helicity $\langle \boldsymbol{u}' \cdot \mathrm{curl}\boldsymbol{u}' \rangle$ of the fluctuations is always negative (Gellert *et al.* 2011). A positive current helicity of the background field results if an axial field and an axial electrical current are parallel. A negative current helicity of the background field results if an axial field and an axial electric current are antiparallel. Hence, the resulting kinetic helicities on the basis of current-driven instabilities have, therefore, opposite signs in opposite hemispheres of the model.

2. Cylindric geometry

We are interested in the stability of a background field $\bar{\boldsymbol{B}} = (0, B_\phi(R), B_0)$ with $B_0 = $ const, and the flow $\bar{\boldsymbol{u}} = (0, R\Omega(R), 0)$ in a dissipative conducting fluid rotating between two rigid cylinders. ν is the kinematic viscosity and η is the magnetic diffusivity, their ratio is the magnetic Prandtl number $\mathrm{Pm} = \nu/\eta$. The stationary background rotation law is $\Omega = a + b/R^2$ with a and b as constants. Ω_{in} and Ω_{out} are the rotation rates of the cylinders and B_{in} and B_{out} are the azimuthal magnetic fields there.

One finds for the current helicity of the background field $\bar{\boldsymbol{B}} \cdot \mathrm{curl}\,\bar{\boldsymbol{B}} \simeq AB_0$, which may be either positive or negative (and of course vanishes for the current-free case $A = 0$).

The inner value B_{in} may be normalized with the uniform vertical field, i.e. $\beta = B_{\mathrm{in}}/B_0$. For a profile with $B_{\mathrm{in}} = B_{\mathrm{out}}$ we have $\bar{\boldsymbol{B}} \cdot \mathrm{curl}\,\bar{\boldsymbol{B}} \propto 1/\beta$ for the normalized current helicity of the background field. The sign of β determines the sign of the current helicity. As usual, the toroidal field amplitude is measured by the Hartmann number $\mathrm{Ha} = B_{\mathrm{in}}D/\sqrt{\mu_0\,\rho\nu\eta}$ and the global rotation by the Reynolds number $\mathrm{Re} = \Omega_{\mathrm{in}}D^2/\nu$ with $D = R_{\mathrm{out}} - R_{\mathrm{in}}$. The Alfvén frequency is $\Omega_{\mathrm{A}} = B_{\mathrm{in}}/\sqrt{\mu_0\rho}D$.

The boundary conditions associated with the perturbation equations are no-slip for the flows and perfectly conducting for the fields. For all computations it is $R_{\mathrm{out}} = 2R_{\mathrm{in}}$.

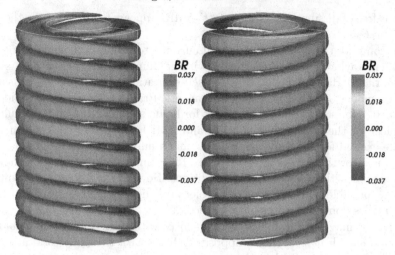

Figure 1. The two modes with opposite helicity values of the same amount which can be excited with different initial conditions. Re $= 0$, Ha $= 200$, $\beta = 0$, Pm $= 1$ (from Gellert *et al.* 2011).

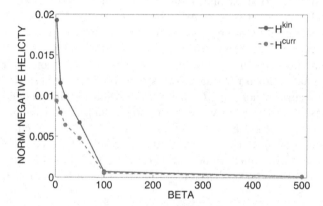

Figure 2. The normalized negative kinetic helicity and the negative current helicity for the nonaxisymmetric perturbations as functions of β. Re $= 200$, Ha $= 100$, $\Omega_{\rm in} = 2\Omega_{\rm out}$, Pm $= 1$.

For purely toroidal fields the expected net helicities vanish as two nonaxisymmetric modes with $m = \pm 1$ exist with identical excitation conditions and the same amount of helicity but of opposite sign (Fig. 1).

We now compute both the helicities for helical fields with various pitch values β which is inverse to the current helicity of the large-scale field. The astrophysically relevant case is $\Omega > \Omega_{\rm A}$ (which is not the classical realization of the Tayler instability). The instability only exists for nonrigid rotation so that we always work with a rotation law $\Omega \propto 1/R$. The radial profile of the toroidal field has been modelled by the above mentioned most simple profile with almost uniform toroidal field. Figure 2 gives the results (in units of $\Omega_{\rm A}^2 D$). Both the kinetic helicity $\langle \boldsymbol{u}' \cdot {\rm curl}\boldsymbol{u}' \rangle$ as well as the current helicity $\langle \boldsymbol{B}' \cdot {\rm curl}\boldsymbol{B}' \rangle$ are negative. They are decreasing functions of β.

We have also calculated the α-effect via the determination of the electromotive force $\langle \boldsymbol{u}' \times \boldsymbol{B}' \rangle$ by the fluctuations. According to the rule that the azimuthal α-effect is anti-correlated with the (kinetic) helicity we expect the azimuthal α-effect as positive for $\beta > 0$. This is indeed the case (see Fig. 3, left). One finds positive and negative signs in the container but the positive values dominate so that in the average the azimuthal α-

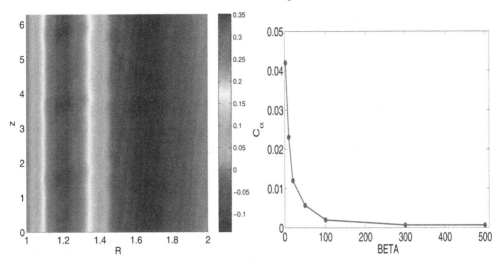

Figure 3. The α-effect for $\Omega > \Omega_\mathrm{A}$. Left: $\beta = 3$. Right: The dimensionless dynamo number C_α of the azimuthal α-effect which decays decays like C/β with $C \simeq 0.05$. Pm $= 1$.

effect is indeed positive. On the other hand, for negative β we expect positive small-scale kinetic helicity and negative α-effect.

The example presented in Fig. 3 refers to the pitch $\beta = 3$. For higher values, i.e. smaller helicity of the background field, the resulting α-effect becomes smaller and smaller. Figure 3 (right) shows that the dynamo number $C_\alpha = \alpha_{\phi\phi}D/\eta$ runs as $C_\alpha \sim C/\beta$ with C of order 10^{-2}. It is easy to show that this value is too small by two orders of magnitudes to allow the operation of a classical $\alpha\Omega$-dynamo (Gellert *et al.* 2011).

3. Spherical geometry

We start with the amplification of fossil magnetic fields by shear and the magnetic-field back-reaction. The two-dimensional, non-linear simulations thus start with an initial differential rotation and a purely poloidal magnetic field in a radiative stellar zone. The early phase of the simulation shows a generation and steep amplification of toroidal magnetic field through differential rotation. The generated Lorentz force redistributes the angular momentum. This is why at the same time of field amplification, the differential rotation starts to decrease, and the toroidal-field growth is thus limited.

Figure 4 shows the maximum magnetic field amplitude in the spherical shell as a function of time. The Reynolds number is Re $= 20\,000$ and the magnetic Prandtl number is Pm $= 1$. The dotted and dot-dashed lines in Fig. 4 are the results of a linear stability analysis of only the toroidal field. By S1 we refer to a velocity perturbation which is symmetric with respect to the equator and has $m = 1$, A1 is the corresponding antisymmetric perturbation. If at any given time both stability lines are above the solid line, the corresponding snapshot is stable against $m = 1$ perturbations. The stability lines cross the solid one at about $t = 0.0023$ (see Arlt & Rüdiger 2011).

The system is now perturbed in a 3D, nonlinear simulation at a somewhat later time ($t_0 = 0.003$). The perturbation is applied to the magnetic field and has an azimuthal wave number of $m = 1$ (and is symmetric with respect to the equator). The resulting flow is thus antisymmetric and can be compared with the A1-mode in the linear stability diagram in Fig. 4.

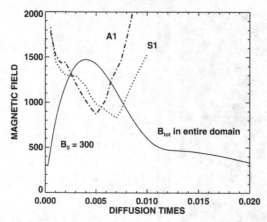

Figure 4. The stability limits of each snapshot of the axisymmetric evolution of a sufficiently strong initial field. The solid line gives the evolution of the toroidal field. The stability limit is given for the nonaxisymmetric perturbation patterns S1 and A1 (from Arlt & Rüdiger 2011).

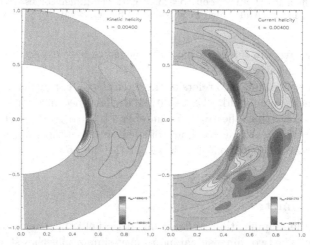

Figure 5. Left: Kinetic helicity, right: Current helicity. Both after averaging over the azimuth. Re = 20 000.

The spatial distribution of the resulting helicity is shown in Fig. 5. Note the concentration of both the helicities near the inner boundary where the tangent cylinder touches the inner sphere. We conclude that a considerable part of the kinetic helicity in the system is due to the presence of an inner sphere. This is slightly different for the current helicity. A considerable amount of *positive* current helicity is measured in the bulk of the northern hemisphere. The situation is unchanged in runs with perfect conductor boundary conditions at the inner radius. Interestingly, Reshetnyak (2006) with convection simulations also finds negative kinetic helicity along the tangent cylinder. We assume that the negative helicity near the tangent cylinder of the northern hemisphere is an inner-boundary effect, neither related to convective nor Tayler instability turbulence.

Figure 6 summarizes the consequences of a numerical experiment. In the unstable domain of Fig. 4 the instability pattern is calculated without global rotation, i.e. for Re = 0. In this case the only existing pseudo-scalar is $\bar{B} \cdot \mathrm{curl} \bar{B}$ like in the above theory in cylindric geometry. The question is whether we can find the same relations between large-scale current helicity (left), kinetic helicity (middle) and α-effect (right). Note first that indeed the dominant role of the tangent cylinder for the kinetic helicity and the

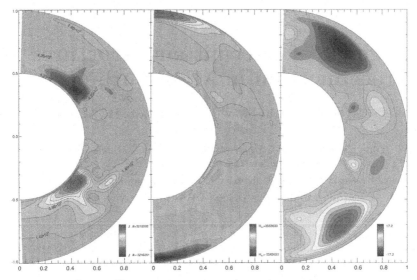

Figure 6. Left: $\bar{B} \cdot \mathrm{curl}\bar{B}$. Middle: kinetic helicity $\langle u' \cdot \mathrm{curl} u' \rangle$. Right: the azimuthal component of the α-effect. Re = 0.

α-effect vanishes so that it is obvious that it is a rotation-induced boundary layer effect. In the southern hemisphere the pseudo-scalar $\bar{B} \cdot \mathrm{curl}\bar{B}$ is positive while the kinetic helicity is negative (but concentrated to the pole). The corresponding α-effect given in the right panel (and computed also by means of the test-field method) proves to be positive (see Arlt & Rüdiger 2011). As it must, at the northern hemisphere all signs are opposite. Hence, the model fulfills the same sign rules as in the above-discussed cylindric setup.

4. Summary

Unstable toroidal fields alone are not able to produce helicity and α-effect. It has been shown, however, that helicity and α-effect are produced by unstable magnetic large-scale field patterns which themselves possess current helicity $\bar{B} \cdot \mathrm{curl}\bar{B}$. This is insofar understandable as helicity and α-effect are both pseudo-scalars which can only be nonvanishing if in the global setup a nonvanishing large-scale pseudo-scalar like $\bar{B} \cdot \mathrm{curl}\bar{B}$ exists. We want to stress, however, that the current helicity of the background field is not the only possible pseudo-scalar existing in magnetized stellar radiation zones on which other forms of helicity and α-effect may base.

References

Arlt, R. & Rüdiger, G. 2011, *MNRAS*, in press
Brott, I., Hunter, I., Anders, P., & Langer, N. 2008, *AIPC*, 990, 273
Eggenberger, P., Maeder, A., & Meynet, G. 2005, *A&A*, 440, L9
Gellert, M., Rüdiger, G., & Hollerbach, R. 2011, *Phys. Rev. E*, in prep.
Reshetnyak, M. Yu. 2006, *Physics of the Solid Earth*, 42, 449
Rüdiger, G. & Kitchatinov, L. L. 1996, *ApJ*, 466, 1078
Rüdiger, G. & Schultz, M. 2010, *Astron. Nachr.*, 331, 121
Rüdiger, G., Gellert, M., & Schultz, M. 2009, *MNRAS*, 399, 996
Spruit, H. C. 2002, *A&A*, 381, 923
Yoon, S.-C., Langer, N., & Norman, C. 2006, *A&A*, 460, 199

Advances in Plasma Astrophysics
Proceedings IAU Symposium No. 274, 2010
A. Bonanno, E. de Gouveia Dal Pino & A.G. Kosovichev, eds.
© International Astronomical Union 2011
doi:10.1017/S1743921311006697

Emergence of intermittent structures and reconnection in MHD turbulence

Antonella Greco[1], Sergio Servidio[1], William H. Matthaeus[2] and Pablo Dmitruk[3]

[1]Dipartimento di Fisica Università della Calabria, Rende (CS), Italy
email: antonella.greco@fis.unical.it

[2]Bartol Research Institute, University of Delaware, Newark DE, United States

[3]Departamento de Física (FCEN-UBA),Buenos Aires, Argentina

Abstract. In recent analyses of numerical simulation and solar wind dataset, the idea that the magnetic discontinuities may be related to intermittent structures that appear spontaneously in MHD turbulence has been explored in details. These studies are consistent with the hypothesis that discontinuity events founds in the solar wind might be of local origin as well, i.e. a by-product of the turbulent evolution of magnetic fluctuations.

Using simulations of 2D MHD turbulence, we are exploring a possible link between tangential discontinuities and magnetic reconnection. The goal is to develop numerical algorithms that may be useful for solar wind applications.

Keywords. (magnetohydrodynamics:) MHD, turbulence, (Sun:) solar wind, methods: numerical

1. Introduction

Solar wind discontinuities are characterized by large and rapid changes in properties of the plasma and magnetic field (Burlaga, 1968; Tsurutani & Smith, 1979). One interpretation of the strong discontinuities is that they are the walls of a filamentary structure of a discontinuous solar-wind plasma (Borovsky, 2008). Another is that some strong discontinuities are fossils from the birth of the solar wind (Burlaga, 1968). We explored an alternative possibility, that observed discontinuities might be the current sheets that form as a consequence of the cascade of MHD turbulence to inertial scales (Greco *et al.*, 2008; Greco *et al.*, 2009a).

In the standard picture of solar wind turbulence where a temporal/spatial cascade of the fluctuations from large to small spatial scales produces the numerous thin current sheets that are a common aspect of the high-speed wind at 1 AU, the small-scale magnetic reconnection occurs relatively frequently at these thin current sheets (e.g., Matthaeus & Lamkin, 1986).

In this work, we start to explore the possibility that discontinuities and local magnetic reconnection events are linked. Magnetic reconnection has been often studied in simplified geometries and boundary conditions, but since it might occur in any region separating topologically distinct magnetic flux structures, it might be expected to be of importance in MHD turbulence. The latter possibility has been recently investigated, leading to the conclusion that in turbulence strong reconnection events locally occur (Servidio *et al.*, 2009; Servidio *et al.*, 2010). Previous studies on discontinuities and theories of reconnection in turbulence are being combined in order to identify possible reconnection events between the intermittent events. In the present report we consider a 2D model in order to simplify the problem.

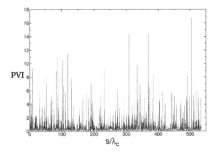

Figure 1. Spatial signal PVI obtained from the simulation by sampling along the trajectory s in the simulation box.

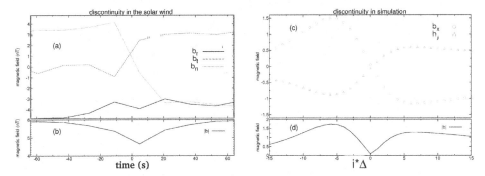

Figure 2. Examples of discontinuities selected by the PVI method. Panel a: the three RTN components of the magnetic field vector and Panel b: magnitude of the magnetic field vector in solar wind data. The discontinuity is centered around zero; Panel c: the two components of the magnetic field vector and Panel d: magnitude of the magnetic field vector in simulation data. Δ is the resolution data.

2. Magnetic discontinuities and intermittent current sheets

To describe rapid changes in the magnetic field, usually people look at the increments $\Delta\mathbf{b}(s, \Delta s) = \mathbf{b}(s + \Delta s) - \mathbf{b}(s)$, being $\mathbf{b}$ the magnetic field, s the 1D coordinate along the trajectory of the spacecraft, and Δs the related spatial increment. By using the magnetic increments, we computed the normalized (squared) Partial Variance of Increments (PVI) (Greco *et al.*, 2008):

$$PVI(\Delta s, \ell, s) = \frac{|\Delta\mathbf{b}(s, \Delta s)|}{\sqrt{\langle|\Delta\mathbf{b}(s, \Delta s)|^2\rangle_\ell}}, \qquad (2.1)$$

where $\langle\bullet\rangle$ denotes a spatial average over the entire domain of total length ℓ. For the numerical analysis performed here $\ell \simeq 535\lambda_C$, being $\lambda_C = 0.18$ the correlation length of turbulence. In Fig. 1, the time series of PVI with $\Delta s = 0.67\lambda_d$, where $\lambda_d = 4.6 \times 10^{-3}$ is the dissipation length of turbulence, is shown, and Fig. 3 shows an example of s-path in the simulation box along with Eq. 2.1 is computed. The PVI dataset is bursty, suggesting the presence of sharp gradients in the magnetic field, localized intermittently in space. These events may correspond to tangential discontinuities (TDs). Imposing a threshold on Eq. 2.1 a collection of discontinuities along s-path can be identified. An example of TD is displayed in Fig. 2. The figure shows that these events are characterized by a rotation of the magnetic field followed by a depression of the magnitude of B (panels c and d). None of these features were present at the initial time (not shown) in which the electric current is not concentrated but rather is randomly distributed by construction.

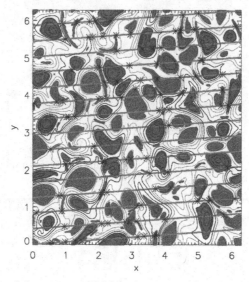

Figure 3. Contour lines of the magnetic field (red areas are the magnetic islands) together with the reconnection (blue) regions. The discontinuities identified by PVI technique are gray stars placed on an a sample path (green line) in the simulation box.

Greco *et al.*, (2008) looked at the distribution of waiting times between discontinuity events identified either by classical methods (e.g. Tsurutani & Smith, 1979) or by applying the thresholding technique based on Eq. 2.1. The authors found that the two methods performed almost interchangeably and the normalized waiting-time (WT) distributions between events were extremely similar in the solar wind and simulation datasets at (inertial range) separations shorter than the correlation scale (Greco *et al.*, 2009a). Fig. 2 shows a directional change, selected by the PVI algorithm in the ACE magnetic field data. Panel a gives the components and panel b the magnitude of the magnetic field vector with 16 s time resolution. More specifically, in Greco *et al.* (2008a) and (2009a) it was found that distributions of WTs between discontinuities for $s < \lambda_C$ were well described by a power law. This conclusion was true both in simulations and in solar wind. Further analysis were employed in Greco *et al.* (2009b) to examine the distribution of WTs for these events, showing that the discontinuities are not distributed without correlations, but rather that non-Poisson correlations are present in the solar wind data at least up to the typical correlation scale. A similar conclusion emerges from Poisson analysis of the simulation dataset.

3. Local reconnection

In previous studies (Servidio *et al.*, 2009; Servidio *et al.*, 2010) it has been confirmed that turbulence lead to the spontaneous formation of local and intermittent reconnection events. We employed a cellular automata's technique (Servidio *et al.*, 2010), applying to a 2D turbulent field, permitting to identify the diffusion regions. Using this cellular automata mapping, the shape and the position of each diffusion region, nearby each X-point of the vector potential a, are defined. The width d, the elongation l, and the respective reconnection rate E_X are computed as well. Note that only the strongest reconnection sites (RSs) are detected by the algorithm (Servidio *et al.*, 2010).

Varying the threshold for Eq. 2.1, we can count how many of the TDs are RSs as follows. Every discontinuity is characterized by a starting and an ending point (see Fig. 2). A set of discontinuities is identified but eventually only a certain number of discontinuities are reconnecting regions. In order to measure this number, we will make use of the cellular automata map. The latter is a 2D matrix that has 0 value out of the diffusion regions or 1 inside them. When at least one point of the discontinuity overlaps with one point of the diffusion region, the event is counted as success, otherwise it is a failure. In the latter case it means that the method is detecting a non reconnecting, high-stress, magnetic field structure. Using for example the threshold 6 in Eq. 2.1, 25 discontinuities have been identified and 18 correspond to reconnection regions. The efficiency of this method can be arbitrarily estimated as proportional to the number of the success over the total number of discontinuities. For this algorithm the goodness is $\simeq 70\%$. An example of discontinuities, together with the reconnecting regions, is shown in Fig. 3.

4. Conclusions

We can draw a firm conclusion for the numerical experiments, that the discontinuity events are formed spontaneously due to nonlinear couplings, cascade and turbulence. They were not present in the initial data. The extension of this conclusion to the solar wind is tempting.. Our studies are consistent with the hypothesis that solar wind intermittent discontinuities are produced by MHD turbulence, even if we have not ruled out that some of these features originate in the lower corona. It is possible that many inertial range structures that contribute to the tails of the PDFs of increments of $\mathbf{B}$ in the solar wind, are formed in situ by local rapid relaxation processes associated with turbulence. A further detailed analysis determined the non-Poisson character of these intermittent events from the shape of waiting time distributions.

Finally, a magnetic discontinuity is a rapid change of the field across a very narrow part of the space, so that strong changes of the magnetic topology are necessarily involved. This implies the possibility that discontinuities and local magnetic reconnection events may be linked. Magnetic reconnection has been often studied in simplified geometries and boundary conditions, but since it might occur in any region separating topologically distinct magnetic flux structures, it might be expected to be of importance in MHD turbulence. Previous studies on discontinuities and theories of reconnection in turbulence could be combined in order to identify possible reconnection events between the intermittent events.

References

Borovsky, J. 2008, *J. Geophys Res.*, 113, A08110.

Burlaga, L. F. 1968, *Solar Physics*, 4, 67.

Greco, A., Chuychai, P., Matthaeus, W. H., Servidio, S., & Dmitruk, P. 2008, *Geophys. Res. Lett.*, 35, L19111, 2008GL035454.

Greco, A., Matthaeus, W. H., Servidio, S., Chuychai, P. & P. Dmitruk 2009, *Astrophys. J.*, 691, L111.

Greco, A., Matthaeus, W. H., Servidio, S., & Dmitruk, P. 2009, *Phys. Rev. E*, 80, 046401.

Matthaeus, W. H., & Lamkin, S. L. 1986, *Physics of Fluids*, 29, 2513.

Servidio, S., Matthaeus, W. H., Shay, M. A., Cassak, P. A., & Dmitruk, P. 2009, *Phys. Rev. Lett.*, 102, 115003.

Servidio, S., Matthaeus, W. H., Shay, M. A., Dmitruk, P., Cassak, P. A., & Wan, M. 2010, *Phys. Plasmas*, 17, 032315.

Tsurutani, B. T. & Smith, E. J. 1979, *J. Geophys. Res.*, 84, 2773.

Advances in Plasma Astrophysics
Proceedings IAU Symposium No. 274, 2010
A. Bonanno, E. de Gouveia Dal Pino & A.G. Kosovichev, eds.
© International Astronomical Union 2011
doi:10.1017/S1743921311006703

Realistic MHD simulations of magnetic self-organization in solar plasma

I. N. Kitiashvili[1,2], A. G. Kosovichev[2], A. A. Wray[1,3], and N. N. Mansour[1,3]

[1] Center for Turbulence Research, Stanford University, Stanford, CA 94305, USA
email: irinasun@stanford.edu

[2] W.W. Hansen Experimental Physics Laboratory, Stanford University, Stanford, CA 94305, USA

[3] NASA Ames Research Center, Moffett Field, Mountain View, CA 94040, USA

Abstract. Filamentary structure is a fundamental property of the magnetized solar plasma. Recent high-resolution observations and numerical simulations have revealed close links between the filamentary structures and plasma dynamics in large-scale solar phenomena, such as sunspots and magnetic network. A new emerging paradigm is that the mechanisms of the filamentary structuring and large-scale organization are natural consequences of turbulent magnetoconvection on the Sun. We present results of 3D radiative MHD large-eddy simulations (LES) of magnetic structures in the turbulent convective boundary layer of the Sun. The results show how the initial relatively weak and uniformly distributed magnetic field forms the filamentary structures, which under certain conditions gets organized on larger scales, creating stable long-living magnetic structures. We discuss the physics of magnetic self-organization in the turbulent solar plasma, and compare the simulation results with observations.

Keywords. Plasmas, turbulence, Sun: granulation, magnetic fields, sunspots, photosphere

1. Introduction

The interaction of turbulent plasma and distributed magnetic fields is presented by various phenomena on the solar surface. Modern observational instruments provide unique detailed data for analysis of the magnetic field topology and convective flows. However, observational limitations and complicated physics of the observed phenomena make of great interest the development of realistic numerical simulations. In these simulations a numerical model is built from first physical principles and takes into account the realistic equation of state, effects of ionization, chemical composition, radiative transfer, turbulence and magnetic field effects. The pioneering numerical investigations by Stein and Nordlund (1989) reproduced quite accurately physical properties of solar convection.

We investigate processes of self-organization of the solar convection in magnetic field by means of the realistic numerical simulations. Our numerical models reproduce the characteristic filamentary structurization and dynamics of magnetoconvection in the sunspot penumbra at different distances from the umbra, and also reveal a mechanism of spontaneous formation of long-living magnetic structures from initially uniform weak magnetic field in the upper convection. We used a 3D radiative MHD code, developed for simulations of top layers of the solar convective zone and low atmosphere (Jacoutot *et al.* 2008). Various sub-grid scale turbulent models are implemented in the code. Here we use a minimal hyperviscosity model.

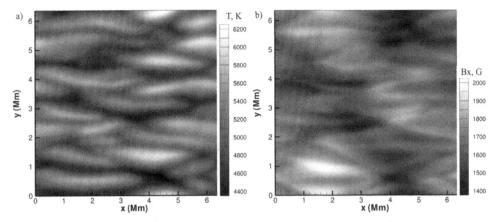

Figure 1. Filamentary structure of magnetoconvection at the solar surface in a highly inclined (along x-axis) strong magnetic field ($B_0 = 1500$ G, $\alpha = 85°$): a) distribution of the temperature surface, T; b) horizontal component of magnetic field, B_x.

2. Magnetic self-organization in a sunspot penumbra

Magnetoconvection in a sunspot penumbra, where convective flows interact with strong, almost horizontal magnetic field, is a source of various observational phenomena. One of them is the Evershed effect, observed as a radial outflows of plasma along magnetic filaments with a speed of 2 – 6 km/s (Evershed 1909). The highly organized behavior of the magnetized plasma flows indicates strong coupling between the dynamics of magnetic field and turbulent convection. The complicated radial distribution of the magnetic field in real sunspots creates difficulties in the numerical penumbra modeling. Therefore, for studying the basic physical properties we simulate separately different penumbra areas, in which the mean magnetic-field strength and inclination are approximately constant. Thus, the various magnetic field strengths, 600 – 2000 G, in our simulation domain, correspond to penumbra areas located at different distances from the sunspot umbra.

Our previous investigation of the influence of magnetic field inclination on solar granulation showed that the granular convective cells become deformed in the direction of the field inclination and form a dynamic filamentary structure of the magnetoconvection (Kitiashvili *et al.* 2009, 2010b). Such stretching of convective granules along the magnetic field lines becomes stronger with the higher field inclination to the surface and with the higher strength of magnetic field. In stronger magnetic field the convective filaments become more narrow, and this causes additional acceleration of the horizontal flows along the field lines. Figure 1 illustrates the surface distributions of temperature (panel a) and the horizontal component of magnetic field (panel b) for the simulations with the field mean strength of 1500 G and the mean inclination angle of 85° towards the surface. The thin elongated granules and the high-speed horizontal flows are located in the magnetized "gaps" with relatively low field strength. The surrounding strong magnetic field controls the direction of the flows and accelerates it by decreasing the space between the magnetic filaments. In addition, the solar convection in inclined magnetic field has properties of magnetoconvection waves traveling in the direction of the field inclination. This results in coherent patches of high-speed flows ($\sim$ 6 km/s), observed as "Evershed clouds". The patches appear quasi-periodically with a characteristic period of 20 – 40 min. The coherent behavior of these flows across several filaments is particularly apparent in the strong field regions (1500 – 2000 G). Such coherent behavior of the filamentary magnetoconvection represents an interesting example of non-linear self-organization of

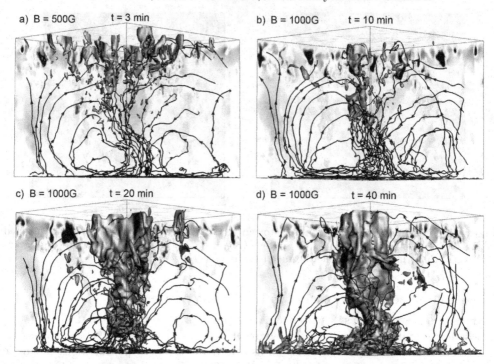

Figure 2. Process of formation of a magnetic structure in the subsurface layers of the Sun. The greyscale isosurface corresponds to 500 G in panel a) and 1000 G in panels b – d). Curves illustrate streamlines of converging subsurface flows.

turbulent magnetized plasma. The partially frozen magnetic field lines are deformed by dragging convective upflows and downflows. Such "sea-serpent" effect can be observed in the form of bipolar magnetic patches, which move in the direction of magnetic field inclination following the convective cell dynamics (Kitiashvili *et al.* 2010a).

3. Effects of weak magnetic fields on the turbulent convection

In regions with relatively weak ($\ll$ 1000 G) magnetic fields on the Sun convection plays the lead role. The magnetic field gets organized by turbulent convective processes. For example, a weak initially uniform vertical magnetic field ($B_0 = 1 - 10$ G) becomes concentrated of the surface and near-surface layer in the intergranular lanes in the form of compact magnetic flux tubes (observed as bright points), in which the field strength reaches $\sim$ 1000 G and which are accompanied downflows. In deeper subsurface layers this field forms diffuse magnetic structures. Thus, the MHD simulations show that the ubiquitous magnetic flux tubes observed in the "quiet"-Sun regions are highly dynamic and can be maintained in the near-surface layer by convective downdrafts.

For a stronger ($B_0 = 100$ G) initial vertical field, distributed uniformly in the simulation domain ($6.4^2 \times 5.5$ Mm), our results reveal a process of magnetic field accumulation into a large-scale (pore-like) magnetic structure. This transient process starts in the vicinity of a large vortex tube ($\sim$ 0.5 Mm in diameter) after the uniform field initiation, and leads to creation of an unstable region. This region is characterized by a low gas pressure, strong downflows ($\sim$ 7 km/s) and supersonic horizontal velocities. The strong converging downflows and a sharp decrease of density cause further amplification of the magnetic field concentration in the region.

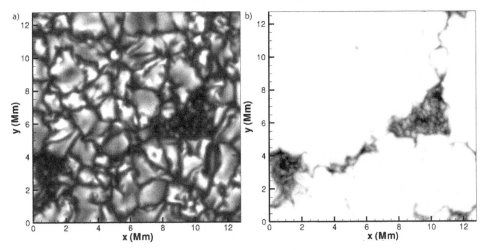

Figure 3. Formation of two magnetic structures in the simulations with a large box domain. The structures are shown at the surface in the blue continuum intensity (panel a) and the vertical magnetic field distribution (panel b), where the range of the grey scale is from 0 (white) to 1500 G (black).

Figure 2 illustrates different stages of formation of the self-organized magnetic structure. The stream lines (with arrows) show converging convective flows that support the process of accumulation of magnetic field. During the first few minutes, the initial vertical uniform 100 G vertical magnetic field is collapsed into the intergranular lanes. Locally the field strength increases up to 500 G (shown as isosurfaces in Fig. 2a). Due to the converging flows and the local pressure cavity the individual magnetic patches move into the region of the structure formation (Fig. 2b). Finally, the complicated diffusive magnetic structure becomes compact and stable (Fig. 2c-d). The local accumulation of the magnetic field increases the magnetic field strength up to $\sim 1400 - 1800$ G at the surface, and up to ~ 6 kG below the surface (Kitiashvili *et al.* 2010c). It is interesting that this compact magnetic pore-like structure has long life-time, at least, for 9 hours computed in our simulations, but also it is highly dynamic and can be moved by occasional strong convective flows.

The process of the structure formation depends on the state of convection at the moment of magnetic field initialization. We repeated the simulations for the same domain ($6.4^2 \times 5.5$ Mm), but for different moments of the magnetic field initialization and found that the self-organized structures can be formed in different places and have different properties, such as size, magnetic field strength and distribution with depth, but qualitatively the process is the same in all cases. In some cases, two separate magnetic structures with very different properties are formed. However, these structures are usually close to each other and merged during the evolution. In the case of a larger simulation domain ($12.8^2 \times 5$ Mm) and a similar grid resolution (25 km in horizontal and 22 km in the vertical directions), the uniformly distributed magnetic field became initially concentrated in five magnetic structures. During the evolution some of these structures merged, and finally we obtained two stable pore-like structures. Figure 3 shows an intensity image of these pores in the blue continuum (panel a) and also the corresponding vertical magnetic field map (panel b). An interesting feature of the intensity map is the internal substructure consisting of umbra-dot-like dots features inside the pores. The umbral dots represent small convective granules suppressed by the strong magnetic field. The distribution of the vertical magnetic field (Fig. 3b) shows that the pore-like structures are magnetically connected. Inside the structures, the magnetic field is mostly concentrated between the bright umbral dots.

4. Conclusion

Convection on the solar surface is characterized by strong turbulent processes. The influence of magnetic field makes the plasma flow behavior more organized. The almost horizontal magnetic field of sunspot penumbrae creates a subsurface shear layer, in which the plasma is accelerated in the direction of magnetic field inclination, leading to the Evershed effect. A weak vertical magnetic field (1 – 10 G) imposed in the convective layer results in formation of magnetic concentrations (bright points) with downdrafts at the solar surface, connected to diffuse subsurface magnetic structures. A stronger 100 G field leads to a process of spontaneous formation of stable magnetic pore-like structures, with the magnetic field strength ~ 1.5 kG at the surface and ~ 6 kG below the surface. These examples of magnetic self-organization of turbulent convection on the Sun illustrate importance of the realistic numerical simulations, based on first principles, for getting insight into complicated non-linear phenomena of plasma astrophysics.

Acknowledgement

We thank the International Space Science Institute (Bern) for the opportunity to discuss these results at the international team meeting on solar magnetism.

References

Evershed, J. 1909, *MNRAS*, 69, 454
Jacoutot, L., Kosovichev, A. G., Wray, A., & Mansour, N. N. 2008, *ApJ*, 682, 1386.
Kitiashvili, I. N., Kosovichev, A. G., Wray, A. A., & Mansour, N. N. 2009, *ApJ*, 700, L178.
Kitiashvili, I. N., Bellot Rubio, L. R., Kosovichev, A. G., Mansour, N. N., Sainz Dalda, A., & Wray, A. A. 2010a, *ApJ*, 716, L181.
Kitiashvili, I. N., Kosovichev, A. G., Wray, A. A., & Mansour, N. N. 2010b, *Proc. of 3rd Hinode Science Meeting*, 1-4 December, 2009, in press.
Kitiashvili, I. N., Kosovichev, A. G., Wray, A. A., & Mansour, N. N. 2010c, *ApJ*, 719, 307.
Stein, R. F. & Nordlund, Å. 1989, *ApJ*, 342, L95

Advances in Plasma Astrophysics
Proceedings IAU Symposium No. 274, 2010
A. Bonanno, E. de Gouveia Dal Pino & A. G. Kosovichev, eds.
© International Astronomical Union 2011
doi:10.1017/S1743921311006715

Helicity transport in a simulated coronal mass ejection

B. Kliem[1,2,3], S. Rust[1] and N. Seehafer[1]

[1]Institute of Physics and Astronomy, University of Potsdam, Karl-Liebknecht-Str. 24-25, 14476 Potsdam, Germany

[2]Mullard Space Science Laboratory, University College London, Holmbury St. Mary, Dorking, Surrey, RH5 6NT, UK

[3]Space Science Division, Naval Research Laboratory, Washington, DC 20375, USA

Abstract. It has been suggested that coronal mass ejections (CMEs) remove the magnetic helicity of their coronal source region from the Sun. Such removal is often regarded to be necessary due to the hemispheric sign preference of the helicity, which inhibits a simple annihilation by reconnection between volumes of opposite chirality. Here we monitor the relative magnetic helicity contained in the coronal volume of a simulated flux rope CME, as well as the upward flux of relative helicity through horizontal planes in the simulation box. The unstable and erupting flux rope carries away only a minor part of the initial relative helicity; the major part remains in the volume. This is a consequence of the requirement that the current through an expanding loop must decrease if the magnetic energy of the configuration is to decrease as the loop rises, to provide the kinetic energy of the CME.

Keywords. magnetic fields, (magnetohydrodynamics:) MHD, Sun: coronal mass ejections

1. Introduction

The helicity of the solar magnetic field obeys a hemispheric preference which is invariant with respect to the sign reversal of the global magnetic field with the activity cycle (Hale 1925; Seehafer 1990). This has led to the suggestion that coronal mass ejections (CMEs) must remove magnetic helicity from the Sun to prevent indefinite accumulation of the helicity in each hemisphere (Rust 1994; Low 1996). It was also found that the accumulation and removal of helicity can control the rate of mean-field dynamo action, so that the evolution of the activity cycle may be related to the flow of helicity through the Sun (Blackman & Field 2001; Brandenburg & Subramanian 2005). Careful studies of the long-term helicity budget of two solar active regions (Démoulin *et al.* 2002; Green *et al.* 2002) appear to confirm the conjecture of efficient helicity shedding by CMEs. However, both investigations used the linear force-free field approximation to estimate the helicity in the active region atmosphere. The accuracy of this estimate is not known. Similarly, the estimates of the helicity in interplanetary CMEs are still subject to considerable uncertainty (Demoulin 2007). In this paper, numerical simulation is used to quantify the transport of magnetic helicity from the source volume of CMEs.

2. Relative magnetic helicity in a simulated CME

We monitor the relative magnetic helicity $H_r = \int (\mathbf{A} + \mathbf{A}_p) \cdot (\mathbf{B} - \mathbf{B}_p)\,dV$ (Berger & Field 1984; Finn & Antonsen 1985) in an MHD simulation of a flux rope CME. The force-free equlibrium of a toroidal current channel partially submerged below the photosphere (Titov & Démoulin 1999) is chosen as initial condition. The twist is set to a supercritical

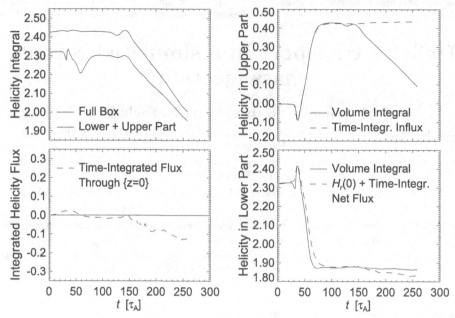

Figure 1. Relative magnetic helicity $H_r(t)$ in the simulation box and in the lower and upper sub-volumes, and time-integrated helicity fluxes trough the bottom and diagnostics planes (bottom right) and through the diagnostics plane only (top right) in a CME simulation. The erupting flux begins to cross the diagnostics plane and the upper boundary of the box at $t \approx 30\tau_A$ and $t \approx 120\tau_A$, respectively. A helicity of about 0.4 (in normalized units) is transported from the lower into the upper sub-volume and then out of the box. This is about 1/5 of the initial helicity $H_r(0) \approx 2.4$. (The minor outflux of helicity through the bottom boundary for $t \gtrsim 150\tau_A$ results from downward propagating perturbations triggered by the ejected flux at large heights and may be a numerical artefact. The simulation was terminated by numerical instability in one of the upper corners of the box, where an open and two closed boundaries meet and numerical stability is more difficult to maintain than in the interior of the box.)

value, so that the kink-unstable magnetic flux rope formed by the current channel spontaneously starts to rise. The ideal MHD equations are integrated, neglecting pressure, with numerical diffusion enabling magnetic reconnection. The simulation is similar to the one in Török & Kliem (2005), except for an open upper boundary and a larger box size (of 40^3 unit lengths, set to be the initial flux rope apex height h_0). Approximations of the instantaneous helicity in the simulation box and of the helicity flux through the bottom boundary are obtained from Equations (1)–(7) in DeVore (2000). These are exact only if the field strength has fallen to zero at the top and lateral boundaries of the considered volume. We have checked that the computed approximate helicity of the initial equilibrium approaches a limit for increasing box size and that it deviates by less then 2% from the apparent limit value for the chosen size.

The approximation degrades as the flux rope passes through the top boundary, although the field strength then still decreases by more than two orders of magnitude from the bottom to the top of the box. To estimate the magnitude of the error, we divide the box by a horizontal "diagnostics plane" at height $z = 6h_0$, such that the field drops with increasing height by a similar factor in each sub-volume. At $t = 0$ the factor is of order 10^2, justifying the use of the approximation in each sub-volume. The top left panel in Figure 1 shows that the summed helicities of the sub-volumes differ by $\lesssim 5\%$ from the helicity of the box as a whole, with the error obviously originating from the lower sub-volume, since $H_r(t = 0) = 0$ is correctly found in the upper sub-volume (top right

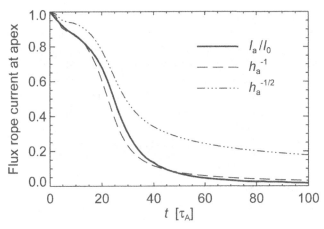

Figure 2. Total current through the apex of the flux rope, normalized by initial current, $I_a(t)/I_0$. For comparison, the inverse of the apex height $h_a(t)$, a proxy for the flux rope length, and $h_a(t)^{-1/2}$ are shown (see text).

panel). When the upper part of the unstable flux rope begins to propagate through the diagnostics plane, the approximation is degraded in the lower sub-volume, but is still of similar quality as before in the upper sub-volume and in the box as a whole. The resulting error is given by the additional difference between the two values for the whole volume in the relevant time interval, $t \sim (30\text{--}80)\tau_A$, where $\tau_A = h_0/V_A$ and V_A is the initial Alfvén velocity in the flux rope. The error reaches a peak value $\lesssim 5\%$ at $t = 60\tau_A$, when the flux rope apex has risen to $z = 17h_0$, and decreases considerably thereafter. The helicity calculation for $t \gtrsim 120\tau_A$, when the CME leaves the box, should even be more precise, since the field strengths of the flux rope at the upper boundary are then smaller, by an order of magnitude, than the field strengths at the passage of the diagnostics plane.

Figure 1 shows that a relative helicity of ≈ 0.4 (in the normalized units of the simulation, which prescribe a field strength of unity at the apex of the initial flux rope, $|\mathbf{B}_0(0,0,1)| = 1$) is transported from the lower into the upper subvolume and then out of the box when the upper part of the erupting flux rope crosses the respective top boundary. (A helicity of ≈ 0.1 is not yet ejected when the simulation is terminated due to numerical instability, but it is clear from the plot that this will be ejected as well.) The top right plot shows a very small rate of helicity flux through the diagnostics plane after the top part of the flux rope has propagated into the upper sub-volume ($t \gtrsim 80\tau_A$). This suggests that the further upward stretching of the flux rope legs and the addition of flux to the rope by reconnection in the vertical current sheet under the rope do not contribute strongly to the ejected helicity, at least not at the scales covered by the simulation, which extend to flux rope apex heights of ≈ 25 times the footpoint distance (several solar radii when scaled to a large solar active region). The upward reconnection outflow velocity, a proxy of the reconnection rate in our ideal MHD simulation, has decreased to one quarter of its peak value, $u_z(t = 52) = 0.8V_A$, by the end of the simulation. The soft X-ray flux of long-duration solar ejective events, an indicator of the energy release by reconnection, decreases strongly from its peak value on such scales. Therefore, although this simulation is terminated by numerical instability, it likely models most of the helicity ejection in a CME. However, this amounts to only about $1/5$ of the initial relative helicity.

The relatively small efficiency of helicity ejection can be related to the evolution of the current distribution in the volume. These currents carry the relative helicity (which vanishes in a potential field). The initial coronal equilibrium consists of a section of a toroidal current ring. The energy of a current ring is given by $W = LI^2/2 \approx I^2 R \left[\ln(8R/a) - 7/4\right]$, where I, R, a, and L are the current, major and minor radius, and the inductance of the ring, respectively. Since magnetic energy must be released in order to accelerate the ejecta and the term in brackets does not vary strongly, the current in the rising flux loop of the CME must decrease faster than $R^{-1/2}$. In the approximation of ideal MHD the current in the loop decreases roughly as R^{-1}, because the number of field line turns in the loop is conserved. The simulation shows such a fast decrease (Figure 2), which results in most of the current staying low in the box. Consequently, only a minor part of the initial relative helicity leaves the system with the ejected flux.

3. Conclusions and Discussion

The simulated CME ejects only a minor part of the initial relative magnetic helicity from its source volume. Although this result requires substantiation through the study of its parametric dependence and of other equilibria, the necessary decrease of the current through an expanding unstable flux loop leads us to expect that it holds generally. The number of CMEs per active region varies within very wide limits. Between 30 and 65 CMEs have been estimated to occur throughout the lifetime of the two very CME-prolific active regions studied in Démoulin *et al.* (2002) and Green et al. (2002). On the other hand, the majority of active regions produces no CME at all, or only one CME in their lifetime. Hence, the shedding of helicity by CMEs may be of lower importance than originally conjectured.

It appears natural to assume that, perhaps generally, much of an active region's helicity submerges when the region disperses and the major part of its flux submerges below the photosphere. The helicity may then follow the slow journey of magnetic flux in the course of the solar cycle. Annihilation of helicity in the interior of the Sun, following the transport of the helicity-carrying flux to the equatorial plane by the meridional flow, is one possibility to prevent the helicity in each hemisphere from accumulating indefinitely. Another possibility, opposite to a common conjecture, is that the helicity in the solar interior, like magnetic energy, undergoes a normal (or direct) turbulent cascade towards small spatial scales, where it is dissipated. Also, the cascade directions may be different for small-scale and large-scale fields (Alexakis *et al.* 2006), with the helicity of active-region magnetic fields, considered to be small-scale fields, subject to a direct cascade.

References

Alexakis, A., Mininni, P. D., & Pouquet, A. 2006, *ApJ* 640, 335
Berger, M. A. & Field, G. B. 1984, *J. Fluid Mechanics* 147, 133
Blackman, E. G. & Field, G. B. 2001, *Phys. Plasmas* 8, 2407
Brandenburg, A. & Subramanian, K. 2005, *Physics Reports* 417, 1
Démoulin, P. 2007, *Adv. Space Res.* 39, 1674
Démoulin, P., Mandrini, C. H., van Driel-Gesztelyi, L., *et al.* 2002, *A&A* 382, 650
DeVore, C. R. 2000, *ApJ* 539, 944
Finn, J. M. & Antonsen, T. M. 1985, *Comm. Plasma Phys. Contr. Fusion* 9, 111
Green, L. M., López Fuentes, M. C., Mandrini, C. H., *et al.* 2002, *Solar Phys.* 208, 43
Hale, G. E. 1925, *Publ. Astron. Soc. Pacific* 37, 268
Low, B.-C. 1996, *Solar Phys.* 167, 217
Rust, D. M. 1994, *Geophys. Res. Lett.* 21, 241
Seehafer, N. 1990, *Solar Phys.* 125, 219
Titov, V. S. & Démoulin, P. 1999, *A&A* 351, 707
Török, T. & Kliem, B. 2005, *ApJ* 630, L97

Advances in Plasma Astrophysics
Proceedings IAU Symposium No. 274, 2010
A. Bonanno, E. de Gouveia Dal Pino & A. G. Kosovichev, eds.
© International Astronomical Union 2011
doi:10.1017/S1743921311006727

Coupled Alfvén and kink oscillations in an inhomogeneous corona

David J. Pascoe[1], Andrew N. Wright[1] and Ineke De Moortel[1]

[1]School of Mathematics and Statistics, University of St Andrews,
St Andrews, KY16 9SS, United Kingdom
email: dpascoe@mcs.st-andrews.ac.uk

Abstract. We perform 3D numerical simulations of footpoint-driven transverse waves propagating in a low β plasma. The presence of inhomogeneities in the density profile leads to the coupling of the driven kink mode to Alfvén modes by resonant absorption. The decay of the propagating kink wave as energy is transferred to the local Alfvén mode is in good agreement with a modified interpretation of the analytical expression derived for standing kink modes. This coupling may account for the damping of transverse velocity perturbation waves which have recently been observed to be ubiquitous in the solar corona.

Keywords. MHD, Sun: atmosphere, Sun: corona, Sun: magnetic fields, Sun: oscillations, waves

1. Introduction

Recent observations of the solar corona by Tomczyk *et al.* (2007) and Tomczyk & McIntosh (2009) with the ground-based coronagraph *CoMP* show transverse oscillations propagating outwards everywhere. The dominance of outward-propagating wave power over inward-propagating wave power along closed loops suggests significant attenuation *in situ*. The propagating waves were initially interpreted as Alfvén waves but this was disputed by Van Doorsselaere *et al.* (2008a,b) who interpret them as kink waves. These different interpretations have consequences for the inferred coronal magnetic field strength and the energy budget calculations for the coronal heating problem.

Ruderman & Roberts (2002) show how resonant absorption can damp coronal loop oscillations. Their work was motivated by the global standing kink modes as seen by *TRACE* and is similar to the Edwin & Roberts (1983) configuration except for the inclusion of an inhomogeneous layer of thickness *l*. Resonant absorption takes place in this inhomogeneous layer, transferring energy from the global kink mode to the Alfvén mode.

Our studies (Pascoe *et al.* 2010) show that the transverse waves we launch from the boundary couple efficiently to Alfvén waves when the medium is non-uniform. This is seen as a decay of the driven wave fields. Once the energy is in the form of Alfvén waves, it is well known that these fields will phase-mix (Heyvaerts & Priest 1983; Mann *et al.* 1995) and lead to the development of small transverse scales.

2. Model

Considering the rapid damping of the global kink standing mode in a zero β cylindrical flux tube with an inhomogeneous layer, Ruderman & Roberts (2002) derived the relationship

$$\frac{\tau}{P} = C\frac{a}{l}\frac{\rho_0 + \rho_e}{\rho_0 - \rho_e} \tag{2.1}$$

where τ is the damping time, P is the period of oscillation, a is the loop radius, l is the inhomogeneous layer thickness, and ρ_0 and ρ_e are the internal and external mass densities, respectively. The constant C depends upon the density profile in the inhomogeneous layer, e.g. $C = 2/\pi$ for a sinusoidal profile, whereas for a linear density profile $C = (2/\pi)^2$ (see e.g., Hollweg & Yang 1988; Goossens *et al.* 1992; Roberts 2008).

In our model we consider a straight, uniform magnetic field in the z direction. We use a zero plasma β approximation. Our density profile describes a cylindrical tube aligned with the z axis and defines three regions; the core region with an internal density ρ_0, the external or environment region with density ρ_e, and the inhomogeneous shell region in between, where the density varies linearly from ρ_0 to ρ_e.

The driving condition is applied to the lower z boundary to simulate excitation by footpoint motions (e.g., De Groof *et al.* 2002) and prescribes the x and y components of velocity to have a time dependence based on a single period displacement of the loop axis in the x direction. The spatial dependence of the driver is based on a 2D dipole. In the core region the velocity is constant $\mathbf{u} = (u_0, 0, 0)$ and only in the x-direction, where $u_0 = 0.002$ is chosen to be small to avoid non-linear effects. In the surrounding environment the driver has a 2D dipole form. This flow corresponds to 2D incompressible dipole flow around a circular tube that moves with velocity $(u_0, 0, 0)$. In cylindrical coordinates, this would be described as the $m = 1$ mode, in which u_r is continuous and u_ϕ discontinuous at the tube boundary. We avoid numerical problems arising from such a discontinuity by smoothly changing in the inhomogeneous layer from the core velocity profile to that in the environment.

The simulations are performed using the MHD code LARE3D (Arber *et al.* 2001). The numerical domain is much larger in the z direction than in x or y, but the resolution is higher in the x and y directions in order to resolve the activity in the inhomogeneous layer, particularly when phase-mixing occurs.

3. Results

In the case of a uniform medium the Alfvén speed is constant everywhere. The perturbation applied at the lower boundary propagates upwards uniformly with no signs of any significant attenuation or dispersion. Although we do not drive the boundary for many cycles, and hence do not have a quasi-monochromatic source, the driver does have a well-defined nodal structure and is dominated by a time scale of $P \approx \frac{2}{3}P_0$.

In the case of an inhomogeneous layer ($\rho_0/\rho_e > 1$) the Alfvén speed now varies across the driven region. The Alfvén speed varies continuously in the inhomogeneous layer and resonant absorption occurs where the phase speed of the kink mode matches the Alfvén speed.

This coupling of the wavetrain to a local Alfvén mode causes a decrease in wave energy in the core (and the environment) and hence appears to damp the tube oscillation. Also, since the Alfvén mode is in an inhomogeneous layer, it is subject to phase mixing. The corresponding characteristic spatial scale becomes smaller as a function of time. The time dependence of the phase-mixing length scale agrees well with the analytical relationship of Mann *et al.* (1995). These effects can be seen in Fig. 1, which shows v_x at the axis of the cylinder at time $t = 1.5P_0$. The density profile, outlined by the vertical lines, is defined by $\rho_0/\rho_e = 2$ and $l/a = 0.5$.

In order to quantify the behaviour in our model we calculate the wave energy density, as in Terradas *et al.* (2006, 2008). Mode coupling in the inhomogeneous shell causes the wave energy to become localised there with a corresponding decrease in energy in the core and environment regions, which was used to calculate a decay time τ.

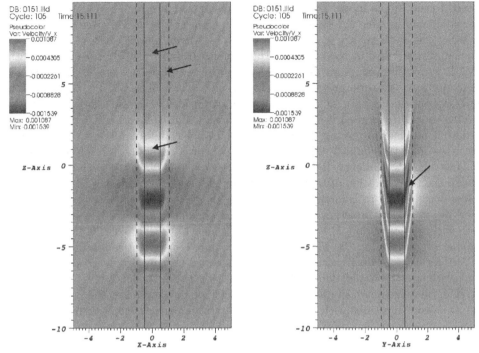

Figure 1. v_x at $y = 0$ (left) and $x = 0$ (right) and at time $t = 1.5P_0$ for $\rho_0/\rho_e = 2$ and $l/a = 0.5$. Energy is transferred from the kink mode in the core region to the Alfvén mode in the inhomogeneous shell region by resonant absorption.

Note that we have a decaying propagating wavepacket, not a decaying standing mode. However, Hood *et al.* (2005) show that standing normal mode calculations can be a good indicator of wavepacket behaviour, when the spatial and temporal scales are similar. With this in mind we compare our wavepacket decay rates with the Ruderman & Roberts (2002) formula for standing modes in eq. [2.1].

For our linear density profile in the inhomogeneous layer, the analysis predicts $C = (2/\pi)^2$ for standing modes. We find for our simulations $C = 0.9$ for all values of ρ and l considered. Choosing this value of C, the Ruderman & Roberts (2002) result in eq. [2.1] can be regarded as providing an empirical fit to our simulation results. That the value of C differs from that in the Ruderman & Roberts (2002) calculation can be attributed to the fact that our model is not accurately described as a thin flux tube with a thin boundary layer.

4. Implications

Our non-monochromatic driver on the lower boundary produced an upward propagating wavetrain. In the case of an inhomogeneous medium the wavetrain is subject to mode coupling through resonant absorption. Here we will discuss these results in the context of the observations of Tomczyk *et al.* (2007) and the interpretation of Van Doorsselaere *et al.* (2008a,b) as kink waves. In this case our driver corresponds to some general photospheric motion and our wavetrain corresponds to the Doppler shift in coronal emission observed by Tomczyk *et al.* (2007).

In the most general case of a continuously non-uniform corona, the behaviour demonstrated by our numerical simulations is that of a quasi-mode composed of the kink mode coupled to the Alfvén mode. Due to resonant absorption and the introduction of a characteristic damping time (τ) the time-dependent nature of the mode must be considered. For $t > \tau$ the wave energy is concentrated in the Alfvén mode, whereas for $t < \tau$ the behaviour is described by the kink mode of Edwin & Roberts (1983). The coupled nature of MHD waves in a non-uniform plasma is also discussed by Goossens et al. (2009).

According to this interpretation, the transverse waves observed by Tomczyk et al. (2007) are in the regime $t \approx \tau$ and so are an intrinsically coupled mode. This will be the case unless the corona is uniform or contains a flux tube whose density profile is discontinuous. The properties of the observed Doppler shifts will resemble a damped kink mode since the coupled Alfvén mode component is generally unresolved by modern solar instruments. However, its presence is necessary to explain the rapid damping and leads to the concentration of wave energy to smaller scales and so may be an important process contributing to coronal heating.

For the typical parameters of our numerical simulations, we estimate that for the observed waves of period $P \approx 300$ seconds (Tomczyk & McIntosh 2009) we expect a damping length of $L_d \approx 750$ Mm. This would require a footpoint separation of $2L_d/\pi \approx 500$ Mm for closed loop structures to give an outward propagating wave signature. This simple estimate of damping length scales is consistent with Tomczyk & McIntosh (2009), who show that the outward directed wave velocity power dominates the inward directed power for loop structures with a footpoint separation greater than 300 Mm.

References

Arber, T. D., Longbottom, A. W., Gerrard, C. L., & Milne, A. M. 2001, JCP, 171, 151

De Groof, A., Paes, K., & Goossens, M. 2002, A&A, 386, 681

Edwin, P. M. & Roberts, B. 1983, Solar Phys., 88, 179

Goossens, M., Hollweg, J. V., & Sakurai, T. 1992, Solar Phys., 138, 233

Goossens, M., Terradas, J., Andries, J., Arregui, I., & Ballester, J. L. 2009, A&A, 503, 213

Heyvaerts, J. & Priest, E. R. 1983, ApJ, 117, 220

Hollweg, J. V. & Yang, G. 1988, JGR, 93, 5423

Hood, A. W., Brooks, S. J., & Wright, A. N. 2005, Proc. R. Soc. Lond. A, 461, 237

Mann, I. R., Wright A. N., & Cally P. S. 1995, JGR, 100, 19441

Pascoe, D. J., Wright, A. N., & De Moortel, I. 2010, ApJ, 711, 990

Roberts, B. 2008, in: R. Erdélyi & C. A. Mendoza-Briceño (eds.), Waves & Oscillations in the Solar Atmosphere: Heating and Magneto-Seismology, Proc. IAU Symposium No. 247 (Cambridge: CUP), p. 3

Ruderman, M. S. & Roberts, B. 2002, ApJ, 577, 475

Terradas, J., Oliver, R., & Ballester, J. L. 2006, ApJ, 642, 533

Terradas, J., Arregui, I., Oliver, R., & Ballester, J. L. 2008, ApJ, 679, 1611

Tomczyk, S. & McIntosh, S. W. 2009, ApJ, 697, 1384

Tomczyk, S., McIntosh, S. W., Keil, S. L., Judge, P. G., Schad, T., Seeley, D. H., & Edmondson, J. 2007, Science, 317, 1192

Van Doorsselaere, T., Nakariakov, V. M., & Verwichte, E. 2008a, ApJ, 676, L73

Van Doorsselaere, T., Brady, C. S., Verwichte, E., & Nakariakov, V. M. 2008b, A&A, 491, L9

Advances in Plasma Astrophysics
Proceedings IAU Symposium No. 274, 2010   © International Astronomical Union 2011
A. Bonanno, E. de Gouveia Dal Pino & A.G. Kosovichev, eds.   doi:10.1017/S1743921311006739

Weak turbulence theory of dispersive waves in the solar corona

Felix Spanier[1] and Rami Vainio[2]

[1] Lehrstuhl für Astronomie, Universität Würzburg, Am Hubland, 97074 Würzburg, Germany
email: `fspanier@astro.uni-wuerzburg.de`

[2] University of Helsinki, Department of Physics, P.O.Box 64, 00014 University of Helsinki, Finland
email: `rami.vainio@helsinki.fi`

Abstract. The interaction of plasma waves plays a crucial role in the dynamics of weakly turbulent plasmas. So far the interaction of non-dispersive waves has been studied. In this paper the theory is extended to dispersive waves. It is well known that dispersive waves may be found in the solar corona, where they contribute to the heating of the corona. Here the possible interactions in the solar corona are described and the interaction rates are derived in the framework of Hall MHD.

Keywords. Sun: corona, plasmas, waves

---

1. Introduction

The sun exhibits a wide range of plasma phenomena from its core to the solar wind. On all scales thermal plasma, turbulent waves and non-thermal particle distributions can be found. Due to their variety not all phenomena can be explained by single model. Here we focus on coronal holes, where the turbulence level is rather low and the weak turbulence model, i.e., the interaction of plasma waves, is an adequate description. Weak turbulence theory allows the calculation of energy transport between different wave modes on different scales. From observation of nonthermal velocity fluctuations of ions (e.g., Laitinen *et al.*, 2003) one can show that magnetic field perturbation amplitudes are well below the background magnetic field, which is necessary prerequisite for the application of weak turbulence theory. We will use this theory to describe the problem of turbulence evolution of coronal holes.

One class of models for the heating of coronal holes is based on the dissipation of low-frequency ($\omega < \Omega_i$) electromagnetic plasma waves. Such waves could be generated, for example, by small reconnection events in the chromospheric magnetic network below coronal holes (Axford & McKenzie, 1992). However, the dissipation and spectral evolution of such waves is still not properly understood, as the results of wave-wave interactions and dispersion are typically neglected. In this paper, we attempt to describe the non-linear interaction of dispersive plasma waves in coronal-hole-type plasmas.

2. Derivation of the wave-wave interaction

Following the approach from Sagdeev & Galeev (1969) plasma waves are considered as quasi particles. Therefore in the interaction of three waves momentum and energy of the waves have to be conserved

$$\vec{k}_{\mathrm{M}} = \vec{k}_{\mathrm{P}} + \vec{k}_{\mathrm{Q}} \tag{2.1}$$

$$\omega_{\mathrm{M}} = \omega_{\mathrm{P}} + \omega_{\mathrm{Q}} \tag{2.2}$$

This system has been studied for Alfvén and sound waves by Chin & Wentzel (1972) and Vainio & Spanier (2005). These studies limited to frequencies well below the ion-gyrofrequency. For dispersive waves the three-wave interactions have been studied by Spanier & Vainio (2009), where L- and R-modes as well as sound modes are considered. An outcome of this study is also that not all interactions allowed by the resonance condition are actually possible: Additionally in all interactions angular momentum has to be conserved, where L waves are assigned angular momentum $\hbar$ and R waves $-\hbar$. One finds in the dispersive regime interactions which are directly linked to non-dispersive waves and also purely dispersive interactions which are found mainly in the ion-gyrofrequency regime.

To derive the interaction rate of dispersive wave modes one may either use the full kinetic theory provided by Melrose & Sy (1972) or use the ansatz of Chin & Wentzel in the Hall MHD regime. Since the full kinetic theory is fairly complex (see Yoon & Fang, 2008) we resorted to the Hall MHD ansatz. Using the set of MHD equations

$$\partial_t \rho + \nabla \cdot (\rho \vec{v}) = 0 \tag{2.3}$$

$$\rho \partial_t \vec{v} + \rho(\vec{v} \cdot \nabla)\vec{v} = -\nabla p - \frac{1}{4\pi}\vec{B} \times (\nabla \times \vec{B}) \tag{2.4}$$

$$\partial_t \vec{B} = \nabla \times (\vec{v} \times \vec{B}) - \frac{m_i c}{4\pi e}\nabla \times (\frac{1}{\rho}(\nabla \times \vec{B}) \times \vec{B}) \tag{2.5}$$

$$\nabla p = c_s^2 \nabla \rho \tag{2.6}$$

one employs perturbation theory to derive the interaction rate. Using the notation of Chin & Wentzel one finds

$$\partial_t \vec{v} = -\frac{i}{2\omega}\vec{S}(\vec{v}_1, \vec{v}_2) \tag{2.7}$$

where all second-order interactions of waves are included. In the course of this paper the interaction rate is usually presented as $\tilde{S} = |\vec{S}|/(|\vec{v}_1||\vec{v}_2|)$, which is normalized such that the amplitude of the incident waves is assumed to be 1. When changing to occupation numbers in the framework of quasiparticles, these numbers are defined as

$$N = \frac{\mathfrak{E}}{\hbar\omega} \tag{2.8}$$

where $\mathfrak{E}$ is the energy of the given wavemode. For occupation numbers $\vec{S}$ has to be transformed into the interaction rate $|V|^2$ (see Eq. I-26]sg). In order to do so, one define $N(k) = |c(k)|^2$, which yields

$$\partial_t c_0 = V c_1 c_2 \Rightarrow \partial_t N_0 \propto |V|^2 N_1 N_2 \tag{2.9}$$

3. Results

The resonance conditions of the dispersive waves corresponding to the non-dispersive three-wave interactions, $A^+ \leftrightarrow S^+ + A^-$ (Fig. 1), show very different behavior depending on the polarization. For the right-hand polarized waves (Fast-wave – Whistler mode), the interaction is qualitatively similar to the non-dispersive case: The frequency difference of the R-mode waves is small and increasing with increasing R-wave frequency. As a result of damping of the ion-sound mode (I-mode, which is used as dispersive equivalent to sound mode S), the standard inverse cascade of R-waves is produced as in the non-dispersive case. However, the left-hand polarized waves (Alfvén – ion-cyclotron mode) behavior is very different: as the frequency of the mother wave approaches the resonance

Low β-Plasma

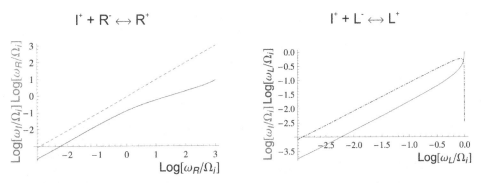

Figure 1. Resonance condition for the standard low-β interactions which resemble the classical interaction $A^+ \leftrightarrow S^+ + A^-$. The green (dashed) curved represents the frequency of R-mode, blue (dashed-dotted) curves the L-mode and the red (solid) curve the I-mode respectively.

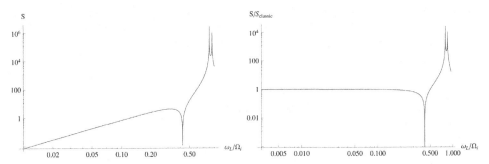

Figure 2. Interaction rate c for the interaction $L^+ \leftrightarrow I^+ + L^-$ and its comparison to the classical rate

at $\omega \to \Omega_i$, the daughter-wave frequency becomes very small. Thus, ion-cyclotron waves are able to decay directly into low-frequency Alfvén waves and at the same time emit high-frequency sound waves that can be easily damped through ion-Landau damping. Most of the energy of the ion-cyclotron waves is, thus, dissipated in the process.

While looking at the interaction rates ($\tilde{S}$), one notes that the R-mode interactions are again pretty similar to the non-dispersive case. The L-mode, on the other hand, experiences a couple of orders-of-magnitude increase in $\tilde{S}$ over the non-dispersive result in the frequency range $\omega \gtrsim 0.6\Omega_i$, where the dispersion relation (and the three-wave resonance condition) is modified. This offers a dissipation channel for waves emitted through kinetic instabilities in this frequency range. Below the high-frequency range (around $\omega \sim 0.5\Omega_i$), there is a short range of frequencies, where the interaction rate is small. However, this range of frequencies will be modified if the resonance takes account of the ^{4}He resonance.

Thus, the inverse cascading of the R mode closely resembles the non-dispersive case, while the L mode interactions are strongly modified. Waves generated in the dispersive range, e.g., through kinetic instabilities involving electron beams, will be cascaded down to lower frequencies through very different channels depending on their polarization.

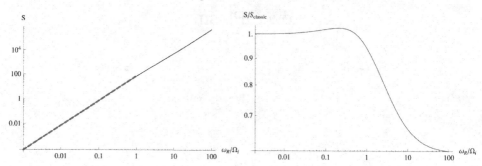

Figure 3. Interaction rate c for the interaction $R^+ \leftrightarrow I^+ + R^-$ and its comparison to the classical rate

4. Discussion and Conclusions

In this paper an extension of three-wave interactions to the dispersive limit is discussed. The possible interactions which are derived from the dispersion relations of L- and R-waves differ with the wave polarization. The present study shows the expected difference of L- and R-modes around the gyrofrequency. While the interaction of R-waves closely resembles the non-dispersive, the L-wave interaction show strong dispersion effects close to the ion-gyrofrequency. This can also be seen in the interaction rate which is derived from Hall-MHD. This modification of L-wave interactions has an influence on the heating in coronal holes.

Wave heating models based on the cyclotron damping mechanism have problems in explaining the proton temperatures see e.g.]mvh, as cyclotron resonance is realized first for minor ions and alphas. The non-linear mechanism relying on the Landau damping of the ions sound waves, however, has the attractive feature that all ions with thermal speeds comparable to the ion sound speeds can Landau damp the waves. As electrons have much higher thermal speeds, their Landau damping rates are smaller than those of the ions.

The presented model is not complete in the sense that only interactions are discussed which resemble classic (non-dispersive) interactions. A full model of non-linear interactions in the solar corona requires also dispersive-only interactions. This model combined with appropriate damping rates would then allow for an accurate prediction of turbulence power spectra.

Due to the obvious difference in the decay of L- and R-modes, which in turn leads to a faster damping of L-modes, one may expect to see changes in the magnetic helicity of the system in the dissipation range ($k > \Omega_i/v_A$).

References

Axford, W. I. & McKenzie, J. F. 1992, *Solar Wind Seven*, 1
Chin, Y.-C. & Wentzel, D. G. 1972, *Ap&SS*, 16, 465
Laitinen, T., Fichtner, H., & Vainio, R. 2003, *JGR (Space Physics)*, 108, 1081
Markovskii, S. A., Vasquez, B. J., & Hollweg, J. V. 2009 *ApJ*, 695, 1413
Melrose, D. B. & Sy, W. 1972, *Ap&SS*, 17, 343
Sagdeev, R. Z. & Galeev, A. A. 1969 *Nonlinear Plasma theory*
Spanier, F. & Vainio, R. 2009, *ASL*, 2, 337
Vainio, R. & Spanier, F. 2005, *A&A*, 437, 1
Yoon, P. H. & Fang, T.-M. 2009 *PPCF*, 50, 8

Advances in Plasma Astrophysics
Proceedings IAU Symposium No. 274, 2010
A. Bonanno, E. de Gouveia Dal Pino & A. G. Kosovichev, eds.
© International Astronomical Union 2011
doi:10.1017/S1743921311006740

The effect of plume/interplume lanes on ion-cyclotron resonance heating

Suzan Doğan and E. Rennan Pekünlü

Department of Astronomy and Space Sciences, Faculty of Science, University of Ege, Bornova, 35100, İzmir, Turkey
email: dogansuzan@gmail.com
email: rennan.pekunlu@ege.edu.tr

Abstract. The effect of the Plume/Interplume Lane (PIPL) structure of the solar North Polar Coronal Hole (NPCH) on the ion-cyclotron resonance (ICR) process is investigated. The ICR process in the interplume lanes is much more effective than in the plumes, agreeing with the observations which show the source of fast solar wind is interplume lanes.

Keywords. Sun: corona, solar wind, acceleration of particles, plasmas, waves.

1. Introduction

Images taken by SOHO/SUMER clearly show that the solar NPCH is structured in radial direction by so-called plume and interplume lanes (PIPL) (see Fig. 1). Plumes are denser and cooler, interplumes are less dense and hotter regions.

Doppler dimming analysis (Kohl *et al.* 1997) showed that the O VI ions display a temperature anisotropy with $T_\perp \sim 10^2 T_\parallel$. Besides, outflow velocities in the solar wind of O VI ions in the interplume lanes are higher than that of the plumes. The origin of the fast solar wind was identified as the interplume lanes (Wilhelm *et al.* 1998; Hollweg 1999a, 1999b, 1999c). These effects could be brought about only by ICR process (Hollweg 1999a, 1999b, 1999c). Therefore, ICR process is regarded as the most efficient heating mechanism for NPCH.

The aim of this study is to investigate the effect of the PIPL structure of the solar NPCH on the propagation characteristics of ion-cyclotron waves (ICW). The gradients of physical parameters both parallel and perpendicular to the magnetic field are considered with the aim of determining how the efficiency of the ICR process varies along the PIPL structure of NPCH.

2. Model

Our model is based on the kinetic theory. We solve the Vlasov equation for O VI ions and obtain the dispersion relation of ICW. O VI ions are considered under the two forces, a) Lorentz force, b) pressure gradient force. The quasi-linearized Vlasov equation is:

$$\frac{df_1}{dt} = \frac{\partial f_1}{\partial t} + \mathbf{v} \cdot \frac{\partial f_1}{\partial \mathbf{R}} + \frac{q_i}{m_i}\left(\mathbf{E}_1 + \frac{1}{c}(\mathbf{v} \times \mathbf{B}_0)\right) \cdot \frac{\partial f_1}{\partial \mathbf{v}} = \left[-\frac{q_i}{m_i}\left(\mathbf{E}_1 + \frac{1}{c}(\mathbf{v} \times \mathbf{B}_1)\right) \right.$$

$$\left. + \frac{k_B}{n_0 m_i}\left(T_{\text{eff}}^\xi \frac{\partial n_0}{\partial \mathbf{R}} + T_{\text{eff}}^\xi \frac{\partial n_0}{\partial \mathbf{x}} + n_0 \frac{\partial T_{\text{eff}}^\xi}{\partial \mathbf{R}} + n_0 \frac{\partial T_{\text{eff}}^\xi}{\partial \mathbf{x}}\right) \right] \cdot \frac{\partial f_0}{\partial \mathbf{v}}$$

(2.1)

Figure 1. SOHO/SUMER image of polar plumes and interplumes in O VI (1032 A). This image was obtained over the North pole in 1996 (Wilhelm *et al.* 1998).

where f_0 and f_1 are the unperturbed and perturbed parts of the velocity distribution function, and $\mathbf{E}_1$ and $\mathbf{B}_1$ are the wave electric and magnetic fields, respectively.

$$f_0 = n_{\mathrm{i}}\, \alpha_\perp^2 \alpha_\parallel\, \pi^{-3/2} \exp\left[-\left(\alpha_\perp^2 v_\perp^2 + \alpha_\parallel^2 v_\parallel^2\right)\right] \tag{2.2}$$

where $\alpha_\perp = (2k_{\mathrm{B}}T_\perp/m_i)^{-1/2}$ and $\alpha_\parallel = (2k_{\mathrm{B}}T_\parallel/m_i)^{-1/2}$ are the inverse of the most probable speeds in the perpendicular and parallel direction to the external magnetic field, respectively. By solving the Vlasov equation analytically and neglecting the residual contribution, we find the dispersion relation of ICW as below,

$$k^2 \left[c^2 - \frac{i\omega_p^2}{2\alpha_\parallel^2\,(\omega - \omega_c)^2}\Pi\right] - \frac{i\omega_p^2\, k}{\sqrt{\pi}\,\alpha_\parallel\,(\omega - \omega_c)}\Pi - \frac{i\omega_p^2\,\omega}{(\omega - \omega_c)} - \omega^2 + \frac{\pi^2\omega q_{\mathrm{i}}\,L}{E_x v_A}\nabla p_1^{\mathrm{r}} = 0 \tag{2.3}$$

where ω_p is the plasma frequency, ω_c is the cyclotron frequency for the O VI ions. For the sake of brevity we replace $[\omega/(\omega - \omega_c)] + (T_\perp/T_\parallel) - 1$ by Π and ∇p_1^{r} is the term designating the pressure gradient:

$$\nabla p_1^{\mathrm{r}} = \frac{k_{\mathrm{B}}}{m_{\mathrm{i}}}\left[\left(T_{\mathrm{eff}}^{\xi}\frac{\partial n_0}{\partial R} + n_0\frac{\partial T_{\mathrm{eff}}^{\xi}}{\partial R}\right)\left(\frac{T_\perp}{T_\parallel}\right)^{1/2} + 2\left(T_{\mathrm{eff}}^{\xi}\frac{\partial n_0}{\partial x} + n_0\frac{\partial T_{\mathrm{eff}}^{\xi}}{\partial x}\right)\right] \tag{2.4}$$

3. Results

Dispersion relation of the ICW given by Eq. (2.3) is solved for two situations, (*i*) the PIPL structure of NPCH is not taken into account, (*ii*) the PIPL structure therefore the plasma gradients in x direction are taken into account. These solutions are compared in Fig. 2. For ω=2500 $rads^{-1}$, all the solutions reveal an infinity in the refractive index, corresponding to a resonance at about 2.38 R. We should emphasize that the waves with frequencies higher than 2500 $rads^{-1}$ resonate with O VI ions at smaller R and vice versa.

When the PIPL structure is considered, the differences in the refractive index of the plumes and interplume lanes also revealed. Fig. 2b and 2d show crests which correspond to the interplume lanes and troughs which correspond to the plumes. As it is apparent from the figures that refractive index in the interplume lanes are about 2 times higher than the ones in the plumes. We may argue that the refractive index of the interplume lanes is readily going to infinity indicating that the resonance process in the interplume lanes is more effective than in the plumes.

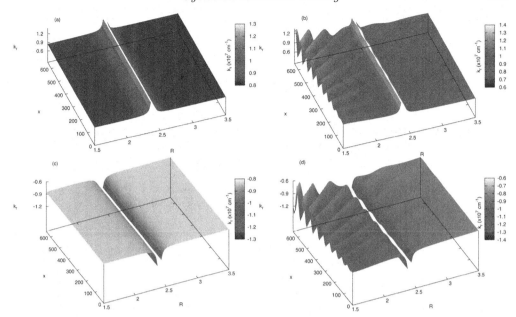

Figure 2. Two roots of the second order dispersion relation given by Eq. (2.3). R is the dimensionless radial distance $(r/R_\odot)$ and x is the distance perpendicular to R. (a) Forward propagating mode when PIPL structure is not taken into account; (b) the same mode with PIPL structure considered; (c) backward propagating mode without PIPL structure and (d) the same mode with PIPL structure considered. The graphs are for ICW with a frequency 2500 rads^{-1}.

When we adopt 10^{-3} for N_{OVI}/N_p, which is the value Vocks (2002) used in his model, the ratio of the wave numbers $k_{\nabla p}/k$ range from 0.2 to 16.8 in the distance range 1.5-3.5 R, where $k_{\nabla p}$ is the wave number of the waves propagating in the presence of gradients considered. For $N_{OVI}/N_p = 1.52 \times 10^{-6}$ which is the value given by Cranmer *et al.* (2008), $k_{\nabla p}/k$ take values between 0.7 and 1.6. We plotted the figures by adopting the later value, i.e. Cranmer *et al.*'s (2008).

4. Conclusion

When the wave frequency is equal to the ion cyclotron frequency, i.e., $\omega = \omega_c$, ICW resonate with and transfer their energy to the O VI ions. The resonance process in the interplume lanes is more effective than in the plumes. This result is confirmed by the observations showing that the seat of the fast solar wind is interplume lanes.

References

Cranmer, S. R., Panasyuk, A. V., & Kohl, J. L. 2008, *ApJ*, 678, 1480
Hollweg, J. V. 1999a, *JGR*, 104, 24781
Hollweg, J. V. 1999b, *JGR*, 104, 24793
Hollweg, J. V. 1999c, *JGR*, 104, 505
Kohl, J. L., Noci, G., Antonucci, E. *et al.* 1997, *SoPh*, 175, 613
Wilhelm, K., Marsch, E., Dwivedi, B. N. *et al.* 1998, *ApJ*, 500, 1023

Advances in Plasma Astrophysics
Proceedings IAU Symposium No. 274, 2010
A. Bonanno, E. De Gouveia Dal Pino & A. G. Kosovichev eds.
© International Astronomical Union 2011
doi:10.1017/S1743921311006752

Small-scale flux emergence events observed by Sunrise/IMaX

S. L. Guglielmino[1,2], V. Martínez Pillet[1], J. C. del Toro Iniesta[3], L. R. Bellot Rubio[3], F. Zuccarello[4], S. K. Solanki[5,6] & the Sunrise/IMaX team

[1]IAC Instituto de Astrofísica de Canarias,
C/ Vía Láctea s/n, La Laguna, Tenerife, E-38200, Spain
email: sgu@iac.es

[2]ULL Departamento de Astrofísica, Univ. de La Laguna,
La Laguna, Tenerife, E-38205, Spain

[3]IAA Instituto de Astrofísica de Andalucía (CSIC),
Apdo. 3004, Granada, E-18080, Spain

[4]Dipartimento di Fisica e Astronomia - Sezione Astrofisica,
Via S. Sofia 78, 95123 Catania, Italy

[5]Max-Planck-Institut für Sonnensystemforschung,
Max-Planck-Str. 2, 37191 Katlenburg-Lindau, Germany

[6]School of Space Research, Kyung Hee University,
Yongin, Gyeonggi 446-701, Republic of Korea

Abstract. Thanks to the unprecedented combination of high spatial resolution ($0''.2$) and high temporal cadence (33 s) spectropolarimetric measurements, the IMaX magnetograph aboard the Sunrise balloon-borne telescope is revealing new insights about the plasma dynamics of the all-pervasive small-scale flux concentrations in the quiet Sun. We present the result of a case study concerning the appearance of a bipole, with a size of about $4''$and a flux content of 5×10^{17} Mx, with strong signal of horizontal fields during the emergence. We analyze the data set using the SIR inversion code and obtain indications about the three-dimensional shape of the bipole and its evolution with time.

Keywords. Sun: photosphere, Sun: magnetic fields, Sun: activity, Techniques: high angular resolution

1. IMaX Observations and Data Analysis

A small-scale emerging bipole was observed in 09 June 2009 near the disk center at high spatial resolution with the Imaging Magnetograph eXperiment (IMaX, Martínez Pillet *et al.*, 2010), mounted on the 1-m aperture telescope flown during the Sunrise mission, a balloon-borne solar observatory (Barthol *et al.*, 2010; Solanki *et al.*, 2010). IMaX took polarization maps at five wavelength positions over the Fe I 525.02 nm line, at -80, -40, +40, +80, +227 mÅ from the line center, with a cadence of 33 s and a pixel size of $0''.055$, recording the full Stokes parameters I, Q, U, and V over the full IMaX field-of-view (FOV) of about $50'' \times 50''$. In this study we analyze non-reconstructed data, as the Phase-Diversity technique used to obtain reconstructed data implies an apodization of the full FOV of IMaX, and the emerging bipole appeared just at its right border.

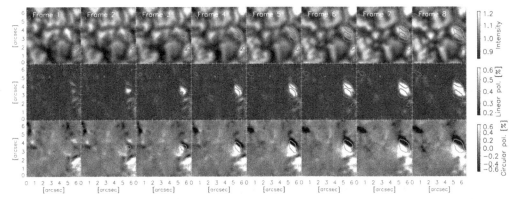

Figure 1. Top row: Time evolution of the continuum maps during the appearance of the bipole; red (blue) contours represent a circular polarization signal of $+0.8$ (0.8) % of the I_c; green contours over the circular polarization maps represent a linear polarization signal of 0.5 % of the I_c. Middle row: the same for the mean linear polarization signal L_s. Bottom row: the same for the mean circular polarization signal V_s. The bipole appears in the frame 1, giving rise to a considerable signal in the L_s map. In the frame 3 we see two small patches of opposite polarity at opposite edges of the structure seen in the L_s map. The evolution of the emerging loop is clearly recognizable in the subsequent frames, where opposite polarities of the bipole separate each other in opposite directions, with a quite strong L_s signal in between them, which coincides with the top of a granule in the continuum map.

We obtained maps of the mean circular polarization signal V_s and of the mean linear polarization signal L_s, given by

$$V_s = 1/4 \langle I_c \rangle \epsilon \sum_{i=1}^{4} |V_i| \quad \text{and} \quad L_s = 1/4 \langle I_c \rangle \sum_{i=1}^{4} \sqrt{Q_i^2 + U_i^2},$$

where I_c is the continuum intensity averaged over the IMaX FOV, $\epsilon = \pm 1$ depends on the sign of the blue lobe of Stokes V, and i runs over the first four wavelength positions. The signal-to-noise ratio in these measurements is about $2 \cdot 10^3$.

We carried out inversions of the observed Stokes vector using the SIR code (Ruiz Cobo & del Toro Iniesta, 1992), which yields the temperature stratification, the line-of-sight (LOS) velocity, the magnetic field strength, inclination and azimuth angles in the LOS reference frame. While these latter are assumed by SIR inversion to be constant throughout the photosphere, the LOS velocity and magnetic field strength have a linear gradient. Such a gradient has been found to be necessary in order to fit some profiles that show a considerable asymmetry between the blue and red lobes in Stokes V. No stray light contamination was taken into account for the inversions.

2. Results

In Figure 1 we display the temporal sequences of the mean linear polarization and mean circular polarization maps, that clearly show the emergence and the evolution of the bipole at the beginning of its appearance: first a patch with a strong linear polarization signal appears, and a pair of structures with opposite polarities is seen at the opposite edges of the patch in the following frame. Then the opposite polarities separate in time. In the continuum map is clearly visible that the polarization signals coincide with a granule.

Summarizing the results obtained from the SIR inversion:

- upflow motions are found in correspondence of the region with strong linear polarization signal;
- magnetic field strength and flux grow in time, reaching a peak about five minutes since the first detection of the bipole;
- footpoints do not reach a vertical orientation, but rather they remain inclined at about 45° with respect to the vertical;
- in the emergence zone, the azimuth angle is quite homogeneous, ranging from 25° to 55°;
- the magnetic flux content of the bipole at its maximum is of about 5×10^{17} Mx, while the size of about 4″, roughly of the same order of a granule.

We have also produced the scatter plots of the magnetic field strength and LOS velocity vs. zenith angle. The footpoints of the loop have a field strength of about 300 G and downflows up to 3 km s^{-1}, while the horizontal parts show upflows of 1.5 km s^{-1}, decreasing with time. In both cases, no evidence of asymmetry between the footpoints of the bipole is found in the data.

3. Conclusions

IMaX/Sunrise observations of the solar photosphere taken at the disk center have revealed a number of small-scale episodes of magnetic flux emergence (Danilovic *et al.*, 2010; Solanki *et al.*, 2010). We have carried out an analysis concerning the emergence of a small magnetic bipole. We have analyzed the polarization maps and then we have inverted the Stokes profiles with the SIR code, to obtain information on the physical parameters of this magnetic structure.

The magnetic flux content locates this small bipole in a halfway point between the ephemeral regions studied by Hagenaar (2001) and the all-pervasive loops of the quiet Sun. The bipole indeed appears to be coincident with a granule in the continuum map, so it would apparently represent a typical case of flux emergence at granular scale. Nevertheless, the footpoints separate up to 4500 km and their remnants remain visible until 15 minutes since the first detection of the magnetic structure. There are also evidences for opposite motions along the LOS at different altitude in the atmosphere in the same pixels: upflows in the higher layers and downflows in the deeper layers at the external border of the negative footpoint might indicate the presence of small-scale jets, due to the interaction of the emerging loop with the ambient magnetic field.

Acknowledgements

This work has been partially funded by the Spanish Ministerio de Educación y Ciencia, through Projects ESP2006-13030-C06-01/02/03/04 and AYA2009-14105-C06-03/06 INTA+GACE, and Junta de Andalucía, through Project P07-TEP-2687, including a percentage from European FEDER funds. Financial support by the European Commission through the SOLAIRE Network (MTRN-CT-2006-035484) is gratefully acknowledged.

References

Barthol, P., *et al.* 2010, *Solar Phys.*, in press
Danilovic, S., *et al.* 2010, *Astrophys. Jour. Lett.*, 723, L149
Hagenaar, H. J., 2001, *Astrophys. Jour.*, 555, 448
Martnez Pillet, V., *et al.* 2010, *Solar Phys.*, in press
Ruiz Cobo, B. & del Toro Iniesta, J.C. 1992, *Astrophys. Jour.*, 398, 375
Solanki, S. K., *et al.* 2010, *Astrophys. Jour. Lett.*, 723, L127

Advances in Plasma Astrophysics
Proceedings IAU Symposium No. 274, 2010
A.Bonanno, E. de Gouveia Dal Pino & A. G. Kosovichev, eds.
© International Astronomical Union 2011
doi:10.1017/S1743921311006764

Cross helicity in stellar magnetoconvection

Manfred Küker and Günther Rüdiger

Astrophysikalisches Institut Potsdam,
An der Sternwarte 16, 14482, Potsdam, Germany

Abstract. Magnetic diffusion is a key ingredient in mean-field dynamo models but neither observations nor theory are able to produce reliable values. Numerical simulations provide an alternative way to determine the turbulent electromotive force. Cross helicity allows us to determine the turbulent magnetic diffusion coefficient in simulations of stellar magnetoconvection.

Keywords. convection, Sun: activity, stars: activity

1. Introduction

Mean-field MHD has successfully modelled the solar differential rotation and many aspects of its activity cycle. In this formulation of MHD, the mean magnetic field is governed by the induction equation

$$\frac{\partial \langle \boldsymbol{B} \rangle}{\partial t} = \nabla \times (\langle \boldsymbol{u} \rangle \times \langle \boldsymbol{B} \rangle - \eta \nabla \times \langle \boldsymbol{B} \rangle + \langle \boldsymbol{u}' \times \boldsymbol{B}' \rangle) \tag{1.1}$$

where η is the magnetic diffusion coefficient, brackets indicate mean (i.e. averaged) quantities and primes the fluctuations. Eq.1.1 is formally identical with the standard induction equation except for the last term on the right hand side, the turbulent electromotive force. The latter can be expressed in terms of the mean field:

$$\langle \boldsymbol{u}' \times \boldsymbol{B}' \rangle \approx \alpha \langle \boldsymbol{B} \rangle - \eta_T \langle \nabla \times \boldsymbol{B} \rangle. \tag{1.2}$$

The first term on the RHS is the field-generating α effect caused by kinetic helicity and the second term is turbulent diffusion. While the α effect is caused by the Coriolis force, the turbulent magnetic diffusivity exists in non-rotating turbulence, too. Both the α and η_T coefficients depend on statistical properties of the fluctuations only.

The kinetic helicity on the solar surface has been measured by Komm *et al.* (2008). The magnetic diffusivity can be derived from the decay of sunspots. However, these are areas with strong magnetic fields and the diffusivity is quenched by the back-reaction of the magnetic field on the fluid. The magnetic diffusivity for the quiet Sun can be determined by measuring the cross helicity, $\langle \boldsymbol{u}' \cdot \boldsymbol{B}' \rangle$. The component involving the radial components of the gas velocity and the magnetic field can be expressed in terms of the radial component of the mean magnetic field (Rüdiger *et al.* 2011):

$$\frac{\langle u_r' B_r' \rangle}{\langle B_r \rangle} \simeq -\frac{\eta_T}{H_\rho}, \tag{1.3}$$

where H_ρ is the density scale height. Eq. 1.3 allows the determination of the turbulent magnetic diffusivity coefficient from measurements of the radial velocity and magnetic field components provided the density stratification is known.

2. Numerical simulations

Rüdiger *et al.* (2011) studied the cross helicity in forced turbulence and found it to be anti-correlated with the radial magnetic field, as required by Eq. 1.3 for a positive magnetic diffusivity. Here we check if this result holds for convection. We therefore run numerical simulations using the Nirvana code (Ziegler 2002), which uses a conservative finite difference scheme in cartesian coordinates. The code solves the equation of motion, the induction equation, and the equations of energy and mass conservation. We assume an ideal, fully ionized gas that is heated from below and kept at a fixed temperature at the top of the simulation box. Periodic boundary conditions apply at the horizontal boundaries. A homogeneous vertical magnetic field is applied. The upper and lower boundaries are impenetrable and stress-free.

In the dimensionless units described there the size of our simulation box is $6 \times 6 \times 2$ in the x, y, and z directions, respectively. As we treat the non-rotating case, the z axis is aligned with the stratification vector, i.e. represents the radial direction in spherical geometry. The x and y coordinates denote the horizontal directions. The stratification of density, pressure, and temperature is piecewise polytropic as described in Ziegler (2002). Similar setups have been used by Cattaneo *et al.* (1991), Brandenburg *et al.* (1996), Chan (2001), and Ossendrijver *et al.* (2001). The initial state is in hydrostatic equilibrium but convectively unstable in the upper half of the box. The z coordinate is negative in our setup, with $z = 0$ at the upper boundary. The stable layer thus extends from $z = -2$ to $z = -1$, the unstable layer from $z = -1$ to $z = 0$. The density varies by a factor 50 over the depth of the box, i.e. the density scale height is 0.5.

The initial magnetic field is vertical and homogeneous. We run the simulations until a quasi-stationary state evolves. Our control parameters are the heat conduction coefficient, κ, the Prandtl number, Pr$= \nu/\kappa$, where ν is the viscosity, the magnetic diffusivity coefficient, η, and the strength of the initial magnetic field, B_0. Convection sets in if the Rayleigh number,

$$\mathrm{Ra} = \frac{\rho g c_p d^4}{T \kappa \nu} \left(\frac{dT}{dz} - \frac{g}{c_p} \right), \tag{2.1}$$

with the density ρ, the specific heat capacity c_p, the gravity force g, and the length scale d, exceeds a critical value. The length scale is defined by the depth of the convectively unstable layer, i.e. $d = 1$.

3. Results

The velocity field shows the asymmetry between upwards and downwards motion characteristic of convection in stratified media. The downward motion is concentrated at the boundaries of the convection cells and particularly at the corners. The upwards motion fills the interior of the convection cells. As it covers a much larger area the gas motion is much slower than in the concentrated downdrafts. The magnetic field shows a similar pattern. The vertical field is concentrated in the areas with downwards motion and weak in the areas with upward motion. As the total vertical magnetic flux is conserved, this is the result of field advection.

Figure 1 shows results from a run with rather weak magnetic field, $B_0 = 10^{-5}$, and Ra$= 6 \times 10^7$. The value of the magnetic Prandtl number, Pm$= \nu/\eta$, is 0.1. The left diagram shows the horizontal average of the cross helicity as a function of the depth, the right diagram shows the vertical (z) component, $\langle u'_z B'_z \rangle$. The difference between the two quantities is very small, with the vertical component actually being slightly larger than the sum of the vertical and horizontal correlations. It is thus a good proxy for

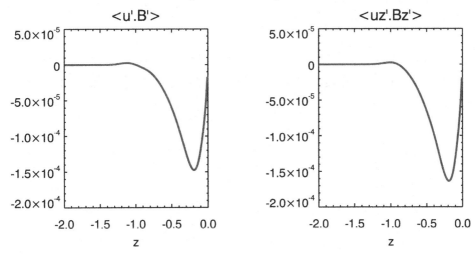

Figure 1. Horizontally-averaged cross helicity as a function of depth. Left: full cross helicity. Center: correlation of the vertical components of the magnetic field and gas velocity.

the full cross correlation. The correlations do not vanish abruptly at the bottom of the unstable layer because of overshoot, which affects the upper half of the stable layer. The correlations are positive and much smaller than in the unstable layer.

As the density scale height and the magnetic diffusivity are both positive quantities, Eq. 1.3 can only be fulfilled if the (vertical) cross correlation and the mean magnetic field have opposite signs. This is indeed the case. For positive values of the mean magnetic field $\langle B_z \rangle$ the cross-helicity is negative in the unstably stratified layer. If the field polarity is reversed and everything else left unchanged the cross correlation becomes positive with the same amplitude as in the case shown.

References

Brandenburg, A., Jennings, R. L., Nordlund, A. *et al.* 1996, *J. Fluid Mech.*, 306, 14370
Cattaneo, F., Brummell, N. H., Toomre, J., *et al.* 1991, ApJ, 370, 282
Chan, K. L., 2001, *ApJ*, 548, 1102
Komm, R., Hill, F., & Howe, R., *J. Phys. Conf. Ser.*, 118, 012035
Ossendrijver, M., Stix, M., Brandenburg, A., 2001, *A&A*, 376, 713
Rüdiger, G., Kitchatinov, L. L., & Brandenburg, A. 2011, *Sol. Phys.*, 30, 490
Ziegler, U. 2002, *A&A*, 386,331

Advances in Plasma Astrophysics
Proceedings IAU Symposium No. 274, 2010   © International Astronomical Union 2011
A. Bonanno, E. de Gouveia Dal Pino & A.G. Kosovichev, eds.   doi:10.1017/S1743921311006776

Some Properties of prominence eruption associated CMEs during 1996-2009

Nishant Mittal[1,2]

[1] Astrophysics Research Group, Meerut College, Meerut-250001, India
email: nishantphysics@yahoo.com

[2] Dept. of Physics, Krishna Institute of Management and Technology, Moradabad-244001,
U. P., India

Abstract. Solar prominences can be viewed as pre-eruptive states of coronal mass ejections (CMEs). Eruptive prominences are the phenomena most related to CMEs observed in the lower layers of the solar atmosphere. We have made a comprehensive statistical study on the CMEs associated with prominence eruptions. We have examined the distribution of CMEs speed and acceleration for prominence eruptions associated CMEs. We also examine the speed-acceleration correlation for these events and there is no correlation between speed and acceleration. The mean angular width is almost similar to normal CMEs. The number variation during solar cycle of prominence activities is similar to the sunspot cycle.

Keywords. Sun; coronal mass ejections; prominence eruptions;

1. Introduction

CMEs are often seen as spectacular eruptions of matter from the Sun, which propagate outward through the heliosphere and often interact with the Earth's magnetosphere (Hundhausen 1997; Mittal & Narain 2010 & references there in). It is well known that these interactions can have substantial consequences on the geomagnetic environment of the Earth, sometimes resulting in damage to satellites (Mittal *et al.* 2010. The "classical" structure of a CME consists of three parts and one of these parts, a bright core, is the remnant of an eruptive prominence (Crifo *et al.* 1983; Illing & Hundhausen 1985; Hundhausen 1999 and Gopalswamy *et al.* 2006, Mittal & Narain 2010).

2. Data and results

Over the past 14-years the SOHO/LASCO instrument has been detecting CMEs. The LASCO instrument (Brueckner *et al.* 1995) consists of three coronagraphs C1, C2 and C3 that span the fields of view 1.1-3$R_\odot$, 2-6$R_\odot$S and 4-30$R_\odot$, respectively. The height-time data of CMEs used in this study are taken from the online SOHO/LASCO CME catalogue (http://cdaw.gsfc.nasa.gov/CME_list). The Prominence Eruption data has been taken from Nobeyama Radioheliograph; which detect 396 events during the period. The distribution of Prominence Eruption associated CMEs has been shown in Figure 1. It seems that the peak of the histogram occur at 350 km s^{-1}; where the tail of large speeds is very short. The overall median (average) speed 447 (511) km s^{-1}; respectively. Figure 2 is the histogram of apparent angular width for the period 1996–2009. The average width from the 14–year data for Prominence Eruption associated CMEs (390) is 670 and the median width is 550. Figure 3 is the histogram of CME acceleration for the period 1996–2009. It is clear from this figure that a majority (55%) of Prominence Eruption associated CMEs are decelerated, about 12% of them move with little acceleration and the

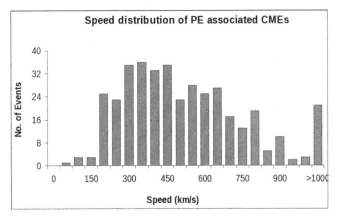

Figure 1. Histogram shows the speed distribution of Prominence Eruption associated CMEs during 1996-2009.

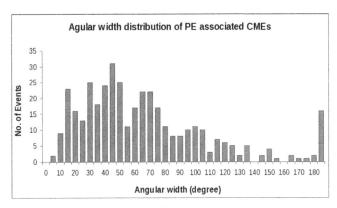

Figure 2. The width distribution of SOHO/LASCO Prominence Eruption associated CMEs from 1996 to 2009. The last bin shows all CMEs with width ¿1800, which amounts to 5-6% of all Prominence Eruption associated CMEs.

remaining 33% have positive acceleration. Thus Prominence Eruption associated CMEs have clear bias towards deceleration. Figure 4 shows a scatter plot between the measured acceleration, a (in m s^{-2}) and speed, V (in km s^{-1}) of Prominence Eruption associated CMEs for which the acceleration estimate was possible.

Figure 5 exhibits total CMEs and Prominence Eruptions. It is clear from the figure that the solar maximum peak occurs at the year 2000 and 2002. Occurrence rate of Prominence Eruptions follows the solar cycle variation whereas CMEs occurrence rate does not follow the solar cycle variation.

3. Conclusion

In this paper, I have made a comprehensive statistical study on Prominence Eruption associated CMEs from 1996 to 2009. During the period 1996-2009 more than 14000 CMEs observed by SOHO/LASCO, whereas Nobeyama Radioheliograph detect 396 Prominence Eruption events, out of which 390 Prominence Eruptions are associated with CMEs.

1. Figure 1 shows that the Prominence Eruption associated CMEs have median (average) speed 447 (511) km s^{-1}. The number of CMEs having speeds greater than 1000 km s^{-1} is quite small (5-6%).

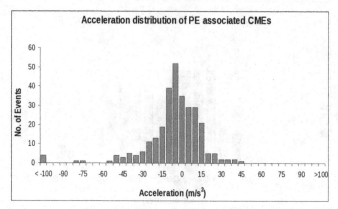

Figure 3. Histogram of acceleration of Prominence Eruption associated CMEs, showing clear bias towards deceleration.

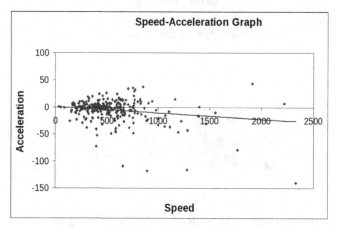

Figure 4. Acceleration as a function of speed of Prominence Eruption associated CMEs from 1996 to 2009. The acceleration has a large scatter, but there is a clear trend that the fast CMEs decelerate, while slow CMEs accelerate.

2. The average width for Prominence Eruption associated CMEs (390) is 670 and the median width is 550.

3. The acceleration distribution shows a conspicuous peak near -5 m s^{-1}, about 55% have negative acceleration, 33% have positive acceleration and the remaining 12% have very little acceleration. The Prominence Eruption associated CME distribution is biased towards deceleration.

4. Rate of Prominence Eruptions varies with solar cycle.

Acknowledgement

Author is highly thankful to IAU and LOC/SOC of IAU S274 for providing financial help to attend the meeting. Author is thankful to KIMT, Moradabad authorities for their help and encouragement. I am also thankful to Pankaj Kumar for his useful suggestions. The CME catalog is generated and maintained at the CDAW Data Center by NASA and The Catholic University of America in cooperation with the Naval Research Laboratory.

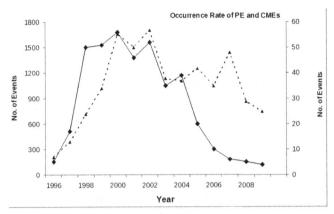

Figure 5. Comparison between annual CME occurrence rate and Prominence Eruptions from 1996-2009.

SOHO is a project of international cooperation between ESA and NASA. Author is also thankful to Nobeyama Radioheliograph for providing the prominence eruption data

References

Mittal, N. & Narain, U. 2010, *Journal of Atmospheric and Solar-Terrestrial Physics*, 72, 643

Crifo, F., Picat, J. P., & Cailloux, M. 1983, *Sol. phys*, 83, 143

Gopalswamy, N., Mikić, Z., Maia, D., Alexander, D., Cremades, H., Kaufmann, P., Tripathi, D., & Wang, Y.-M. 2006, *SSR*, 123, 303

Hundhausen, A. 1999, The many faces of the sun: a summary of the results from NASA's Solar Maximum Mission., 143

Hundhausen, A. J., Burkepile, J. T., & St. Cyr, O. C. 1994, *JGR*, 99, 6543

Illing, R. M. E. & Hundhausen, A. J. 1985, *JGR*, 90, 275

Advances in Plasma Astrophysics
Proceedings IAU Symposium No. 274, 2010
A. Bonanno, E. de Gouveia Dal Pino & A. G. Kosovichev, eds
© International Astronomical Union 2011
doi:10.1017/S1743921311006788

Separation of solar radio bursts in a complex spectrum

Hana Mészárosová[1], Ján Rybák[2], Marian Karlický[1], and Karel Jiřička[1]

[1]Astronomical Institute of the Academy of Sciences,
CZ-25165 Ondřejov, Czech Republic
email: hana@asu.cas.cz

[2]Astronomical Institute, Slovak Academy of Sciences,
SK-05960 Tatranská Lomnica, Slovak Republic
email: choc@ta3.sk

Abstract. Radio spectra, observed during solar flares, are usually very complex (many bursts and fine structures). We have developed a new method to separate them into individual bursts and analyze them separately. The method is used in the analysis of the 0.8–2.0 GHz radio spectrum of the April 11, 2001 event, which was rich in drifting pulsating structures (DPSs). Using this method we showed that the complex radio spectrum consists of at least four DPSs separated with respect to their different frequency drifts (-115, -36, -23, and -11 MHz s^{-1}). These DPSs indicate a presence of at least four plasmoids expected to be formed in a flaring current sheet. These plasmoids produce the radio emission on close frequencies giving thus a mixture of superimposed DPSs observed in the radio spectrum.

Keywords. Sun: corona, Sun: flares, Sun: radio radiation, Sun: oscillations

1. Introduction

The drifting pulsating structures (Karlický 2004, Reiner *et al.* 2008) observed at the beginning of the eruptive solar flares have been found to be radio signatures of the plasmoid ejection (Karlický & Odstrčil 1994, Ohyama & Shibata 1998, Karlický *et al.* 2002). Based on the MHD numerical simulations, Kliem *et al.* (2000) suggested that the drifting pulsation structure is generated by superthermal electrons, trapped in the magnetic island (plasmoid) in the bursty regime of the magnetic field reconnection. The global slow negative frequency drift of the structure was explained by a plasmoid propagation upwards in the solar corona towards lower plasma densities. This model of DPSs was further developed and verified in the papers by Karlický & Bárta (2007), Bárta *et al.* (2008), Karlický *et al.* (2010).

2. Observations and analysis

Observed radio spectra usually consist of many radio bursts and fine structures superimposed in the same time and frequency intervals. Therefore, it would be highly desirable to separate these individual radio features and analyze them separately. For this purpose we developed a new method (Mészárosová *et al.* 2010) that is able to separate the bursts and fine structures according to their frequency widths and temporal scales. The method is based on the wavelet analysis technique. In order to present capabilities of our method we have tested it using artificial radio spectra (Mészárosová *et al.* 2010).

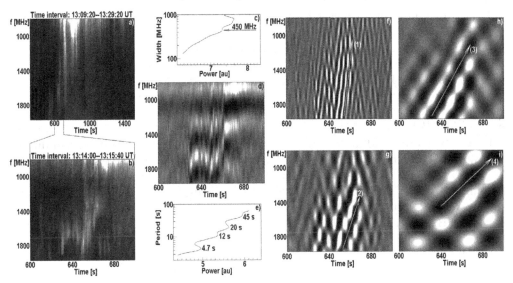

Figure 1. Separation of DPSs in the radio dynamic spectrum of the April 11, 2001 event: (a) original radio spectrum with DPSs, (b) part of the original radio spectrum with DPSs, (c) averaged global spectrum (ASf) with a peak for the frequency width = 450 MHz, (d) filtered radio spectrum in the frequency width range = 350–500 MHz, (e) averaged global spectrum (ASt) made from the spectrum in the panel *d* (f) separated DPSs in the period range 3–6 s, (g) separated DPSs in the period range 6–15 s, (h) separated DPSs in the period range 15–29 s, and (i) separated DPSs in the period range 29–52 s. Arrows in panels show frequency drifts of DPSs. The positive and negative parts of amplitudes (in the relation to their mean values) are in white and black colors, respectively (panels *d* and *f–i*).

In the present paper we applied this method on the 0.8–2.0 GHz radio spectrum observed during the April 11, 2001 event (GOES M2.3, NOAA AR9415) by the Ondřejov radiospectrograph (Jiřička *et al.* 1993). We analyzed this spectrum in the time interval 13:09:20–13:29:20 UT (i.e. 1200 s) and in the frequency range 0.8–2.0 GHz (see Figure 1, panel *a*). Its time resolution is 0.1 s. As an example of our analysis, we present here the analysis of the spectrum in the time subinterval 13:14:00–13:15:40 UT. As can be seen here, this part of the radio spectrum consists of several superimposed DPSs. We used the same method as in the paper by Mészárosová *et al.* (2010). We took the spectrum (Figure 1, panel *b*) as an input data set for the separation according to frequency widths of DPSs. Then we computed the averaged wavelet spectrum (ASf, panel *c*). It shows a peak for the frequency width 450 MHz. It means that the most of DPSs in the radio spectrum (panel *b*) have the frequency width of about 450 MHz. Therefore, we computed a new filtered radio spectrum in the frequency widths range 350–500 MHz, where the values 350 and 500 MHz correspond to local minima around the peak at 450 MHz. This filtered spectrum is presented in the panel *d*. Then, the data set in panel *d* became the input data for the separation according to temporal periods of DPSs. The averaged global spectrum (ASt) shows four peaks (panel *e*) for the periods $P=$ 4.7, 12, 20, and 45 s. It means that the DPSs with frequency width of about 450 MHz have the characteristic periods 4.7, 12, 20, and 45 s. Then, using the inverse wavelet transform we computed the final spectra filtered in the period ranges 3–6 s (panel *f*), 6–15 s (panel *g*), 15–29 s (panel *h*), and 29–52 s (panel *i*) where these range values correspond to local minima around the period peaks in panel *e*. The positive and negative parts of amplitudes are in white and black colors, respectively (panels *d* and *f–i*). This type of displaying makes possible to see individual DPSs in very good contrast.

At least four different frequency drifts can be recognized (-115, -36, -23, and -11 MHz s^{-1}, see the arrows 1, 2, 3, and 4 in the panels f, g, h, and i, respectively). It means that at least four DPSs are present in the complex radio spectrum observed in the April 11, 2001 event. We have obtained similar results also for other selected subintervals during the time interval 13:15:40–13:29:20 UT.

3. Conclusions

Using new method (Mészárosová *et al.* 2010), the complex spectrum observed during the April 11, 2001 flare with many drifting pulsating structures (DPSs) was analyzed. We found that the characteristic bandwidth of these DPSs is about 450 MHz in whole time interval under study (13:09:20–13:29:20 UT, Figure 1, panel a). The DPSs in time subinterval (13:14:00–13:15:40 UT, panel b) revealed the frequency drifts -115, -36, -23, and -11 MHz s^{-1}, respectively (panels f–i). The DPSs of the others time subintervals (13:15:40–13:29:20 UT) show similar frequency drifts.

Usually only one DPS is observed as a signature of the ejection of one dominant plasmoid. However, in the flaring current sheet a series of plasmoids of different sizes can be formed (Shibata & Tanuma 2001). If these plasmoids produce the radio emission on close frequencies then on the radio spectrum we can see a mixture of different and superimposed DPSs. The radio spectrum of the April 11, 2001 event shows such superimposed DPSs. Our analysis of this spectrum revealed that it consists of at least four different DPSs (panels f–i), which means that at least four plasmoids (magnetic islands) were generated in the flaring current sheet. But the velocities of these plasmoids, oriented in the upward direction in the solar atmosphere, differ.

Acknowledgements

H. M., M. K., and K. J. acknowledge support from the Grant IAA300030701 of the Academy of Sciences of the Czech Republic and the research project AVOZ10030501 of the Astronomical Institute AS CR. The work of J. R. was partly supported by the Slovak Grant Agency VEGA (project 2/0064/09). The program of mobility between the academies of the Czech Republic and Slovakia is also acknowledged. The wavelet analysis was performed using the software based on tools provided by C. Torrence and G. P. Compo at http://paos.colorado.edu/research/wavelets.

References

Bárta, M., Karlický, M., & Žemlička, R. 2008, *Solar Phys.*, 253, 173
Jiřička, K., Karlický, M., Kepka, O., & Tlamicha, A. 1993, *Solar Phys.*, 147, 203
Karlický, M. & Odstrčil, D. 1994, *Solar Phys.*, 155, 171
Karlický, M., Fárník, F., & Mészárosová, H. 2002, *A&A*, 395, 677
Karlický, M. 2004, *A&A*, 417, 325
Karlický, M. & Bárta, M. 2007, *A&A*, 464, 735
Karlický, M., Bárta, M., & Rybák, J. 2010, *A&A*, 514, A28
Kliem, B., Karlický, M., & Benz, A. O. 2000, *A&A*, 360, 715
Mészárosová, H., Rybák, J., & Karlický, M. 2010, *A&A*, submitted
Ohyama, M. & Shibata, K. 1998, *ApJ*, 499, 934
Reiner, M. J., Klein, K. L., Karlický, M., *et al.* 2008, *Solar Phys.*, 249, 337
Shibata, K. & Tanuma, S. 2001, *Earth, Planets and Space*, 53, 473

Advances in Plasma Astrophysics
Proceedings IAU Symposium No. 274, 2010
A. Bonanno, E. de Gouveia Dal Pino & A.G. Kosovichev, eds.

© International Astronomical Union 2011
doi:10.1017/S174392131100679X

An hybrid neuro-wavelet approach for long-term prediction of solar wind

Christian Napoli[1], Francesco Bonanno[2] and Giacomo Capizzi[2]

[1] Dept. of Physics and Astronomy, University of Catania,
Via S. Sofia, 95125, Catania - ITALY
email: chnapoli@gmail.com

[2] Dept. of Electrical, Electronic and Systems Engineering, University of Catania,
Viale A. Doria, 95125, Catania - ITALY
email: gcapizzi@diees.unict.it

Abstract. Nowadays the interest for space weather and solar wind forecasting is increasing to become a main relevance problem especially for telecommunication industry, military, and for scientific research. At present the goal for weather forecasting reach the ultimate high ground of the cosmos where the environment can affect the technological instrumentation. Some interests then rise about the correct prediction of space events, like ionized turbulence in the ionosphere or impacts from the energetic particles in the Van Allen belts, then of the intensity and features of the solar wind and magnetospheric response. The problem of data prediction can be faced using hybrid computation methods so as wavelet decomposition and recurrent neural networks (RNNs). Wavelet analysis was used in order to reduce the data redundancies so obtaining representation which can express their intrinsic structure. The main advantage of the wavelet use is the ability to pack the energy of a signal, and in turn the relevant carried informations, in few significant uncoupled coefficients. Neural networks (NNs) are a promising technique to exploit the complexity of non-linear data correlation. To obtain a correct prediction of solar wind an RNN was designed starting on the data series. As reported in literature, because of the temporal memory of the data an Adaptative Amplitude Real Time Recurrent Learning algorithm was used for a full connected RNN with temporal delays. The inputs for the RNN were given by the set of coefficients coming from the biorthogonal wavelet decomposition of the solar wind velocity time series. The experimental data were collected during the NASA mission WIND. It is a spin stabilized spacecraft launched in 1994 in a halo orbit around the L1 point. The data are provided by the SWE, a subsystem of the main craft designed to measure the flux of thermal protons and positive ions.

Keywords. solar wind, magnetic fields, Sun: activity, methods: data analysis

1. Introduction

The space environment is a dynamic system for the most part driven by the solar radiation and emission of plasma and particles from the star surface. The main effect of the solar activity is of course the solar wind, the set of emissions which voyage from the sun to us, and that can be measured providing informations on the energy and intensity. Solar emissions and electromagnetic disturbances (which could hamper the communication and satellite network), fast particles and flares bursts reaches Earth in few hours or even minutes from the solar event, as far as big clouds of plasma and charged particles hit the globe in the range of few days, without any early warning. The problem of correct and precise prediction of the solar wind then rises starting from the available space weather data survey. The problem of data prediction can be faced by using soft computing techniques as wavelet decompositions, recurrent neural networks or joint hybrid methods (Capizzi, Bonanno & Napoli 2010, Goh & Mandic 2003).

2. The basic of wavelet and neural networks theory

As shown by (Cybenko 1989), a finite weighted sum of continuous discriminating functions $\sigma(a_i^T x + b_i)$, where b_i are real scalar and a_i n-dimensional real vectors, is dense in the continuous functions space on $[0,1]^n$, so any continuous function f can be approximated by this kind of finite sum. Analogously according to wavelet theory we can state that the finite weighted sum of $det(D_i^{1/2})\psi(d_i x - t_i)$ is dense in the $L^2(\mathbb{R}^n)$, where $\{d_i\}$ is the dilatation vector, $\{t_i\}$ the translation vector, $D_i = diag(d_i)$ and ψ the mother wavelet function whose translates and dilates forms the basis for the $L^2(\mathbb{R}^n)$ space. This statement shows how the wavelets offer high advantages over other kinds of activation functions. In particular the performance of the wavelet decomposition can be noticed for nonlinear dynamical systems and predictors(Daubechies 1990). The wavelet transform, in fact, packs the energy of the signal, and in turn its relevant information in few significant non zero coefficients reducing the redundancies and showing the intrinsic structure in time and frequencies. The so obtained representation with wavelet decomposition offer advantages while used with neural networks. Typically a neural network is composed by the input and output neural layer and generally at least two hidden layers which are the core of the computational process which will assign a set of weights to the connections between different neurons. As reported in literature, in presence of a temporal memory in the dataset recurrent neural networks are used. The recurrent neural networks have a dynamic structure, which evolves during time because these networks contain feedback lines. In particular to obtain the best performance, even in terms of computational complexity and overall error, these neural networks are trained by an Adaptive Amplitude Real Time Recurrent Learning algorithm.

3. Proposed data analysis and simulations

The solar wind speed data were provided by the WIND SWE (Kasper 2006) and their analysis was done by using wavelet decomposition in order to pack the relevant informations in few significant coefficients. The approach lead to design a neural network which adapt himself to the signal through the significant wavelet coefficients set. In order to achieve this goal we used biorthogonal wavelets, (which have more freedom degrees respect the orthonormal ones) as result of a multiple evaluation of behaviors for an extended set of different kind of wavelets. The analysis is mainly based on the RMS performance which leaded to our final choose. Biorthogonal wavelets allow more degrees of freedom respect to traditional orthogonal (and of course the not orthogonal) wavelets, and because these have two different scaling functions, which can generate twin resolution analysis by two different wave functions ψ and ψ'. The scaling sequence must satisfy the biorthogonality condition and the wavelet coefficients in the general form are $b_n = (-1)^n a'_{N-1-n}$ and $b_n' = (-1)^n a_{M-1-n}$. By using the MATLAB package we developed a series of routines performing a biorthogonal wavelet analysis of the data. The coefficients set resulting from the wavelet analysis was used as input for the designed neural network. A biorthogonal wavelet decomposition was used to extract a shortened number of non-zero coefficients from a signal representative of solar wind velocity sampled trough time. These proposed calculations allowed us to drastically cut down the useful data set. The subbands of the decomposition were used to predict the future values of wavelet coefficients. The correlation between different subbands and temporal trends allow the subdivision of the neural network in serial subnets. In this adaptive conception the network can be modularized in relation with the time bands in order to cut the computational process based on the prediction goal at the selected temporal window. The

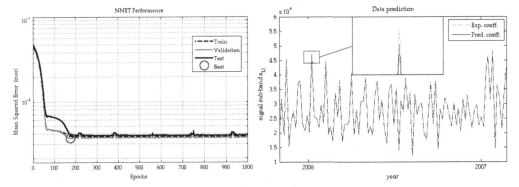

Figure 1. Performance of the NN   **Figure 2.** Solar wind velocity prediction

solar wind speed were decomposed in 13 subbands with the described method reducing the 99% of the energy in less than 4% of non zero coefficients. We trained the network to the 12th temporal sub-band to predict the rate of change in the solar wind velocity for the next 6 days. According the proposed modular strategy, for a signal sub band a_n, and in order to predict the data at a time interval τ_1 from a time serie of a previous interval τ_0, then the input for the NN is given in the form of $[a_{n+1}(\tau_0)|d_n(\tau_0)]$ so that the output will be the sub band $a_n(\tau_1)$. The selected NN is a full RNN NARX, trained by a gradient descent with momentum and adaptive amplitude learning rule backpropagation algorithm. The selected NN have 3 hidden layer: the first of 15 neuron with tribas transfer function, the second 11 neurons with tansig transfer function, and the third of 5 neurons with a competitive transfer function. Finally time vectors were used to create 5 delay lines for the input and 3 for the output.

4. Conclusions and results

The problem of solar wind data prediction, here faced using hybrid computations methods so as wavelet decomposition and recurrent neural networks, has been demostrated to be efficient. In fact the simulations for long-term prediction of solar wind velocity show the power of the proposed neuro-wavelet method and the obtained good performance in terms of RMS. The predicted solar wind speed accurately matches the experimental data within an RMS of 6%. The performance of the selected neural network is shown in fig. 1. The most significant results of solar wind velocity prediction are shown in fig. 2.

References

Capizzi G., Bonanno F., & Napoli C. 2010, Proc. Speedam 2010, p. 586
Goh S. L. & Mandic D. P. 2003, *Neural Networks*, Vol.16, p. 1095
Cybenko G. 1989, *Math. of Control, Signals, and Syst.*, Vol.2, No.4, p. 303
Daubechies I. 1990, *IEEE Trans. Inf. Theory*, Vol.36, No.5, p. 961
Kasper J. *et al.* 2006 *J. of Geophys. Res.*, Vol.1
Gleisner H., Lundstedt H., & Wintoft P. 1996, *Ann. Geophys.*, Vol.4, No.7, p. 679

Advances in Plasma Astrophysics
Proceedings IAU Symposium No. 274, 2010
A. Bonanno, E. de Gouveia Dal Pino & A.G. Kosovichev, eds.

© International Astronomical Union 2011
doi:10.1017/S1743921311006806

Exploiting solar wind time series correlation with magnetospheric response by using an hybrid neuro-wavelet approach

Christian Napoli[1], Francesco Bonanno[2] and Giacomo Capizzi[2]

[1] Dept. of Physics and Astronomy, University of Catania,
Via S. Sofia, 95125, Catania - ITALY
email: chnapoli@gmail.com

[2] Dept. of Electrical, Electronic and Systems Engineering, University of Catania,
Viale A. Doria, 95125, Catania - ITALY
email: gcapizzi@diees.unict.it

Abstract. The studies about the Sun rise a strong interest regarding modifications caused by the solar activity on the Earth. For almost a century in literature was discussed the problem of forecasting and analysis of the space weather, which in his definition covers both the near-earth space and the biospheric affection due to the environmental interaction with the Sun. In particular in the last years increased the attention for magnetospheric response in conjunction with the technological infrastructure and the biosphere itself. This to prevent i.e. spacecraft failures or possible treats to human health. Since the main effect of the activity of the Sun is the solar wind, rises the aim to found a correlation between itself and the localized variations induced on the magnetosphere being the purpose to predict long-term variation of the magnetic field from solar wind time series. As recently proposed for solar wind forecasting, an hybrid approach will be here used than joining the wavelet analysis with the prediction capabilities of recurrent neural networks with an adaptive amplitude activation function algorithm in order to avoid the need to standardize or rescaling the input signal and to match the exact range of the activation function.

Keywords. solar wind, magnetic fields, Sun: activity, methods: data analysis

1. Introduction

Modern industry, army and science operations have expanded their area of interests to the ultimate high ground of the cosmos where space based communications, navigation, scientific and surveillance systems are all affected by the operating environment. It includes the ionized turbulences in the ionopshere, which can degrade the communication signal between Earth and orbital devices and full impacts range from the energetic particles in the Van Allen radiation belts which can disrupts the satellite microelectronic, and even fluctuations in the magnetosphere induced by the solar emissions. The major availability of data of solar wind respect of magnetic field measures, such as the minor cost, even in terms of infrastructures, technological effort and site-related problems, aim to correlate the magnetospheric response with the velocity of solar wind. As recently proposed (Napoli, Bonanno & Capizzi 2010) the problem can be faced by using hybrid computing techniques as wavelet and recurrent neural networks.

2. The basic of neuro-wavelet theory

Neural networks (NNs) are a promising technique to exploit the complexity of non-linear data correlation, but the main problem in connectionist theory regards the development of learning algorithms that can optimize the computational capabilities of these networks. A wide variety of different approaches have been proposed during the last years, in particular a branch of research have focused on learning algorithms that use recurrent connections to deal with time series (Williams & Zipser 1989). In a general framework for this kind of problems it is possible to unfold a recurrent neural network (RNN) into a multilayer feedforward network (FF) that grows of one layer on each time step. This approach is called backpropagation through time (BPTT). The main advantages of BPTT relies on his generality, then is possible to improve the predictive capability and reduce the training time of this kind of RNNs using adaptive amplitude of activation functions based on a real time recurrent algorithm (AARTRL) which can be employed as nonlinear adaptive filter for fully connected RNNs. The inputs for the RNN were given by the set of coefficients coming from the biorthogonal wavelet decomposition of the solar wind velocity time series. Wavelet analysis was used in order to reduce the data redundancies so obtaining representation which can express their intrinsic structure. In fact the main advantage of the use of wavelet is the ability to pack the energy of a signal, and then the relevant informations carried by it, in few significant uncoupled coefficients. This characteristic is very useful to optimize the performances of neural networks (M. M. Gupta, L. Jin, N. Homma 2003). As recently proposed a biorthogonal wavelet decomposition was used due to his freedom degrees. The biorthogonal wavelets utilize a particular transform, which is invertible but not necessarily orthogonal, based on coupled filters, and allowing more degrees of freedom respect to traditional orthogonal (and of course the not orthogonal) wavelets (S. Mallat 2009).

3. Proposed data analysis

In this paper the authors propose hybrid fully recurrent neuro-wavelet RNNs, to correlate the magnetopsheric response to the solar wind and forecast these variations from predicted solar wind data. To obtain a correct prediction of magnetic field from solar wind velocity data an RNN was designed starting on the data series. The inputs for the RNN were given by the set of coefficients coming from the biorthogonal wavelet decomposition of the solar wind velocity time series. By using the MATLAB package we developed a series of routines performing a biorthogonal wavelet analysis of the data of solar wind and magnetic field extract a shortened number of non-zero coefficients from a representative time series. This method allowed to drastically reduce the coefficients data sets aiming to correlate them. The subbands of the decomposition of solar wind data then were used to predict the future values of wavelet coefficients for the magnetic field. As advanced features of the process the temporal trend was associated with the different sub-bands of the decomposition, allowing to subdivide the neural network in modular serial subnets in order to gain computation time and drop down the complexity. The available data, regarding a time period from january 1998 to may 2010, were collected during the WIND mission of the NASA administration. WIND is a spin stabilized spacecraft launched in November 1, 1994 and placed in a halo orbit around the L1 Lagrange point, more than 200 Re upstream of Earth to observe the unperturbed solar wind that is about to impact the magnetosphere of Earth. The solar wind data are provided, as done in a previous paper by the SWE. While the magnetic field data are provided by the Magnetic Field Investigation, a subsystem consistent of dual triaxial fluxgate magnetometers on a 12 m radial boom, designed to provide accurate, high resolution vector magnetic field measurements in near real time on a continuous basis in a wide dynamic measuring range. A detailed description of the experimental system can be found in the (PhD dissertation of J. Kasper 2002). The solar wind data were deceomposed as reported in (Napoli, Bonanno

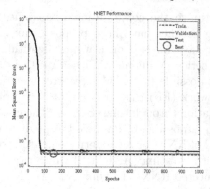

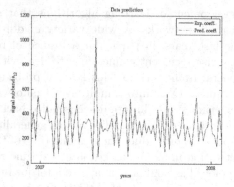

Figure 1. Performance of the NN   **Figure 2.** Magnetic field prediction

& Capizzi 2010), while the magnetic field data were also decomposed in 13 subbands with the described method reducing the 99.75% of the energy in 0.05% of non zero coefficients. We trained the network to the 12th temporal sub-band to predict the rate of change in the magnetic field for the next 6 days by using solar wind data. For a magnetic field time sub band b_n, and in order to predict the data at a time interval τ_1 from a time serie of a previous interval τ_0, then the input for the NN is given in the form of solar wind time series $[a_{n+1}(\tau_0)|d_n(\tau_0)]$ so that the output will be the sub band $b_n(\tau_1)$. The NN was a full RNN NARX, trained by a gradient descent with momentum and adaptive amplitude learning rule backpropagation algorithm and composed by 3 hidden layer: the first of 13 neuron with tribas transfer function (tf), the second of 7 neurons with tansig tf, and the third of 5 neurons with a competitive tf. Finally time vectors were used to create 5 delay lines for the input and 3 for the output.

4. Conclusions and results

The simulation results show the power and the good performance obtained with the proposed method. The predicted magnetic field accurately matches the experimental data within an RMS of less than 0.4% as shown in fig. 1 and 2. The high degree of accuracy highlights the correlation between magnetic field measurements in the L1 point and solar activity in the form of solar wind creating the basis for further works about prediction of solar manifestations (such coronal holes or flares) by mapping the magnetic field near the Earth orbit.

References

Napoli C. *et al.* 2010, in: Advances in Plasma Astrophysics, Proc. IAU Symposium No. 274
Capizzi G., Bonanno F., & Napoli C. 2010, Proc. Speedam 2010, p. 586
Williams R. J. & Zipser D. 1989, *Neural Comput.*, Vol.1, p. 270
M. M. Gupta *et al.* 2003, *Static and Dynamic Neural Networks*, J. Wiley & Sons Inc.
S. Mallat 2009, *A Wavelet Tour of Signal Processing: The Sparse Way*, Academic Press
Kasper J. 2002, *Excerpts from PhD dissertation*, Cap.2, p. 49
Gleisner H., Lundstedt H., & Wintoft P. 1996, *Ann. Geophys.*, Vol.4, No.7, p. 679
Eselevich V. G. *et al.* 2009, *Cosmic Res.*, Vol.47, No.2, p. 95

Advances in Plasma Astrophysics
Proceedings IAU Symposium No. 274, 2010   © International Astronomical Union 2011
A. Bonanno, E. de Gouveia Dal Pino & A. G. Kosovichev eds.   doi:10.1017/S1743921311006818

A shell model for turbulent dynamos

G. Nigro, D. Perrone and P. Veltri

Università della Calabria, Dipartimento di Fisica and Centro Nazionale Interuniversitario
Struttura della Materia, Unità di Cosenza, I-87030 Arcavacata di Rende, Italy
email: giusy.nigro@gmail.com

Abstract. A self-consistent nonlinear dynamo model is presented. The nonlinear behavior of the plasma at small scale is described by using a MHD shell model for fields fluctuations; this allow us to study the dynamo problem in a large parameter regime which characterizes the dynamo phenomenon in many natural systems and which is beyond the power of supercomputers at today. The model is able to reproduce dynamical situations in which the system can undergo transactions to different dynamo regimes. In one of these the large-scale magnetic field jumps between two states reproducing the magnetic polarity reversals. From the analysis of long time series of reversals we infer results about the statistics of persistence times, revealing the presence of hidden long-time correlations in the chaotic dynamo process.

Keywords. Dynamo models; magnetic reversals; MHD turbulence; shell models

1. Introduction

The understanding of dynamo problem is fundamental to explain the origin and the self-sustained of large scale magnetic field observed in natural systems like planets, stars, galaxies, black holes etc.

Many researchers (e.g., Kagemayama *et al.*, 2008) dealt with this problem using direct numerical simulation (DNS), even if realistic parameter regimes are beyond the power of actual supercomputers. Difficulties arise from the realistic description of both large-scales and small-scale (high Reynolds numbers) turbulence. Actual DNS are able to simulate only some few polarity reversals of the magnetic field. In order to overcome these difficulties we have built a model which takes into account very large Reynolds numbers and is able to reproduce very long time series of reversals which can be statistically analyzed giving the possibility to make a comparison with paleomagnetic data.

2. Turbulent Shell Dynamo Model

The starting point of our model is the decomposition of the fields in an average part, varying only on the large scale L, and a turbulent fluctuating part, varying at small-scales $\sim \ell$, with the assumption $\ell \ll L$ (Parker, 1955). Performing this scale separation we obtain, in the induction equation at large scale, a term which describes the action of small scales on the large one consisting in a turbulent e.m.f. that can be written in terms of the Fourier modes of velocity ($\mathbf{u}(\mathbf{k}, t)$) and magnetic field ($\mathbf{b}(\mathbf{k}, t)$) small scale fluctuations as follow:

$$\epsilon - -\sum_{\mathbf{k}} \mathbf{u}(\mathbf{k}, t) \times \mathbf{b}^*(\mathbf{k}, t) . \tag{2.1}$$

Introducing a basis in the spectral space: $\widehat{e_1}(\mathbf{k})$, $\widehat{e_2}(\mathbf{k}) = \widehat{e_3}(\mathbf{k}) \times \widehat{e_1}(\mathbf{k})$, $\widehat{e_3}(\mathbf{k}) = i\mathbf{k}/|\mathbf{k}|$; and writing expression (2.1) in a form symmetric with respect to the change of $\mathbf{k}$ in $-\mathbf{k}$

we finally find:

$$\epsilon = -\sum_{\mathbf{k}(k_z>0)} \hat{e}_3 \left[(u_1^* \, b_2 - u_2 \, b_1^*) + (u_2^* \, b_1 - u_1 \, b_2^*)\right] \tag{2.2}$$

where u_1 and u_2 (b_1 and b_2) are the components of $\mathbf{u}(\mathbf{k},t)$ ($\mathbf{b}(\mathbf{k},t)$), along $\hat{e}_1$ and $\hat{e}_2$.

We describe the dynamics of the system at large scale by integrating the induction equation in an axisymmetric situation and local approximation: we approximate the toroidal ($\widehat{e_\varphi}$) and the poloidal ($\widehat{e_p}$) unit vectors with the cartesian unit vectors, $\hat{e}_x$ and $\hat{e}_z$. Hence the field at large scale are: $\mathbf{u}_0 = V(y,z)\,\hat{e}_x$; $\mathbf{b}_0 = B_\phi(y,t)\hat{e}_x + B_p(y,t)\hat{e}_z$.

The dynamics at small scales is described by a shell model. At variance with the original MHD shell model (Frick and Sokoloff, 1998; Giuliani & Carbone, 1998), here the nonlinear interactions avoid unphysical correlations of phases for each interacting triad. We can write the set of self-consistent equations for our dynamo model coupling the Eqs. for the small scales and the Eqs. for the field at large scale in which the e.m.f. is in a form consistent with the shell model and the spatial derivative associated with large scale is estimated dividing by the typical large scale L:

$$\frac{dB_\phi}{dt} = \frac{i}{L}\sum_n (u_n^* b_n - u_n b_n^*) + B_p\frac{V}{L} - \eta\frac{B_\phi}{L^2}\,, \tag{2.3a}$$

$$\frac{dB_p}{dt} = \frac{i}{L}\sum_n (u_n^* b_n - u_n b_n^*) - \eta\frac{B_p}{L^2}\,, \tag{2.3b}$$

$$\frac{du_n}{dt} = k_n(B_\phi + B_p)b_n + ik_n\left[(u_{n+1}^* u_{n+2} - b_{n+1}^* b_{n+2}) + \right.$$
$$\left. -\frac{1}{4}(u_{n-1}^* u_{n+1} - b_{n-1}^* b_{n+1}) + \frac{1}{8}(u_{n-2}u_{n-1} - b_{n-2}b_{n-1})\right] - \nu k_n^2 u_n + f_n, \tag{2.3c}$$

$$\frac{db_n}{dt} = ik_n(B_\phi + B_p)u_n + \frac{ik_n}{6}\left[(u_{n+1}^* b_{n+2} - b_{n+1}^* u_{n+2})\right.$$
$$\left. +(u_{n-1}^* b_{n+1} - b_{n-1}^* u_{n+1}) - (u_{n-2}b_{n-1} - b_{n-2}u_{n-1})\right] - \eta k_n^2 b_n; \tag{2.3d}$$

where ν is the viscosity and η is the diffusivity of the MHD flow; n is the shell number ($n = 0, ...N$); f_n is an external forcing term applied only on the first shell $k_0 \sim 2\pi/\ell$ ($n = 0$). This is an exponentially correlated Gaussian noise, characterized by a second moment $< f_1^2 >= \sigma^2/\ln 10$ and a correlation time $\tau_c = 1$, which has the property to preserve the energy flux to small scales (Giuliani & Carbone, 1998).

We solve the model Eqs. assuming $V = 0$, that is equivalent to solve the dynamo problem for Rossby number $\mathrm{Ro} = \frac{\delta u}{V}\frac{\delta b}{B_p} \gg 1$. Therefore $B_p(t) = B_\phi(t) = \mathrm{B}(t)$ and the model describes α^2 dynamo problem.

3. Numerical results

The model Eqs. are numerically solved by a fourth order Runge-Kutta scheme. The results are in dimensionless units: the field fluctuations are measured in Alfvén velocity unit c_A, the time in eddy-turn-over time ($1/(k_0 u_0)$), the lengths are normalized to $1/k_0$, and finally the dissipative coefficients are normalized to c_A/k_0.

At the beginning of each simulation, we let the system become turbulent at small scales. After that, we introduce a magnetic field seed of amplitude 10^{-10} at large scale and we check whether the dynamo effect starts to develop.

The numerical results reveal a strong sensitivity of the system with respect to the magnetic Reynolds number $\mathrm{Rm} \simeq \delta u/k_0\eta$ and a dependence on the hydrodynamic Reynolds number $\mathrm{Re} \simeq \delta u/k_0\nu$, where δu is the r.m.s. of the turbulent velocity fluctuations.

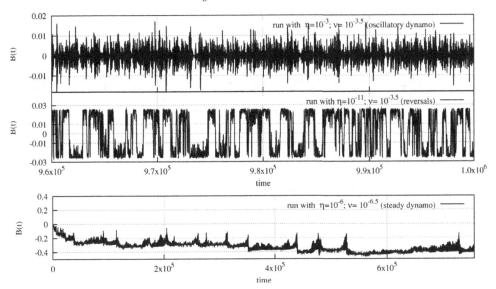

Figure 1. Time evolution of the large scale magnetic field in dimensionless units in different simulations.

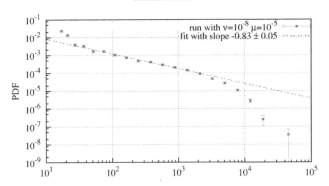

Figure 2. PDF of persistence times Δt in logarithmic scale.

Depending on these parameters, the system evolves towards different scenarios: i) no dynamo; ii) oscillatory dynamo; iii) magnetic reversals; iv) steady dynamo (see Fig. 1).

The model give us the capability to reproduce a long series of magnetic polarity reversals which can be statistically analyzed: the PDFs display a power law behavior (see Fig. 2), revealing the presence of hidden long-time correlations in the chaotic dynamo process. This is an argument in favor of some degree of memory in the chaotic dynamo as observed from analysis on the CK95 dataset of paleomagnetic inversions (Jonkers, 2003; Carbone *et al.*, 2006; Sorriso *et al.*, 2007; Nigro & Carbone, 2010).

References

Kageyama A., Miyagoshi T., & Sato T. 2008, *Nature letters* 454, 1106

Parker E. N. 1955, *ApJ* 122, 293

Frick P. & Sokoloff D. 1998, *Phys. Rev. E* 57, 4155

Giuliani P. & Carbone V. 1998, *Europhys. Lett.* 43, 527

Jonkers A. R. T. 2003, *Phys. Earth and Planet. Int.* 135, 253

Carbone V. *et al.* 2006, *Phys. Rev. Lett.* 96, 128501

Sorriso-Valvo L. *et al.* 2007, *Phys. Earth Planet. Inter.* 164, 197

Nigro G. & Carbone V. 2010, *Phys. Rev. E* 82, 016313

Advances in Plasma Astrophysics
Proceedings IAU Symposium No. 274, 2010   © International Astronomical Union 2011
A. Bonanno, E. de Gouveia Dal Pino, & A. G. Kosovichev, eds.   doi:10.1017/S174392131100682X

Energetic particle acceleration and transport by Alfvén/acoustic waves in tokamak-like Solar flares

Martin Obergaulinger[1] and Manuel García-Muñoz[2]

[1] Max-Planck-Institut für Astrophysik,
Karl-Schwarzschild-Straße 1, D-85741 Garching, Germany
email: mobergaulinger@mpa-garching.mpg.de

[2] Max-Planck-Institut für Plasmaphysik,
Boltzmannstraße 2, D-85748 Garching, Germany
email: mgmunoz@ipp.mpg.de

Abstract. Alfvén/acoustic waves are ubiquitous in astrophysical as well as in laboratory plasmas. Their interplay with energetic ions is of crucial importance to understanding the energy and particle exchange in astrophysical plasmas as well as to obtaining a viable energy source in magnetically confined fusion devices. In magnetically confined fusion plasmas, an experimental phase-space characterisation of convective and diffusive energetic particle losses induced by Alfvén/acoustic waves allows for a better understanding of the underlying physics. The relevance of these results in the problem of the anomalous heating of the solar corona is checked by MHD simulations of Tokamak-like Solar flare tubes.

Keywords. acceleration of particles, waves, (magnetohydrodynamics:) MHD, Sun: flares

1. Introduction

The interplay of Alfvén/acoustic waves with energetic ions is crucial for energy and particle exchange in astrophysical plasmas (e.g., the Solar corona and wind) as well as to obtaining a viable energy source in magnetically confined fusion devices where the excitation of shear Alfvén waves such as Alfvén cascades (ACs), toroidal Alfvén eigenmodes (TAEs) and Alfvén-acoustic eigenmodes is of important for the fast-ion transport across field lines because of their potential to eject fast ions before their thermalization. Despite the very different values of plasma parameters, most of the normalized characteristic lengths and frequencies are often of the same order in Solar flares and Tokamaks.

2. Particle acceleration processes

A large wave-particle exchange of energy and momentum takes place if the resonance condition $\omega - k_\parallel \cdot v_\parallel - l\Omega_c \approx 0$ is fulfilled. Here, ω is the mode frequency, Ω_c the ion cyclotron frequency and $k_\parallel$ and $v_\parallel$ the parallel components of the wave vector and particle velocity. A high phase-space density of resonances leads to a phase mixing in a stochastic phase-space with non-linear exchanges of energy and momentum.

In Tokamaks, radial chains of overlapping Alfven/acoustic modes have been observed to cause strong coherent and incoherent fast-ion losses. Core-localized acoustic fluctuations have a severe impact on the overall fast-ion loss if the particle phase-space is covered by overlapping/crossing main and sideband resonances. Detailed *in situ* measurements of these losses, in particular of the MHD internal structures and particle distribution functions, performed at the Tokamak ASDEX Upgrade (Garcia-Munoz *et al.* 2008) can

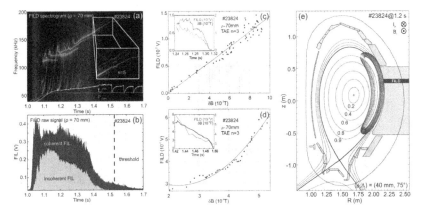

Figure 1. Left panel: Fourier analysis of the phase-space (energy and pitch angle) of fast-ion losses. Central panel: rate of the coherent and incoherent fast-ion losses as a function of the respective AE magnetic fluctuation amplitude. Right panel:Typical orbit of an escaping hydrogen ion with E ≈ 200 keV.

be used to study the underlying physical mechanisms. A Fourier analysis of the fast-ion loss signal Garcia-Munoz *et al.* 2010a allows us to identify the MHD fluctuations responsible for these losses (left panels of Fig. 1). The spectrogram reproduces the fast-ion losses correlated in frequency and phase with the related MHD fluctuation. The strongest coherent losses appear at the TAE frequencies and at the AC frequencies during their interaction with the acoustic branch near the f_{AC}^{min} (geodesic frequency) and during the AC-TAE transition. Diffusive fast-ion losses have been observed with a single TAE above a certain threshold in the local fluctuation amplitude.

The rate of the coherent and incoherent fast-ion losses as a function of the respective AE magnetic fluctuation amplitude is shown in Fig. 1, central panel. We can identify clear differences between particle ejection by single ACs and TAEs (directly proportional to the fluctuation amplitude, δB) and by the overlapping of multiple AC and TAE spatial structures and wave-particle resonances leading to a large diffusive loss that scales as $(\delta B/B)^2$. Simulations using the HAGIS drift orbit code (Pinches, 1998 show that the entire fast-particle phase-space in the energy range measured by the fast-ion loss detectors appear virtually covered by wave-particle resonances, enabling a stochastic fast-ion phase-space (Garcia-Munoz *et al.* 2010b). The right panel of Fig. 1 shows the typical trajectory of a detected escaping hydrogen ion with ≈ 200 keV.

3. Solar flares

We follow the evolution of a simplified model for Solar flares in MHD simulations (Mikic & Linker, 1994); Fig. 2 shows the model at different times. The initial conditions are a simple model of the solar atmosphere in magnetohydrodynamic equilibrium, permeated by a magnetic field emerging from a bipolar group of spots (left panel). A combined converging and shearing motion on the solar surface triggers the rise of magnetic loops (central panel) until an intense current sheet develops, and reconnection sets in, accelerating the emergence of the gas (right panel). A magnetic island disconnects from the surrounding field lines. In 3d, this corresponds to an extended flux rope anchored in the solar surface–a configuration resembling the geometry of the plasma in a tokamak.

To study resonant interactions between waves and particles, we compare Alfvén wave frequencies and particle orbital frequencies (left and central panels of Fig. 2) in the background field of the flare. We integrate the guiding-centre motion of particles gyrating

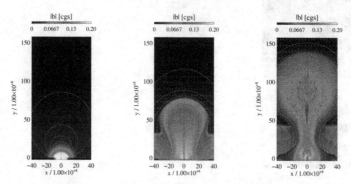

Figure 2. Snapshots of the MHD simulation of a solar flare showing the magnetic field strength (colours), field lines (white), and velocity vectors at (left to right) $t = 0$, $t = 3000$ s, $t = 4000$ s.

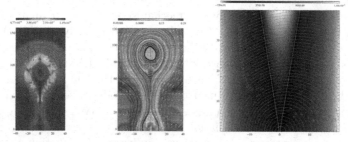

Figure 3. Left panel: frequencies of Alfvén (colour coded) propagating along sample field lines. Central panel: particle orbits in the flare model. Right panel: a resonance diagram for particles orbiting in the flare.

around the field lines of the model including the $\vec{E} \times \vec{B}$ drift; the vectors plotted show the velocity field of the guiding-centre drift.

We compute resonance diagrams (Karimabadi *et al.* 1990) in the phase space of particle momentum parallel and perpendicular to the wave vector, $p_{\parallel}$ and $p_{\perp}$. For a given particle energy, pitch angle and wave frequency, we compute the intersections between R surfaces fulfilling the condition for resonance, and H (Hamiltonian) surface defined by an energy equal to the particle energy. Resonance is possible if these surfaces intersect; if the intersections lie close to each other, the particle can gain energy by successive interactions in adjacent resonances. In the right panel of Fig. 3, the colours show the Hamiltonian function H for a proton of 1 MeV energy; white lines show the H surfaces where the particle is located. Green lines show R surfaces. The densely distributed interaction points indicate a high potential for resonant acceleration of the particle.

References

Garcia-Munoz M. *et al.*, 2008, *Phys. Rev. Lett.* **100** 055005.
Garcia-Munoz M. *et al.*, 2010a, *Phys. Rev. Lett.* **104** 185002.
Garcia-Munoz M. *et al.*, 2010b, *Nucl. Fusion* **50** 084004.
Pinches S. D. *et al.*, 1998, *Comput. Phys. Commun.* **111** 131.
Mikic Z. & Linker J.-A., 1994, *ApJ* **430**.
Karimabadi H. *et al.*, 1990, *Phys. of Fluids B* **2**.

Advances in Plasma Astrophysics
Proceedings IAU Symposium No. 274, 2010
A. Bonanno, E. de Gouveia Dal Pino & A.G. Kosovichev, eds.
© International Astronomical Union 2011
doi:10.1017/S1743921311006831

CME evolution and 3D reconstruction with STEREO Data

A. Orlando[1,2]**, F. Zuccarello**[1]**, P. Romano**[2]**, F.P. Zuccarello**[3]**,
M. Mierla**[4,5,6]**, D. Spadaro**[2] **and R. Ventura**[2]

[1]Dipartimento di Fisica e Astronomia, Sezione Astrofisica, Via S. Sofia 78, 95123 Catania, Italy

[2]INAF Osservatorio Astrofisico di Catania, Italy

[3]Centre for Plasma Astrophysics, K.U. Leuven, Belgium

[4]Royal Observatory of Belgium, Brussels, Belgium

[5]Institute of Geodynamics of the Romanian Academy, Bucharest, Romania

[6]Research Center for Atomic Physics and Astrophysics, University of Bucharest, Romania

Abstract. We describe a CME event, occurred in NOAA 11059 on April 3 2010, using STEREO and MDI/SOHO data. We analyze the CME evolution using data provided by SECCHI-EUVI and COR1 onboard STEREO satellites, and we perform a 3D reconstruction of the CME using the LCT-TP method. Using MDI/SOHO line-of-sight magnetograms we analyze the magnetic configuration of NOAA 11059 and we determine the magnetic helicity trend.

Keywords. Sun: coronal mass ejections (CMEs), Sun: magnetic fields

1. Introduction

Coronal mass ejections (CMEs) are eruptions of large amounts of mass (10^{16} g) and embedded magnetic fields from the Sun. Using data provided by the SECCHI corona-graphs onboard the STEREO mission (Kaiser *et al.*, 2008), we can infer the propagation direction and the 3-D structure of such events.

The SECCHI experiment (Howard *et al.*, 2008) on the STEREO mission is a suite of remote sensing instruments consisting of an extreme UV imager (EUVI), two white light coronographs (COR1, observing from 1.4 to 4 $R_\odot$, and COR2, characterized by a field of view from 2.5 to 15 $R_\odot$), and two Heliospheric Imager (HI1 and HI2) observing the heliosphere from 15 to 318 $R_\odot$.

We have analyzed a CME occurred on April 3 2010 at 09:05 UT in NOAA 11059 (S25 W03). The GOES 14 satellite (Stern *et al.*, 2004) recorded a B7.4 flare, beginning at 09:04 UT, with peak at 09:54 UT and ending at 10:58 UT, occurring in the same active region. In Fig. 1 (a) - (b) we report STEREO EUVI-A and B 195 Å images of the event.

2. CME observation and reconstruction

To obtain the images of the CME event with SECCHI COR1, a minimum intensity background over the images of the day when the CME was observed is subtracted from each frame containing the CME; the images are co-aligned in STEREO mission plane. Fig. 1 (c) - (d) show the CME observed by COR1 A and B, respectively.

Using the LCT (local correlation tracking) method we identify the same feature in COR1-A and COR1-B images (this technique is described in Mierla *et al.*, 2009). Once the association is found, making use of the position of the two spacecrafts, we obtain the 3D coordinates of our feature by a triangulation technique (sometimes referred as tie-pointing (TP), see Inhester, 2006).

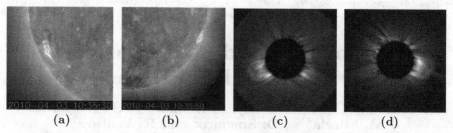

| (a) | (b) | (c) | (d) |

Figure 1. (a) STEREO-EUVI A image of NOAA 11059 acquired at 10:35 UT: a system of post-flare loops is visible at the flaring site; (b) STEREO-EUVI B image, showing the same event at the west-southern limb; (c) STEREO-COR1 A image, showing the CME at 09:45 UT; (d) COR1 B image of the CME at the same time. The separation angle between STEREO A and B on 3 April 2010 was 138.6 degrees.

Fig. 2 shows the LCT-TP reconstruction from three different points of view. The coordinate system is HEEQ (Heliocentric Earth Equatorial), with the origin in the Sun center, X pointing towards Earth and Z towards the solar north. The greyscale colors indicate the distance from the Sun center: black closest to the Sun and grey-white closest to the Earth.

3. Magnetic Configuration

To analyze the magnetic configuration of NOAA 11059 we used MDI/SOHO line-of-sight magnetograms acquired between 00:00 UT on April 1 to 04:00 UT on April 5 2010. We found that the magnetic flux has values within 4.2 - 6.0 $\times$ 10^{21} Mx, and is characterized by an initial decreasing trend, followed by a modest increase in the hours preceding the CME. After the CME, the magnetic flux shows an abrupt decrease and later on it remains almost constant.

We performed a force-free field extrapolation on the MDI magnetogram acquired at 07:59 UT using the method introduced by Alissandrakis (1981) and the value for the force-free field parameter $\alpha = 0.02$. The results show several systems or bundles of field lines. These bundle of field lines have the same shape of the loops observed by TRACE at 195 Å and some of them outline the magnetic arcades supporting and confining some filaments observed in Hα images. These field lines most probably correspond to the post-flare loop system observed in STEREO A images (Fig. 1 (a)).

To determine the magnetic helicity in the active region we determined the horizontal motions of the field line footpoints using the differential affine velocity estimator (DAVE) method (Schuck, 2005), with a window size of 20 arcsec and a time interval between two successive frames of 96 min. Subsequently we estimated the magnetic helicity rate and helicity accumulation using the method described by Pariat *et al.*, (2005). The results indicate that the rate of magnetic helicity is extremely variable, with an initial predominance of negative values, followed by positive values after about one day from the initial time (00:00 UT on April 1 2010). Concerning the accumulated magnetic helicity, the maximum negative value is reached at about 00:00 UT on April 2, and later the helicity increases almost continuously. However, immediately after the CME occurrence, there is an abrupt change in the accumulated magnetic helicity.

4. Discussion and Conclusion

Using STEREO-EUVI data we could visualize a system of post-flare loops developing after more than one hour from the beginning of a B7.4 flare associated to the CME.

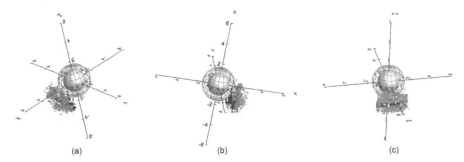

Figure 2. (a) 3D reconstruction from STEREO A view point at 09:45 UT, as seen edge-on; (b) CME 3D reconstruction as seen by STEREO B; (c) CME 3D reconstruction as seen by the Earth.

The analysis of Hα images showed the presence of some filaments along the magnetic neutral line: we believe that the activation of parts of these filaments gave rise to the B7.4 flare and to the CME. This hypothesis is corroborated by the results obtained from the force-free field extrapolation, showing the presence of several systems or bundles of field lines, characterized by different heights and connectivity domains, some of which clearly resembles the post-flare loops system seen by EUVI on STEREO A.

Using the LCT-TP method to COR1 data acquired at 09:45 UT, we carried out the 3D reconstruction of the CME: due to the optically thin properties of the CME plasma, it is not possible to get a full 3D geometry of the CME, but an estimate of the direction of propagation can be accurately determined (from Fig. 2 (c) it is clear that the CME is directed towards the Earth, and indeed, the CME hit the Earth on 5 April).

The analysis of the MDI magnetograms indicates that the magnetic flux showed a small increase in the hours preceding the CME and an abrupt decrease immediately after it. The helicity accumulation changed abruptly after the CME, indicating a torque unbalance between the subphotospheric and the coronal domain of the magnetic field (see, e.g., Smyrli *et al.*, 2010).

Acknowledgements

This work was supported by the European Commission through the SOLAIRE Network (MTRN-CT-2006-035484), by the Istituto Nazionale di Astrofisica (INAF), by the Catania University, by the Agenzia Spaziale Italiana (contract I/015/07/0).

References

Alissandrakis, C. E. 1981, *A & A*, 100, 197
Howard R. A., Moses J. D., Vourlidas A. *et al.* 2008, *Space Sci. Rev.*, 136, 67
Kaiser M. L., Kucera T. A., Davila J. M., *et al.*, 2008, *Space Sci. Rev.*, 136, 5
Inhester B. 2006, *arXiv:astro-ph/0612649*
Pariat E., Démoulin P., & Berger M. A. 2005, *A & A*, 439, 1191
Mierla, M., Inhester, B., Marqu/'e, C., Rodriguez, L., Gissot, S., Zhukov, A. N., Berghmans, D., & Davila, J. 2009, *Solar Phys.*, 259, 123
Schuck P. W. 2005, *ApJ*, 632, L53
Smyrli A., Zuccarello, F., Romano, P., Zuccarello, F. P., Guglielmino, S. L., Spadaro, D., Hood, A. W., & Mackay, D. 2010, *A & A*, in press
Stern, R. A., *et al.* 2004, *Proc. SPIE*, 5171, 77

Advances in Plasma Astrophysics
Proceedings IAU Symposium No. 274, 2010
A. Bonanno, E. de Gouveia Dal Pino & A.G. Kosovichev, eds.
© International Astronomical Union 2011
doi:10.1017/S1743921311006843

Hybrid Vlasov simulations for alpha particles heating in the solar wind

Denise Perrone[1], Francesco Valentini[2,1] and Pierluigi Veltri[1]

[1]Università della Calabria, Dipartimento di Fisica and CNISM,
Unità di Cosenza, I-87030 Arcavacata di Rende, Italy
email: denise.perrone@fis.unical.it; pierluigi.veltri@fis.unical.it

[2]Università di Pisa, Dipartimento di Fisica and CNISM, 56127 Pisa, Italy
email: francesco.valentini@fis.unical.it

Abstract. Heating and acceleration of heavy ions in the solar wind and corona represent a long-standing theoretical problem in space physics and are distinct experimental signatures of kinetic processes occurring in collisionless plasmas. To address this problem, we propose the use of a low-noise hybrid-Vlasov code in four dimensional phase space (1D in physical space and 3D in velocity space) configuration. We trigger a turbulent cascade injecting the energy at large wavelengths and analyze the role of kinetic effects along the development of the energy spectra. Following the evolution of both proton and α distribution functions shows that both the ion species significantly depart from the maxwellian equilibrium, with the appearance of beams of accelerated particles in the direction parallel to the background magnetic field.

Keywords. plasmas, turbulence, waves.

Helium is the second most abundant element in the Sun and plays an important and not yet understood role in the dynamics of the solar wind, corona and interior. In situ measurements in the solar wind have clearly shown that the heavy minor ions in this essentially collisionless space plasma are heated and accelerated preferentially as compared to protons and electrons. These observations indicate that there are sources of ion heating and momentum exchange which operate differently on protons and α throughout solar wind (Kasper *et al.*, 2008, Hansteen *et al.*, 1997). The investigation of these preferential effects has pushed several workers to study the quasi-linear resonant cyclotron interaction of solar wind ions with parallel-propagating ion cyclotron waves (Dusenbery & Hollweg, 1981; Marsch *et al.*, 1982, Isenberg & Hollweg, 1983; Isenberg, 1984; Hollweg & Isenberg, 2002; Xie & Ofman, 2004; Ofman, 2010).

The fast technological development of supercomputers gives the possibility of using kinetic Eulerian Vlasov codes that solve the Vlasov-Maxwell equations in multidimensional phase space. The use of these "zero-noise" codes is crucial since Eulerian algorithms (Mangeney *et al.*, 2002, Valentini *et al.*, 2007) allow for the first time the analysis of kinetic effects in the small-scale tail of the turbulent cascade, where the energy level of the fluctuations is typically very low. In this spectral region, Lagrangian PIC algorithms fail due to their intrinsic noise. In recent years, thank to the use of this code (4D or 5D), significant steps forward are made in the analysis of the evolution of solar wind turbulence towards dissipation (Valentini *et al.*, 2008; Valentini & Veltri, 2009; Valentini *et al.*, 2010; Valentini *et al.*, 2010). These numerical results have shown that, at least for $T_e/T_p > 3 - 4$, newly identified electrostatic (acoustic-like) modes, longitudinal with respect to the average magnetic field and driven by particle distribution functions far from local thermodynamic equilibrium, represent a privileged channel for the turbulence to carry the energy towards small disordered scales.

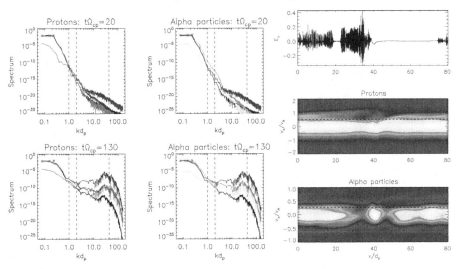

Figure 1. Energy spectra at two different times: magnetic (black line) and electric energy (blue line). For protons (left): kinetic energy (purple line) and density (red line). For α particles (middle): kinetic energy (yellow line) and density (green line). (Right) Parallel electric field versus x at $t = 150$ (top plot). $x - v_x$ level lines of the reduced distribution for protons (middle plot) and for α (bottom plot) at $t = 150$.

In the present paper, we propose the use of an updated version of the hybrid-Vlasov code, which includes the kinetic dynamics of heavy ions. Electrons are treated as a fluid and a generalized Ohm equation, that retains Hall effect and electron inertia terms, is considered. Faraday equation, Ampère equation (in which the displacement current is neglected) and an equation of state for the electron pressure close the system. Quasi-neutrality is assumed. The solutions of the above equations are obtained through a numerical hybrid-Vlasov code in 4D (1D in space and 3D in velocity) phase space. Periodic boundary conditions are imposed in physical space. Times are scaled by the proton cyclotron frequency Ω_{cp}, lengths by the proton skin depth $d_p = v_A/\Omega_{cp}$, v_A being the Alfvén speed, and masses by the proton mass m_p. We simulate a plasma embedded in a background magnetic field $\mathbf{B} = B_0 \mathbf{e}_x$, x being the direction of wave propagation. The system is perturbed at $t = 0$ by a circularly left hand polarized Alfvén wave in the perpendicular plane. We choose an isothermal equation of state for the electron pressure $P_e = (\beta/2)n_e T_e/T_p$, with $\beta = 2v_{tp}^2/v_A^2 = 0.5$, $T_e/T_p = 10$ and $T_\alpha = T_p$. The mass ratios are $m_e/m_p = 1/1836$ and $m_\alpha/m_p = 4$, the charge number is $Z_\alpha = 2$ and the density ratio is $n_\alpha^{(0)}/n_p^{(0)} = 5\%$. The length of the physical domain is $L_x = 12.8d_p$, while the limits of the velocity domain in each direction are fixed at $v_i^{max} = 5v_{th_i}$. We use 4096 gridpoints in physical space and 51^3 in velocity space. At $t = 0$, the first three modes in the spectrum of velocity and magnetic perturbations are excited with amplitude $\epsilon = 0.5$. No density disturbance is imposed at $t = 0$. The simulation is carried up to $t = 200$.

Figure 1 (left and middle) reports the numerical spectra of particle density ($|n_k^{(p)}|^2$, red line, $|n_k^{(\alpha)}|^2$, green line), kinetic ($|U_k^{(p)}|^2$, purple line; $|U_k^{(\alpha)}|^2$, yellow line), magnetic ($|B_k|^2$, black line) and electric ($|E_k|^2$, blue line) energies. The four plot refer to two different times in the simulations for protons and α particles. The same phenomenology described for the protons by Valentini *et al.*, 2008, is observed: the energy transfer to small scales is not driven by a continuos cascade, but for $t > 60$ a well defined group of wavenumbers are excited. The energy spectra are evidently peaked around $k \simeq 30$ and the magnetic energy is sensibly lower than electric one, meaning that electrostatic

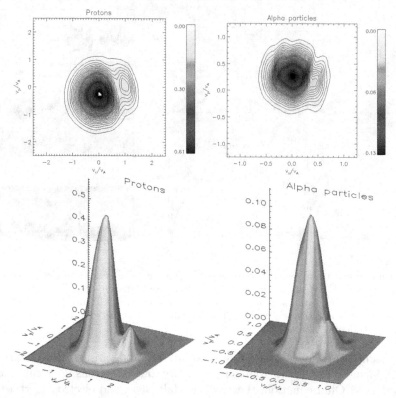

Figure 2. Level lines (at the top) and surface plot (at the bottom) of protons (left) and α particles (right) distributions in the velocity plane $v_x - v_y$ at $t = 150$.

activity is observed. The top right plot of Fig. 1 shows E_x as a function of x at $t = 150$; the middle and the bottom plots show the $x - v_x$ level lines of the reduced distribution for protons and α respectively. Phase space holes are generated in correspondence of the regions of highly impulsive behavior of E_x. These vortices are the typical signature of particles trapping. Another important phenomenon, recovered in correspondence of the maximum intensity of electrostatic activity in solar wind, is the generation of a double-stream ion distribution. These secondary beams, that move along the ambient field with different velocities, are shown in Fig. 2. The top plot displays the $v_x - v_y$ contour lines of f at $t = 150$ integrated over v_z and for a given point x_M in the physical domain, with x_M being the point where the $x-v_x$ trapped region has maximum velocity width. The bottom To identify the short-wavelength fluctuations observed in the simulations, we considered the κ-ω spectrum of the parallel electric field energy. This Fourier analysis displays the presence of only one branch of electrostatic waves, driven by kinetic trapping effects. The ion-acoustic waves, discussed in Valentini & Veltri (2009), are not present. These simulations show that the gross features of the kinetic effects occurring across the ion skin depth are not affected by the presence of α particles. Nevertheless this introduction allows to study the evolution of their distribution function and gives rise to some peculiar behavior.

References

Dusenbery, P. B. & Hollweg, J. V. 1981, *J. Geophys. Res.* 86, 153
Hansteen, V. H., Leer, E., & Holzer, T. E. 1997, *ApJ* 482, 498

Hollweg, J. V. & Isenberg, P. A. 2002, *J. Geophys. Res.* 107, A7, 1147

Isenberg, P. A. & Hollweg, J. V. 1983, *J. Geophys. Res.* 88, 3924

Isenberg, P. A. 1984, *J. Geophys. Res.* 89, A4, 2133

Kasper, J. C., Lazarus, A. J., & Gary, S. P. 2008, *Phys. Rev. Lett.* 101, 261103

Mangeney, A., Califano, F., Cavazzoni, C., & Travnicek, P. 2002, *J. Comput. Phys.* 179, 405

Marsch, E., Goertz, C. K., & Richter, K. 1982a, *J. Geophys. Res.* 87, 5030

Ofman, L. 2010, *J. Geophys. Res.* 115, A04108

Valentini, F., Trávníček, P., Califano, F., Hellinger, P., & Mangeney, A. 2007 *J. Comput. Phys.* 225, 753

Valentini, F., Veltri, P., Califano, F., & Mangeney, A. 2008, *Phys. Rev. Lett.* 101, 025006

Valentini, F. & Veltri, P. 2009, *Phys. Rev. Lett.* 102, 225001

Valentini, F., Califano, F., & Veltri, P. 2010, *Planetary and Space Science* doi:10.1016/j.pss.2009.11.007 (in press)

Valentini, F., Califano, F., & Veltri, P. 2010, *Phys. Rev. Lett.* 104, 205002

Xie, H. & Ofman, L. 2004, *J. Geophys. Res.* 109, A08103

Advances in Plasma Astrophysics
Proceedings IAU Symposium No. 274, 2010
A. Bonanno, E. de Gouveia Dal Pino & A.G. Kosovichev, eds.
© International Astronomical Union 2011
doi:10.1017/S1743921311006855

Solar dynamo in two-layer medium

Helen Popova

Faculty of Physics, M.V.Lomonosov Moscow State University
Leninskie Gory, Moscow, Russia
email: popovaelp@hotmail.com

Abstract. We studied the problem of the behavior of the magnetic field in the case of two-layer medium. We included of the meridional circulation in this model and investigated the influence of the meridional circulation on the nature of distribution and configuration of the dynamo-waves.

Keywords. waves, turbulence, MHD

1. Introduction

The cycles of solar magnetic activity associated with the action of the solar dynamo mechanism that is based on the combined effect of the differential rotation and alpha-effect. Such representation provides a solution in the form of oscillating waves of toroidal field, extending from middle latitudes to the equator. Scheme of the dynamo was proposed in the fundamental work by Parker (1955). The toroidal magnetic field is generated from the poloidal field by the differential rotation occurring in the solar convective zone. The reverse transformation of toroidal into poloidal field arises due to the absence of mirror symmetry in the convection occurring in the rotating body. In Parker (1955) it assumed that the generation of a dynamo occurs in one spherical shell, where the alpha-effect and differential rotation acting together. This scheme gives the duration of the cycle smaller than actually observed (Kuzanyan & Sokoloff (1995)). Accounting for the meridional circulation provided the necessary lengthening of the cycle (Popova *et al.* (2008), Popova & Sokoloff (2008), Popova (2009)).

However, consideration of a single-layer medium is described by one-way flow of matter and it does not allow describing return of matter. Parker suggested a way to describe an interface dynamo with dynamo generators distributed in two radial layers, using for each layer equations that are very similar to that of the Parker (1955) migratory dynamo Parker (1993). We developed a WKB method for the asymptotic solution of the corresponding dynamo equations. We consider two-layer medium, in which the layers have oppositely directed movement of substances and different diffusion coefficients. In this work we studied the problem of the behavior of the magnetic field in the case of two-layer medium with meridional circulation.

2. Basic Equations

We included the meridional flows in each layer of Parker equations

$$\frac{\partial B}{\partial t} + \frac{\partial (VB)}{\partial \theta} = \beta \Delta B, \qquad \frac{\partial A}{\partial t} + V\frac{\partial A}{\partial \theta} = \alpha B + \beta \Delta A, \qquad (2.1)$$

$$\frac{\partial b}{\partial t} + \frac{\partial (vb)}{\partial \theta} = D\cos\theta\frac{\partial a}{\partial \theta} + \Delta b, \qquad \frac{\partial a}{\partial t} + v\frac{\partial a}{\partial \theta} = \Delta a, \qquad (2.2)$$

for a dynamo with the α-effect present in one radial layer and shear of differential rotation present in the other. Here β is the ratio of the diffusion coefficients in first and second layers, $B(r, \theta, t), b(r, \theta, t)$ are the corresponding toroidal magnetic fields in these layers, and $A(r, \theta, t), a(r, \theta, t)$ are proportional to the corresponding toroidal components of the vector potential (which determines the poloidal magnetic field). Here $V(\theta), v(\theta)$ are the meridional flows in the respective layers, D is the dynamo number. Parker assumed the differential rotation to dominate in one layer and to vanish in the other, and, conversely, the α-effect to prevail in the second layer and to vanish in the first.

Following Parker we prescribe $r = 0$ for the radial boundary between two layers and use boundary conditions:

$$b = B, \qquad a = A, \qquad \frac{\partial b}{\partial r} = \beta \frac{\partial B}{\partial r}, \qquad \frac{\partial a}{\partial r} = \frac{\partial A}{\partial r}. \tag{2.3}$$

In view of the symmetry conditions $\alpha(-\theta) = -\alpha(\theta), V(-\theta) = -V(\theta)$ the above system of equations can be considered in only one (e.g., the northern) hemisphere using anti-symmetry (dipolar symmetry) or symmetry (quadrupolar symmetry) conditions at the equator.

We obtained Hamilton-Jacobi equation for Eqs. (2.1) and (2.2) by a method similar to the method described in Popova *et al.* (2010).

Remind that we represent the solution of Eqs. (2.1) and (2.2) in the form of waves travelling in the θ-direction with an appropriate parametrization of the r-dependence:

$$B = \mu e^{iD^{1/3}S\theta + \gamma D^{2/3}t - iD^{1/3}m_1 r}, \qquad A = (\nu + D^{2/3}\nu_1 r)e^{iD^{1/3}S\theta + \gamma D^{2/3}t - iD^{1/3}m_1 r}, \tag{2.4}$$

$$a = \zeta e^{iD^{1/3}S\theta + \gamma D^{2/3}t - iD^{1/3}m_2 r}, \qquad b = (\chi + D^{-2/3}\chi_1 r)e^{iD^{1/3}S\theta + \gamma D^{2/3}t - iD^{1/3}m_2 r}. \tag{2.5}$$

Here $\gamma, \nu, \nu_1, \varsigma, \chi, \chi_1, m_1, m_2$ are slowly varying functions of θ and $S(\theta) = \int k d\theta$.

S is similar to the action in quantum theory and $k = S'$ corresponds to the wave vector or impulse, which is complex in our case. The complex quantity γ determines the eigenvalue, with its real part giving the growth rate and its imaginary part the duration of the activity cycle.

Substituting the desired form of solution in Eqs. (2.1) and (2.2) and using boundary conditions we obtained Hamilton-Jacobi equation for eqs. (2.1) and (2.2):

$$(\beta\sqrt{-(\gamma + iVk)/\beta - k^2} + \sqrt{-\gamma - ivk - k^2})(\sqrt{-(\gamma + iVk)/\beta - k^2} + \tag{2.6}$$

$$+\sqrt{-\gamma - ivk - k^2}) = -\frac{4\hat{a}k}{i\beta\sqrt{-\gamma - ivk - k^2}\sqrt{-(\gamma + iVk)/\beta - k^2}}.$$

For simplicity, we assume that the meridional circulation does not depend on latitude. Ways to study Hamilton-Jacobi equations with the meridional circulation are similar to the methods described in Popova (2009).

3. Results

In result we obtained the ratio of the diffusion coefficients affect to duration activity cycle. The growth of the intensity of the meridional flow of matter slows the spread of the dynamo waves. Minimum of the magnetic solar activity may occur in the case of large intensity of the meridional circulation in both layers, as well as substantial differences in physical characteristics between the layers.

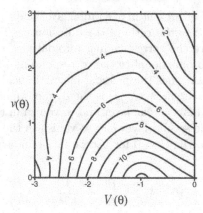

Figure 1. The ratios of β, $V(\theta)$, and $v(\theta)$ are for case when the duration of cycle is equal 11 years

Fig. 1 presents case when the meridional circulation is directed against spread of the dynamo waves in layer where differential rotation acting. Herewith, the meridional circulation is directed along spread of the dynamo waves in the layer where differential rotation acting. $V(\theta)$ is the meridional circulation in the layer with alpha-effect acting; $v(\theta)$ is the meridional circulation in the layer with differential rotation acting; isolines are *beta*. The ratios of β, $V(\theta)$, and $v(\theta)$ are for case when the duration of cycle is equal 11 years. To cycle duration was 11 years, as we can see from the figure, it must either increase the intensity of the meridional circulation in both layers or an increase β.

The more the meridional motion of matter is intense in the outer layer in comparison with the inner layer, the less β can be to achieve the 11-years cycle. If the meridional flows in the inner layer of a substance is stronger than in the external layer it will require greater value β to achieve the 11-years cycle than in the previous case.

If the values of $V(\theta)$ and $v(\theta)$ are interchanged the dependence of the cycle duration of β will not be changed.

For $V(\theta) = v(\theta) = 0$ the minimum of the solar activity ($\sim$ 100 years) is achieved in case when $\beta \approx 30$. If β tends to infinity the velocity of the dynamo wave spreading tends to zero. For $V(\theta) \neq v(\theta) \neq 0$ the minimum of the solar activity ($\sim$ 100 years) is achieved in case when either $\beta \approx 10$ and difference in the values of the velocities of the meridional flow from the different layers can be several tens or the order of β is several tens and difference in the values of the velocities can be about 10.

Obtained results results can be compared with observation data for β, $V(\theta)$, and $v(\theta)$.

Financial support from RFBR under grant 07-02-00127, 10-02-09688 and 09-02-01010 is acknowledged. Author is grateful to D. Sokoloff and M. Artyushkova for useful discussion.

References

Parker, E. N. 1955, *Astrophys. J.*, 122, 293-314.
Kuzanyan, K. M. & Sokoloff, D. D. 1995, *Geophys. Astrophys. Fluid Dyn.*, 81, 113-129.
Popova, H. P., Reshetnyak, M. Yu., & Sokoloff, D. 2008, *Astron. Rep.*, 52, 157-163.
Popova, E. P. & Sokoloff D. 2008, *Astron. Nachr.*, 329, 766-768.
Popova, E. P. 2009, *Astron. Rep.*, 53, 863-868.
Parker, E. N. 1993, *Astrophys. J.*, 408, 707-719.
Popova, H. P., Artyushkova M. E., & Sokoloff, D. 2010, *Geophys. Astrophys. Fluid Dyn.*, ID: 507201 (in print).

Advances in Plasma Astrophysics
Proceedings IAU Symposium No. 274, 2010
A. Bonanno, E. de Gouveia Dal Pino & A.G. Kosovichev, eds.

© International Astronomical Union 2011
doi:10.1017/S1743921311006867

Modeling circumstellar envelope with advanced numerical codes

P. Procopio[1], A. De Rosa[1], C. Burigana[1], G. Umana[2], and C. Trigilio[2]

[1]INAF - IASF Bologna, Via P. Gobetti 101, 40129 Bologna
[2]INAF - OACt, Via S. Sofia 78, Catania

Abstract. We propose a modeling study on the formation and evolution of the Circumstellar Envelopes (CSEs) of a sample of selected radio-loud objects, based on an innovative interaction between two codes widely used by the scientific community, but in different fields. CLOUDY† (Ferland *et al.* 1998) is a widely used code to model the spectral energy distribution (SED) of the several objects characterized by clouds of gas heated and ionized by a central object. CosmoMC‡ (Lewis & Bridle 2002) instead is usually used for exploring cosmological parameter space. We investigate here on the exploitation of the sampling performance of the Markov-Chain Monte-Carlo (MCMC) engine of CosmoMC to search for a best fit model of the considered objects through the spectral synthesis capacity of CLOUDY.

Keywords. circumstellar matter, planetary nebulae, MCMC

1. Introduction

Detailed observations of (CSE) are of crucial importance in determining fine details of different aspects related to the physical properties of plasma and dust around various classes of stars in their late evolution stages. The modeling and characterization of the CSEs are of extreme relevance for understanding the formation and evolution of relatively fast evolving objects, like e.g., Planetary Nebulae (PN), on which this contribution is focused. On the other side, the enrichment of the Interstellar Medium (ISM) depends on the chemistry and the composition of gas and dust that are coming out from PNs and Supernovae, through stellar winds or shockwaves. In this study, the spectral emission is predicted through CLOUDY, a performing public C++ based code, while the core of the optimization of the model parameters is CosmoMC, a FORTRAN90 MCMC engine, supplied with additional codes to analyze specific output data of the sampler.

2. The Data-set

The sample of PNs has been extracted from the Condon & Kaplan catalogue (1998). Only objects characterized by a flux density higher than 100 mJy at 1.4 GHz were selected (in order to guarantee a good detection with the Noto radiotelescope). Some observations has been already performed at 8.4 and 43 GHz (Umana *et al.* 2008). To consider the dust emission associated to the PNs, data from near (2 μm) to far (100 μm) infrared were retrieved as well: photometric measurements have been extracted from IRAS, 2MASS, and MSX archives.

---

† CLOUDY is available for free download at http://www.nublado.org
‡ ComoMC is available for free download at http://cosmologist.info/cosmomc/

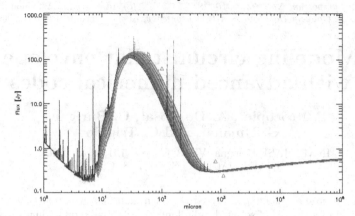

Figure 1. The effect of the variation of the density parameter of the SED model for NGC6543: increasing the parameter the peak of the model shifts to higher frequencies. The observed data are indicated with blue triangles.

3. Modeling through CLOUDY

We make wide use of the spectral synthesis code CLOUDY in order to fit models to observational data. We assume a reasonable set of initial parameters (central star radiation field, CSE density distribution, elements abundances and properties, geometric structure, filling factor, distance of the PN) describing the ionizing star of the considered PN. Once the initial model is obtained, it is possible to perform a fine tuning of some critical parameters in order to achieve their best estimate and to derive a fiducial model for the considered object. The variation of the parameters could be done inside CLOUDY itself, through its own optimization method, but it is necessary to provide a good initial guess of the parameters.

Astrophysical parameters - We focus in particular on the parameters essential for the modelization of the observed SED. Among them, we remember: the inner radius and the extension of the ionized plasma region, with typical values, for the inner radius, that may span between 10^{16} and 10^{19} cm; the filling factor, that takes into account the small scale structure of the nebula; the temperature of the central star, that may range from about ten to many tens of thousands Kelvin, and the total density of H (HII + HI + H$_2$) in the PN, that may range from some thousands to a million of atoms per cm^3. Also, the dust composition (astronomical silicate and/or graphite) and the grain dimensions play a fundamental role in shaping of the SED. These parameters act differently on the SED, but a certain degree of degeneracy persists.

Optimization process - The optimization process in CLOUDY is performed by the PHYMIR algorithm (van Hoof 1997), a non-standard fit procedure for minimizing the χ^2. The observables are divided in four categories: emission lines ratio, infrared emission, radio flux or H-β line, and angular diameter. The total χ^2 of the model is computed considering the contribution of each observable: $\chi_i^2 = [(O_i^m - O_i^o)/(min(O_i^m, O_i^o)\sigma_i)]^2$. Here the suffix o identifies the observed quantity, while m the value obtained through the model for the considered observable, σ the error on the observed quantity i. Anyway, some problems can be encountered: possibility of a non-converging optimization phase; long lasting runs; errors due to chance of convergence to a secondary minimum of the χ^2. We develop an alternative method for the optimization of the models exploiting the power of the MCMC sampling, having also the possibility to test different minimization criteria.

4. The Markov-Chain Monte-Carlo sampler

The extraction of the parameters needed to run CLOUDY is performed through the code CosmoMC. Being originally designed for cosmological applications, CosmoMC can also be compiled as a generic sampler, providing a likelihood function, through which behavior, the sampler drives the extractions in order minimize the differences between models and observations.

5. Interface between the codes

Unfortunately, making the codes talking is not that simple. The principal difficulty lies in the different programming language of the codes. The parameters exchange between the codes is then a crucial phase and several modifications have to be made on CosmoMC. To set up properly the sampler it is necessary to define and introduce the new parameters in more than one file of the CosmoMC package. Through the MCMC sampler it is possible to work in two ways: *i*) finding the best fit of the fiducial model; *ii*) probing the entire space defined by the parameters one wants to recover. The first option is the faster, but no additional informations are available on the sampling. The second is of course slower, but allows one to deeply analyze the extraction history, checking, for example, for secondary minima in the distribution of the models.

6. Discussion and conclusion

The computational time depends strongly on the required number of iterations, on the choice of the sampling method, and on the number of chains in the sampling: for about twenty objects, considering about 3×10^4 iterations and 8 chains, the computational time estimate is around 2×10^5 CPU hours, in line with computational resources available on new generation of extreme multi-core platforms. Our first tests indicate that the fiducial model recovered through the interactions of these two codes, could be more solid. In addition, a complete scan of the parameter space can be performed, allowing to accurately analyze the behavior of each crucial parameter. Furthermore, an additional tuning could come from a proper definition of the likelihood function.

7. Acknowledgement

We acknowledge the use of the computational facilities of INAF-IASF Bologna. We acknowledge partial support by the ASI/INAF Agreement I/072/09/0 for the *Planck* LFI Activity of Phase E2 and the ASI contract I/016/07/0 COFIS. We acknowledge the use of the CosmoMC code Lewis & Bridle 2002. Calculations were performed with version 08.00 of CLOUDY, last described by Ferland *et al.* 1998.

References

Condon, J. J. & Kaplan, D. L. 1998, *ApJ* Supp. Series, 117, 361
Ferland, G. J., Korista, K. T., Verner, D. A., Ferguson, J. W., Kingdon, J. B., & Verner, E. M. 1998, *PASP*, 110, 761
Lewis, A. & Bridle, S. 2002, *Phys. Rev. D*, 66, 103511
Umana, G., Leto, P., Trigilio, C., Buemi, C. S., Manzitto, P., Toscano, S., Dolci, S., & Cerrigone, L. 2008, *A&A*, 482, 529
van Hoof, P., PhD Thesis, http://homepage.oma.be/pvh/thesis.html

Advances in Plasma Astrophysics
Proceedings IAU Symposium No. 274, 2010
A. Bonanno, E. de Gouveia Dal Pino & A.G. Kosovichev, eds.

© International Astronomical Union 2011
doi:10.1017/S1743921311006879

Energetic Solar Electrons – Whistler Bootstrap, Magnetic Knots and Small-scale Reconnection

Ilan Roth

Space Sciences, University of California at Berkeley,
Berkeley, CA 94720 , USA
email: ilan@ssl.berkeley.edu

Abstract. The (near) relativistic electrons, emanating from the solar corona in long-lasting, gradual events, are generally observed at 1 AU as delayed vs the less energetic, type-III beams. The observations are consistent with the delayed electrons being energized along the stretched post-CME coronal field lines, when the tail of an anisotropic seed population, which is injected in conjunction to the observed radioheliograph bursts, interacts with the self-excited whistler waves (bootstrap mechanism). These bursts indicate efficient processes where suprathermal seed electrons are injected as a result of magnetic reconnection at the marginally stable coronal configuration left behind the emerging CME. The dependence of the bootstrap mechanism on the electron injection raises the general question of the MHD description and its deviation over the small electron skin-depth scale. The similarity between MHD and knot theories allows one to characterize any turbulent magnetic configuration through topological invariants, while deviation over electron skin-depth scale, characterized by the generalized vorticity, which is enhanced due to density inhomogeneity, creates the conditions for the potential injection sites.

Keywords. relativistic electrons, solar flares, knots, reconnection.

1. Introduction

Formation of non-thermal electron populations in solar environment has been investigated experimentally through (a) remote sensing of electromagnetic waves emitted by the accelerated electrons via their local interaction with plasma or with magnetic field and (b) in situ measurements by satellites traversing various heliospheric regions during active solar periods. In this paper we address briefly two physical aspects of electron energization with potential implications beyond solar physics: (1) the injection of a suprathermal seed population through reconnection process and (2) the successive energization of the electron tail to relativistic energies. We discuss the bootstrap acceleration process, specify the MHD description in terms of knot theory invariants and show its violation over small scales, where the electrons are frozen into the generalized vorticity.

2. Bootstrap model

The near-relativistic electrons are observed at 1 AU with a solar injection timing significantly delayed with respect to the lower energy electrons, pointing out to different inception and energization mechanisms. Precise timing of the observed electron fluxes (Krucker *et al.*, 1999) showed that the injected electrons could be characterized into low-energy, injected instantaneously with the electromagnetic radiation, and more energetic with a delay of up to 30 minutes. Klassen *et al.* (2005) investigated the intense Halloween 2003 event and found long-lasting high energy relativistic electron oppulation with an onset of 25 minutes after the type III initialization. Maia et al (2007) concluded from the April 15 2001 event that the energization process operates at the very low corona and the energetic electrons, observed through illuminated loops, are formed behind the CME, in the disturbed, turbulent, marginally unstable corona. These observations lead to the bootstrap energization model described below.

At active solar times, after CME uplift, intense spectroscopic signals are observed by radio-heliographs (RH), indicating coronal injection. The energization occurs behind the CME leading edge, often at low solar altitudes, on closed field lines, without direct correlation to the CME. It has been therefore conjectured that reconnection process injects non-isotropic electrons, which destabilize whistler waves, resulting in an efficient tail energization (Roth, 2008). The energization proceeds via resonant interaction between the waves and the electrons: $k_\parallel v_\parallel = \omega - n\Omega/\gamma$. The bootstrap model requires injection of a seed population in tandem with the RH emissions, which is facilitated by unstable, turbulent magnetic configurations; their description is presented in the next sections.

3. Magnetized plasma descriptions

General description of plasma consists of a fluid or Boltzmann model. Since we are interested in magnetic structures which are supported mainly by the bulk distributions, and because various thermal effects can be incorporated a posteriori, we shall adhere here to a fluid approximation. Plasma can then be divided into electron fluid which follows changes due to electric field, Lorentz and Hall forces, electron pressure gradient and plasma resistivity (with the standard notation)

$$-(m_e/e)[du/dt] + J \times B/nec - \nabla p_e/ne + \eta j = E + u \times B/c \qquad (3.1)$$

and center of mass fluid which is advanced under the effects of the (total) kinetic and magnetic pressures and magnetic tension,

$$\rho[dV/dt] = J \times B/c - \nabla p = -\nabla(p + B^2/8\pi) + (B.\nabla)B/4\pi \qquad (3.2)$$

while the electromagnetic fields are related via Maxwell equation:

$$\nabla \times E = c^{-1}\partial B/\partial t. \qquad (3.3)$$

The set of the above equations allows one to consider approximations at different scales and with different physical implications:

a. On a scale larger than ion skin depth the lowest order approximation for the electron fluid ignores all the terms on the lhs of Eq 3.1 and substitutes this result into the Maxwell equation. The electrons now are tied almost entirely to the ions, $u \sim V$, and the magnetic field is frozen into the plasma: $\partial_t B = \nabla \times (V \times B)$. This constitutes the evolutionary equation for the (turbulent) magnetic field in the MHD approximation where both ion fluids move together, dragging the magnetic field (Section 4).

b. On the scale smaller than the ion skin depth one chooses various approximations to the "ion" fluid in eq 3.2 and attempts to include as many effects as possible in the electron dynamics, Eq. 3.1. Over scale much below ion skin depth, in the lowest order, the ion motion is completely neglected and using vector identities Eq 3.1 becomes

$$\partial_t G = \nabla \times (u \times G); \quad G = B - (mc/e)\nabla \times u. \qquad (3.4)$$

Here the electrons are dragging the G field, termed generalized vorticity, indicating that on the small scale they decouple from the magnetic field. The electron drift takes the role of a vector potential.

Figure 1. (a)-(b) Writhe operations; (c) Trefoil knot; (d) Figure Eigth knot.

4. MHD Knot description

MHD is an approximate description of the magnetic field evolution as it is immersed in a plasma fluid; although the field may form complicated structures, the magnetic flux through surface intersecting B lines is conserved. The plasma flow drags the magnetic field lines such that they may only stretch and bend. MHD turbulence forms then a collection of non-intersecting, entangled fields. Similarly, a mathematical knot is depicted as a closed loop of a non-self-intersecting curve, whose evolution is determined via continuous deformation in R^3, following laws of knot topology - smooth changes in the surrounding viscous fluid, allowing only stretching or bending. Hence, MHD field evolution may be viewed as a topological deformation and its dynamics forms equivalent configurations with a set of invariants; similarly, knots are distinguished by a variety of topological invariants. Such invariants are crucial in obtaining topological information about the knot or a link (collection of non-intersecting, entangled knots), and equivalently about the (turbulent) magnetic field configuration. The topological information about knots may be obtained from their diagrams - 2D projections which preserve the over/under crossing of the 3D curve. The general deformations which satisfy this equivalency were described in the three link moves R_j, j = 1,3 (Reidemeister, 1926, Hass and Lagarias, 2001). Then, to each knot or link one can attribute a set of mathematical operations and check if the result is preserved under the R_j moves. Figures 1a-b show two basic configurations with assigned value of a writhe at crossing p, $\epsilon(p)$, which takes the values of ±1. More general knot characterization is obtained where each intersection with undercrossing arcs a, b and overcrossing arc c is assigned algebraic relation c(t) = ta + (1-t)b for variable t; summarizing over the whole knot forms a consistency matrix whose determinant results in a Polynomial P(t) (Alexander, 1928). This characteristic feature becomes an important invariant of each knot. For instance, the Trefoil and Figure Eight Knot have the Polynomial invariants $t^{-1} - 1 + t$ and $t^{-1} - 3 + t$, respectively, showing their inequivalent character (Fig 1c-d). For 60 years Alexander Polynomial was the only known invariant until Jones (1985), using skein relations discovered more powerful invariant in the form of Laurent polynomial, which is able to distinguishes between a knot and its mirror. Hence, the topological information contained in the knot invariant may be useful in description of the MHD (turbulent) field. The description of magnetic configuration through its topological invariants is valuable in characterizing the complexity of the magnetic field and in assessing possibility of conversion between the various magnetic structures in the realm of MHD. Magnetic helicity was shown to satisfy the R_j moves.

Space observations and laboratory measurements indicate that some physical processes violate the smooth knot evolution. To allow pinching and reconfiguration the knot theory invoked a procedure of the Connected Sum of Knots, which joins two knots near a chosen point on each one of them. This mathematical operation is commutative and includes the unit element (unknot) forming a semigroup (group without inverse), while in physics this process violates the frozen-in condition and requires a modified approach.

5. Electron-MHD

At time scales much shorter than the ion gyrofrequency and on spatial scale smaller than ion skin depth, plasma dynamics is determined mainly by electrons. This one-species regime, commonly designed as EMHD, is dominated by a helicon/whistler mode, which replaces the role of Alfven wave in MHD. Ampere's law relates the electron velocity u to the magnetic field where ($\alpha = (c/4\pi e \hat{n})$ and $\hat{n}$ denotes an average density,

$$u = -[c/4\pi e n_o(x)]\nabla \times B = -(\alpha/\nu)\nabla \times B \tag{5.1}$$

which casts the evolution of the magnetic field into Eq 3.4 with an extended expression for the generalized vorticity

$$G = [1 - (d_e^2/\nu)\nabla^2]B + (d_e^2/\nu)(\nabla \times B) \times \nabla ln\nu \tag{5.2}$$

where $\nu = n_o(x)/\hat{n}$, $d = c/\omega_e$ (electron skin depth). Eq (5.2) combines the effects of the current concentration on the small electron skin depth scale with density dips, which are observed in data (Mozer *et al.*, 2003; Mozer, 2005). When the generalized vorticity differs significantly from the magnetic field, the violation of the frozen-in condition becomes conducive to the formation of electron vortices; it is conjectured, based on preliminary simulations, that regions with large field and density gradients are susceptible to form sites of enhanced current modifications which lead to electron injection.

6. Summary

The bootstrap model of energization in solar gradual events requires injection of an electron seed population into the marginally stable corona left behind the emerging CME. Formation of this seed population in magnetic configuration raises a general question of turbulent magnetic field description, which is here given as a collection of knots moving smoothly in a viscous fluid, preserving their topological invariants, which characterize well a complex MHD magnetic field. It is suggested that this classification may be fruitful in comparing various magnetic configurations and, with physical input, assessing the timescales for their modifications. It is shown that on the electron skin depth scale the electron fluid, which decouples from the stationary ions and the magnetic field, is frozen in the generalized vorticity; it is conjectured that regions of largest deviations of G from B form the sites of local current enhancements and electron seed injection.

Acknowledgments

Partial support for this work is acknowledged to NASA grant NNX09AE41G-1/11.

References

Alexander, J. W., *Trans Amer Math Society*, 30, 275, 1928.
Hass, J. and J. Lagarias, *J. Amer. Math. Soc.* 14 , no. 2, 399, 2001.
Jones, V., *Bull. Am. Math. Soc.* 12, 103-111, 1985.
Klassen, A.,S. *et al.*, *Geoph. Res.*, 110, AO9S04, 2005.
Krucker, S., *et al.*, *Astrophys. J.*, 519, 864, 1999.
Maia, D. J. F. *et al.*, *Astrophys. J.*, 660, 874, 2007.
Mozer, F. S. *et al.*, *Phys Rev Lett*, 91, 245002, 2003.
Mozer, F. S., *Jour, Geoph, Res.*, 1190, A12222, 2005.
Reidemeister, K., *Abh. Math. Sem. Univ. Hamburg* 5 24-32, 1926.
Roth, I., *Jour Atmospheric Solar-Terrestrial Phys.*, 70, 490, 2008.

Advances in Plasma Astrophysics
Proceedings IAU Symposium No. 274, 2010
A. Bonanno, E. de Gouveia Dal Pino & A.G. Kosovichev, eds.

© International Astronomical Union 2011
doi:10.1017/S1743921311006880

Solar flares: observations vs simulations

Fatima Rubio da Costa,[1] Francesca Zuccarello,[1] Nicolas Labrosse,[2] Lyndsay Fletcher,[2] Tomáš Prosecký[3] and Jana Kašparová[3]

[1]Dipartimento di Fisica e Astronomia, Via S. Sofia, 78. 95123. Catania, Italy
[2]University of Glasgow, School of Physics and Astronomy
Kelvin Building, University of Glasgow. G12 8QQ, Glasgow, UK
[3]Astronomical Institute of the Academy of Sciences of the Czech Republic, v. v. i.
Fričova 298, 25165, Ondřejov, Czech Republic
email: frdc@oact.inaf.it

Abstract. In order to study the properties of faint, moderate and bright flares, we simulate the conditions of the solar atmosphere using a radiative hydrodynamic model (Abbett & Hawley, 1999). A constant beam of non-thermal electrons is injected at the apex of a 1D coronal loop and heating from thermal soft X-ray emission is included. We compare the results with some observational data in Ly-α (using TRACE 1216 and 1600 Å data and estimating the "pure" Ly-α emission) and in Hα (data taken with a Multichannel Flare Spectrograph, at the Ondrejov Observatory).

Keywords. Methods: numerical, radiative transfer, Sun: atmosphere, Sun: flares

1. Introduction

The NLTE Radyn code models a 1D plane-parallel atmosphere using the plane-parallel equation of radiative hydrodynamics (Carlsson & Stein, 1997).

We obtain the non-LTE solution of the population equations for hydrogen, helium, single ionized calcium and single ionized magnesium atoms, assuming complete redistribution. We calculate the evolution of the atmospheric plasma parameters, and selected radiated line intensities as a function of time and we compare the results obtained for the hydrogen atom with the intensity measured in Ly-α and Hα during a solar flare to test agreement between observations and the simulations.

2. Simulations

A constant beam of non-thermal electrons is injected at the apex of a 1D coronal loop and heating from thermal soft X-ray and UV emission is included (Abbett & Hawley, 1999). Cooling resulting from bremsstrahlung and collisionally excited metal transitions is also added. The non-thermal electron beam is assumed to have a power-law distribution with a spectral index $\delta = 4$ and a low-energy cutoff of $E_c = 20$ keV.

To investigate the properties of the lower atmosphere during the impulsive phase of a solar flare, the code takes into account three levels of non-thermal electron energy flux:

- $\mathcal{F} = 10^9$ ergs cm^{-2} s^{-1}, hereafter called run F9, which corresponds to *weak* flares. The quiescent atmosphere is heated for 67.1 s.
- $\mathcal{F} = 10^{10}$ ergs cm^{-2} s^{-1}, run F10, which simulates *moderate* flares, for 45.2 s.
- $\mathcal{F} = 10^{11}$ ergs cm^{-2} s^{-1}, run F11, associated to *strong* flares. This case is computationally much more problematic and the run is only for 1.62 s.

Fig. 1 shows the evolution of the temperature stratification for the three different flare cases; the dotted line is the temperature of the initial preflare atmosphere (at t = 0

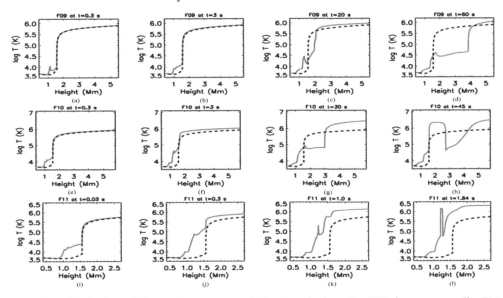

Figure 1. Evolution of the temperature stratification during the F09 (upper pannel), F10 (middle pannel) and F11 (bottom pannel) runs. The dotted line represents the temperature of the initial preflare atmosphere and the solid line reports the temperature during the flare.

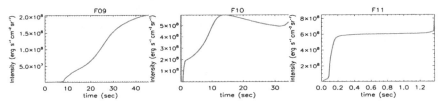

Figure 2. Ly-α integrated intensity as a function of time for the three different flare models.

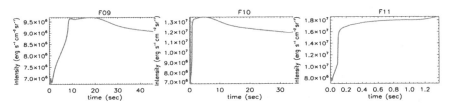

Figure 3. Hα integrated intensity as a function of time for the three different flare models.

seconds) and the solid line is the temperature stratification at each time during the flare. Moderate and Strong flares have an initial *gentle* phase (see panels (f) and (j)) and a subsequent *explosive* phase (see panels (h) and (l)), characterized by large material flows, while Weak flares do not have enough energy to produce the *explosive* phase.

We calculate the radiative emission in Hα and Ly-α from the chromospheric footpoints of the flare. The line intensity is obtained by integrating the line profiles over an interval of $\Delta\lambda = 5$ Å. The intensity of the non-thermal electron beam affects the Ly-α intensity in time (Fig. 2), being only a factor 4 higher for F11 than for F09. Fig. 3 shows the Hα light curve during the flare for each run. The Hα intensity varies by less than a factor of 2 between F09 and F11.

	M6.6 Flare	M1.4 Flare
I_{L_α} (erg s^{-1} cm^{-2} sr^{-1})	9.9×10^5	1.4×10^7

Table 1. Intensity estimated at the beginning of the impulsive phase at the flare footpoints for two flares observed by TRACE.

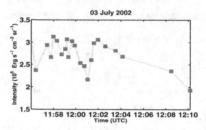

Figure 4. Hα light curve obtained integrating the intensity along the line profile ($\Delta\lambda = 5$ Å).

3. Observations

We studied two different flares using TRACE data: an M1.4 class flare that occurred on 08 September 1999 at 12:08 UT and an M6.6 class flare that occurred on 28 February 1999 at 16:31 UT. A cleaned version of the Ly-α image is obtained by correcting the influence of the 1600 Å TRACE channel on the 1216 Å TRACE channel using: $I_{Ly_\alpha} = 0.97 \times I_{1216} - 0.14 \times I_{1600}$ (Kim *et al.*, 2006). Measuring the count rate in a region of 250×250 pixels^2 at the footpoints of the flare in the corrected Ly-α image (using a threshold range between 1200 and 4090 DN), we estimate the intensity in the flare in Ly-α (see Table 1) (Rubio da Costa *et al.*, 2009).

A C.3 class flare that occurred on 03 July 2002 at 11:43 UT was observed at the Ondrejov Observatory, with the Multichannel Flare Spectrograph (MFS), from 11:56:24 UT to 12:10:42 UT. We calibrated the data obtaining the intensity at different wavelengths along the Hα line profile at different times, determining therefore the evolution of the Hα line profile during the flare for $\Delta\lambda = 5$ Å (see Fig. 4). This can be directly compared to calculated values.

4. Conclusions

The Ly-α intensity estimated from TRACE images is comparable with the theoretical one obtained from the RADYN code even if during the simulations the energy flux of the non-thermal electron beam is constant in time.

The values of the Hα light curve obtained from observations are similar to the simulated ones. The behaviour of the two light curves is different (compare Fig. 3 with Fig. 4): this might be explained by the time variation of the beam flux.

Acknowledgements

Financial support by the European Commission through the SOLAIRE Network (MTRN-CT-2006-035484) is gratefully acknowledged.

References

Abbett, W. P. & Hawley, S. L. 1999, *ApJ*, 521, 906
Carlsson, M. & Stein, R. F. 1997, *ApJ*, 481, 500
Kim, S. S., Roh, H.-S., Cho, K.-S., & Shin, J. 2006, *A&A*, 456, 747
Rubio da Costa, F., Fletcher, L., Labrosse, N., & Zuccarello, F. 2009, *A&A*, 507, 1005

Advances in Plasma Astrophysics
Proceedings IAU Symposium No. 274, 2010   © International Astronomical Union 2011
A. Bonanno, E. de Gouveia Dal Pino & A.G. Kosovichev, eds.   doi:10.1017/S1743921311006892

A way to detect the magnetic helicity using the observable polarized radio emission

Rodion Stepanov and Antonina Volegova

Institute of Continuous Media Mechanics, Ac. Koroleva str. 1, 614013 Perm, Russia
email: rodion@icmm.ru

Abstract. We discuss inverse problem of detection turbulence magnetic field helical properties using radio survey observations statistics. In this paper, we present principal solution which connects magnetic helicity and correlation between Faraday rotation measure and polarization degree of radio synchrotron emission. The effect of depolarization plays the main role in this problem and allows to detect magnetic helicity for certain frequency range of observable radio emission. We show that the proposed method is mainly sensitive to a large-scale magnetic field component.

Keywords. ISM: magnetic fields, methods: data analysis

1. Introduction

Magnetic fields exist not only in compact astrophysical objects such as planets and stars but also are observed everywhere in the Universe, interstellar space and can be attributed to galaxies and galactic clusters (Zweibel & Heiles, 1997). Substantially the dynamo theory explains the nature and evolution of cosmic magnetic fields (Parker, 1979). The dynamics specific property of magnetohydrodynamics (MHD) systems are turbulent motions of media. The mean field theory (Krause & Rädler, 1980) predicts magnetic field generation, as a result of α-effect which can appear under the condition of helical turbulent flows. As a result magnetic field becomes helical too (Brandenburg, 2001) and can be described by magnetic helicity $H = \mathbf{A} \cdot \mathbf{B}$, where $\mathbf{A}$ - is the vector potential of magnetic field $\mathbf{B} = \text{curl } \mathbf{A}$.

Recently the special role of magnetic helicity in space magnetic fields evolution processes is noted by Sokoloff, (2007). Total magnetic helicity of a system is integral of motion and conserved in the nondissipative limit. The results of theoretical and numerical researches show that the magnetic helicity can accumulate in the system and suppress the generation mechanisms (Mininni, 2007). This put into question a possibility of the turbulent dynamo and has demanded construction of the adequate model describing dynamics H.

The model of dynamo in the galactic disk has been added by equations describing outflux magnetic helicity that has allowed to overcome catastrophic suppression dynamo processes (Shukurov *et al.*, 2006). Thus, mechanisms of solar dynamo have been considered and it is shown that allowance for the helicity of the small-scale magnetic fields is of crucial importance in limiting the energy of the generated large-scale magnetic field (Pipin, 2007). Using the results of Shukurov *et al.*, (2007), it is proved necessity of coronal ejections for the strong large-scale solar magnetic field generation (Brandenburg, 2007).

The development of the models that describe evolution of magnetic helicity requires understanding of nonlinear processes in multi-scale systems as well as notions about magnetic energy and helicity spectral distributions, non-uniformity and anisotropy properties of the spatial distributions. It is extremely important to have factual material confirming presence of helicity and its connections with other components of the media.

The observations of helicity in the solar convective zone indicate the existence of connection between the intensity of current helicity and dynamo processes (Zhang et al., 2006). Study of MHD turbulence in laboratory conditions is extremely difficult (for review, see Stefani et al., 2008). Single successful experiments for measuring the turbulent magnetic fields (Denisov et al., 2008; Frick et al., 2010) considerably differ in values of the characteristic parameters, primarily the magnetic Reynolds number. Analysis of astrophysical observations remains the most promising direction of research in this question.

In current astrophysical researches there is no general approach to derivation helicity of interstellar magnetic fields. It is discussed a possible way to detect magnetic helicity from cosmic microwave background (CMB) fluctuations data (Kahniashvili, 2006). In some cases (Kahniashvili & Vachaspati, 2006), information about helicity can be extracted from the properties of cosmic rays if their source is known. The authors of these researches noted that their approaches requires presence of high-accuracy observation data, which we don't have at present, therefore the practical application of these approaches is limited.

New generation of radio telescopes (SKA and LOFAR) offers great opportunities (Beck, 2007) because new high accuracy and resolution data about the space magnetism will be available in the near future. Magnetic fields of the interstellar medium are the most suitable object for derivation MHD turbulence properties. Due to relatively large scales, we can neglect the contribution of regular magnetic fields, stars and planets and assume that the continuous electrically conductive interstellar medium is in a state of turbulent motion, which is excited by explosions of supernovae (Ruzmaikin et al., 1988). Another significant feature of the interstellar medium is in its "transparency", i.e. depth distribution of radio sources allows to make an analysis in all three dimensions.

The aim of the paper is to show the possibility of magnetic helicity detection in the ionized interstellar plasma by statistical analysis of radio polarized observations. We consider the model distribution of the magnetic field with given properties and determine relation of magnetic helicity and correlation coefficient between Faraday rotation measure and polarization degree of radio emission.

2. Interstellar medium model

The simulation domain is a cube of side L. Let the coordinates x, y describe the sky plane and the axis z corresponds to the line of sight. For generation of artificial polarized radio data distributions it is necessary to define some distributions of the ISM components such as magnetic field $\mathbf{B}$, densities of relativistic n_c and free thermal electrons n_e. An indicator of the magnetic field in the interstellar medium is synchrotron emission resulting from relativistic electrons passing through the magnetic field.

At the first step of work we calculate three-dimensional distribution of uniform and isotropic magnetic field. The input parameters of the model are power low of spectrum for magnetic energy distribution α, the turbulent energy scale l and magnetic helicity value. Also, the resulting magnetic field is solenoidal, i.e. $\nabla \cdot \mathbf{B} = 0$. We can satisfy this conditions conveniently using the Fourier representation of the magnetic field $\hat{\mathbf{B}}$ expressed via the vector potential $\mathbf{A}$

$$\hat{\mathbf{B}}(\mathbf{k}) = i\mathbf{k} \times \hat{\mathbf{A}}(\mathbf{k}), \quad \hat{\mathbf{A}}(\mathbf{k}) = \frac{\mathbf{c}}{|\mathbf{k} \times \mathbf{c}|}|\mathbf{k}|^{\alpha/2-1}, \tag{2.1}$$

where $\mathbf{k}$ is the wave-vector, $\mathbf{c} = \mathbf{a} + i\mathbf{b}$ is the random complex vector, whose distribution determines magnetic helicity value. If the random vectors $\mathbf{a}$ and $\mathbf{b}$ have uniform

(a)   (b)   (c)

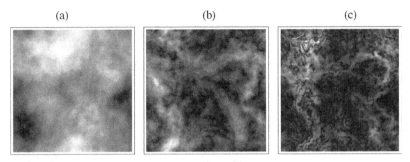

Figure 1. Radio maps. Distributions are represented in the observation plane (x, y): (a) Faraday rotation measure distribution $\mathrm{RM}(x, y, z = L)$, (b) Polarization degree p for $\lambda = 0.05\,\mathrm{m}$ and (c) for $\lambda = 0.2\,\mathrm{m}$. Image resolution is 256x256 pixels. Black colour corresponds to minimum values and white corresponds to maximum values.

distribution the mean value of helicity $\langle H \rangle$ will be close to zero. If we choose only those pairs of vectors which give the same sign of $\mathbf{k} \cdot (\mathbf{a} \times \mathbf{b})$ then $\langle H \rangle$ will be positive or negative, respectively. The extreme value of magnetic helicity for given magnetic energy will be obtained with

$$\mathbf{b} = \pm \frac{\mathbf{k} \times \mathbf{a}}{|\mathbf{k} \times \mathbf{a}|} |\mathbf{a}|. \tag{2.2}$$

The turbulent cells number along line of sight is defined by $N = [Lk_0]$, where $k_0 = l^{-1}$ is the length of the wave-vector till which magnetic energy is equal to zero. Starting from k_0, the energy spectrum obeys the Kolmogorov law $\alpha = -5/3$.

The next step of solving the problem is calculation of artificial polarized radio emission maps and RM images. The total intensity of synchrotron emission is given by

$$I(x, y) = \int_0^L \epsilon(x, y, z)dz, \tag{2.3}$$

where $\epsilon(x, y, z)$ is the synchrotron emissivity. Using simplified representations about spectral distribution of n_c we can consider that $\epsilon \sim n_c(B_x^2 + B_y^2)$. The synchrotron emissivity initially has some degree of polarization γ and the polarization angle is defined by a perpendicular direction to $\mathbf{B}$ in the sky plane (x, y). The intrinsic polarization angle at the point of emission (x, y, z) is given by

$$\psi_0(x, y, z) = \arctan(B_y/B_x) + \pi/2. \tag{2.4}$$

When polarized radio emission propagates thought magnetized plasma, the polarization plane rotated by Faraday effect. Thus, the polarization angle at some point with

$$\psi(x, y, z) = \psi_0(x, y, z) + \lambda^2 \mathrm{RM}(x, y, z), \tag{2.5}$$

where λ is the wavelength of observed radio emission, and RM is Faraday rotation measure which determined by the integral with variable upper limit

$$\mathrm{RM}(x, y, z) = K \int_0^z n_e B_z(x, y, z')dz'. \tag{2.6}$$

Note that Faraday rotation of the polarization plane depends on the magnetic field component along the line of sight, while polarized intensity and polarization angle are defined by the perpendicular component. The observed Stocks parameters Q and U can be used to determine the complex intensity of the polarized emission $P = Q + \mathrm{i}U$ which

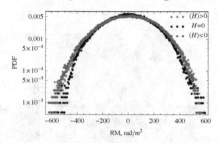

Figure 2. Probability distribution function of Faraday rotation measure for different levels of magnetic helicity

given as

$$P(x,y) = \gamma \int_0^L \epsilon(x,y,z) \exp\{2\psi(x,y,z)\}dz. \tag{2.7}$$

Superposition of the electromagnetic waves with different polarization angles causes depolarization, thus that the observed polarization degree $p = |P|/I$ varies from 0 to γ. Depolarization may be caused not only by physical reasons and also by the limited radio telescope resolution. In this work instrumental effects are not considered. The physical size of the simulation domain is $L = 0.5$ kpc (1 kiloparsec $\approx 3 \cdot 10^{19}$) that corresponds to the half-thickness of the galactic disk. The dimensional constant in (2.6) $K = 0.81$ if λ is measured in meters, z in parsecs, and n_e in cm^{-3}. We accepted the mean value of the magnetic field $\overline{B} = 1$ μG and the thermal electron density $n_e = 1$ cm^{-3} as a typical values for the ISM.

Figure 1 shows typical view of calculated Faraday rotation measures and polarization degree distributions for wavelengths of radio emission 0.05 m, 0.2 m. These distributions of the radio data contain information about all magnetic field components. The changes of distribution details p for long λ are explained by Faraday depolarization. It is possible to see formation of thin black structures (Fig. 1c) in domains corresponding to the maximal values of Faraday rotation measure, this structures are typical for real astrophysical observations. The analysis of these structures called "canals", allows to identify some properties of interstellar turbulence (Fletcher & Shukurov, 2006). Noted canals appearance reflects the fact of reliability for the chosen ISM model.

3. Statistical analysis

Figure 2 shows probability distribution functions of RM for three levels of magnetic helicity. These distributions have symmetric shapes which are sufficiently approximated by the normal law. The difference between distributions is insignificant that cannot be used for diagnostics of helicity.

The situation changes substantially if we consider joint probabilities. The joint probability distribution density of RM and p depending on sign of magnetic helicity is shown in Fig. 3. Magnetic helicity destroys the symmetry of the distribution function. For $\langle H \rangle > 0$ (Fig. 3a) low degree of polarization is most likely corresponds to negative values of RM, and for $\langle H \rangle < 0$ (Fig. 3c) – positive values. The quantitative estimation of revealed statistical characteristics can be derived using the correlation coefficient

$$C = \frac{\langle RMp \rangle - \langle RM \rangle \langle p \rangle}{\sqrt{(\langle RM^2 \rangle - \langle RM \rangle^2)(\langle p^2 \rangle - \langle p \rangle^2)}}, \tag{3.1}$$

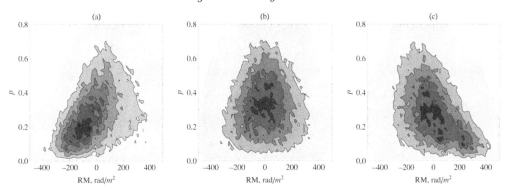

Figure 3. Joint probability distribution of pair RM and p for three levels of magnetic helicity: (a) positive, (b) zero-order, (c) negative

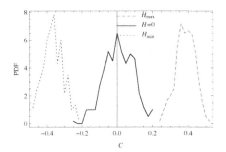

Figure 4. Probability distribution function of C with $\lambda = 0.2$ m for three levels of magnetic helicity

where mean value is taken in the observation plane (x, y) for $z = L$. The probability distribution function of C is defined by repeated calculations of random magnetic field realizations with given level of magnetic helicity. For construction of the results that shown in Fig. 4 we used 300 realizations for each level of magnetic helicity, respectively. The ranges of C values for each magnetic helicity level practically doesn't intersect that allows to identify the sign of magnetic helicity. As mentioned previously, the main reason for connection of RM and p is probably the depolarization effect caused by Faraday rotation, that confirms the relationship of mean correlation $\overline{C}$ with the wavelength λ (see Fig. 5).

For wavelengths less than 6 cm we have low Faraday depolarization that $\overline{C}$ is almost equal to zero. The most successful wavelength for given parameters of the interstellar medium model is $\lambda \approx 15$ cm, which produces the extremum of $\overline{C}$. The connection between $\overline{C}$ and the number of turbulent cells along the line of sight N also was investigated. Our calculations show that the value of $\overline{C}$ suddenly decreases with increasing number of turbulent cells (see Fig. 6).

According to the mathematical modeling results, the magnetic helicity initiates a correlation between polarization degree and Faraday rotation measure. Thus strong correlation is approximately equal to 0.4 and achieved for a definite wavelength. Optimum wavelength for the observations will depend on the parameters of the interstellar medium: the domain size L, magnetic field B, the thermal electron density n_e. However, it is found that for maximum effect, the interstellar medium should provide specific Faraday

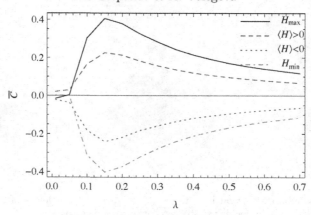

Figure 5. Mean of correlation $\overline{C}$ depending on wave length λ for different levels of magnetic helicity

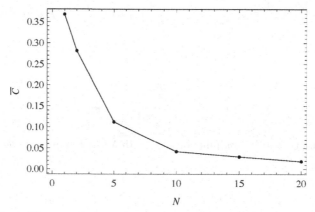

Figure 6. Mean of $\overline{C}$ as a function of turbulent cells number N along the line-of-sight for $\lambda = 0.2$ m

rotation, rotation of the polarization plane through angle of 2π. And then relation

$$\mathrm{RM}\,\lambda^2 \approx K\,L\,B\,n_e\,\lambda^2 \approx 2\pi. \tag{3.2}$$

The correlation decreases with increasing of the number of turbulent cells along the line of sight. It means that the proposed method for diagnosis of the magnetic helicity is mainly sensitive to large-scale magnetic field component. The noise appears during radio polarized observations process, probably, also produces accuracy problems for helicity detection. This influence can be assessed within the proposed model, but it makes sense to do so, when we have real data with known signal-to-noise ratio and other features.

This work was supported by the Russian Foundation for Basic Research and the German Research Foundation (projects 08-02-92881, 11-01-96031-ural) and by the Council of the President of the Russian Federation (grant YD-4471.2011.1).

References

Beck, R. 2007, *Adv. Radio Sc.*, 5, 399
Brandenburg, A. 2001, *ApJ*, 550, 824

—. 2007, *Highlights of Astronomy*, 14, 291

Denisov, S. A., Noskov, V. I., Stepanov, R. A., & Frick, P. G. 2008, *JETP Lett.*, 88, 167

P. Frick, V. Noskov, S. Denisov, & R. Stepanov. 2010, PRL, 105, 184502

Fletcher, A. & Shukurov, A. 2006, *MNRAS*, 371, L21

Kahniashvili, T. 2006, *New Astronomy Review*, 50, 1015

Kahniashvili, T. & Vachaspati, T. 2006, *Phys. Rev. D*, 73, 063507

Krause, F. & Rädler, K.-H. 1980, Mean-field magnetohydrodynamics and dynamo theory, ed. L. O. Williams

Mininni, P. D. 2007, *Phys. Rev. E*, 76, 026316

Parker, E. N. 1979, Cosmical magnetic fields: Their origin and their activity, ed. E. N. Parker

Pipin, V. V. 2007, *AZh*, 51, 411

Ruzmaikin, A. A., Sokolov, D. D., & Shukurov, A. M. 1988, Magnetic fields of galaxies (M.: Nauka), 280

Shukurov, A., Sokoloff, D., Subramanian, K., & Brandenburg, A. 2006, *A&A*, 448, L33

Sokoloff, D. 2007, *Plasma Physics and Controlled Fusion*, 49, 447

Stefani, F., Gailitis, A., & Gerbeth, G. 2008, *Zamm-Zeitschrift Fur Angewandte Mathematik Und Mechanik*, 88, 930

Zhang, H., Sokoloff, D., Rogachevskii, I., et al. 2006, *MNRAS*, 365, 276

Zweibel, E. G. & Heiles, C. 1997, *Nat*, 385, 131

Advances in Plasma Astrophysics
Proceedings IAU Symposium No. 274, 2010
A. Bonanno, E. de Gouveia Dal Pino & A. G. Kosovichev, eds.
© International Astronomical Union 2011
doi:10.1017/S1743921311006909

Magnetic helicity evolution inside a hexagonal convective cell

Aimilia Smyrli[1,2], Duncan Mackay[2] and Francesca Zuccarello[1]

[1]Dipartimento di Fisica e Astronomia - Sezione Astrofisica,
Via S. Sofia 78, 95123 Catania, Italy
email: `emilia@oact.inaf.it`

[2]School of Mathematics and Statistics, University of St. Andrews,
The North Haugh, St. Andrews,
Fife KY 169SS, Scotland, U.K.

Abstract. Magnetic helicity has received considerable attention in the area of fluid dynamics. Recently, this quantity is attracting the interest of solar physicists and much research has been carried out related to magnetic helicity generation and transport through different solar layers, starting from the interior and the convection zone, towards the photosphere, the corona and finally into the heliosphere. Taking into account the global importance of supergranular cells in convection theories, we study the motion of magnetic features into such a geometrical element simplified as hexagonal cell and we analyse the results in terms of the accumulated magnetic helicity. We compute the emergence of a bipole inside the hexagonal cell and its motion from the centre of the cell towards its sides and its vertices, where the magnetic elements are considered to be sinking down. Multiple bipoles are also considered and phenomena such as cancellation, coalescence and fragmentation are also investigated. We find that the most important process for the accumulation of magnetic helicity is the shear motion between the polarities. The magnetic helicity accumulation changes its trend when one polarity reaches the side of the hexagon, and later the vertex. It has zero value when there is no shear motion inside the hexagonal cell, and it is constant when there is no shear between the two polarities during their motion along the cell sides.

Keywords. Sun: photosphere, Sun: interior, convection, Sun: magnetic fields

1. Introduction

The study of magnetic helicity is an important topic in solar physics, as it can describe the evolution of a magnetic field showing its complex geometry and mixing properties (such as the twist, linkage and braiding of the magnetic field lines Berger & Field 1986).

Observationally, the study of magnetic helicity can be applied to solar eruptive phenomena such as flares (Park 2010), filaments (Mackay & Gaizuaskas 2003; Romano *et al.* 2003) and coronal mass ejections (Smyrli *et al.* 2010). We can obtain information on the topological properties of the magnetic field from the study of magnetic helicity. It is important to investigate the build-up of this quantity in the photospheric layer of both the large and the small scale field . Here we only consider the small scale field so-called magnetic carpet configuration and in a first attempt only a small portion of it, i.e inside a single supergranular cell.

The aim of this work is to study the magnetic helicity behaviour in processes occurring inside convective (hexagonal) cells that are the result of energy transport through convection from the solar interior and lead to the evolution of the magnetic carpet and consequently drive coronal interactions (Priest *et al.* 2002) and rearrangements in the outer solar atmosphere.

In particular, we study the following processes related to the motions of magnetic features :

• flux emergence (two magnetic features of opposite but equal amounts of flux appearing simultaneously).

• fly-by or advection (two polarities are moving relative to one another without any interaction between them).

• flux cancellation (two magnetic concentrations of opposite sign disappearing while approaching and interacting each other).

• coalescence (two adjacent magnetic features of the same polarity moving towards each other and merging into a larger polarity).

• fragmentation (a large magnetic concentration splitting into two smaller polarities of equal flux)

2. The Simulation

In order to investigate the influence of different motions of magnetic features on magnetic helicity evolution, we perform a 2-D simulation in cartesian coordinates. The computational domain is a square box of size 256×256 pixels (24000×24000 km^2), and the resolution is 0.129 arcsec/pixel. We represent the supergranular cell with an idealized hexagon of length $L = 7200$ km, in accordance with the typical values of the diameter of a supergranular cell (Parnell 2001, Mackay *et al.* 2002) and each magnetic element by a circle of radius $r_0 = 0.10\ L$. We assume that each feature is characterized by a magnetic field with a Gaussian profile and peak strength of 100 G.

Initially we study the case of a single bipole emerging at a random position inside a box, located at the centre of the supergranular cell, having a length $l = L/2$. Each magnetic polarity moves radially inside the hexagonal cell and as soon as it reaches the edge it moves towards the nearest vertex, which represents a stagnation point, and here it stops. We consider that the flow towards and along the cell boundaries is uniform with a constant speed for each polarity equal to 0.5 km/sec. The computational run has 800 time steps and each step corresponds to 0.5 min. We consider all the possible relative motions for a single bipole by dividing the hexagonal cell into six equilateral triangles and keeping one polarity (positive) inside one triangle while allowing the other (negative) to move in each one of the six triangles. In a second step, we consider the multiple bipole case, where new bipoles emerge simultaneously in the same manner as described above. For this case we can study different processes (such as the magnetic flux cancellation and coalescence) for different numbers of bipoles (maximum 5). For each computational case we calculate the accumulated amount of magnetic helicity using the method developed by Chae (2001).

3. Discussion and Conclusions

• Single bipole. The fly-by case gives a larger amount of accumulated helicity (for total, positive and negative concentrations) compared to the cancellation, emergence, fragmentation or coalescence cases. When the vertical distance between the two polarities is smallest (a condition that in our simulation means strongest shear between the two magnetic polarities), the enhancement of magnetic helicity accumulation reaches a maximum (Fig. 1, top row).

• Multiple bipoles. When the number of emerging bipoles increases, the amount of accumulated helicity at the end of the computational time increases. The rate of change of magnetic helicity, shows some spikes when the number of bipoles is greater than one

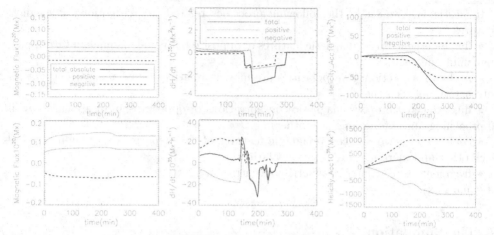

Figure 1. Top row: the bipole fly-by case, (from left to right) magnetic flux, magnetic helicity rate and magnetic helicity accumulation as a function of time. Second row: the 5-bipole case.

and the total magnetic flux shows an abrupt change. The magnetic helicity accumulation trend for positive and negative concentrations is similar for all the cases of bipoles until they reach the border of the hexagon; later on the behavior may change, depending on each case. The total accumulated amount of helicity shows sudden changes (from positive to negative inclination and reverse) almost at the same time as changes in the total magnetic flux occur (Fig. 1, bottom row). Moreover, this study shows that fragmentation is the process that mainly affects the helicity evolution.

We have studied the 2D motion of magnetic concentrations inside a hexagonal cell. This is a first attempt to interpret the behaviour of magnetic helicity for the solar magnetic carpet configuration which can later be used as a tool for interpreting helicity in higher levels of the solar atmosphere for eruptive events. The results show that the shearing motions of flux elements on the borders of the hexagonal cell provokes a large amount of helicity to be accumulated. However, since magnetic fields are three dimensional, a full magnetohydrodynamic (MHD) code should be applied to fully interpret the convective flows and the dynamics of the processes of emerging and cancelling magnetic features.

Acknowledgements

This work was supported by the European Commission through the SOLAIRE Network (MTRN-CT-2006-035484).

References

Berger, M. A. & Field, G. B. 1984, *J. Fluid Mech.*, 147, 133
Chae, J. 2001, *ApJ*, 560, L95
Mackay, D. H., Priest, E. R., & Lockwood, M. 2002, *Sol. Phys.*, 209, 287
Mackay, D. H. & Gaizauskas, V. 2003, *Sol. Phys.*, 216, 121
Park, J., Chae, J., Haimin, W. 2010, *ApJ*, in press
Parnell C. E. 2001, *Sol. Phys*, 200, 23
Priest, E. R., Heyvaerts, J. F., & Title, A. M. 2002, *ApJ*, 576, 533
Romano P., Contarino L., & Zuccarello F. 2003, *Sol.Phys.*, 218, 137
Smyrli A., Zuccarello, F., Romano, P., Zuccarello, F. P., Guglielmino, S. L., Spadaro, D., Hood, A. W., & Mackay, D. H., 2010, *A & A*, in press

Advances in Plasma Astrophysics
Proceedings IAU Symposium No. 274, 2010
A. Bonanno, E. de Gouveia Dal Pino & A.G. Kosovichev, eds.

© International Astronomical Union 2011
doi:10.1017/S1743921311006910

Unveiling the butterfly diagram structure

Maurizio Ternullo

INAF – Osservatorio Astrofisico di Catania
v. S. Sofia 78, 95123 Catania, Italia
email: maurizio.ternullo@oact.inaf.it

Abstract. A Butterfly Diagram showing the spotted area distribution is presented. The diagram reveals that most of the spotted area is concentrated in few, small portions ("knots") of the butterfly wings. A knot may appear at either lower or higher latitudes than previous ones, in a seemingly random way; accordingly, the spot mean latitude abruptly drifts equatorward or even poleward at any knot activation, in spite of any smoothing procedure. The description, assuming that spots scatter around the "spot mean latitude" steadily drifting equatorward, is questioned. In a relevant number of cases, knots appear to be arranged in two roughly parallel, oblique streams, the "spot mean latitude" being located in the underspotted band lying between these streams.

Keywords. Sun: activity, Sun: magnetic fields, Sun: sunspots

Recent works have shown that the trace of the spot zone centroid in the Butterfly Diagram (BD) results – in any hemisphere – from the quasi-biennial alternation of high-speed prograde phases with stationary or even retrograde phases, the average total duration of the latter phases amounting to $\approx 35\%$ of the cycle total duration (Ternullo 2007a,b,c). Moreover, Ternullo (2008, 2010a, 2010b) has shown that most of the spotted area is concentrated in small portions ("*knots*") of the BD, giving it a *"leopard skin"* aspect. An attempt to describe the order governing the knot seemingly random distribution is made with the present work. The data set compiled at the Royal Greenwich Observatory, integrated with data compiled by the US Air Force Solar Observing Optical Network and the NOAA has been used (Hathaway *et al.* 2003).

For each of the 1696 Carrington rotations (from the 330th to the 2025th) and for each of the 84 1°–wide latitude strips in the interval $-42, +42°$, the spotted area average values have been computed. The resulting figures are the elements of an 84×1696 array. This array has been smoothed by a triangular running window covering 5 Carrington rotations and visualized by means of level curves. A portion of the resulting diagram is shown in Figure 1 (cycles 15-17). Here, the aforementioned *"leopard skin"* aspect is apparent. Knots are the signature of *sunspot nests* (Castenmiller et al. 1986). A knot may appear at a latitude either higher or lower than that of previous ones. Figure 1 suggests that in some cycles (e.g., 15, 16 and 17 [s.h.]), knots are arranged into two oblique, discontinuous, roughly parallel streams, between which an underspotted band may be recognized. The existence of such a pattern may be proved by decomposing any wing into a set of oblique, 1°-wide "elementary bands", whose slope is allowed to vary at small steps, and thereinafter visualizing the sums of the array elements falling inside each elementary band by a histogram (a histogram for any slope). If an underspotted band actually crosses a wing with the unknown slope $\sigma°\,\mathrm{yr}^{-1}$ (measured in degrees per year units), we expect to see a depression in the central portion of the histogram plotted for such a slope value.

This approach has quantitatively confirmed that depletion channels actually exist even in cases where – due to the knot pattern complexity – it was not trivial to single out

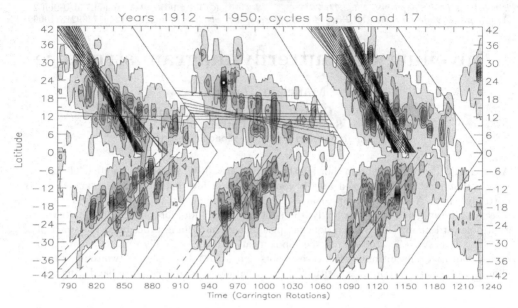

Figure 1. Butterfly Diagram for Carrington rotations 790–1300 (years 1912–1950). The sunspot area distribution is represented by means of 10 level curves, dividing the range of spotted area values into equal intervals. The levels of grey qualitatively correspond to the spotted area density. The Butterfly Diagram appears as a cluster of small, highly concentrated spotted area aggregations (= *knots*): this is the *leopard skin* pattern. In any butterfly wing, it is possible to single out in many different ways a triplet of channels such that the spotted area density in the central channel (the *"depletion channel"*) is significantly lower (at a level of significance not lower than $7\,\sigma$) than in both adjacent comparison channels. As regards the southern hemisphere (s.h.), only one triplet of channels – namely, the one characterized by the maximal level of significance – has been depicted for any wing, As regards the n.h., any triplet of channels fulfilling the same statistical requirements is schematically represented by the line passing through its center.

them at a glance. Moreover, their slopes vary in a restricted range (≈ 4 to $8°\,\mathrm{y}^{-1}$). This finding enables us to conclude that any typical spot cycle actually splits into two activity waves, in any hemisphere; the cycle begins with the activation of the first wave, at a latitude usually not larger than $24 \approx 30°$; it generates the knots lying by the butterfly wing equatorward boundary; the second wave starts a couple of years after the first, at a latitude higher than the first one. Accordingly, spotgroups belonging to different waves lie in belts ≈ 6 through $10°$ apart. The depletion channel marks the separation between these two waves of activity; its slope amounts to the equatorward drift rate of each wave. The sequence of activations and extinctions of knots belonging to either stream accounts for the zigzag displacements of the spot zone noticed by Norton & Gilman (2004) and extensively described by Ternullo (1997, 2001, 2007a,b,c): indeed, the activation of the second wave mimics the first poleward drift of the spot zone centroid; afterwards, other retrograde phases occur because of the extinction of a low latitude knot followed by the activation of a high latitude one.

Acknowledgements

Ms. L. Santagati has reviewed the English form of this paper. This work was partially supported by Agenzia Spaziale Italiana (Accordo num. I/023/09/0 "Attività Scientifica per l'Analisi Dati Sole e Plasma - Fase E2" siglato fra INAF ed ASI).

Cycle 16 s.h.   Cycle 16 n.h   Cycle 17 s.h.   Cycle 17 n.h.

slope (°/y)

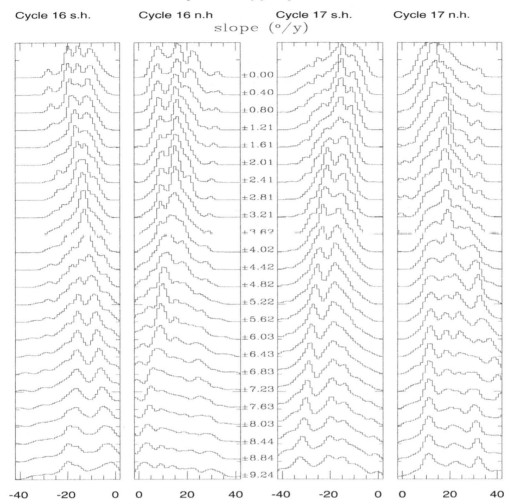

-40   -20   0   0   20   40   -40   -20   0   0   20   40

Figure 2. Histograms showing the spotted area distribution in sets of parallel, oblique, 1°-wide elementary bands scanning the Butterfly Diagram (cycles 16 and 17). The slope scans the range $[0, \pm 9.24°\mathrm{yr}^{-1}]$ (slopes are positive for the s.h. and negative for the n.h.). Each band is assigned to the latitude it crosses at the maximum activity epoch. By sequentially examining histograms related to a given semicycle, it is easy to find that a small dip becomes a sharper and sharper depression until a special slope value is attained; for further slopes, the inverse process occurs.

References

Castenmiller, M. J. M. *et al.* 1986, *Solar Phys.*, 105, 237

Hathaway, D. H., *et al.* 2003, *ApJ*, 589, 665

Ternullo, M. 1990, *Solar Phys.*, 127, 29

Ternullo, M. 1997 *Solar Phys.*, 172, 37

Ternullo, M. 2001a *MemSAI*, 72, 565

Ternullo, M. 2007a *Solar Phys.*, 240, 153

Ternullo, M. 2007b *MemSAI*, 78, 596

Ternullo, M. 2007c *AN*, 328, 1023

Ternullo, M. 2010a, *Ap&SS*, 328, 301; DOI: 10.1007/s10509-010-0270-9

Ternullo, M. 2010b, *MemSAI Suppl.*, 14, 202

Ternullo, M. 2010c, in: D. P. Choudhary & K. G. Strassmeier (eds.), *Physics of the Sun and Star Spots*, Proc. IAU Symposium No. 273

Advances in Plasma Astrophysics
Proceedings IAU Symposium No. 274, 2010
A. Bonanno, E. de Gouveia Dal Pino & A.G. Kosovichev, eds.
© International Astronomical Union 2011
doi:10.1017/S1743921311006922

Superdiffusive and ballistic propagation of protons in solar energetic particle events

Enrico M. Trotta[1] and Gaetano Zimbardo[2]

[1] [2] Department of Physic, University of Calabria,
Via P.Bucci Cubo 31 C, 87036 Arcavacata di Rende (CS), Italy
email: gaetano.zimbardo@fis.unical.it

Abstract. In this work we show that protons can exhibit both superdiffusive and ballistic propagation, at variance with standard diffusion. We carry out an analysis of impulsive solar energetic particle (SEP) events, for which the observed time profile of energetic particle fluxes represent the propagator of the corresponding transport equation. We show that in the case of superdiffusive or ballistic transport the propagator in the time asymptotic regime has a power law form, and that a fit of the observed time profiles allows to determine the transport regime. Using data obtained from ACE and SoHO spacecraft, two proton and electron events, which exhibit both superdiffusive and ballistic transport, will be shown. The finding of these anomalous regimes implies that no finite mean free path can be defined.

1. Introduction

The solar corona is a powerful particle accelerator, being able to accelerate ions to energies of the order of 1 GeV, and electrons to energies of the order of tens of MeV. These particles escape the coronal plasma and propagate in the solar wind along the spiral magnetic field.In order to find anomalous transport regimes, i.e., superdiffusive-and ballistic regimes, we carry out an analysis of impulsive solar energetic particle(SEP) events. We selected events from SOHO and ACE data set (e.g. Krucker & Lin, 2000) where it is possible to assume that the energetic particles fluxesare accelerated in impulsive events like a flare. The characterization of the impulsive events follows the criteria proposedby Reames (1999): Intensity-time profiles of electrons and protons with a prevalenceof electron intensity and a duration of tens of hours; 3He-rich events. In the case of super diffusive or ballistic transport the propagator in thetime asymptotic regime has a power law form (Zumofen & Klafter, 1993; Metzler & Klafter, 2000) and a fit of the observed time profilesallows to determine the transport regime.

2. Impulsive events data and transport analysis

The transport analysis assumes that the impulsive events correspond to an injection of particles localized in space and time, so the transport of particles is described by the shape of the propagator. The propagator form related to super diffusive transport is used (Zumofen & Klafter, 1993 Perri & Zimbardo, 2007; Perri & Zimbardo, 2008), in particular its form calculated in real space in the approximation of long times:

$P(x,t) \sim a_0 / \left[(t)^{1/(\mu-1)} \right]$ which corresponds to the diffusion law $\langle \Delta x^2(t) \rangle \simeq 2D_\alpha t^\alpha$ with $\alpha = 4 - \mu$. In the case of transport in 3-D, assuming normal diffusion in the directions perpendicular to the magnetic field, the long time propagator scales as (Trotta & Zimbardo in progress)

$$P(t - t') \simeq \frac{N_0}{(t - t')^{\mu/(\mu-1)}} \tag{2.1}$$

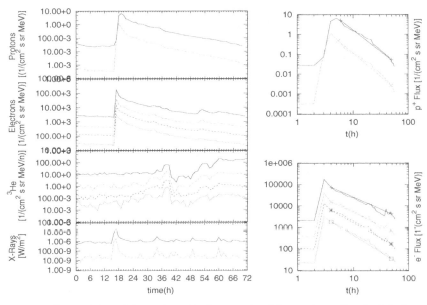

Figure 1. 2000 June 10 impulsive event. Lin-log graph of fluxes data. On the left from top to bottom: proton flux from SOHO, electron flux from ACE/EPAM, ^3He flux from ACE/ULEIS and X-Rays from GOES/SEM. On the right protons (top) and electrons (bottom) fit

The value of μ can be determinate using linear fit in log-log scale with slope give by $m = \mu/(\mu - 1)$: different values of μ identify different transport regimes: normal diffusion for $\mu > 3$, super diffusive transport for $2 < \mu < 3$, and ballistic (e.g. scatter-free) for $\mu < 2$.

Table 1. Summary of 2000 June 10 impulsive event

Data	$E_{max}(MeV)$	Satellite	Duration (h)	m	μ	χ^2	α	Flare-coords
Protons	16.40	SOHO	50.00	-2.12	1.89	0.20	2.00	N22W38
	33.00	SOHO	50.00	-2.28	1.78	0.22	2.00	
Electrons	0.05	ACE	50,00	-1,18	6.70	0.36	1.00	N22W38
	0.10	ACE	50.00	-1.34	3.96	0.24	1.00	
	0.18	ACE	50.00	-1.51	2.94	0.12	1.06	
	0.32	ACE	50.00	-1.62	2.61	0.09	1.39	

The flare is shown by the peak present in X-rays, that is also synchronized with the peak flux of protons and electrons. This allows us to characterize the event as impulsive event (Reames, 1999).The slope and the calculation of the exponent in transport equation, indicates ballistic (e.g. scatter-free) transport for protons (Fig. 1, Tab.1) and a superdiffusive transport for electrons. Table 1 also gives the energy channels, the duration of the event, and the χ^2 of the fit, and the flare coordinates. The next event (Fig. 2) is characterized by the presence of different transport type for protons in different channels of energy. Electron transport is superdiffusive. As previous events, proton and electron fluxes are compared with ^3He flux and SXR to characterize the event as impulsive event. Even for this event, the slope and the calculation of the exponent μ, indicates superdiffusive transport for both electrons and protons (Fig. 2, Tab. 2).

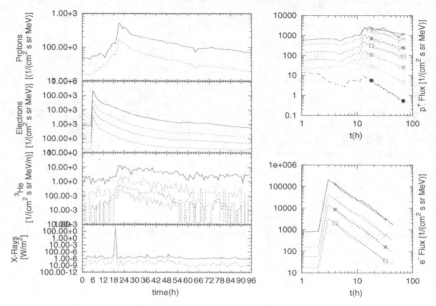

Figure 2. Same as Fig. 1, but for 2002 February 20 event

Table 2. Summary of 2002 February 20 impulsive event

Data	$E_{max}(MeV)$	Satellite	Duration (h)	m	μ	χ^2	α	Flare-coords
Protons	1.06	ACE	72.00	-1.05	21.00	0.03	1.00	N12W72
	1.91	ACE	72.00	-1.27	4.70	0.02	1.00	
	4.75	ACE	72.00	-1.66	2.52	0.03	1.48	
Electrons	0.05	ACE	52.00	-1.70	2.43	0.09	1.57	N12W72
	0.10	ACE	52.00	-1.81	2.23	0.07	1.77	
	0.18	ACE	52.00	-1.88	2.14	0.04	1.86	
	0.32	ACE	52.00	-1.85	2.18	0.03	1.82	

3. Conclusions

Using data obtained from ACE and SOHO spacecraft, we have shown that proton and electron events exhibit both superdiffusive and ballistic transport. These anomalous transport regime are important for the mechanisms of energetic particle acceleration and for space weather predictions.

References

Krucker, S. & Lin R. P. 2000, *AJ*, 542, L61
Metzler, R. & Klafter, J. 2000, *Physics Reports*, 339, 1
Perri, S. & Zimbardo, G. 2007, *AJ*, 671, L000
Perri, S. & Zimbardo, G. 2008, *J. Geophys. Res.*, 113, A03107
Reames, D. V. 1999, *Space Sci. Rev.*, 90, 413
Zumofen, G. & Klafter, J. 1993, *Phys. Rev. A*, 47, 2

Advances in Plasma Astrophysics
Proceedings IAU Symposium No. 274, 2010
A. Bonanno, E. de Gouveia Dal Pino & A. G. Kosovichev, eds.

© International Astronomical Union 2011
doi:10.1017/S1743921311006934

Electron acceleration by a wavy shock front: details on angular distribution

Marek Vandas and Marian Karlický

Astronomical Institute, Academy of Sciences, 25165 Ondřejov, Czech Republic
email: vandas@ig.cas.cz, karlicky@asu.cas.cz

Abstract. We studied numerically electron acceleration by a perpendicular wavy shock. Distribution function of accelerated electrons is highly anisotropic, with many sharp peaks. The peaks are caused by (usually single) reflections of electrons by the shock and subsequent transmission.

Keywords. acceleration of particles, shock waves, Sun: corona

1. Introduction

Electrons accelerated at a nearly perpendicular shock wave have been observed in 1970s and the process first theoretically treated by Wu (1984) and Leroy & Mangeney (1984). Vandas (1989) and Krauss-Varban & Wu (1989) have identified the process as a drift acceleration of electrons inside a shock layer. Zlobec *et al.* (1993) suggested qualitatively that the acceleration of electrons could be more efficient when a shock front was wavy. In a series of papers, Vandas & Karlický (2000, 2005, 2010) examined this possibility quantitatively. They found that accelerated electrons had an unexpectedly anisotropic angular distributions with many sharp peaks. Here we investigate causes of this peculiarity.

2. Results

Fig. 1 shows a model of a wavy shock. The shock has a sine-like shock front with the amplitude A and "wavelength" L (amplitude is exaggerated in the figure). Upstream magnetic field is parallel to a smoothed shock front (perpendicular shock wave). To calculate motion of electrons inside the shock layer, the wavy shock is converted into a plane shock wave with B_n, the normal component of the upstream homogeneous magnetic field to the wavy shock, varying. Electron trajectories are calculated numerically in guiding centre approximation. The shock geometry causes that all upstream electrons are eventually transmitted downstream through the shock after an interaction with it. The model and calculation method are described in detail by Vandas & Karlický (2000). Parameters of the shock were taken the same as in the cited paper: the upstream magnetic field strength $B_1 = 0.5$ mT (5 G), the shock velocity $V_1 = 1000$ km s^{-1}, the shock thickness $d = 50$ m, the magnetic field jump at the shock $\nu = 1.6$ (corresponding to the Mach number 1.5 and $\beta = 0.035$), $L = 1000$ km, $A = 100$ km ($A/L = 0.1$). These parameters describe a coronal shock. The initial distribution function of electrons (seed population) is used the same as in Vandas & Karlický (2000), a kappa function describing halo electrons (suprathermal tail).

Fig. 2a shows an angular distribution of accelerated electrons at a particular place in the downstream region. Their distribution function $f(v, \alpha)$ is plotted relatively to the initial distribution function $f_i(v)$ and this value f/f_i is displayed as a function of the pitch angle α; the angular resolution is 0.5°. The velocity v in the plots corresponds to electron energy of 3 keV. Fig. 2a demonstrates a very anisotropic distribution function,

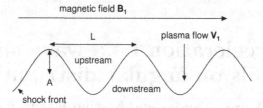

Figure 1. Model shock wave

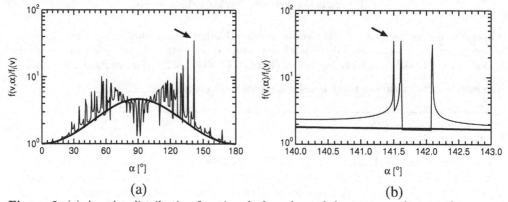

Figure 2. (a) Angular distribution function f of accelerated downstream electrons for energy 3 keV with dependence on the pitch angle α. (b) Detailed view on a selected interval of pitch angles. The angular distribution is plotted relatively to the initial distribution function f_i which is isotropic. The smooth thick lines show the angular distribution of accelerated downstream electrons at a plane perpendicular shock.

with many sharp peaks. The highest peak is labeled by an arrow (at $\alpha = 141.5°$) and a part of angular distribution around it is displayed in Fig. 2b with the higher angular resolution of $0.01°$. The mentioned peak is again labeled by the arrow, and apart of it, additional peaks appear which were not present in Fig. 2a with the lower resolution. It points to a high sensitivity to electron initial values. Due to the wavy form of the shock, electrons may interact with the shock many times. So one can speculate that sharp peaks in Fig. 2, indicating a significantly higher energy gain of electrons, are caused by a sudden increase in a number of interactions for particularly suited combinations of initial parameters. However, a detailed analysis shows that it is not the case. The peaks occur when conditions are suitable for a reflection at the shock. Usually an electron undergoes many transmissions through the shock where the energy gain is small, the decisive role plays a reflection. Fig. 3 demonstrates the situation. Fig. 3a (its right part) shows a trajectory of an electron with the final energy 3 keV and pitch angle $141.50°$. It is the case labeled by the arrows in Fig. 2. The electron is one time reflected at the shock (denoted by "r_1") and eleven times transmitted (denoted by "t" with consecutive numbers in its subscript). Its initial energy was 0.73 keV and pitch angle $93.52°$. The trajectories of the electron within the shock are plotted in the left part of the figure and also labeled by "r" and "t". As it has been described, the shock layer is modeled by a plane shock wave. Mutual positions of trajectories do not correspond to reality, they are shifted to approximately match crossings of the shock front in the right part of the figure. Vertical and horizontal directions are not in scale, the latter one is many times zoomed in and differently for the left (much more times) and right parts. Fig. 2b shows a sharp decrease of the labeled peak for a higher α. And indeed, only a tiny increase of it changes the situation. Fig. 3b displays an electron with the final energy 3 keV and pitch

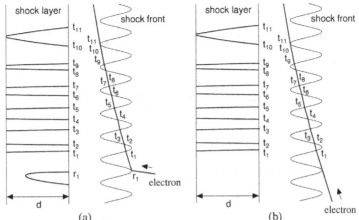

Figure 3. Trajectories (thick lines) of an electron with the final energy 3 keV and pitch angle (a) $\alpha = 141.50°$ and (b) $\alpha = 141.51°$ around and inside the shock. More description is in the text.

angle 141.51°. The electron missed the reflection and it is only eleven times transmitted. Consequently, its energy gain is significantly lower, it had the initial energy 2.13 keV and pitch angle 144.27°. The second peak in Fig. 2b is a result of only one reflection and one transmission. Number of interactions plays a secondary role but it can explain decrease of values in Fig. 2a around $\alpha \approx 90°$. Electrons with such pitch angles move slowly along the shock front and it suppresses the number of interactions (units in contrast to tens for other pitch angles).

3. Conclusions

Distribution function of accelerated electrons at a wavy shock front is highly anisotropic, with many sharp peaks. The peaks are not caused by many interactions of electrons with the shock front, but by their rare reflections at the shock; for specific values of pitch angles and energies. It follows that a regularly oscillating wave pattern of the shock front (as in our model) is not a necessary condition for these anisotropic distributions. One can expect that similar results would be obtained for an irregular wavy shock.

Acknowledgment

This research was supported by grants 205/09/0170 from GA ČR, IAA300030805 and IAA300030701 from AV ČR, and by ESA PECS contract 98068.

References

Krauss-Varban, D. & Wu, C. S. 1989, *J. Geophys. Res.*, 94, 15367
Leroy, M. & Mangeney, A. 1984, *Ann. Geophys.*, 2, 449
Vandas, M. 1989, *Bull. Astron. Inst. Czech.*, 40, 189
Vandas, M. & Karlický, M. 2000, *Solar Phys.*, 197, 85
Vandas, M. & Karlický, M. 2005, in: B. Fleck, T. H. Zurbuchen & H. Lacoste (eds.), *Proc. Solar Wind 11 — SOHO 16 "Connecting Sun and Heliosphere"*, ESA SP-592 (Noordwijk: ESA), p. 453
Vandas, M. & Karlický, M. 2010, *A&A*, submitted
Wu, C. 1984, *J. Geophys. Res.*, 89, 8857
Zlobec, P., Messerotti, M., Karlický, M., & Urbarz H. 1993, *Solar Phys.*, 144, 373

Advances in Plasma Astrophysics
Proceedings IAU Symposium No. 274, 2010
A. Bonanno, E. de Gouveia Dal Pino & A.G. Kosovichev, eds.

© International Astronomical Union 2011
doi:10.1017/S1743921311006946

Spatio-temporal variability of the photospheric magnetic field

A. Vecchio[1,2], M. Laurenza[3], D. Meduri[2], V. Carbone[2,4], and M. Storini[3]

[1]Consorzio Nazionale Interuniversitario per le Scienze Fisiche della Materia (CNISM), unità di ricerca di Cosenza - Ponte P. Bucci, Cubo 31C, 87036 Rende (CS) - Italy
email: antonio.vecchio@fis.unical.it

[2]Dipartimento di Fisica, Università della Calabria, Ponte P. Bucci, Cubo 31C, 87036 Rende (CS), Italy

[3]INAF/IFSI-Roma, Via del Fosso del Cavaliere, 100, 00133 Roma, Italy

[4]LICRYL/CRN, Ponte P. Bucci, Cubo 31C, 87036 Rende (CS)

Abstract. The spatio-temporal dynamics of the solar magnetic field has been investigated by using NSO/Kitt Peak synoptic magnetic maps covering $\sim$ 28 yr. For each heliographic latitude the field has been analyzed through the Empirical Mode Decomposition, in order to investigate the time evolution of the various characteristic oscillating frequencies. Preliminary results are discussed.

Keywords. Sun: magnetic fields, Sun: activity

1. Introduction

The 11 yr cyclic behavior of magnetic activity is driven by the dynamo action, usually related to the emergence of magnetic field in active regions. Apart from the main 11 yr cycle, or 22 yr considering the field polarity reversals, the Quasi-Biennial Oscillations (QBOs) on timescale of $\sim$ 2 yr have been identified in many solar parameters. QBOs are better detected in correspondence of cycle maxima and suffer, as the 11-yr cycle, of variations in the period length (Vecchio & Carbone 2009). The QBO origin is still unknown even if it seems to be related to the dynamo action in the inner solar layers (Benevolenskaya 1998, Fletcher *et al.* 2010) being also detected in phenomena directly connected with the solar interior. In fact, the equatorial rotation rate close to the tachocline varies with a 1.3 yr period (Howe *et al.* 2000), as detected from GONG and MDI observations, the solar neutrino flux shows a significant modulation at the QBO rate (Vecchio *et al.* 2010) and a 2 yr periodicity has been observed in the global solar oscillation frequencies (Fletcher *et al.* 2010).

2. Data analysis

In order to establish the role played in the magnetic field at the solar surface by the QBOs, we study synoptic maps of the magnetic flux density. We analyze the NSO/Kitt Peak maps as function of heliographic longitude (in the range $-70° \div +70°$, with a resolution of 1°) and latitude (with a resolution of 0.01 in sine latitude), consisting of 363 maps from August 1976 to September 2003†. In panel a of Figure 1 we report the latitude-time evolution of the magnetic field $B(\theta, t)$. The raw data set shows the typical 11 yr butterfly picture of equatorward magnetic flux with occasional strong poleward

† the data set is available at ftp://nsokp.nso.edu/kpvt/synoptic/mag/. NSO/Kitt Peak data are produced cooperatively by NSF/NOAO, NASA/GSFC, and NOAA/SEL.

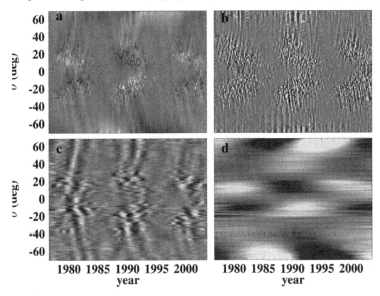

Figure 1. (a) Butterfly diagram of the net magnetic flux density, averaged over longitude for each Carrington rotation. Positive polarities appear white, negative polarities black. The map is saturated at ± 20 G. (b, c, d) Reconstruction through partial sums of IMFs $b_j(\theta, t)$ for the three different base periods, namely $P_j \leqslant 1$ yr (panel a), $1 < P_j \leqslant 4.5$ yr (panel b), and $18 < P_j$ yr (panel c). The maps are saturated at ± 3 G.

surges (Wang, Nash & Sheeley 1989, Knaack & Stenflo 2005). We tried to identify the periodicities present in the data set, and their relative amplitude, through the Empirical Mode Decomposition (EMD), a technique developed to process non stationary data (Huang *et al.* 1998). Through the EMD, a time series $B(\theta, t)$ for each latitude θ_k, is decomposed into a finite number m of oscillating Intrinsic Mode Functions (IMFs) as

$$B(\theta_k, t) = \sum_{j=1}^{m-1} b_j(\theta_k, t) + r_m(\theta_k, t). \tag{2.1}$$

The IMFs $b_j(\theta_k, t)$ are a set of empirical basis functions obtained from the data set under analysis (Huang *et al.* 1998). They represent zero mean oscillations with characteristic timescale P_j. The IMFs are not restricted to a particular frequency but can experience both amplitude and frequency modulation. The residue $r_m(t)$ describes the mean trend. This kind of decomposition is local, complete and orthogonal; in particular, the orthogonality can be exploited to reconstruct the signal through partial sums in (2.1) (Huang *et al.* 1998).

3. Results

Three independent spatio-temporal patterns (panels b, c, d of Figure 1) have been obtained by summing up, in (2.1), significant IMFs (see Wu & Huang (2004) for the significance test) oscillating with timescales P_j in the following ranges: $P_j \leqslant 1$ yr, $1 < P_j \leqslant 4.5$ yr, and $P_j > 18$ yr. The high frequency pattern (panel b; $P_j \leqslant 1$ yr) traces out the butterfly diagram, thus indicating the magnetic flux emergence, on monthly basis, progressively toward the equator during the Schwabe cycle. This result suggests that the "magnetic butterfly" is associated to the emergence of magnetic flux concentration at active region scales. The spatio-temporal pattern, corresponding to the QBOs (panel c), shows a displacement of magnetic flux in poleward and equatorward directions, from latitudes $|\theta| \sim 25°$. This represents an evidence that the polar and equatorial activity belts, in which the magnetic tracers migrate along in opposite directions, do exist and are associated to the QBO periodicity of the magnetic field. Moreover, this finding confirms that magnetic flux periodically migrates poleward, even if just some few strong poleward surges can be directly identified in the raw data. The periodic character of this pattern could be associated to oscillating dynamo solutions which, starting at $|\theta| \sim 25°$, propagate both polewards and equatorwards. This should reflect the existence of a dynamo wave over a ~ 2 yr period developed in the tachocline given that the same periodicity is found in several processes taking place in the solar interior. Panel d of Figure 1, showing the spatio-temporal pattern for periods longer than 18 yr, is associated with the periodic polarity reversals of the Sun and clearly reveals some well known features of the Hale cycle. The reconstruction highlights that the polarity reversal at the polar regions takes place around the maximum of the solar activity. On the other hand, an antisymmetry with respect to the equator is apparent in the belt $\pm 25°$; each polarity is constant for ~ 11 yr with inversion at the activity minima. Moreover, at intermediate latitudes ($30° < |\theta| < 70°$) the inversion of the poloidal field seems to happen shifted in time with increasing latitude while butterfly like structures can be recognized in the active latitude bands. The observed spatio-temporal pattern is consistent with $\alpha - \omega$ dynamo solutions, including meridional circulation obtained by Dikpati & Gilman (2001), supporting the hypothesis that the large scale flux transport sets the period of ~ 22 yr.

4. Conclusion

The spatio-temporal dynamics of the photospheric magnetic field has been investigated by applying the EMD technique to synoptic maps for about three solar cycles. Both the 22 yr and QBO spatio-temporal patterns show regular magnetic flux migration toward the poles and signatures of migrations toward the equator. The main features of these patterns indicate that the QBO could be due to a second dynamo action.

Acknowledgements

This work was partially supported by the Italian Space Agency for the BepiColombo mission.

References

Benevolenskaya, E. E. 1998, *ApJ* (Letters), 509, L49
Dikpati, M. & Gilman, P. 2001, *ApJ*, 559, 428
Fletcher, S. T., Broomhall, A., Salabert, D., Basu S., Chaplin, W. J., Elsworth, Y., Garcia, R. A., & New, R. 2010, *ApJ*, 718, L19
Howe, R., Christensen-Dalsgaard, J., Hill, F., Komm, R. W., Larsen, R. M., Schou, J., Thompson, M. J., & Toomre, J. 2000, *Science*, 287, 2456
Huang, N. E., *et al.* 1998, *Royal Society of London Proceedings Series A*, 454, 903
Knaack, R. & Stenflo, J. O. 2005, *A&A*, 438, 349
Vecchio, A. & Carbone, V. 2009, *A&A*, 502, 981
Vecchio, A., Laurenza, M., Carbone, V., & Storini, M. 2010, *ApJ* (Letters), 709, L1
Wang, Y., Nash, A. G., & Sheeley, Jr., N. R. 1989, *ApJ*, 347, 529
Wu, Z. & Huang, N. E. 2004, *Royal Society of London Proceedings Series A*, 460, 1597

Plasma around compact objects

Advances in Plasma Astrophysics
Proceedings IAU Symposium No. 274, 2010
A. Bonanno, E. de Gouveia Dal Pino & A.G. Kosovichev, eds.

© International Astronomical Union 2011
doi:10.1017/S1743921311006958

Plasma processes in pulsar magnetospheres

D. B. Melrose

SIfA, School of Physics, The University of Sydney
NSW 2006, Australia
email: melrose@physics.usyd.edu.au

Abstract. It is pointed out that the standard model for pulsar electrodynamics is based on a false premise, related to neglecting the displacement current, and the associated need for current screening. Wave dispersion in the standard model is reviewed, and its relation to the interpretation of pulsar radio emission and its polarization is discussed. Inclusion of the displacement current results in large-amplitude oscillations; some of the implications of these oscillations on the interpretation of the radio emission are discussed.

Keywords. pulsars: general, plasmas, magnetic fields, polarization

1. Introduction

Pulsar radio emission is poorly understood. In one sense, our failure to identify the emission mechanism unambiguously seems surprising. There is an enormous body of observational evidence on radio pulsars, and one would expect this to severely constrain the emission mechanism. Moreover, the number of possible mechanisms is relatively small (I comment on four) and one would expect to identify signatures that could distinguish between them. However, the difficulties in identifying the mechanism uniquely are formidable. On the theoretical side, our understanding of pulsar electrodynamics contains serious deficiencies, and it is unrealistic to suppose that we can predict the emission mechanism from first principles. On the observational side, although there are many rules that describe the huge variety of features in pulsar radio emission, there are exceptions to every rule. Furthermore, it is strongly believed that the emission is generated by relativistic particles, and for highly relativistic particles many features of the emission depend on the Lorentz factor, γ, and are insensitive to differences between different emission mechanisms. One might hope to identify the mechanism from the observed polarization, but evidence on orthogonally polarized modes (OPMs) strongly suggests that the observed polarization is determined as a propagation effect, rather than being intrinsic to the emission mechanism. When all the difficulties are taken into account, the concern is whether it is even possible in principle to identify the pulsar emission mechanism unambiguously.

There is a widely accepted standard model for pulsar magnetospheres, but this is based on a false premise. Criticism of the standard model is far from new (Michel 2004), and the assumption that I criticize specifically is that electric fields can be screened by charges: this is possible only if one neglects the displacement current and such neglect is justified only for an aligned rotator, which cannot produce any pulses. Available alternative models have other difficulties, and none has received wide acceptance. One can hope that an acceptable theory for pulsar electrodynamics will ultimately emerge, and that it will contain many of the features in the standard model. One new feature in such an acceptable model should be large-amplitude electric oscillations (LAEWs). As argued by Sturrock (1971) and confirmed by numerical calculations (Levinson *et al.* 2005, Beloborodov & Thompson 2007), when the displacement current is included, the magnetosphere is unstable to the development of LAEWs. Models for wave dispersion and pulsar radio emission are affected substantially by LAEWs. Here I make a distinction between a "standard" model without LAEWs and an "oscillating" model with LAEWs.

I discuss pulsar electrodynamics in §2, wave dispersion in a standard model in §3, pulsar radio emission mechanisms in §4, and some implications of an oscillating model in §5.

2. Pulsar electrodynamics revisited

Pulsar electrodynamics involves attempting to reconcile two incompatible models: a rotating magnetized star in vacuo, and a corotating magnetosphere. The standard model is based on an aligned rotator in which the electric field is not a function of time, but such a model does not pulse. For an oblique rotator, the electric field is intrinsically time-dependent, and the displacement current cannot be neglected.

Rotating dipole model: A rotating point dipole, $\mathbf{m}$, has time derivatives $\dot{\mathbf{m}} = \boldsymbol{\omega} \times \mathbf{m}$, $\ddot{\mathbf{m}} = \boldsymbol{\omega} \times (\boldsymbol{\omega} \times \mathbf{m})$, where $\boldsymbol{\omega}$ is the angular velocity. Using labels dip = dipolar, ind = inductive, rad = radiative, with $r_L = c/\omega$ the light cylinder radius, the magnetic field has terms with three different dependences of the radial distance, r. These are $\mathbf{B}_{\mathrm{dip}} \propto 1/r^3$, $\mathbf{B}_{\mathrm{ind}} \propto 1/r_L r^2$, $\mathbf{B}_{\mathrm{rad}} \propto 1/r_L^2 r$. The associated electric field has two such terms: $\mathbf{E}_{\mathrm{ind}} \propto 1/r_L r^2$, $\mathbf{E}_{\mathrm{rad}} \propto 1/r_L^2 r$. The leading term in the displacement current is $\varepsilon_0 \partial \mathbf{E}_{\mathrm{ind}}/\partial t \propto 1/r_L^2 r^2$.

Magnetic dipole radiation: The radiation field $\mathbf{E}_{\mathrm{rad}}, \mathbf{B}_{\mathrm{rad}}$ dominates at $r \gg r_L$, and leads to electromagnetic radiation at frequency ω, which carries away energy and angular momentum. The radiative energy loss is used to infer the slowing down, and hence to determine the dipolar component of the magnetic field at the stellar surface, $B_{\mathrm{dip}} \propto (P\dot{P})^{1/2}$ with $\omega = 2\pi/P$, and the age of the pulsar, $P/2\dot{P}$.

Quadrupolar electric field: Assuming that the interior of the star is a perfect conductor, there is a corotation electric field, $\mathbf{E}_{\mathrm{cor}} = -(\boldsymbol{\omega} \times \mathbf{x}) \times \mathbf{B}$, inside the star, $r < R_*$. The boundary conditions at the surface of the star imply a "quadrupolar" field $\mathbf{E}_{\mathrm{quad}} \propto R_*^2/r_L r^4$ at $r > R_*$. $\mathbf{E}_{\mathrm{quad}}$ has a component along $\mathbf{B}_{\mathrm{dip}}$, which rips charges off the surface of the star, populating the surrounding region with charges of one sign, and invalidating the vacuum model.

The rotating-dipole-in-vacuo model is unacceptable: the radiation cannot escape to infinity, and even if it did, the model predicts that the magnetic and rotation axes become aligned on the spin down time, which is inconsistent with observations.

Corotating model: In a corotating model, it is assumed that the corotation electric field, $\mathbf{E}_{\mathrm{cor}}$, is the only electric field in the magnetosphere. The divergence of $\mathbf{E}_{\mathrm{cor}}$ implies the Goldreich-Julian charge density, $\rho_{\mathrm{GJ}} = -2\varepsilon_0 \boldsymbol{\omega} \cdot \mathbf{B}_{\mathrm{dip}} \propto 1/r_L r^3+$ other terms. In the standard model, $\mathbf{E}_{\mathrm{quad}}$ rips charges off the star, setting up ρ_{GJ} immediately above the star, but an additional source of charge is needed to provide ρ_{GJ} elsewhere in the magnetosphere. Further acceleration of the "primary" charges ripped off the star trigger a pair cascade, localized in a pair formation front (PFF). These "secondary" pairs provide the additional charges needed to screen $\mathbf{E}_{\mathrm{quad}}$ above the PFF.

Current screening: In an aligned rotator, the only electric field outside the star is $\mathbf{E}_{\mathrm{quad}}$, and because it is a potential field it can be screened by charges. In general, for $\mathbf{E}_{\mathrm{cor}}$ to be the only electric field, one must also screen $\mathbf{E}_{\mathrm{ind}}$, and this requires a screening current, $\mathbf{J}_{\mathrm{screen}} = \varepsilon_0 \partial[\mathbf{E}_{\mathrm{ind}} - \mathbf{E}_{\mathrm{cor}}]/\partial t \propto 1/r_L^2 r^2$. The usual assumption made in the literature is that the magnetosphere is stationary in a corotating frame, which involves neglecting

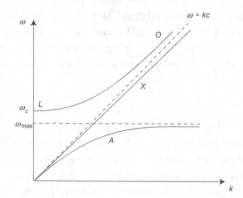

Figure 1. Dispersion curves for a pulsar plasma well below the cyclotron resonance: the cutoff is in the L-O mode, and the resonances is in the Alfvén (A) mode. Not shown is a small region near $\omega_c\gamma$ for nearly parallel propagation, where the O-mode dispersion curve crosses the light line (shown dashed), cf. (Melrose *et al.* 1999).

the displacement current, so that this screening current is zero by hypothesis. However, the displacement current is the proverbial "elephant in the room" and its neglect is not justified, and when it is included LAEWs develop.

The existence of the screening current has not been pointed out previously, and its implications have yet to be thought through in detail. The screening current is similar to the current, $\rho_{\rm GJ}\boldsymbol{\omega}\times\mathbf{x}$, due to the rotating charge density, in that it has the same radial dependence, $\propto 1/r_L^2 r^2$, but unlike this current it is not in the azimuthal direction, and it has a component across the magnetic field lines. For example, one implication is that if the density of pairs is not uniform across the field the required relative flow of electrons and positrons to produce the screening should set up a charge separation.

3. Wave dispersion in pulsar plasma

The standard, or polar-cap, model (Goldreich & Julian 1969, Ruderman & Sutherland 1975) is based on an aligned rotator, with $\mathbf{E}_{\rm quad}$ confined to a "gap" below a PFF. Above the PFF, the only electric field is $\mathbf{E}_{\rm cor}$. The secondary pairs, created in the PFF, radiate away their perpendicular energy, forming a 1D pair plasma with $p_\perp = 0$. The primary particles, with $\gamma \gtrsim 10^6$, that trigger the cascade form a beam propagating through this pair plasma.

Wave dispersion in a pulsar plasma is relatively simple low in the polar cap, where radio frequencies are much smaller than the cyclotron frequency, $\omega \ll \Omega_e/\gamma$. The natural modes are then linearly polarized (Barnard & Arons 1986). The X mode has vacuum-like properties. The O mode has a cutoff at $\omega \approx \omega_p/\gamma^{1/2}$. A beam instability is possible in the O mode at $\omega \approx \omega_p\gamma^{1/2}$, where its refractive index is greater than unity for small angles of propagation (Melrose *et al.* 1999). The Alfvén mode exist at $\omega \leqslant \omega_{\rm max} \approx \omega_p/\gamma^{1/2}$. These properties are illustrated in Fig. 1.

The radio evidence for orthogonally polarized modes (OPMs) is strongly indicative that the polarization is affected by the wave properties near the cyclotron resonance. Indirect evidence on the heights of the source of the radio emission (Gupta & Gangadhara 2003) favors heights well below where the wave frequency is equal to the cyclotron frequency. A plausible interpretation is that the observed polarization is the result of propagation effects modifying the polarization as the radiation passes through the cyclotron resonance. To discuss the interpretation of OPMs one needs a model for the wave dispersion near the cyclotron resonance. The simplest model is to assume that the plasma is cold in its rest frame and that electrons and positrons stream with the same Lorentz factor. This model may be treated by solving for the wave dispersion in the rest frame, and Lorentz transforming to the pulsar frame (Melrose & Luo 2004). Near the cyclotron resonance itself, the spread in Lorentz factors of the electrons smears out the resonance. The natural modes are elliptically polarized near the cyclotron resonance.

The interpretation of the OPMs requires mode coupling in two stages (Wang, Lai & Han 2010). Assuming that the emission is in a single mode, one stage is to produce a mixture of two modes, due to twisting of the magnetic field. The other is to produce the observed polarization, involving elliptically polarized modes, which requires a polarization limiting region near the cyclotron resonance.

4. Radio emission mechanisms

Pulsar radio emission has an extremely high brightness temperature, T_B, and an acceptable emission mechanism must be "coherent" in the sense that it can account for very high T_B. There are three coherence mechanisms: emission by bunches (particles localized in $\mathbf{x}$ and $\mathbf{p}$); a reactive instability (particles localized in $\mathbf{p}$), and a maser growth, which is equivalent to negative absorption. Emission by bunches requires an effective bunching mechanism, and none has been identified; moreover, the back reaction to the coherent emission tends to disperse a bunch (Melrose 1981), so that any bunching instability rapidly evolves into a reactive instability. A reactive instability evolves very rapidly, and the back reaction to it causes a spread in $\mathbf{p}$, so that a reactive instability evolves into a maser instability. For coherent emission from an astrophysical source to be observable requires emission from a relatively large volume for a relatively long time, and without a compelling argument to the contrary, a maser mechanism is implied. I outline four maser emission mechanisms that have been proposed for pulsars.

Plasma-like emission: A relative streaming motion in the ouflowing relativistic particles can lead to a beam instability that generates O mode waves in the small range where the condition $n_O^2 > 1$ is satisfied (Melrose *et al.* 1999). Unlike the Langmuir waves in a nonrelativistic plasma, the dispersion curve for the O mode allows these waves to escape. One problem with this mechanism is in identifying an effective relative streaming motion, to give the energy inversion, $df(\gamma)/d\gamma > 0$, required to drive the instability.

Curvature emission: In the simplest treatment, curvature emission is synchrotron-like in that maser emission is not possible. Maser emission is possible when the curvature drift motion is included or when the magnetic field is twisted. The mechanism also requires $df(\gamma)/d\gamma > 0$ (Luo & Melrose 1995).

Linear acceleration emission: LAE is due to acceleration by $E_\parallel$. It is particularly relevant in an oscillating model when $E_\parallel$ is present as a LAEW (Luo & Melrose 2008). Maser emission is possible for LAE (Melrose & Luo 2009); as for plasma-like emission and maser curvature emission, maser LAE requires $df(\gamma)/d\gamma > 0$. Maser emission is not favorable for LAEWs that cause the electrons and positrons to oscillate with highly relativistic amplitudes.

Anomalous cyclotron emission: Anomalous cyclotron resonance causes electrons (or positrons) in their ground state to jump to their first excited state while emitting a photon, and this requires that the refractive index be greater than unity. The resonance conditions is $\omega - s\Omega - k_\parallel v_\parallel = 0$, with $s = -1$ (Lyutikov 1999). For the frequency to be in the radio range requires that the magnetic field be relatively weak, so that the source region would be far from the star.

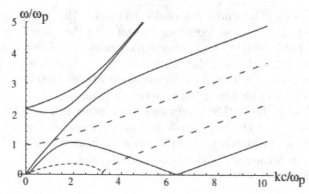

Figure 2. Dispersion curves ω vs k in normalized units for parallel propagation in a cold counter-streaming plasma with $\beta = 0.3$, $\Omega_e = 2\omega_p$. Dotted lines show imaginary parts, dashed lines show longitudinal real parts and solid lines show transverse real parts (Verdon & Melrose 2008).

An argument against maser emissions, and hence against any of these specific maser mechanisms, is that the most extreme examples of pulsar radiation, giant pulses, have T_B seemingly too large to be explained by any of them. A suggested alternative is that the emission is associated with an intrinsically nonlinear instability, similar to the collapse of Langmuir turbulence (Hankins & Eilek 2007).

5. Implications of an oscillating model

For an oblique rotator, inclusion of the displacement current leads to an instability, causing a LAEW to develop. As the (electric) amplitude of the LAEW increases, the maximum Lorentz factor of the oscillating particles increases proportional to it. Once the amplitude exceeds the threshold for effective pair creation, mass loading by pairs limits the amplitude of the LAEW.

In my opinion, the development of oscillations should be an essential feature of any acceptable model for an oblique rotator, and hence for any radio pulsar. It is implausible that the oscillations are coherent over large distances across field lines. Along any individual field lines the LAEW has a dominant effect on the distribution of electrons and positrons, causing them to counter-stream with an oscillating counter-streaming velocity corresponding to the Lorentz factor $\sim 10^6$ needed to generate pairs.

The effect of counter-streaming on wave dispersion in the simplest case of a cold plasma for parallel propagation is illustrated in Fig. 2. Dispersion curves for counter-streaming cold electrons and positrons are illustrated in Fig. 2. The parameters chosen in Fig. 2 are for convenience in illustrating all branches in a single figure. For small but non-zero angle of propagation, the dispersion curves do not cross, as they do in Fig. 2.

Counter-streaming leads to instability, which can result in plasma-like emission. The dispersion curves include an intrinsically growing mode at low frequencies, and in the neighborhood of the crossing points for parallel propagation (Verdon & Melrose 2008). The growth rate of the low-frequency mode decreases with increasing γ, and the most effective growth is at the phase of the LAEW where the counter-streaming is mildly relativistic. The periodic variation causes the wave frequency to change periodically, and a given wave can experience a burst of growth at the same phase over many periods of the LAEW. This is similar to plasma-like emission in the standard model, with the growth of radiation at a specific frequency occurring at a specific phase of the LAEW rather than at a specific height in the magnetosphere. An obvious advantage of the oscillating model is that the required streaming motions are intrinsic to the model.

Mode coupling associated with the cyclotron resonance occurs at a phase of the LAEW where γ is such that Ω_e/γ equals the frequency of the radiation. A theory for mode coupling in a time-dependent magnetized plasmas is needed to discuss such mode coupling in detail.

6. Conclusions

Electrostatic screening of the vacuum field through the Goldreich-Julian charge density is adequate only in an aligned model. In a realistic case a current density is also needed to screen the inductive electric field of the obliquely rotating magnetic dipole, and to set up the postulated corotation electric field. The neglect of the displacement current in models for pulsar electrodynamics obscures an important aspect of the physics: the magnetosphere of an oblique rotator is violently unstable to the development of LAEWs. The properties of wave dispersion in the standard model based on an aligned rotator are reviewed, as are four maser emission mechanisms that have been considered as the pulsar radio emission mechanism.

In an oscillating model, which assumes LAEWs are present, various new possibilities arise for the interpretation of pulsar radio emission. In particular, plasma-like emission and mode coupling to produce OPMs occurs in a similar way to the standard model. The important change is that these effects occur in time, at specific phases of the LAEW, rather than at specific locations as in the standard model. Effective growth of a counter-streaming instability occurs at the phase where the counter-streaming speed is comparable with the intrinsic spread in velocities, when the growth rate is maximum. The mode coupling required to produce observed OPMs is most effective at the phase where the cyclotron frequency, Ω_e/γ is minimum. These effects are currently being investigated.

References

Barnard, J. J. & Arons, A. 1986, *ApJ*, 302, 138

Beloborodov, A. M. & Thompson, C. 2007, *ApJ*, 657, 967

Goldreich, P. & Julian, W. H. 1969 *ApJ*, 157, 869

Gupta, Y. & Gangadhara, R. T. 2003 *ApJ*, 584, 418

Hankins, T. H. & Eilek, J. A. 2007 *ApJ*, 670, 693

Levinson, A., Melrose, D., Judge, A., & Luo, Q. 2005, *ApJ*, 631, 456

Luo, Q. & Melrose, D. B. 1995, *MNRAS*, 276, 372

Luo, Q. & Melrose, D. B. 2008, *MNRAS*, 387, 1291

Lyutikov, M. 1999, *ApSS*, 264, 411

Melrose, D. B. 1981, in W. Sieber, R. Wielebinski (eds) *Pulsars*, IAU Symp. 95, D. Reidel (Dordrecht), p. 133

Melrose, D. B., Gedalin, M. E., Kennett, M. P., & Fletcher, C. S. 1999 *J. Plasma Phys.*, 62, 233

Melrose, D. B. & Luo, Q. 2004 *MNRAS*, 352, 519

Melrose, D. B. & Luo, Q. 2009 *ApJ*, 698, 124

Michel, F. C. 2004 *Adv. Space Res.*, 33, 542

Ruderman, M. A. & Sutherland, P. G. 1975 *ApJ*, 196, 51

Sturrock, P. A. 1971 *ApJ*, 164, 529

Verdon, M. W. & Melrose, D. B. 2008 *Phys. Rev. E*, 77, 046403

Wang, C., Lai, D. & Han, J. 2010 *MNRAS*, 403, 569

Advances in Plasma Astrophysics
Proceedings IAU Symposium No. 274, 2010
A. Bonanno, E. de Gouveia Dal Pino & A.G. Kosovichev, eds.
© International Astronomical Union 2011
doi:10.1017/S174392131100696X

Collisionless shocks and particle acceleration: Lessons from studies of heliospheric shocks

Toshio Terasawa[1]

[1]Institute for Cosmic Ray Research, University of Tokyo,
5-1-5 Kashiwa-no-ha, Kashiwa, Chiba 277-8582, Japan
email: terasawa@icrr.u-tokyo.ac.jp

Abstract. Acceleration processes at astrophysical collisionless shocks are reviewed with a special emphasis on the importance of *in situ* observations of heliospheric shocks. Topics to be included are nonlinear reaction of shock acceleration process, effect of neutral particles, and electron acceleration.

Keywords. heliosphere, diffusive shock acceleration, cosmic rays, nonlinearity, supernova remnants, whistler waves

1. Introduction

Astrophysical collisionless shock waves are thought to be efficient accelerators for nonthermal particles. In 1977-1978 Axford *et al.* (1977), Bell (1978), Blandford & Ostriker (1978), and Krymsky (1977) published the idea of 'diffusive shock acceleration' (DSA) process, where nonthermal particles, cosmic rays particles for their discussions, are accelerated around shock fronts through the scattering processes repeating to occur both in the shock upstream and downstream regions. While shocks within the heliosphere largely differ from astrophysical shocks in their speeds or spatial scales, there are certain commonalities among them in basic physical mechanisms, so that our understanding of heliospheric shocks based on *in situ* observations could lead to comprehensive understanding of physical processes of distant astrophysical shocks.

In the heliosphere shocks are formed ahead of coronal mass ejections (CME), ahead of planetary/cometary magnetospheres/ionospheres, around corotating interaction regions (CIR), and ahead of the heliopause (see, e.g., Schwartz, 2006; Terasawa, 2011). While the usual coverage of acceleration parameters obtainable from these heliospheric observations is rather limited, we could have some exceptional values at least a few times per solar cycle. For example, while velocities of heliospheric shocks are usually of the order of several hundred km/s or less, the historically fastest record of interplanetary shock velocities reached ∼4000 km/s (Figure 1a) which is almost comparable to the shock velocities found around young supernova remnants.

Hillas (1984) argued that the maximum energy $E_{\max}$ of particles of charge Ze accelerated in the region with a characteristic velocity v, a magnetic field B, and a spatial size L, is regulated by a simple relation,

$$E_{\max} = ZevBL \tag{1.1}$$

and applied (1.1) to search possible astrophysical sites for the acceleration of ultra high energy cosmic rays ($E_{\max} > \sim 10^{19}$ eV). In Fig. 1b and Table 1, we compare the results of (1.1) with the observed maximum energies in several heliospheric acceleration phenomena, for which we include not only due to shock acceleration but also second-order stochastic acceleration as well as reconnection-related acceleration. As seen in Fig. 1b,

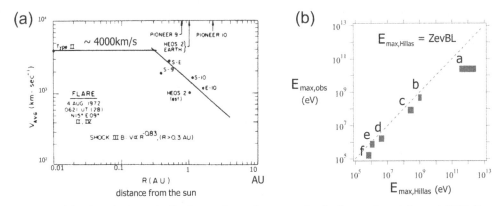

Figure 1. (a) left: The historically fastest interplanetary shock observed on 4 August 1972 (from Smart & Shea (1985)). The horizontal and vertical axes show the distance from the sun and the estimated shock velocity, respectively. (b) right: Heliospheric Hillas relation. The horizontal and vertical axes show E_{max} expected from the Hillas relation (1.1) and E_{max} from observations, respectively. For the items, **a-f**, see corresponding lines in Table 1.

Table 1. Heliospheric Hillas relation vs. Observations (Terasawa, 2001))

	sites	$B_1 \equiv B/1G$	$L_{10} \equiv L/10^{10}$ cm	v/c	E_{max} by Hillas relation (1.1)	observed E_{max}
a	solar flares (CME[1] shocks)	10^2	1	0.001-0.01	300 GeV - 3 TeV	p<~ 30 GeV
b	Van Allen Belt[2] *earth* *Jupiter*	0.1 1	0.1 (1.4 R_E) 1 (1.4 R_J)	0.03 0.03	1 GeV 100 GeV	p~0.6 GeV ?
c	heliospheric shocks <~ 1AU: $B \propto L^{-2}$ >~ 1AU: $B \propto L^{-1}$	6×10^{-5} at 1 AU	10^3 (1 AU)	0.001-0.002	0.2-0.4 GeV	ESP[3] <~ 0.1 GeV ACR[4] <~ 0.1 GeV
d	earth's bow shock + IMF[5] kink	6×10^{-5}	3 (50 R_E)	0.001-0.002	3-6 MeV (for $Z = 6$)	O^{+6} ~2 MeV
e	earth's magnetotail	10^{-4}	1 (15 R_E)	0.003-0.005	0.9-1.5 MeV	e,p~1 MeV
f	earth's foreshock	6×10^{-5}	3 (50 R_E)	0.001-0.002	0.5-1 MeV	p~0.1-0.2 MeV

Notes:
[1] CME $\equiv$ Coronal Mass Ejection. For observed $E_{max} \sim 30$ GeV, see Kahler (1994).
[2] CRAND(=cosmic ray albedo neutron decay) + stochastic betatron acceleration.
[3] ESP$\equiv$Energetic Storm Particles.
[4] ACR$\equiv$Anomalous Cosmic Ray components.
[5] IMF$\equiv$Interplanetary Magnetic Field. Freeman & Parks (2000).

there is a close correlation between the heliospheric observations and the expectations based on the Hillas relation (1.1).

2. Nonlinear reaction of shock acceleration

It is expected that the acceleration process at strong astrophysical shocks is so efficient as the energy densities of accelerated particles, being comparable to the background thermal/magnetic energy densities, contribute to modify the shock structure. This modification has been treated in terms of 'cosmic-ray-modified' or 'cosmic-ray-mediated' shocks (CRMS) (e.g., Drury & Völk, 1981, Malkov & Drury, 2001). Observational studies of

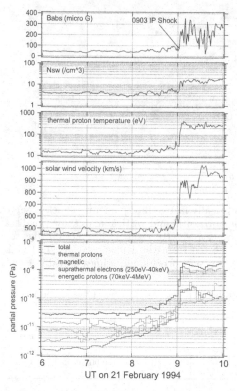

Figure 2. Observational evidence of the CRMS nature at an interplanetary shock detected at 09:03 UT on 21 February 1994. From the top panel, the magnetic field amplitude, the solar wind density, the proton temperature, the solar wind velocity, and subpressures are shown.

CRMS can be made at some of strong heliospheric shocks. Figure 2 shows an example (Terasawa, 2005, Terasawa *et al.*, 2006), where the role of 'cosmic rays' is played by energetic protons of energy $<\sim$ several MeV and suprathermal electrons of energy $<\sim$ 40 keV, whose energy densities reached $\sim$15 % of the upstream solar wind kinetic energy density (the bottom panel). Correspondingly, the increase of the solar wind velocity in the observer's frame† was seen 30 minutes-1 hour before the arrival of the shock front itself (the fourth panel). This is an expected feature of the CRMS. Other examples of heliospheric CRMS can be found at the earth's bow shock (e.g., Zhang *et al.*, 1995), and at the solar wind termination shock (e.g., Florinski *et al.* 2009. Also see the next section).

3. Neutral particles and shock acceleration process

It is known that the heliosphere is surrounded by the partially ionized interstellar gas with an ionization rate of $\sim$50%. The neutral component of interstellar gas (mainly H and He) can penetrate deeply into the upstream region of the solar wind termination shock. After penetration these interstellar H and He are eventually ionized either through the charge exchange process with the solar wind protons, through the photo ionization process by solar UV photons, and through the impact ionization process by energetic electrons (The third process is thought to be negligible in the outer heliosphere). Right after the ionization, these newly born ions are picked-up by the solar wind electromagnetic

† Decrease of the upstream plasma flow in the shock rest frame.

field and start contributing to the solar wind dynamics importantly both in microscopic and macroscopic ways. The solar wind termination shock (TS) was crossed by Voyager 1 and 2 spacecraft successively in December 2004 and August-September 2007. For the latter crossing of TS, the plasma parameter changes were observed. At the transition layer of TS, most of the upstream solar wind flow energy is transferred not to their own thermal energy but to the interstellar pickup ions, so that the downstream solar wind ions are kept relatively cool and the flow speed after the TS crossing was supersonic (Richardson *et al.* 2008; Florinski *et al.* 2009). In this sense, TS is a 'pickup ion mediated' shock.

Interestingly, Helder *et al.* (2009) obtained a conclusion similar to the above TS case from the analysis of the ion temperature downstream of the shock around the supernova remnant RCW 86, which is also surrounded by the partially ionized interstellar gas. The downstream ion temperature of the RCW 86 shock is ~2.3 keV which is by a factor ~18 colder than the value expected from the Rankine-Hugoniot relation in the standard gas shock calculation, $(3/16)\mu m_p v_s^2$ with μ:cosmic abundance (~ 0.6), m_p:proton mass, and v_s:shock speed (~ 6000 km/s). It is noted that RCW 86 is a TeV gamma ray source (Aharonian *et al.*, 2009), where efficient particle acceleration is expected to occur. Ohira and Takahara (2010) have made a detailed calculation of the shock structure mediated by pickup ions as well as by accelerated particles, and concluded that the overall shock compression ratio can be changed significantly.

4. Shock acceleration of electrons

The direct measurement of cosmic ray particles at the earth shows that the electron to proton ratio is about 1% around 10 GeV. How to interpret this ratio is an important unsolved issue. Since the resonance conditions with the MHD turbulence for ions and electrons become identical above the rigidity of ~10 GV, the differentiation between them should occur in the low rigidity regime where their initial accelerations occur (the 'injection' process). While the ion injection process is well understood in terms of self-excitation of resonant MHD waves by ions themselves, how to realize the electron injection process is an unsolved issue (see, e.g., Levinson, 1992). Difficulty for the electron injection is explained in Fig. 3a: The efficient interaction between electrons and scattering waves should occur along the 'resonance line',

$$\omega = kV_b - \Omega_{ce} \qquad (4.1)$$

where ω is the wave angular frequency, k the wavenumber, V_b the velocity of electrons beaming away from the shock along the background magnetic field ($V_b < 0$), and Ω_{ce} the electron cyclotron frequency. On this resonance line there are two interaction points, A and B. At the point A the energy exchange between electrons and whistler waves is possible, but the wave energy is *absorbed* by the beaming electrons, so that the wave self-excitation is not possible. At the point B the electron beam energy could be transferred to left-handed ion cyclotron waves. However, unless the electron energy is sufficiently high (namely, unless the wavenumber k at B is sufficiently low), thermal ions tends to suppress the wave amplification via cyclotron damping interaction.

Recent progress in the electron injection problem is made by Amano & Hoshino (2010) who showed that at shocks with sufficiently high Alfvén Mach number M_A, whistler waves can be self-excited at the point A from the loss-cone pitch angle distribution of the electrons, a natural byproduct of their shock drift acceleration. According to Amano

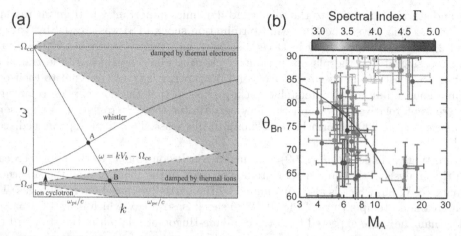

Figure 3. (a) left: Schematic dispersion diagram for circularly polarized electromagnetic waves in an electron-ion plasma, where $\omega > 0$ ($\omega < 0$) is for right-hand (left-hand) polarization. Waves in the shaded regions are strongly damped through the cyclotron resonant interaction with thermal particles (from Amano & Hoshino (2010)). (b) right: The spectral index Γ is obtained by fitting the power law function $E^{-\Gamma}$ to the observed electron energy spectrum (0.05-15 keV). The observed Γ values are shown by the color code (red-yellow-green-blue) and plotted on a M_A-θ_{Bn} plane. A black curve shows the condition (4.2) where β_e is set 1 (from Oka *et al.* (2006)).

& Hoshino (2010), the condition for this mechanism can be approximated as,

$$M_A \geqslant \frac{\cos\theta_{Bn}}{2}\sqrt{\frac{m_p}{m_e}}\beta_e \equiv M_A^{inj} \tag{4.2}$$

where θ_{Bn} is the angle between the shock normal direction and the upstream background magnetic field, m_e the electron mass, β_e the ratio between the electron and magnetic pressures. The authors further argue that for shocks satisfying (4.2) whistler waves grow quickly within the shock transition region and contribute to energize electrons there. This argument consistently explains the bow shock observation by Oka *et al.* (2006) that electrons with hard spectral index ($\Gamma < 3.5$) is mainly found when the condition (4.2) is satisfied (Figure 3b).

It should be noted, however, that the whistler wave generation by the above mechanism is possible only in the region of spatial scale of a few mean free paths, beyond which electrons are isotropized and cannot excite whistler waves. For the DSA process to accelerate low energy electrons, whistler waves should spread over the entire foreshock region. Based on the observations of diffusive electron acceleration at interplanetary shocks (rare but there being a few events per one solar cycle), Nakata *et al.* (2003) suggested a possibility that low frequency MHD waves excited by ions nonlinearly cascade to high frequency whistler wave regime. Since the former waves are known to distribute over the entire foreshock regions (e.g., Bamert *et al.*, 2004), such a nonlinear driving process of whistler waves, if realized, could supply the necessary electron scatterers there. Recent developments in the nonlinear theory of whistler wave turbulence (Saito *et al.*, 2008; Gary *et al.*, 2008; Narita & Gary, 2010) make it possible to test Nakata's suggestion (Oka *et al.*, in preparation).

5. Concluding remarks

As described in this article, the heliosphere has been playing the role of an astrophysical laboratory for the study of collisionless shocks and relating acceleration phenomena. It is interesting to note that the historically fastest interplanetary shock seems to have arrived at the Mercury's orbit (0.3 AU) within its 'free expansion' phase (Figure 1a). For a future Mercury mission or a solar-orbiter mission, we should be ready to observe such a very fast shock environment and relating acceleration phenomena.

References

Aharonian, F., *et al.* 2009, *Astrophys. J.*, 692, 1500

Amano, & Hoshino, M. 2010, *Phys. Rev. Lett.*, 104, 181102

Axford, I. A., Leer, E., & Skadron, G. 1977, *Proc. 15th Int. Cosmic Ray Conf.*, 11, 132

Bamert, K., *et al.* 2004, *Astrophys. J. Lett.*, 601, L99

Bell, A. R. 1978, *Mon. Not. R. Astron. Soc.*, 182, 147

Blandford, R. D. & Ostriker, J. P. 1978, *Astrophys. J.*, 221, L29

Drury, L. O. C., & Völk, H. J. 1981, *Astrophys. J.*, 248, 344

Florinski, V., Decker, R. B., le Roux, J. A., & Zank, G. P. 2009, *Geophys. Res. Lett.*, 36, L12101

Freeman, T. J. & Parks, G. K. 2000, *J. Geophys. Res.*, 105, 15715

Gary, S. P., Saito, S., & Li, H. 2008, *Geophys. Res. Lett.*, 35, L02104

Helder, E. A., *et al.* 2009, *Science*, 325, 719

Hillas, A. M. 1984, *Ann. Rev. Astron. Astrophys.*, 22, 425

Kahler, S. 1994, *Astrophys. J.*, 428, 837

Krymsky, G. F. 1977, *Sov. Phys. Dokl.*, 22, 327

Levinson, A. 1992, *Astrophys. J.*, 401, 73

Malkov, M. A. & Drury, L. O. C. 2001, *Rep. Prog. Phys.*, 64, 429

Nakata, K., *et al.* 2003, *Proc. 28th Int. Cosmic Ray Conf.*, 6, 3697

Narita, Y. & Gary, S. P. 2010, *Ann. Geophys.*, 28, 597

Ohira, Y. & Takahara, F. 2010, *Astrophys. J. Lett*, 721, L43

Oka, M., *et al.* 2006, *Geophys. Res. Lett.*, 33, L24104

Richardson, J. D., *et al.*, 2008, *Nature*, 454, 63

Saito, S., Gary, S. P., Li, H., & Narita, Y. 2008, *Phys. Plasma*, 15, 102305

Schwartz, S. J. 2006, *Space Science Rev.*, 124, 333

Smart, D. F. & Shea, M. A. 1985, *J. Geophys. Res.*, 90, 183

Terasawa, T. 2001, *Science Tech. Adv. Materials*, 2, 461

Terasawa, T. 2005, *COSPAR Colloquium Ser.*, 16, 267

Terasawa, T., Oka, M., Nakata, K., *et al.* 2006, *Adv. Space Res.*, 37, 1408

Terasawa, T. 2011, Chapter 12 of *The Sun, the solar wind, and the heliosphere*, IAGA special Sopron book series, 4, eds. Miralles, M. P. & Almeida, J. S., in press.

Zhang, T. L., Schwingenschuh, K., & Russell, C. T. 1995, *Adv. Space Res.*, 15, 137

Advances in Plasma Astrophysics
Proceedings IAU Symposium No. 274, 2010 © International Astronomical Union 2011
A. Bonanno, E. de Gouveia Dal Pino & A.G. Kosovichev, eds. doi:10.1017/S1743921311006971

Special relativistic magnetohydrodynamic simulation of two-component outflow powered by magnetic explosion on compact stars

Jin Matsumoto[1,2], Youhei Masada[3,4], Eiji Asano[1] and Kazunari Shibata[1]

[1]Kwasan and Hida Observatories, Kyoto University, Kyoto, Japan
email: jin@kusastro.kyoto-u.ac.jp

[2]Department of Astronomy, Kyoto University, Kyoto, Japan

[3]Graduate School of System Informatics, Department of Computational Science,
Kobe University, Kobe, Japan

[4]Hinode Science Project, National Astronomical Observatory of Japan, Tokyo, Japan

Abstract. The nonlinear dynamics of the outflow driven by magnetic explosion on the surface of compact object is investigated through special relativistic magnetohydrodynamic simulations. We adopt, as an initial equilibrium state, a spherical stellar object embedded in the hydrostatic plasma which has a density $\rho(r) \propto r^{-\alpha}$ and is threaded by a dipole magnetic field. The injection of magnetic energy at the surface of compact star breaks the dynamical equilibrium and triggers two-component outflow. At the early evolutionary stage, the magnetic pressure increases rapidly in time around the stellar surface, initiating a magnetically driven outflow. Then it excites a strong forward shock, shock driven outflow. The expansion velocity of the magnetically driven outflow is characterized by the Alfvén velocity on the stellar surface, and follows a simple scaling relation $v_{\mathrm{mag}} \propto v_{\mathrm{A}}^{1/2}$. When the initial density profile declines steeply with radius, the strong shock is accelerated self-similarly to relativistic velocity ahead of the magnetically driven component. We find that the evolution of the strong forward shock can be described by a self-similar relation $\Gamma_{\mathrm{sh}} \propto r_{\mathrm{sh}}$, where Γ_{sh} is the Lorentz factor of the plasma measured at the shock surface r_{sh}. It should be stressed that the pure hydrodynamic process is responsible for the acceleration of the shock driven outflow. Our two-component outflow model, which is the natural outcome of the magnetic explosion, would deepen the understanding of the magnetic active phenomena on various magnetized stellar objects.

Keywords. relativistic, MHD, neutron stars, numerical

---

1. Introduction

Plasma outflow from gravitationally bounded stellar objects is a universal phenomenon in astrophysics although it has different spatial and energetic scales. The energetic origin of plasma outflows would be different in the various astrophysical systems. However, there should be a characteristic common to all the outflow phenomena. The magnetohydrodynamic processes play essential roles in powering and regulating the outflow from a lot of stelar and black-hole systems. It has been a central issue in the outflow study to identify the MHD process which controls the outflow dynamics (Hayashi *et al.* 1996; Beskin & Nokhrina 2006; Spruit 2010).

In this context, the physical properties and the acceleration mechanism of the plasma outflow powered by the magnetic energy stored in compact stars have not been studied sufficiently yet although they are important not only in understanding active phenomena

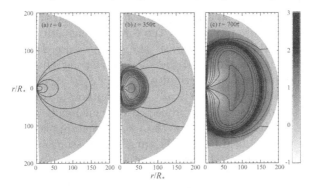

Figure 1. The time evolution of the density distribution (log scale) and magnetic field lines in the meridional plane for the case where $\alpha = 3.50$ and $\beta_0 = 3.2$. The density is normalized by its initial value: $\rho(r, \theta, t)/\rho(r, \theta, 0)$. The left, middle and right panels are corresponding to those in $t/\tau = 0$, 350, and 700 respectively. The normalization unit in time is the light crossing time $\tau \equiv R_*/c$ where R_* and c are central star radius and light speed, respectively.

on the neutron star/magnetar (Thompson & Duncan 1995), but also in unifying the driving mechanisms of the astrophysical outflow. It is interesting from a physical point of view that the outflow from such compact object can be accelerated to the relativistic velocity (Komissarov & Lyubarsky 2004; Komissarov 2006; Takahashi *et al.* 2009).

We focus on the plasma outflows powered by a magnetic explosions on a dense compact object. We investigate their nonlinear dynamics using axisymmetric special relativistic MHD simulations. As an initial setting, we set a spherical compact object embedded in the hydrostatic gaseous plasma which has a density $\rho(r) \propto r^{-\alpha}$ and is threaded by a dipole magnetic field where r and α are the distance from the centroid of the compact star and the power-law index, respectively. In order to initiate magnetically driven outflows, an azimuthal shearing motion is assumed around the equatorial surface, generating the azimuthal component of the magnetic field (Mikic & Linker 1994). It is physically modeling the suppling process of the helical component of the magnetic field and magnetic energy from the compact star interior into the magnetosphere of the star.

2. Results

Figure 1 depicts the time evolution of the expanding outflow for the case where $\alpha = 3.5$ and $\beta_0 = 3.2$. Here β_0 is the plasma beta on the equatorial surface. The density and the magnetic field line projected on the meridional plane are demonstrated at the time (a) $t = 0$, (b) 300τ and (c) 700τ respectively. The normalization unit in time is the light crossing time $\tau \equiv R_*/c$ where R_* and c are central star radius and light speed, respectively. Note that the density is normalized by its initial value and filled by the gray contour in a logarithmic scale.

The magnetic pressure resulted from the azimuthal field component, which is generated by the shearing motion around the stellar equatorial surface, initiates and accelerates the quasi-spherical outflow. The driven outflow is strongly magnetized, in which the magnetic pressure mainly supported by newly generated toroidal field dominates the gas pressure, and carries dense plasma away from the compact object.

The magnetically driven outflow excites a strong forward shock as another outflow component, shock driven outflow. The forward shock expands supersonically into the surrounding medium and is accelerated to sub-relativistic velocity, $\sim 0.3c$, in the case where $\alpha = 3.5$ and $\beta_0 = 3.2$.

We find that there exists a tangential discontinuity behind the propagating forward shock. The magnetic pressure provided by the toroidal magnetic field becomes predominant behind the discontinuity. In contrast, the gas pressure dominates the magnetic pressure between the discontinuity and the shock surface at all the evolutionary stage except a transit region near the discontinuity.

The physical properties of the magnetically driven outflow are examined with changing α-parameter that controls the density profile. The magnetically driven outflow velocity v_{mag} would be characterized by the strength of the initial dipole field which is the seed of the predominant toroidal field behind the tangential discontinuity.

Figure 2a shows the magnetically driven outflow velocity v_{mag} in the relation with initial Alfvén velocity $v_{A,0} = B_0/\sqrt{4\pi\rho_0}$ at the equatorial surface of the central star which is the representative of the seed field strength in the cases $\alpha = 3.50, 4.50, 5.50,$ $6.50,$ and 7.00 respectively. Here B_0 and ρ_0 are the magnetic field strength and density on the compact star surface. The initial Alfvén velocity is measured at the equatorial surface of the star, and varies from $3.7 \times 10^{-3}c$ to $0.15c$. The logarithmic fitting of the numerical data provides a simple scaling relation $v_{\mathrm{mag}} \propto v_{A,0}^{1/2}$.

We focus on the dynamic balance of the system to draw a physical picture for accounting the scaling relation. Our numerical models suggest that at the final steady state, the strength of the toroidal magnetic field behind the tangential discontinuity does not significantly change in time. Hence the radial advection of the azimuthal magnetic field should counterbalance with the generation of the azimuthal field by the shearing motion near the stellar surface. Then we can give an induction equation in the steady form

$$\frac{\partial B_\phi}{\partial t} = [\nabla \times (\mathbf{v} \times \mathbf{B})]_\phi \tag{2.1}$$

$$\sim \frac{B_0 \Delta v_\phi}{L} - \frac{B_{\phi,\mathrm{near}} v_{r,\mathrm{near}}}{R_*} \sim 0, \tag{2.2}$$

where $B_{\phi,\mathrm{near}}$ and $v_{r,\mathrm{near}}$ are the toroidal magnetic field and radial fluid velocity in the region above the stellar surface where the toroidal magnetic flux is stored. Δv_ϕ represents the shearing velocity. The first term in the right hand side of eq (2.2) denotes the generation of the toroidal magnetic field by the shearing motion with the typical latitudinal width L. The second term shows the radial advection of the toroidal magnetic field. The toroidal magnetic field $B_{\phi,\mathrm{near}}$ near the stellar surface gives

$$B_{\phi,\mathrm{near}} \sim \left(\frac{R_*}{L}\right)\left(\frac{\Delta v_\phi}{v_{r,\mathrm{near}}}\right) B_0 . \tag{2.3}$$

The magnetic pressure due to the azimuthal component of the magnetic field near the stellar surface should drive the outflow. Hence the kinetic energy of the outflow is powered by the magnetic energy mainly contributed from the toroidal field component,

$$\frac{1}{2}\rho_{\mathrm{mag}} v_{\mathrm{mag}}^2 = \frac{B_{\phi,\mathrm{near}}^2}{8\pi} , \tag{2.4}$$

where ρ_{mag} and v_{mag} are the typical density and velocity of the magnetically driven outflow, respectively. Since the typical velocity of the outflow is roughly provided by the radial advection velocity,

$$v_{\mathrm{mag}} \sim v_{r,\mathrm{near}} . \tag{2.5}$$

we can obtain, with combining equations (2.4) and (2.5) with (2.3),

$$v_{\mathrm{mag}} \sim v_{A,0}^{1/2} \Delta v_\phi^{1/2} \left(\frac{L}{R_*}\right)^{-1/2} \left(\frac{\rho_{\mathrm{mag}}}{\rho_0}\right)^{-1/4} . \tag{2.6}$$

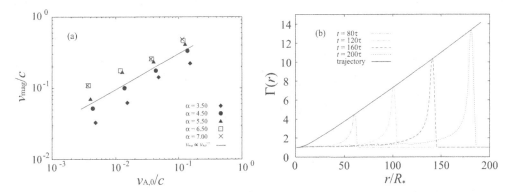

Figure 2. Panel (a): The relation between the Alfvén velocity at the equatorial surface and the velocity of the tangential discontinuity measured at $r = 70R_*$ when $\alpha = 3.50$, 4.50, 5.50, 6.50, and 7.00. There is a simple scaling relation between these two values, $v_{\mathrm{mag}} \propto v_{\mathrm{A},0}^{1/2}$. Panel (b): The time evolution of the Lorentz factor of the expanding gas in the case with $\alpha = 7.25$ and $\beta_0 = 3.2$. The dotted, dashed, dot-dashed, and dashed-two dotted curves indicate the cases where $t/\tau = 80, 120, 160$, and 200 respectively. A solid line represents the time trajectory of the Lorentz factor of the fluid velocity at the shock surface measured in the laboratory frame.

Note that the parameters $v_{\mathrm{A},0}$, Δv_ϕ, L and R_* are the parameters initially given. This indicates that the velocity of the magnetically driven outflow has a scaling relation $v_{\mathrm{mag}} \propto v_A^{1/2}$. We can exhibit the thick line in Figure 2a when the typical parameters which describes our numerical model are adopted. It is noted again that the two balancing equations, which are reduced from the induction and energy equations, account for this scaling relation.

When the initial density profile declines steeply with radius, the outflow-driven shock is accelerated to relativistic velocity ahead of the magnetically driven outflow. Figure 2b illustrates the time evolution of the radial velocity profile in the case where $\alpha = 7.25$ and $\beta_0 = 3.2$. It is found from this figure that the evolution of the relativistic shock has self-similar property. The fitting of numerical data provides a scaling relation between the Lorentz factor Γ_{sh} and the equatorial radius r_{sh} at the shock surface in the laboratory frame

$$\Gamma_{\mathrm{sh}} \propto r_{\mathrm{sh}}. \tag{2.7}$$

The Lorentz factor of the plasma velocity at the shock surface increases linearly with the equatorial radius. The self-similar property of the relativistic shock can be also reproduced by one-dimensional hydrodynamic model (Matsumoto *et al.* 2011). This thing suggests that the pure hydrodynamic process is responsible for the self-similar acceleration of the relativistic shock we found in the 2D MHD model.

References

Beskin, V. S. & Nokhrina, E. E. 2006, *MNRAS*, 367, 375

Hayashi, M. R., Shibata, K., & Matsumoto, R. 1996, *ApJ* (Letters), 468, L37

Komissarov, S. S. & Lyubarsky, Y. E. 2004, *MNRAS*, 349, 779

Komissarov, S. S. 2006, *MNRAS*, 367, 19

Mikic, Z. & Linker, J. A. 1994, *ApJ*, 430, 898

Spruit, H. C. 2010, *Lecture Notes in Physics, Berlin Springer Verlag*, 794, 233

Takahashi, H. R., Asano, E., & Matsumoto, R. 2009, *MNRAS*, 394, 547

Thompson, C. & Duncan, R. C. 1995, *MNRAS*, 275, 255

Advances in Plasma Astrophysics
Proceedings IAU Symposium No. 274, 2010
A. Bonanno, E. de Gouveia Dal Pino & A.G. Kosovichev, eds.

© International Astronomical Union 2011
doi:10.1017/S1743921311006983

Alfvén resonance absorption in electron-positron plasmas

N. F. Cramer

School of Physics, The University of Sydney,
NSW 2006, Australia
email: cramer@physics.usyd.edu.au

Abstract. Waves propagating obliquely in a magnetized cold pair plasma experience an approximate resonance in the wavevector component perpendicular to the magnetic field, which is the analogue of the Alfvén resonance in normal electron-ion plasmas. Wave absorption at the resonance can take place via mode conversion to the analogue of the short wavelength inertial Alfvén wave. The Alfvén resonance could play a role in wave propagation in the pulsar magnetosphere leading to pulsar radio emission. Ducting of waves in strong plasma gradients may occur in the pulsar magnetosphere, which leads to the consideration of Alfvén surface waves, whose energy is concentrated in the region of strong gradients.

Keywords. plasmas, waves, pulsars

1. Introduction

Because the time and spatial scales associated with nonrelativistic electrons-positron pair plasmas are the same in each species, such pair plasmas have dispersive properties quite different to electron-ion plasmas. A distinctive feature of linear wave propagation in magnetized plasmas with equal numbers of same-mass pair species is that there is no Faraday rotation, so that waves propagating parallel to the magnetic field are linearly polarized, rather than circularly polarized, and that there is no analogue of the ion-acoustic wave (Stewart & Laing, 1992; Iwamoto, 1993 and Zank & Greaves, 1995).

If a cold plasma of equal-mass species is considered, but with different number densities of the species (uncompensated), the wave properties are altered, in that circularly polarized waves propagate along the magnetic field. Such an overall charge neutral plasma can be achieved if some of the charge resides on another species, for example consisting of ions or relatively massive dust particles. If the third species is effectively immobile, it can be thought of a charge sink, and the wave properties are determined by the dynamics of the (relatively) light pair species. Another circumstance of an uncompensated pair plasma occurs in the rotating pulsar magnetosphere, where different overall charges exist in different regions of the magnetospheric plasma in pulsar models.

An important feature of real astrophysical plasmas is that they are in general spatially nonuniform on the large scale and inhomogeneous on shorter scales. Waves in such plasmas experience reflection, transmission and absorption processes related to the local spatial variation of the plasma parameters such as the plasma frequencies and cyclotron frequencies. A particular process of interest in this paper is the Alfvén resonance absorption of wave energy, where, in the context of a normal electron-ion plasma, a fast magnetoacoustic wave propagates obliquely to the magnetic field in a density gradient, and at some point the local value of the Alfvén speed equals the wave phase velocity, or in other words the magnetoacoustic wave couples to the Alfvén wave. Energy absorption, via collisional or mode-conversion processes, occurs at this resonance point where the local value of the wavenumber perpendicular to the magnetic field becomes large (Cramer, 2001), and this process in normal plasmas has been invoked as a mechanism for the heating of the solar corona.

The refraction and ducting of waves, launched by emission processes in the nonuniform pulsar magnetosphere, have been invoked as important processes influencing the properties of the emergent pulsar radio waves escaping the magnetosphere, such as pulse profiles (Arons & Barnard, 1986 and Weltevrede *et al.*, 2003). Analyses of refraction of waves in a nonuniform plasma do not usually take into account resonance absorption processes such as the Alfvén resonance, which might play a role in wave propagation in the pulsar magnetosphere.

2. The Dispersion Equation

We consider the case of a cold magnetized plasma, composed of electrons and positrons, plus an effectively immobile species that retains part of the charge, positive or negative. The magnetic field is in the z-direction, and the dielectric tensor components are then:

$$K_{11} = K_{22} = 1 - \frac{\omega_{p+}^2 + \omega_{p-}^2}{\omega^2 - \Omega^2} = 1 - \frac{\omega_p^2}{\omega^2 - \Omega^2}, \tag{2.1}$$

$$K_{12} = -K_{21} = -i\frac{(\omega_{p+}^2 - \omega_{p-}^2)\Omega}{\omega(\omega^2 - \Omega^2)}$$

$$= -i\eta\frac{\omega_p^2\Omega}{\omega(\omega^2 - \Omega^2)}, \tag{2.2}$$

$$K_{33} = 1 - \frac{\omega_p^2}{\omega^2}, \tag{2.3}$$

where ω_{p+} and ω_{p-} are the positron and electron plasma frequencies, and ω_p is the plasma frequency of the combined positron-electron fluid. Ω is the common cyclotron frequency, and η is a measure of the charge imbalance of the two light species:

$$\eta = (n_+ - n_-)/(n_+ + n_-).$$

All the other dielectric tensor components are zero. We define a function

$$A = K_{11} - \frac{c^2}{\omega^2}k_z^2. \tag{2.4}$$

The vanishing of this function for normal electron-ion plasmas defines the position of the Alfvén resonance, because for low frequency it corresponds to the point where the phase velocity along the magnetic field equals the Alfvén speed, where a resonance in the perpendicular wavenumber occurs (Cramer, 2001). A plays a similar role, under certain conditions, for a pair plasma.

In uniform plasma, the wave has wavenumber **k**. If $k_y = 0$, the dispersion equation is:

$$k_x^4(c^2/\omega^2)^2 K_{11} + k_x^2(c^2/\omega^2)(-A(K_{33} + K_{11}) - K_{12}^2) + (A^2 + K_{12}^2)K_{33} = 0. \tag{2.5}$$

3. Alfvén Resonance

We now fix ω and k_z, and vary the plasma density in the x-direction. There are 2 distinct modes. The local k_x^2 may approach zero, i.e. cutoff or strong reflection, or it may become very large, i.e. resonance or strong absorption, as shown in Figure 1.

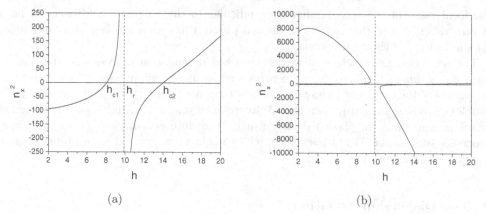

(a)   (b)

Figure 1. (a) Square of the perpendicular refractive index $n_x^2 = c^2 k_x^2 / \Omega^2$ plotted against $h = \omega_p / \Omega$, for case $f \ll h$. Here $f = 0.1$, $n_z = 10$ and $\eta = 0.05$. The Alfvén resonance at h_r is shown by the dashed line, and the two cutoffs $h_{c1,2}$ are indicated. (b) As for (a), but the scale of n_x^2 is expanded, to reveal the second high $|n_x|$ mode on each side of the Alfvén resonance

If $\omega_p \gg \omega$, $|K_{33}| \gg 1$, there are two approximate solutions:

$$c^2 k_{x1}^2 / \omega^2 = A + K_{12}^2 / A, \tag{3.1}$$

$$c^2 k_{x2}^2 / \omega^2 = A K_{33} / K_{11}. \tag{3.2}$$

If $A \approx 0$, there is no longer a true Alfvén resonance where $|n_x^2| \to \infty$, but $|n_x^2|$ acquires large values

$$n_{x1,2}^2 \approx \pm K_{12} (-K_{33}/K_{11})^{1/2} \approx \pm i \eta \frac{h^3}{f^2 (1 + h^2)^{1/2}} \tag{3.3}$$

If $\omega_p \lesssim \omega$, then $|K_{33}|$ is no longer much greater than unity, and there is no longer a pronounced resonance: the moduli of the refractive indices of the two modes are of comparable size in the vicinity of the resonance point.

The larger the value of $|K_{33}|$, the narrower the stop-band about the resonance point, and the sharper the resonance. Equation (3.1), representing a (single) mode with a true resonance at $A = 0$ enclosed by two cutoffs, becomes a better approximation for the actual dispersion relation. This mode is equivalent to the fast compressional wave in cold normal ion-electron plasmas with the electron inertia neglected, which also experiences a cutoff-resonance-cutoff triplet (Cramer, 2001), with the role of the finite ion-cyclotron frequency due to Hall currents in the case of the normal plasma now being played by the imbalance in the pair plasma densities.

Equation (3.1) can be written for $f \ll 1$ in terms of the wavenumbers as

$$v_A^2 k_x^2 = \omega^2 - v_A^2 k_z^2 - \frac{\eta^2 \Omega^2 \omega^2}{\omega^2 - v_A^2 k_z^2}, \tag{3.4}$$

which shows explicitly the Alfvén resonance ($k_x^2 \to \infty$) occurring where the parallel wave phase velocity equals the Alfvén velocity, with the dispersive term depending on the pair species imbalance factor η.

The large $|n_x|$ mode has the approximate dispersion relation

$$\omega^2 = \frac{v_A^2 k_z^2}{1 + \delta_e^2 k_x^2} \tag{3.5}$$

where $\delta_e = c/\omega_p$ = electron inertial length. This is the analogue of the Inertial Alfvén

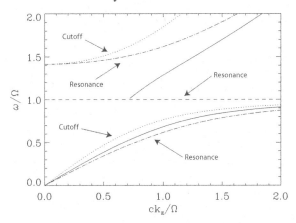

Figure 2. Surface wave dispersion relation (solid curves). Here $h = 1$, $k_y = k_z$ and $\eta = 0$.

wave that occurs in normal plasmas. Surface waves exist on a pair-plasma/vacuum interface for $k_y \neq 0$ (Figure 2). The surface waves may be damped due to Alfvén resonance damping, or coupling into the Inertial Alfvén wave.

4. Conclusion

It has been shown that the Alfvén resonance absorption process occurs in a pair plasma. Analogies are drawn between the resonance-cutoff structures of noncompensated pair plasmas and those of normal electron-ion plasmas with finite ion-cyclotron frequency, with the role of the imbalanced Hall currents in the latter case being played by the imbalanced species densities in the pair plasma.

A true Alfvén resonance can be identified only in the limit of wave frequency much less than the plasma frequency. A long wavelength wave propagating into the "Alfvén resonance" point will partially mode convert into a short wavelength mode which is the pair-plasma version of the "Inertial Alfvén Wave" which occurs in the Alfvén resonance heating of fusion plasmas, and in the form of solitary waves in the Earth's magnetosphere. On a sharp interface between plasmas of differing densities, such as a magnetic flux tube or density duct in the pulsar magnetosphere, a localized surface wave exists. The analogues of Alfvén surface waves on an electron-positron plasma/vacuum interface have been demonstrated. Future work could allow for thermal effects (Stewart and Laing, 1992) and relativistic effects (Luo, Melrose & Fussel, 2002) on the pair plasma dielectric tensor.

References

Arons, J. & Barnard, J. J. 1986, *ApJ*, 302, 120

Cramer, N. F. 2001, *The Physics of Alfvén Waves* (Berlin: Wiley-VCH)

Iwamoto, N. 1993, *Phys. Rev. E*, 47, 604

Luo, Q., Melrose, D. B., & Fussell, D. 2002, *Phys. Rev. E*, 66, 026405.

Stewart, G. A. & Laing, E. W. 1992, *J. Plasma Phys.*, 47, 295

Weltevrede, P., Stappers, B. W., van den Horn, L. J., & Edwards, R.T. 2003, *A&A*, 412, 473

Zank, G. P. & Greaves, R. G. 1995, *Phys. Rev. E*, 51, 6079

Advances in Plasma Astrophysics
Proceedings IAU Symposium No. 274, 2010
A. Bonanno, E. de Gouveia Dal Pino & A. G. Kosovichev, eds.

© International Astronomical Union 2011
doi:10.1017/S1743921311006995

Theory of quasi-stationary kinetic dynamos in magnetized accretion discs

Claudio Cremaschini[1,2], John C. Miller[1,2,3] and Massimo Tessarotto[4,5]

[1]International School for Advanced Studies, SISSA, Trieste, Italy

[2]INFN, Trieste Section, Trieste, Italy

[3]Department of Physics (Astrophysics), University of Oxford, Oxford, U.K.

[4]Department of Mathematics and Informatics, University of Trieste, Trieste, Italy

[5]Consortium for Magnetofluid Dynamics, University of Trieste, Trieste, Italy

Abstract. Magnetic fields are a distinctive feature of accretion disc plasmas around compact objects (i.e., black holes and neutron stars) and they play a decisive role in their dynamical evolution. A fundamental theoretical question related with this concerns investigation of the so-called gravitational MHD dynamo effect, responsible for the *self-generation* of magnetic fields in these systems. Experimental observations and theoretical models, based on fluid MHD descriptions of various types support the conjecture that accretion discs should be characterized by coherent and slowly time-varying magnetic fields with both *poloidal* and *toroidal* components. However, the precise origin of these magnetic structures and their interaction with the disc plasmas is currently unclear. The aim of this paper is to address this problem in the context of kinetic theory. The starting point is the investigation of a general class of Vlasov-Maxwell kinetic equilibria for axi-symmetric collisionless magnetized plasmas characterized by temperature anisotropy and mainly toroidal flow velocity. Retaining finite Larmor-radius effects in the calculation of the fluid fields, we show how these configurations are capable of sustaining both toroidal and poloidal current densities. As a result, we suggest the possible existence of a *kinetic dynamo effect*, which can generate a stationary toroidal magnetic field in the disc even without any net radial accretion flow. The results presented may have important implications for equilibrium solutions and stability analysis of accretion disc dynamics.

Keywords. kinetic theory, accretion disks, plasmas, magnetic fields

1. Introduction: magnetic fields in accretion discs

In this paper basic issues concerned with the origin and the structure of magnetic fields in accretion disc (AD) plasmas are discussed, with particular reference to the dynamo phenomenon which leads to the self-generation of the magnetic field appearing in these systems. More precisely, we address the problem of the generation of both the *poloidal* and *toroidal* components of the AD magnetic field as a consequence of plasma currents produced purely by *collisionless* and *quasi-stationary kinetic mechanisms*.

Experimental observations and theoretical models based on fluid MHD descriptions suggest that accretion discs should be characterized by coherent and slowly time-varying magnetic fields with both *poloidal* and *toroidal* components (see for example Frank *et al.*, 2002 and Szuszkiewicz & Miller, 2001). An interesting development within this context has been the work by Coppi (2005) and Coppi & Rousseau (2006), who showed that stationary magnetic configurations in AD plasmas, for both low and high magnetic energy densities, can exhibit complex magnetic structures characterized locally by plasma rings with closed nested magnetic surfaces. However, even the most sophisticated fluid models are still not able to give a satisfactory explanation for all of the complexity of the phenomena arising in these systems. In particular, the precise origin of these magnetic structures and their interaction with the disc plasmas remains unclear. This is especially true for what concerns the toroidal magnetic field. Usually it is thought that such fields are the result of non-stationary processes associated with some ongoing instabilities in the plasma (possibly the same ones responsible for the accretion), but the actual mechanisms for their generation are still a matter of debate. In contrast with these fluid descriptions, the aim of this paper is to address this problem in the context of kinetic theory, which can provide a description of the plasma dynamics at a more fundamental level (Cremaschini *et al.*, 2008 and 2010). The work is based on results presented in a recent paper by Cremaschini *et al.*, (2010), concerning the investigation of a general class of asymptotic (i.e., *quasi-stationary*) Vlasov-Maxwell kinetic equilibria for axi-symmetric collisionless magnetized plasmas characterized by temperature anisotropy and mainly toroidal flow velocity. Note that what is meant here by the term "equilibrium" is in general a stationary-flow solution. Retaining finite Larmor-radius effects in the calculation of the fluid fields, we show how these configurations are capable of sustaining both toroidal and poloidal current densities. As a result, for these configurations of the magnetic field, we conjecture the possible existence of a *quasi-stationary kinetic dynamo effect* which can generate a stationary toroidal magnetic field in the disc even without any net radial accretion flow. The theory presented is of interest for improving our understanding of magnetic fields in accretion discs, concerning both their local structure and generation mechanisms. These results may also have important implications for further studies of the equilibrium solutions and the stability analysis of accretion disc dynamics.

2. Basic assumptions and notation

For what concerns the notation adopted and the basic assumptions about the AD plasma and the magnetic field configuration, we refer to the paper by Cremaschini *et al.*, (2010). The same holds also for the meaning of the dimensionless parameters ε, ε_M, δ and δ_{Ts} used in constructing the asymptotic kinetic theory and its relevant expansions. For the sake of clarity, we recall that the present analysis is restricted to the particular situation where the equilibrium magnetic field $\mathbf{B}$ admits, at least locally, a family of nested axi-symmetric closed toroidal magnetic surfaces $\{\psi(\mathbf{r})\} \equiv \{\psi(\mathbf{r}) = const.\}$, where ψ denotes the poloidal magnetic flux of $\mathbf{B}$ (see Coppi (2005), Coppi & Rousseau (2006) for a proof of the possible existence of such configurations in the context of astrophysical accretion discs; see also Cremaschini *et al.*, (2010) for further discussion of this). In this situation, a set of magnetic coordinates $(\psi, \varphi, \vartheta)$ can be defined locally, where ϑ is a curvilinear angle-like coordinate on the magnetic surfaces $\psi(\mathbf{r}) = const.$ In particular, we shall assume that the magnetic field is slowly varying in time and of the form

$$\mathbf{B} \equiv \nabla \times \mathbf{A} = \mathbf{B}^{self}(\mathbf{r}, \varepsilon_M t) + \mathbf{B}^{ext}(\mathbf{r}, \varepsilon_M t), \tag{2.1}$$

where $\mathbf{B}^{self}$ and $\mathbf{B}^{ext}$ denote the self-generated magnetic field produced by the AD plasma and a non-vanishing external magnetic field produced by the central object. We also take the self field to be the dominant component, while $\mathbf{B}^{self}$ and $\mathbf{B}^{ext}$ are defined as

$$\mathbf{B}^{self} = I(\mathbf{r}, \varepsilon_M t)\nabla\varphi + \nabla\psi_p(\mathbf{r}, \varepsilon_M t) \times \nabla\varphi, \tag{2.2}$$

$$\mathbf{B}^{ext} = \nabla\psi_D(\mathbf{r}, \varepsilon_M t) \times \nabla\varphi, \tag{2.3}$$

where $\mathbf{B}_T \equiv I(\mathbf{r}, \varepsilon_M t)\nabla\varphi$ and $\mathbf{B}_P \equiv \nabla\psi_p(\mathbf{r}, \varepsilon_M t) \times \nabla\varphi$ are the toroidal and poloidal components of the self-field.

3. Asymptotic stationary solution and analytical expansion

In this section we briefly summarize the solution for the asymptotic stationary kinetic distribution function (KDF) for the case considered here, and its appropriate analytic expansion for describing strongly magnetized collisionless AD plasmas in which the temperature is anisotropic (Cremaschini *et al.*, 2010). An extended treatment can be found in Cremaschini *et al.* (2010). A convenient solution for the KDF in this configuration is given by

$$\widehat{f}_{*s} = \frac{\eta_s}{(2\pi/M_s)^{3/2} \left(T_{\|*s}\right)^{1/2} \widehat{T}_{\perp s}} \exp\left\{-\frac{H_{*s}}{T_{\|*s}} - \frac{m_s' B'}{\widehat{\Delta}_{T_s}}\right\}, \tag{3.1}$$

which we refer to as the *Generalized bi-Maxwellian KDF*, where $\frac{1}{\widehat{\Delta T_s}} \equiv \frac{1}{\widehat{T}_{\perp s}} - \frac{1}{T_{\parallel *s}}$, with the following *kinetic constraints* (Cremaschini, 2010): $\frac{n_s}{\widehat{T}_{\perp s}} = \widehat{\beta}_{*s}(\psi_{*s})$, $T_{\parallel *s} = T_{\parallel *s}(\psi_{*s})$, $\frac{B'}{\widehat{\Delta T_s}} = \widehat{\alpha}_{*s}(\psi_{*s})$ and $H_{*s} = E_s - \frac{Z_s e}{c}\psi_{*s}\Omega_s(\psi_{*s})$. Here ψ_{*s} is proportional to the canonical momentum, E_s is the total particle energy and m'_s is the gyrokinetic magnetic moment. Then, for *strongly magnetized AD plasmas* a convenient analytical expansion for $\widehat{f}_{*s}$ can be made in terms of the small dimensionless parameter ε. Retaining only the leading-order expression for the guiding-center magnetic moment $m'_s \simeq \mu'_s = \frac{M_s w'^2}{2B'}$, the following relation holds to first order in ε: $\widehat{f}_{*s} = \widehat{f}_s[1 + h_{Ds}] + O(\varepsilon^n)$, $n \geqslant 2$. Here, the zero order distribution $\widehat{f}_s$ is expressed as

$$\widehat{f}_s = \frac{n_s}{(2\pi/M_s)^{3/2}\left(T_{\parallel s}\right)^{1/2}T_{\perp s}} \exp\left\{-\frac{M_s\left(\mathbf{v} - \mathbf{V}_s\right)^2}{2T_{\parallel s}} - \frac{M_s w'^2}{2\Delta T_s}\right\}, \qquad (3.2)$$

which we will call the *bi-Maxwellian KDF*, with $\frac{1}{\Delta T_s} \equiv \frac{1}{T_{\perp s}} - \frac{1}{T_{\parallel s}}$ being related to the temperature anisotropy, while the quantity h_{Ds} represents the *diamagnetic part* of the KDF $\widehat{f}_{*s}$, which depends on the thermodynamic forces associated with the gradients of the fluid fields (Cremaschini *et al.*, (2010)). As a consequence of the asymptotic expansion, fluid moments associated to the stationary KDF can be computed analytically. In particular, the total flow velocity $\mathbf{V}_s^{tot}$ takes on the form $n_s^{tot}\mathbf{V}_s^{tot} \equiv \int d\mathbf{v}\mathbf{v}\widehat{f}_{*s} \simeq n_s[\mathbf{V}_s + \Delta\mathbf{U}_s]$, where by definition $\mathbf{V}_s = \Omega_s(\psi)Re_\varphi$ is the leading order term, which is purely toroidal, while $\Delta\mathbf{U}_s$ represents the *self-consistent finite Larmor-radius (FLR) velocity corrections of first order, with both toroidal and poloidal components* (Cremaschini, 2010).

4. The Ampere equation and the "kinetic dynamo"

In this section we investigate the consequences of the kinetic treatment developed here concerning magnetic field generation, showing that, besides a self-generated poloidal magnetic field, the kinetic equilibrium can also sustain a quasi-stationary toroidal field (*kinetic dynamo*). This is diamagnetic in origin and is due to the combined effects of FLR corrections and temperature anisotropies. For seeing this, consider the Ampere equation. Using the analytic calculation of the fluid fields discussed in the previous section, this can be written as follows for the self-generated magnetic field:

$$\nabla \times \mathbf{B}^{self} = \frac{4\pi}{c}\sum_{s=i,e} q_s n_s\left[\mathbf{V}_s + \Delta\mathbf{U}_s\right]. \qquad (4.1)$$

The toroidal component of this equation gives the generalized Grad-Shafranov equation for the poloidal flux function ψ_p, which in this approximation becomes:

$$\frac{\partial b_{\psi_p}}{\partial\vartheta} + \frac{\partial b_\vartheta}{\partial\psi} = -\frac{4\pi}{Rc}J\sum_{s=e,i} q_s n_s\left[\Omega_s\left(\psi\right)R + \Delta_{\varphi s}\right], \qquad (4.2)$$

where $b_{\psi_p} \equiv (\frac{J}{R^2}\nabla\psi_p \cdot \nabla\vartheta)$ and $b_\vartheta \equiv (\frac{J}{R^2}|\nabla\psi_p|^2)$, while $J \equiv \frac{1}{|\nabla\psi \times \nabla\varphi \cdot \nabla\vartheta|}$ is the Jacobian of the coordinate transformations. The remaining terms in Eq.(4.1) give the equation for the toroidal component of the magnetic field $\frac{I(\psi,\vartheta)}{R}$. In the same approximation, this is:

$$\nabla I(\psi,\vartheta) \times \nabla\varphi = \frac{4\pi}{c}\sum_{s=i,e} q_s n_s \frac{\Delta_{3s}}{B}\nabla\psi \times \nabla\varphi, \qquad (4.3)$$

where Δ_{3s} contains the contributions of the species temperature anisotropies. For consistency with the approximation introduced, in the small inverse aspect ratio ordering, it follows that $\frac{\partial I(\psi,\vartheta)}{\partial\vartheta} = 0 + O(\delta^k)$, i.e., to leading order in δ: $I = I(\psi) + O(\delta^k)$, with $k \geqslant 1$. This is the only constraint imposed on the kinetic solution by the Ampere equation

(Cremaschini *et al.* 2010), which in turn also requires that the corresponding current density in Eq. (4.3) is necessarily a flux function. Then, correct to $O(\varepsilon)$, $O(\varepsilon_M^0)$ and $O(\delta^0)$, the differential equation for $I(\psi)$ becomes:

$$\frac{\partial I(\psi)}{\partial \psi} = \frac{4\pi}{c} \sum_{s=e,i} q_s n_s \frac{\Delta_{3s}}{B}, \tag{4.4}$$

which uniquely determines an approximate solution for the toroidal magnetic field. This result is remarkable because it shows that there can exist a *stationary kinetic dynamo effect* which generates an equilibrium toroidal magnetic field *without requiring any net accretion* and *without any possible instability/turbulence phenomena*. This new mechanism results from poloidal currents arising due to the FLR effects and temperature anisotropies which are characteristic of the equilibrium KDF for collisionless plasmas. The self-generation of the stationary magnetic field is purely diamagnetic. In particular, the toroidal component is associated with the drifts of the plasma away from the flux surfaces. In the present formulation, possible dissipative phenomena leading to a non-stationary self field have been ignored. Such dissipative phenomena probably do arise in practice and could occur both in the local domain where the equilibrium magnetic surfaces are closed and nested, and elsewhere. Temperature anisotropies are therefore an important physical property of collisionless AD plasmas, giving a possible mechanism for producing a stationary toroidal magnetic field. We stress that this effect disappears altogether in the case of isotropic temperatures.

5. Conclusions

In this paper we have presented a kinetic formulation for the quasi-stationary dynamo effect, responsible for the self-generation of magnetic fields in accretion discs around compact objects. The theory, arrived at within the framework of the Vlasov-Maxwell description, is applicable to non-relativistic axi-symmetric collisionless AD plasmas immersed in both gravitational and magnetic fields, the latter being assumed to admit locally closed nested poloidal flux surfaces. In particular, it has been shown that a collisionless AD plasma with temperature anisotropy can produce both poloidal and toroidal asymptotic stationary magnetic fields. For the magnetic field configuration considered here, this may occur even without any net radial accretion flow. This remarkable conclusion can cast further light on the physical mechanisms responsible for the generation of magnetic fields in accretion discs and their structure. Finally, these results can also have important implications for our understanding of the equilibrium properties of accretion discs and their dynamical stability properties.

6. Acknowledgments

Work developed in the framework of the Consortium for Magnetofluid Dynamics, Trieste, Italy and the European Research Group on Applied Magnetoscience (GAMAS PDRE), CNRS, France.

References

Frank J., King A., & Raine D. 2002, *Accretion power in astrophysics*, CUP
Cremaschini C., Beklemishev A., Miller J., & Tessarotto M. 2008, *AIP Conf. Proc.* 1084, 1067
Cremaschini C., Beklemishev A., Miller J., & Tessarotto M. 2008, *AIP Conf. Proc.* 1084, 1073
Szuszkiewicz E. & Miller J. C. 2001, *MNRAS* 328, 36
Coppi B. 2005, *Phys. Plasmas* 12, 057302
Coppi B. & Rousseau F. 2006, *ApJ* 641, 458
Cremaschini C., Miller J. C., & Tessarotto M. 2010, *Phys. Plasmas* 17, 072902
Catto P. J., Bernstein I. B., & Tessarotto M. 1987, *Phys. Fluids B* 30, 2784

Advances in Plasma Astrophysics
Proceedings IAU Symposium No. 274, 2010
A. Bonanno, E. de Gouveia Dal Pino & A. G. Kosovichev, eds.
© International Astronomical Union 2011
doi:10.1017/S1743921311007009

Stationary and axisymmetric configurations of compact stars with extremely strong and highly localized magnetic fields

Kotaro Fujisawa*, Shin'ichiro Yoshida and Yoshiharu Eriguchi

Department of Earth Science and Astronomy,
Graduate School of Arts and Sciences, University of Tokyo,
Komaba, Meguro-ku, Tokyo 153-8902, Japan
*email: `fujisawa@ea.c.u-tokyo.ac.jp`

Abstract. Using a new formulation to compute structures of stationary and axisymmetric magnetized barotropic stars in Newtonian gravity, we have succeeded in obtaining numerically exact models of stars with extremely high interior magnetic fields. In this formulation, there appear four arbitrary functions of the magnetic flux function from the integrability conditions among the basic equations. Since in our new formulation these arbitrary functions appear in the expression of the current density, configurations with different current distributions can be specified by choosing the forms of the arbitrary functions.

By choosing appropriate forms for the four arbitrary functions, we have solved many kinds of equilibrium configurations both with poloidal and toroidal magnetic fields. Among them, by choosing special form for the *toroidal current density*, we have been able to obtain magnetized stars which have extremely strong poloidal magnetic fields deep inside the core region near the symmetric axis. By adopting the appropriate model parameters for the neutron stars, the magnetic fields could be $10^{14} \sim 10^{15}$ G on the surfaces and be about 10^{17} G in the deep interior regions. For other model parameters appropriate for white dwarfs, the magnetic fields could be around $10^{7} \sim 10^{8}$ G (surface regions) and $10^{9} \sim 10^{10}$ G (core regions). It is remarkable that the regions with very strong interior magnetic fields are confined to a very narrow region around the symmetric axis in the central part of the stars. The issues of stability of these configurations and of evolutionary paths to reach such configurations need to be investigated in the future work.

Keywords. Stars: interior - Stars: magnetic fields - Stars: neutron - Stars: white dwarfs

---

1. Introduction

According to the recent observations, there are many strong magnetized white dwarfs and neutron stars. Using observations of Zeeman effect and cyclotron absorption or assuming magnetic dipole spin-down, the strength of their magnetic fields at the stellar surfaces has been estimated. For magnetized white dwarfs, the observed magnetic fields on the surfaces range $\sim 3 \times 10^{4} - 10^{9}$ G. On the other hand, for neutron stars, they are estimated to be $10^{9} - 10^{15}$ G. However, we cannot observe the inner magnetic fields directly and almost nothing has been known about them. Thus, it is a very interesting and important task to know the inner magnetic fields by some other means. One of the possibilities is to construct and study theoretical models in the possible domain of parameter space. To this end, we have extended the Tomimura-Eriguchi formulation (Tomimura & Eriguchi 2005) to a more general formulation and have developed a new numerical scheme. We have calculated the magnetized equilibria consistently using this new scheme. As a result, we have succeeded in obtaining configurations with extremely strong and highly localized poloidal magnetic fields deep inside the core region near the symmetric axis.

2. Formulation

In this section, we will explain our new formulation briefly. Our formulation is based on Tomimura-Eriguchi scheme (Tomimura & Eriguchi 2005, Yoshida & Eriguchi 2006, Yoshida *et al.* 2006, and Otani *et al.* 2009).

Assumptions for the magnetized stars are as follows. (1) They are in stationary states, i.e. $\partial/\partial t = 0$. (2) They are axisymmetric about the magnetic or the rotational axis, i.e. $\partial/\partial\varphi = 0$. (3) When stars are rotating and have magnetic fields, the rotational axis and the magnetic field axis coincide. (4) The stars are self-gravitating. (5) Gravity is Newtonian. (6) The ideal MHD approximation applies. (7) No currents appear in the vacuum region. (8) Barotropic equation of state ($p = p(\rho)$). For neutron stars, we employ the polytropic equation ($p = K\rho^{1+1/N}$) with $N = 1$ (Fujisawa *et al.* 2011a). For white dwarfs, we use the equation of state for completely degenerate gas with zero temperature (Fermi gas) (Fujisawa *et al.* 2011b).

Under these assumptions, we reduce Maxwell equations, Poisson equation for the gravitational potential ϕ_g and the equation of motion in the stationary state which reads as follows:

$$\frac{1}{\rho}\nabla p = -\nabla\phi_g - \frac{1}{2}\nabla|\boldsymbol{v}|^2 + \boldsymbol{v}\times\boldsymbol{\omega} + \frac{1}{\rho}\left(\frac{\boldsymbol{j}}{c}\times\boldsymbol{H}\right), \tag{2.1}$$

where variables have their usual meanings. In order to obtain axisymmetric and stationary states of magnetized stars, we introduce the magnetic flux function Ψ as follows:

$$H_R \equiv -\frac{1}{R}\frac{\partial\Psi}{\partial z}, \quad H_z \equiv \frac{1}{R}\frac{\partial\Psi}{\partial R}. \tag{2.2}$$

Here, we use the cylindrical coordinates (R, φ, z). Usually, using this flux function Ψ, the Grad-Shafranov (GS) equation has been solved (see e.g. Eq.(19) in Lovelace *et al.* 1986). However, in our new formulation, instead of solving the GS equation, we derive the elliptical type partial differential equation for the φ-component of the vector potential $A_\varphi (= \Psi/R)$ from the Maxwell equation, i.e. the Ampére's law as follows:

$$\Delta(A_\varphi\sin\varphi) = -4\pi\frac{j_\varphi}{c}\sin\varphi. \tag{2.3}$$

Thus, we have two Poisson-type equations. Using Green's functions for the Laplacian, we can take account of the appropriate boundary conditions at infinity and obtain the following integral equations:

$$\phi_g = -G\int\frac{\rho(\boldsymbol{r}')}{|\boldsymbol{r}-\boldsymbol{r}'|}d^3\boldsymbol{r}', \quad A_\varphi\sin\varphi = \frac{1}{c}\int\frac{j_\varphi(\boldsymbol{r}')}{|\boldsymbol{r}-\boldsymbol{r}'|}d^3\boldsymbol{r}'. \tag{2.4}$$

In order to integrate these equations, we need the distributions of source terms ρ and j_φ. We calculate the density ρ from the first integral of the equation of motion:

$$\int\frac{dp}{\rho} = -\phi_g - \frac{1}{2}|\boldsymbol{v}|^2 + \int\mu(\Psi)\,d\Psi + Rv_\varphi\Omega(\Psi) + C, \tag{2.5}$$

and the current density j_φ from the integrability of the equation of motion:

$$\frac{\boldsymbol{j}}{c} = \left(\frac{\kappa'(\Psi)}{4\pi} + \frac{Q'(\Psi)}{4\pi}Rv_\varphi\right)\boldsymbol{H} + \frac{Q(\Psi)}{4\pi}\boldsymbol{\omega} + \rho R\left(\mu(\Psi) + \Omega'(\Psi)Rv_\varphi\right)\boldsymbol{e}_\varphi. \tag{2.6}$$

Where, κ, Q, μ and Ω are arbitrary functions of the flux function Ψ. These four arbitrary functions are essentially the same ones as shown in Lovelace *et al.* (1986). Choosing forms of these arbitrary functions appropriately, we first calculate Eq. (2.5) and Eq. (2.6) and then obtain ϕ_g and Ψ from Eqs. (2.4). This iteration is cycled until the system

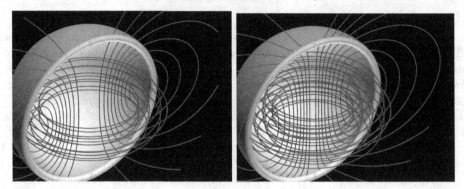

Figure 1. Common features appear in all configurations with both poloidal and toroidal magnetic fields. Configurations with almost dipole distributions of the magnetic fields (left, $m = 0.0$) and configurations with highly localized and extremely strong magnetic fields (right, $m = -3.0$).

settles down to a converged set of physical quantities and we obtain equilibrium states of magnetized stars.

We choose the arbitrary function form as $\mu(\Psi) = \mu_0(\Psi + \epsilon)^m$ (Fujisawa *et al.* 2011a). It should be noted that previous investigations such as Tomimura & Eriguchi (2005) treated only $\mu = \mu_0$ (constant) configurations. Employing this functional form, we have obtained various magnetized configurations by changing the values of the parameter m.

3. Results and Discussion

Fig. 1 shows the magnetic field distributions for the configurations with both poloidal and toroidal magnetic fields. Since almost all of our numerically obtained results have the similar helical structures of the magnetic fields, it is very natural to consider that this kind of magnetic structures are commonly realized in axisymmetric and stationary barotropic equilibrium configurations. Although the equilibrium state in the right panel ($m = -3.0$ solution) shows a helical field distribution, its poloidal magnetic field is highly localized near the symmetry axis. Thus we will call this highly localized configuration.

In Figure 2, shown are the structure (left panel) and the relative strength of the magnetic multipole moments (right panel) for neutron stars with the parameter $m = 0$. In the left panel, the stellar surface, the field lines of the poloidal magnetic fields and the continuous distributions of $\log_{10}|\boldsymbol{H}|$ are shown. In the right panel, the ratios of the strength of the magnetic multipolar moments in the vacuum region to that of the dipole component are displayed. In this figure, the values of the averaged magnetic fields on the stellar surface and in the central region are $H_{sur} \sim 8.0 \times 10^{14}$G and $H_c \sim 5.0 \times 10^{15}$G, respectively. As seen from this figure, higher multipole components are not so large, the magnetic filed of this configuration is described mostly by the dipole moment.

In Figure 3, the same quantities as those in Figure 2 are shown for configurations with the parameter $m = -2.2$. Although we have computed the model so that the value of the surface magnetic field can be almost the same as that for the configuration in Figure 2, i.e. $H_{sur} \sim 8.0 \times 10^{14}$G, the poloidal magnetic field of this model is highly localized near the symmetry axis and H_c reaches 10^{17} G.

It is also remarkable that the contribution of the higher multipole components to the total magnetic fields is appreciable. For instance, the $n = 3$ component (oct-pole component) is about 15 % of the dipole component. Moreover, the $n = 5$ and $n = 7$ components are few percents of the dipole component. Therefore, this configuration

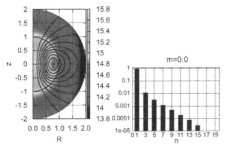

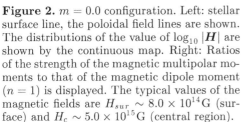

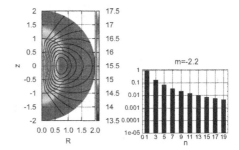

Figure 2. $m = 0.0$ configuration. Left: stellar surface line, the poloidal field lines are shown. The distributions of the value of $\log_{10} |\boldsymbol{H}|$ are shown by the continuous map. Right: Ratios of the strength of the magnetic multipolar moments to that of the magnetic dipole moment ($n = 1$) is displayed. The typical values of the magnetic fields are $H_{sur} \sim 8.0 \times 10^{14}$G (surface) and $H_c \sim 5.0 \times 10^{15}$G (central region).

Figure 3. $m = -2.2$ configuration. Left: stellar surface line, the poloidal field lines are shown. The distributions of the value of $\log_{10} |\boldsymbol{H}|$ are shown by the continuous map. Right: Ratios of the strength of the magnetic multipolar moments to that of the magnetic dipole moment ($n = 1$) is displayed. The typical values of the magnetic fields are $H_{sur} \sim 8.0 \times 10^{14}$G (surface) but $H_c \sim 1.0 \times 10^{17}$G (central region).

cannot be represented by a simple dipole moment configuration but should be considered as a configuration with an extremely strong and highly localized magnetic field.

We have also calculated equilibrium structures of white dwarfs and obtained similar highly localized configurations whose magnetic fields are around $10^7 \sim 10^8$ G on the surfaces and $10^9 \sim 10^{10}$ G in the central core regions (see Fujisawa *et al.* 2011b).

It would be a very interesting problem to investigate stability natures of our newly obtained magnetized configurations. Magnetic field structures of our configurations are similar to the Twisted-Torus configurations (Braithwaite & Spruit 2004) which are stable and stationary state reached after a long time evolution. Therefore there is a possibility that our configurations might be stable. However, since we do not have a suitable method to analyze the stability natures of our configurations, we cannot tell exactly whether our configurations are stable or not, at the present time.

4. Summary

We have succeeded in obtaining a wide range of axisymmetric and stationary configurations with magnetic fields – from configurations with nearly dipole magnetic fields to those with extremely strong and highly localized magnetic fields. Using the models obtained, we have estimated the strengths of the magnetic fields at the central regions of the stars. We have also shown that the components of magnetic multipolar moments are not always represented by a simple dipole moment model.

References

Braithwaite, J. & Spruit, H. C. 2004, *Nature*, 431, 819
Fujisawa, K., Yoshida, Si., & Eriguchi, Y., 2011a, in preparation
Fujisawa, K., Yoshida, Si., & Eriguchi, Y., 2011b, in preparation
Lovelace, R. V. E., Mehanian, C., Mobarry, C. M., & Sulkanen, M. E., 1986, *ApJS*, 62, 1
Otani, J., Takahashi, R., & Eriguchi, Y., 2009, *MNRAS*, 396, 2152
Tomimura, Y. & Eriguchi, Y., 2005, *MNRAS*, 359, 1117
Yoshida, Sj. & Eriguchi, Y., 2006, *ApJ*, 164, 156
Yoshida, Sj., Yoshida, Si., & Eriguchi, Y., 2006, *ApJ*, 651, 462

Advances in Plasma Astrophysics
Proceedings IAU Symposium No. 274, 2010
A. Bonanno, E. de Gouveia Dal Pino, A. G. Kosovichev, eds.

© International Astronomical Union 2011
doi:10.1017/S1743921311007010

Kinetic closure conditions for quasi-stationary collisionless axisymmetric magnetoplasmas

Claudio Cremaschini[1,2]**, John C. Miller**[1,2,3] **and Massimo Tessarotto**[4,5]

[1]International School for Advanced Studies, SISSA, Trieste, Italy

[2]INFN, Trieste Section, Trieste, Italy

[3]Department of Physics (Astrophysics), University of Oxford, Oxford, U.K.

[4]Department of Mathematics and Informatics, University of Trieste, Trieste, Italy

[5]Consortium for Magnetofluid Dynamics, University of Trieste, Trieste, Italy

Abstract. A characteristic feature of fluid theories concerns the difficulty of uniquely defining consistent closure conditions for the fluid equations. In fact it is well known that fluid theories cannot generally provide a closed system of equations for the fluid fields. This feature is typical of collisionless plasmas where, in contrast to collisional plasmas, asymptotic closure conditions do not follow as a consequence of an H-theorem This issue is of particular relevance in astrophysics where fluid approaches are usually adopted. On the other hand, it is well known that the determination of the closure conditions is in principle achievable in the context of kinetic theory. In the case of multi-species thermal magnetoplasmas this requires the determination of the species tensor pressure and of the corresponding heat fluxes. In this paper we investigate this problem in the framework of the Vlasov-Maxwell description for collisionless axisymmetric magnetoplasmas arising in astrophysics, with particular reference to accretion discs around compact objects (like black holes and neutron stars). The dynamics of collisionless plasmas in these environments is determined by the simultaneous presence of gravitational and magnetic fields, where the latter may be both externally produced and self-generated by the plasma currents. Our starting point here is the construction of a solution for the stationary distribution function describing slowly-varying gyrokinetic equilibria. The treatment is applicable to non-relativistic axisymmetric systems characterized by temperature anisotropy and differential rotation flows. It is shown that the kinetic formalism allows one to solve the closure problem and to consistently compute the relevant fluid fields *with the inclusion of finite Larmor-radius effects*. The main features of the theory and relevant applications are discussed.

Keywords. kinetic theory, accretion disks, plasmas, magnetic fields

1. Introduction: fluid equations and kinetic closure conditions

In this paper the issue concerned with the determination of kinetic closure conditions for collisionless magnetized axisymmetric accretion disc (AD) plasmas is discussed. The treatment of this problem is a prerequisite of primary importance for getting correct descriptions for the dynamics of collisionless plasmas in terms of suitable fluid equations and the corresponding fluid fields (Cremaschini *et al.* 2008). The result presented here is obtained within the framework of the Vlasov-Maxwell description and concerns, in particular, collisionless magnetoplasmas in astrophysical accretion discs around compact objects (like black holes and neutron stars). In the present context we focus our attention on the specific case of *asymptotic stationary configurations*, in the sense defined in Cremaschini *et al.*, (2010). We also assume that the magnetic field has both poloidal and toroidal components and we restrict our analysis to configurations in which the poloidal field admits locally a family of closed nested magnetic surfaces $\psi = const.$, ψ denoting the poloidal magnetic flux function (Cremaschini *et al.* 2010 and references cited there). The starting point of the work is the construction of a stationary solution to the Vlasov equation for the equilibrium kinetic distribution function (KDF) $\widehat{f}_{*s}$ describing asymptotic (i.e., slowly-varying in time) gyrokinetic equilibria for AD plasmas, as presented in Cremaschini *et al.* 2008 and 2010. This investigation is applicable to non-relativistic axisymmetric systems characterized by temperature anisotropy and flows with differential rotation. By Taylor expanding the KDF in a suitable asymptotic ordering up to a prescribed order of accuracy we wish to compute *analytically* the relevant stationary fluid fields corresponding to collisionless AD plasmas, retaining finite Larmor-radius (FLR) effects in the calculation. It is shown that the kinetic formalism then allows one both to solve consistently the closure problem of the fluid equations and, at the same time, to derive important conclusions about the physical properties of the stationary collisionless AD plasma, as implied by the kinetic analysis (see also Cremaschini *et al.* 2010 for some discussion related to this point). The theory presented is of interest for those studies aimed at investigating stationary AD plasmas from the point of view of fluid MHD theories.

2. Tensor pressure

Concerning the notation adopted, the basic assumptions about the system and the magnetic field configuration, and the details of the construction of the kinetic theory for AD plasmas used here, we refer to the paper by Cremaschini *et al.* 2010. The species number densities and flow velocities have been computed as fluid moments in the same paper. For this reason, in the following we shall focus attention on the calculation of the tensor pressure, which is required as a closure condition in the Euler fluid equation. The total species tensor pressure is defined as $\underline{\underline{\Pi}}_s^{tot} \equiv \int d\mathbf{v} M_s \left(\mathbf{v} - \mathbf{V}_s^{tot} \right) \left(\mathbf{v} - \mathbf{V}_s^{tot} \right) \widehat{f}_{*s}$. Since the AD plasma is collisionless, the KDF is not Maxwellian and we expect to recover some sort of anisotropy in the final form of the pressure tensor. For example, this can be due to the temperature anisotropy, whose origin is related to the conservation of the magnetic moment as an adiabatic invariant. Parallel and perpendicular temperatures are defined with respect to the local direction of the magnetic field. For this reason, it is convenient to introduce the set of orthogonal unitary vectors given by $(\mathbf{b}, \mathbf{e}_1, \mathbf{e}_2)$, where $\mathbf{b} \equiv \frac{\mathbf{B}}{B}$ is the tangent vector to the magnetic field while $\mathbf{e}_1$ and $\mathbf{e}_2$ are two orthogonal vectors in the plane perpendicular to the magnetic field line. Then, from this basis, we can construct the following unitary tensor: $\underline{\underline{I}} \equiv \mathbf{b}\mathbf{b} + \mathbf{e}_1\mathbf{e}_1 + \mathbf{e}_2\mathbf{e}_2$. The tensor pressure has its simplest representation when expressed in terms of the tensor $\underline{\underline{I}}$ and the vectors $(\mathbf{b}, \mathbf{e}_1, \mathbf{e}_2)$.

By Taylor expanding the equilibrium KDF (Cremaschini *et al.* 2010), the species pressure tensor $\underline{\underline{\Pi}}_s^{tot}$ for magnetized plasmas can be written as follows: $\underline{\underline{\Pi}}_s^{tot} \simeq \underline{\underline{\Pi}}_s + \Delta\underline{\underline{\Pi}}_s$, where $\underline{\underline{\Pi}}_s$ is the leading-order term (with respect to all of the expansion parameters), while $\Delta\underline{\underline{\Pi}}_s$ represents the first-order (i.e., $O(\varepsilon)$) correction term. In particular, $\underline{\underline{\Pi}}_s$ is given by

$$\underline{\underline{\Pi}}_s \equiv T_{\perp s} n_s \underline{\underline{I}} + n_s \left[T_{\| s} - T_{\perp s} \right] \mathbf{b}\mathbf{b}, \tag{2.1}$$

which is expressed in terms of the parallel and perpendicular temperatures. On the other hand, $\Delta\underline{\underline{\Pi}}_s$ can be written as

$$\Delta\underline{\underline{\Pi}}_s \equiv \Delta\Pi_s^1 \underline{\underline{I}} + \Delta\Pi_s^2 \mathbf{b}\mathbf{b} + \Delta\underline{\underline{\Pi}}_s^3, \tag{2.2}$$

in which $\Delta\Pi_s^1$ and $\Delta\Pi_s^2$ are the diagonal first-order anisotropic corrections to the pressure tensor, while $\Delta\underline{\underline{\Pi}}_s^3$ contains all of the non-diagonal contributions. More precisely, $\Delta\Pi_s^1$ is given by

$$\Delta\Pi_s^1 \equiv \gamma_1 \Omega_s RT_{\perp s} n_s - \gamma_2 \frac{2\Omega_s RT_{\perp s}^2 n_s}{B} + \gamma_3 \frac{\Omega_s RT_{\perp s}^2 n_s}{M_s} \left(5 + \frac{T_{\| s}}{\Delta T_s} \right)$$
$$+ \gamma_3 \left(\Omega_s R \right)^3 T_{\perp s} n_s + \gamma_3 \frac{2\Omega_s I^2 T_{\| s} T_{\perp s} n_s}{RB^2 M_s} + \gamma_3 \frac{8\Omega_s RT_{\perp s}^2 n_s}{M_s} \left[\frac{3}{4} - \frac{I^2}{R^2 B^2} \right], \tag{2.3}$$

while $\Delta\Pi_s^2$ is defined as

$$\Delta\Pi_s^2 \equiv \gamma_1 \Omega_s R n_s \left[T_{\|s} - T_{\perp s}\right] - \gamma_2 \frac{\Omega_s R T_{\|s} T_{\perp s} n_s}{B}$$

$$+ \gamma_3 \frac{\Omega_s R n_s T_{\|s} T_{\perp s}}{M_s} \left(5 + 3\frac{T_{\|s}}{\Delta_{T_s}}\right) + + \gamma_3 \left(\Omega_s R\right)^3 n_s \left[T_{\|s} - T_{\perp s}\right] +$$

$$+ \gamma_3 \frac{2\Omega_s R n_s T_{\|s} T_{\perp s}}{M_s} - \gamma_3 \frac{8\Omega_s R T_{\perp s}^2 n_s}{M_s} \left[\frac{3}{4}\frac{I^2}{R^2 B^2} + 1\right] + \gamma_2 \frac{2\Omega_s R T_{\perp s}^2 n_s}{B} +$$

$$- \gamma_3 \frac{\Omega_s R T_{\perp s}^2 n_s}{M_s} \left(5 + \frac{T_{\|s}}{\Delta_{T_s}}\right) + \gamma_3 \frac{6\Omega_s I^2 n_s T_{\|s}^2}{R B^2 M_s} - \gamma_3 \frac{4\Omega_s I^2 T_{\|s} T_{\perp s} n_s}{R B^2 M_s}. \qquad (2.4)$$

Finally, $\Delta\underline{\underline{\Pi}}_s^3$ is symmetric and is given by

$$\Delta\underline{\underline{\Pi}}_s^3 \equiv \gamma_3 \frac{16\Omega_s R T_{\perp s}^2 n_s}{M_s} \left(\mathbf{e}_1 \mathbf{e}_2 : \mathbf{e}_\varphi \mathbf{e}_\varphi\right) \left[\mathbf{e}_1 \mathbf{e}_2 + \mathbf{e}_2 \mathbf{e}_1\right] +$$

$$+ \gamma_3 \frac{4\Omega_s I T_{\|s} T_{\perp s} n_s}{B M_s} \left(\left(\mathbf{e}_2 \cdot \mathbf{e}_\varphi\right) \left[\mathbf{b} \mathbf{e}_2 + \mathbf{e}_2 \mathbf{b}\right] + \left(\mathbf{e}_1 \cdot \mathbf{e}_\varphi\right) \left[\mathbf{b} \mathbf{e}_1 + \mathbf{e}_1 \mathbf{b}\right]\right).$$

The following comments about the solution are in order:
- The total tensor pressure $\underline{\underline{\Pi}}_s^{tot}$ is symmetric in the system defined by the vectors $(\mathbf{b}, \mathbf{e}_1, \mathbf{e}_2)$.
- The leading-order pressure tensor $\underline{\underline{\Pi}}_s$ calculated in this approximation is diagonal but non-isotropic. We notice that the source of this anisotropy in Eq. (2.1) is just the temperature anisotropy. In the limit of isotropic temperature, the leading-order pressure tensor becomes diagonal and isotropic, as can be easily verified.
- The first-order correction $\Delta\underline{\underline{\Pi}}_s$ is non-diagonal and non-isotropic. In particular, non-diagonal terms are carried by the tensor $\Delta\underline{\underline{\Pi}}_s^3$ given in Eq. (2.5). Two main properties of the solution contribute to generating the non-isotropic feature of $\Delta\underline{\underline{\Pi}}_s$. The first is the temperature anisotropy, while the second is the existence of the *diamagnetic part* of the KDF obtained from the Taylor expansion of $\widehat{f}_{*s}$ and depending on the thermodynamic forces associated with the gradients of the fluid fields (Cremaschini *et al.* 2010). Indeed, we notice that taking the limit of isotropic temperature is not enough to make the tensor $\Delta\underline{\underline{\Pi}}_s$ isotropic as well. In fact, even if the parallel and perpendicular temperatures are equal, since the plasma is magnetized and collisionless the KDF will not be perfectly Maxwellian and deviations carried by the diamagnetic part act as a source of anisotropy.

3. Conclusions

In this paper a kinetic solution to the problem concerning the determination of closure conditions for fluid MHD equations has been presented. The theory has been developed within the framework of the Vlasov-Maxwell description and is applicable to non-relativistic axisymmetric collisionless AD plasmas. The specific case of asymptotic stationary configurations has been considered, in which the magnetic field is assumed to admit locally a family of closed nested magnetic surfaces. Some important features of collisionless magnetized plasmas have been included in the treatment. These concern the anisotropies induced by the existence of the magnetic field, like the temperature anisotropy, and the deviation of the equilibrium KDF away from the exact Maxwellian case, expressed through the diamagnetic part of the KDF. In particular, in this study, the calculation of the tensor pressure and the discussion of its physical properties have been considered.

References

Cremaschini, C., Beklemishev, A., Miller, J. C. & Tessarotto, M. 2008, *AIP Conf. Proc.* 1084, 1067

Cremaschini, C., Beklemishev, A., Miller, J. C. & Tessarotto, M. 2008, *AIP Conf. Proc.* 1084, 1073

Cremaschini, C., Miller, J. C. & Tessarotto, M. 2010, *Phys. Plasmas* 17, 072902

Advances in Plasma Astrophysics
Proceedings IAU Symposium No. 274, 2010
A. Bonanno, E. de Gouveia Dal Pino & A.G. Kosovichev, eds.

© International Astronomical Union 2011
doi:10.1017/S1743921311007022

Ponderomotive barrier for plasma particles on the boundary of astrophysical jets

Anna A. Dubinova[1] and Vladimir V. Kocharovsky[1]

[1] Institute of Applied Physics of the Russian Academy of Sciences, Nizhny Novgorod, Russia
email: anndub@gmail.com, kochar@appl.sci-nnov.ru

Abstract. We study kinetics of plasma particles on internal inhomogeneities and on the boundary of astrophysical jets in the presence of intensive low-frequency electromagnetic (surface and bulk waves). These waves with energy density exceeding or of the same order as the particle kinetic energy density can be generated due to the non-equilibrium state of plasma and lead to the jet stratification at various scales. The main reason is a ponderomotive force which is able to change dramatically the particle behavior, the plasma density cross-section profile and the wave collimation. We present results obtained on the basis of our simple ponderomotive model of the self-consistent analysis of the electromagnetic wave propagation and the formation of the plasma density profile.

Keywords. galaxies: jets, plasmas, waves.

1. Introduction

We describe the structure of Poynting-dominated jets (Blandford R., 2003, Guthmann A. W., 2002) in which particle collimation and boundary formation are strongly influenced by low-frequency electromagnetic waves propagating in the jet body and along the boundaries of the jet plasma waveguide. Originated from the guided waves the gradient ponderomotive force might be responsible for the formation of self-consistent plasma density profile $\rho = mn$. We expect the role of the ponderomotive force to be as important as the role of plasma pressure, p, inhomogeneous particle motion with bulk velocity, u, and quasistatic magnetic field pressure, $B_0^2/8\pi$. Previously, the ponderomotive effects were usually discussed with respect to particle acceleration along jets or with respect to the bulk velocity profile formation (e.g. Lundin R. *et al.*, 2006).

This can be explained by means of the generalized Bernoulli law with a corresponding term which stands for the ponderomotive potential w_{pond}

$$\frac{\rho u^2}{2} + p + \frac{B_0^2}{8\pi} + w_{\mathrm{pond}} \cong const. \tag{1.1}$$

At large scales (of order of the jet radius R_{jet}) the MHD approach is usually applicable (e.g. Komissarov S. S., 2010 *et al.*, and McKinney J.C. *et al.*, 2009). At small scales up to the plasma skin depth and the electromagnetic wavelength the kinetic approach is inevitable to describe, for example, the steepening of jet boundaries and inhomogeneitics inside jets. Generally speaking, we should take into consideration also wave and plasma turbulence (e.g. Sagdeev R. Z., 1979; Litvak A. G., 1986) but we restrict ourselves to studying steady-state configurations Fig. 1.

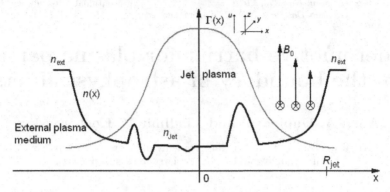

Figure 1. Scheme of the jet cross-section profile ($n(x)$ is the plasma density inside the jet and n_{ext} the plasma density outside the jet, $\Gamma(x)$ the bulk Lorents factor).

2. Features of ponderomotive forces

The well-known non-relativistic expression for the gradient ponderomotive (Miller) force in the presence of a plane monochromatic wave $\sim \boldsymbol{E}(\boldsymbol{r})e^{i\omega t}$ at frequency ω reads (Lundin R. *et al.*, 2006)

$$\boldsymbol{F}_{\text{pe}}(\boldsymbol{r}_{\text{d}}) = -\nabla\Phi_{\text{pe}}, \quad \Phi_{\text{pe}}(\boldsymbol{r}_{\text{d}}) = \frac{e^2}{4m\omega^2}|\boldsymbol{E}(\boldsymbol{r}_{\text{d}})|^2; \quad \frac{m|\dot{\boldsymbol{r}}_{\text{d}}|^2}{2} + \Phi_{\text{pe}}(\boldsymbol{r}_{\text{d}}) = const, \quad (2.1)$$

where e and m are charge and mass of a plasma particle (electron), $\boldsymbol{r}_{\text{d}}$ is radius-vector of a particle position averaged over the wave period $2\pi/\omega$. Taking into consideration the effects caused by an external quasistatic magnetic field, B_0, and a finite rate of particle collisions (or turbulence scattering), ν, one arrives at the more general expression for the ponderomotive force

$$\boldsymbol{F}_{\text{pe}}(\boldsymbol{r}_{\text{d}}) = -\frac{e^2}{4m\omega^2}\left(\frac{1}{1+\nu^2/\omega^2}\nabla|\boldsymbol{E}_{\|}|^2 + \frac{1+\omega_{\text{B}}/\omega}{(1+\omega_{\text{B}}/\omega)^2+\nu^2/\omega^2}\nabla|\boldsymbol{E}_{\perp}^+|^2\right.$$
$$+ \frac{1-\omega_{\text{B}}/\omega}{(1-\omega_{\text{B}}/\omega)^2+\nu^2/\omega^2}\nabla|\boldsymbol{E}_{\perp}^-|^2 + \frac{2\nu/\omega}{1+\nu^2/\omega^2}|\boldsymbol{E}_{\|}|^2\nabla\Psi_{\|}$$
$$\left.+ \frac{2\nu/\omega}{(1+\omega_{\text{B}}/\omega)^2+\nu^2/\omega^2}|\boldsymbol{E}_{\perp}^+|^2\nabla\Psi_{\perp}^+ + \frac{2\nu/\omega}{(1-\omega_{\text{B}}/\omega)^2+\nu^2/\omega^2}|\boldsymbol{E}_{\perp}^-|^2\nabla\Psi_{\perp}^-\right), \quad (2.2)$$

where $\Psi_{\|}, \Psi_{\perp}^+, \Psi_{\perp}^-$ are phases of the corresponding electric field components - longitudinal and transverse circular-polarized ones (with respect to $\boldsymbol{B}_0$), $\omega_{\text{B}} = eB_0/mc$ is the cyclotron frequency (Litvak A. G., 1986). A cyclotron resonance at $\omega = \omega_{\text{B}}$ results in strong dependence of the ponderomotive force on the effects of particle collisions and particle scattering. These effects is also crucial for the dynamics of waves at frequencies at the vicinity of the cyclotron frequency as well as the plasma frequency. In the general case of non-monochromatic electromagnetic field contributions of all the harmonics should be summed incoherently. For a continuous spectrum of incoherent waves it is possible to replace summing by integrating.

We have analyzed this force and the importance of the effective ponderomotive potential for various types of low-frequency waves and their simple possible spectra and found out that there is a wide range of wave and plasma parameters for which the ponderomotive separation of rarefied an dense plasma layers could exist at relatively small scales. Roughly speaking, it happens when the ponderomotive barrier exceeds the average particle energy. That requires the wave energy density of the order of the equipartition value.

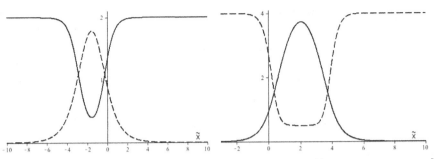

Figure 2. Typical self-consistent solutions of the system (3.2): $\tilde{\omega}_p^2(\tilde{x})$ (solid line) and $\tilde{E}(\tilde{x})$ (dashed line) at $\tilde{k}_z^2 = 0$, $\kappa_\infty^2 = 1$ (the left graph) and at $\tilde{k}_z^2 = 0.5$, $\kappa_\infty^2 = 3$ (the right graph).

3. A simple model

For illustration, let us consider a s-polarized monochromatic wave incident on the one-dimensional inhomogeneous plasma with the density profile determined by the pressure balance. The net pressure at certain point x is a sum of the thermal plasma pressure and the ponderomotive pressure created by the incident electromagnetic wave. We are eager to obtain a stationary self-consistent structure of the electromagnetic field and the plasma density. Combining the Helmholtz equation for a real amplitude of the electric field y-component and the pressure balance equation we arrive at the pair of normalized self-consistent equations with boundary conditions

$$\frac{\partial \kappa^2}{\partial \tilde{x}} + (\kappa^2 + 1 - \tilde{k}_z^2)\tilde{E}\frac{\partial \tilde{E}}{\partial \tilde{x}} = 0, \qquad \frac{\partial^2 \tilde{E}}{\partial \tilde{x}^2} - \kappa^2 \tilde{E} = 0, \tag{3.1}$$

$$\kappa^2(x \to +\infty) = \kappa_\infty^2, \quad \tilde{E}(x \to +\infty) = 0, \quad \frac{\partial \tilde{E}}{\partial \tilde{x}}(x \to +\infty) = 0, \tag{3.2}$$

where we ignore the magnetic field and introduce the following denotations

$$\tilde{E} = \frac{Ee}{\omega\sqrt{2mk_B T}}, \quad \tilde{\omega}_p^2 = \frac{\omega_p^2}{\omega^2}, \quad \tilde{x} = \frac{x\omega}{c}, \quad \tilde{k}_z^2 = \frac{k_z^2 c^2}{\omega^2}, \quad \kappa^2 = \tilde{\omega}_p^2 - 1 + \tilde{k}_z^2. \tag{3.3}$$

Here ω_p is the local plasma frequency proportional to the plasma density, k_z the wave number along z axis, T the plasma temperature which is assumed constant, c the speed of light in vacuum and k_B the Boltzman constant.

System (3.2) can be easily reduced to the non-linear Schödinger equation

$$\partial^2 \tilde{E}/\partial \tilde{x}^2 - \left[(\kappa_\infty^2 + 1 - \tilde{k}_z^2)\exp(-\tilde{E}^2) - (1 - \tilde{k}_z^2)\right]\tilde{E} = 0. \tag{3.4}$$

It gives soliton-like solutions which can be found numerically. They are demonstrated on the graphs on Fig. 2. It is clear to see that the wave field is localized and well-collimated and that the jet boundary width may be of the order of the electromagnetic wavelength. The higher the external plasma frequency and the larger longitudinal wavelength k_z the deeper and wider the plasma channel is.

4. Problems and discussions

Similar self-consistent 1D analysis for p-polarized waves (both bulk and surface) is more complicated because of the presence of the electric field component along the plasma density gradient. This fact leads to the resonant growth of the electric field near the wave reflection point and the resonant excitation of strong plasma waves at the vicinity of the point where the dielectric permittivity is equal to zero. However, these effects do not change the general conclusion that the small-scale plasma stratification and the formation of the sharp boundary could be typical for the Poynting-dominated jets due to the ponderomotive influence of low-frequency electromagnetic waves in the process of their self-channeling. There are also a number of qualitative arguments indicating that scattering, dissipation and various wave instabilities could affect the bulk acceleration, the effective particle temperature, the structure of boundary layers, the intrajet turbulence, the anomalous plasma conductivity, the magnetic field diffusion, the evolution of jet opening angle, etc. The rigorous analysis of all these problems requires additional (usually unavailable) information on jets including spectra of low-frequency powerful waves, their instability rates and scales, features of nonlinear stabilization and slow evolution of small-scale plasma turbulence, and so on.

5. Conclusions

In the presence of intensive electromagnetic surface and bulk waves as well as intensive plasma waves the gradient ponderomotive force can have strong influence on the kinetics of weakly collisional plasma particles and as a consequence on the plasma stratification inside jets and on the boundary plasma layer between the jet and the external medium. The electromagnetic waves are able to create a channel which they cannot escape if their frequency is below the plasma frequency of the external medium.

We point out the vital importance of the gradient ponderomotive force for the formation of internal inhomogeneities and the boundary of Poynting-dominated astrophysical jets. Our preliminary results make it clear that further development of the ponderomotive model of the jet boundary confinement is essential for understanding the processes of jet formation and collimation. The present work is the first step on the way to create a self-consistent ponderomotive model describing the collimation of astrophysical jets along with due analysis of the self-channeling of the low-frequency electromagnetic waves. There is no doubt that the advanced study of the problem will require numerical simulations which combine large-scale MHD and small-scale kinetic approaches.

References

Litvak, A. G. 1986, *Review of Plasma Physics*, 10, 293
Sagdeev, R. Z. 1979, *Rev. Mod. Phys.*, 51, 1
McKinney, J. C. & Bladford, R. D. 2009, *Mon. Not. R. Astron. Soc.*, 394, L126
Guthmann, A. W., Georganopoulos, M. *et al.* 2002, *Relativistic flows in astrophysics*, Springer
Komissarov, S. S., Vlahakis, N. & Königl A. 2010, *Mon. Not. R. Astron. Soc.*, 104, 17
Blandford, R. 2003, *ASP Conference series*, 290, 267
Lundin, R. & Guglielmi, A. 2006, *Space Sci. Rev.*, 127, 1

Advances in Plasma Astrophysics
Proceedings IAU Symposium No. 274, 2010
A. Bonanno, E. de Gouveia Dal Pino & A.G. Kosovichev, eds.

© International Astronomical Union 2011
doi:10.1017/S1743921311007034

GRB spectral parameter modeling

Gregory D. Fleishman[1,2] and Fedor A. Urtiev[3]

[1]New Jersey Institute of Technology, Newark, NJ 07102, USA

[2]Ioffe Institute, St. Petersburg 194021, Russia
email: gfleishm@njit.edu

[3]State Polytechnical University, St.Petersburg, 195251, Russia
email: zigzagworld@rambler.ru

Abstract. Fireball model of the gamma-ray bursts (GRBs) predicts generation of numerous internal shocks, which efficiently accelerate charged particles and generate relatively small-scale stochastic magnetic and electric fields. The accelerated particles diffuse in space due to interaction with the random waves and so emit so called Diffusive Synchrotron Radiation (DSR) in contrast to standard synchrotron radiation they would produce in a large-scale regular magnetic fields. In this contribution we present key results of detailed modeling of the GRB spectral parameters, which demonstrate that the non-perturbative DSR emission mechanism in a strong random magnetic field is consistent with observed distributions of the Band parameters and also with cross-correlations between them.

Keywords. acceleration of particles, shock waves, turbulence, galaxies: jets, radiation mechanisms: non-thermal, magnetic fields

1. Introduction

The fireball model of the gamma-ray burst (GRB) suggests that a central engine produces a number of interacting relativistic shock waves, whose interactions, in the collisionless case, result in generation of fluctuating electromagnetic fields and acceleration of charged particles up to high energies. It is well established by now that the magnetic and electric fields produced in the shock interactions have often a significant random component at various spatial scales. As a result, the shock-accelerated charged particles moving through a plasma with random electromagnetic fields experience random Lorenz forces and so follow random paths representing a kind of spatial diffusion. Accordingly, the particles produce a "diffusive radiation" whose spectra depend on the type of the field (magnetic or electric) and on spectral energy distribution of the field over the spatial scales (Toptygin & Fleishman 1987, Fleishman 2006, Fleishman & Toptygin 2007a, Fleishman & Toptygin 2007b, Reville & Kirk 2010).

Individual spectra of the prompt GRB emission are typically well fitted by a phenomenological Band function (Band *et al.* 1993), which consists of low-energy (spectral index α) and high-energy (spectral index β) power-law regions smoothly linked at a break energy E_{br}. The DSR was shown (Fleishman 2006) to produce spectra consistent with those observed typically from the GRBs (Band *et al.* 1993). It had yet been unclear, however, if the DSR spectra are naturally consistent with observed *distributions* of the GRB spectral parameters (Preece *et al.* 2000; Kaneko *et al.* 2006) and what ranges of physical GRB parameters are needed to reconcile the theoretical spectra with the observed ones. In this conference contribution we present results of the modeling of GRB prompt emission generation by DSR in relativistically expanding GRB jets presented in greater detail by Fleishman & Urtiev (2010).

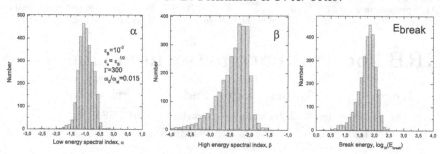

Figure 1. Example of the model Band parameter (α, β, and E_{br}) distributions obtained within the DSR model with strong random magnetic field.

2. Model

Formulation. Adopting a general internal shocks/fireball concept we accept that a single binary collision of relativistic internal shocks results in a single episode of the GRB prompt emission (Fleishman & Urtiev 2010). Microscopically, this shock-shock interaction produces high levels of random magnetic and/or electric fields and accelerates the charged particles up to large ultrarelativistic energies; these particles interact with the random fields to generate the gamma-rays. Although there are some common general properties of all cases of relativistic shock interactions, each shock-shock collision is, nevertheless, unique in terms of combination of the physical parameters involved. Accordingly, we adopt a set of standard (mean) parameters appropriate to account for the most global GRB properties, and then consider if a reasonable scatter of those standard parameters is capable of reproducing more detailed properties of the considered class of events as a whole — the statistical distributions of the GRB spectral parameters and cross-correlations between them. To do so, we considered a number of different emission models including the standard synchrotron radiation and DSR regimes in case of either weak or strong random magnetic field. The spectral slopes and breaks depend on both the emission mechanism and combination of physical parameters affecting the radiation spectra within a given mechanism. Thus, the goal of the modeling is to establish if there exists a parametric space making one or another theoretical model compatible with the observational data on the GRB spectral properties.

Results. Fleishman & Urtiev (2010) conclude that the DSR model with the weak random magnetic field (jitter regime), either perturbative or non-perturbative, cannot offer a consistent fit to the observed α histogram. This complies with independent criticisms of the jitter regime: Kumar & McMahon (2008) noticed that it may imply an unrealistically high level of inverse Compton emission, while Kirk & Reville (2010) argued that the jitter case seems to be in contradiction with the required high efficiency of the particle acceleration at the shocks, so strong magnetic fluctuations are needed to self-consistently accelerate electrons up to the gamma-ray producing energies.

Thus, having the weak random field model (jitter regime) rejected, we turn now to analysis of the strong random field case. According to Fleishman (2005), Fleishman & Bietenholz (2007), and Reville & Kirk (2010), new asymptotes arise in this case, which can yield broader α distribution. In this strong-field regime, the model α distribution depend on adopted ν distribution, which, within the adopted model, is straightforwardly linked to the β distribution, because $\beta = -\nu - 1$. The corresponding model results (Fig. 1) are in a remarkable agreement with the observations. Indeed, the α histogram is a symmetric one, it displays a peak at the right place, $\alpha = -1$, and its bandwidth is comparable to that of the observed histogram. The β histogram almost repeats the observed one, displaying the correct asymmetric shape and the peak at the right place,

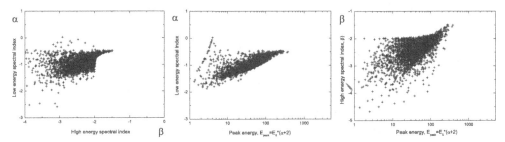

Figure 2. Cross-correlation of the model Band spectral parameters α and β (left); α and $E_{\text{peak}} = (\alpha + 2)E_{\text{br}}$ (middle); and β and $E_{\text{peak}} = (\alpha + 2)E_{\text{br}}$ (right).

$\beta = -2.2$. The E_{br} histogram also agrees with the observed one rather well displaying correct shape and bandwidth.

The cross-correlations between the spectral parameters derived from the model (Fig. 2) are to be compared with fig. 31 from Kaneko *et al.* (2006). Like in the observation, the spectral indices α and β are not highly correlated, although in the model plot the region of $-0.5 < \alpha < 0$ is underpopulated compared with the observed plot (Kaneko *et al.* 2006). Two other plots are in remarkable agreement with the observed cross-correlation plots, presented in Kaneko *et al.* (2006).

3. Conclusions

We conclude that the developed model is naturally capable of reproducing both the Band parameter histograms and their cross-correlations, which is a remarkable success of the non-perturbative DSR model in the presence of strong random magnetic field.

Acknowledgments

This work was supported in part by the Russian Foundation for Basic Research, grants No. 08-02-92228, 09-02-00226, 09-02-00624. We have made use of NASA's Astrophysics Data System Abstract Service.

References

Band, D. *et al.* 1993, *ApJ*, 413, 281
Fleishman, G. D. 2005, ArXiv e-prints: astro-ph/0510317
—. 2006, *ApJ*, 638, 348
Fleishman, G. D. & Bietenholz, M. F. 2007, *MNRAS*, 376, 625
Fleishman, G. D. & Toptygin, I. N. 2007a, *MNRAS*, 381, 1473
—. 2007b, *Phys. Rev. E*, 76, 017401
Fleishman, G. D. & Urtiev, F. A. 2010, *MNRAS*, 406, 644
Kaneko, Y. *et al.* 2006, *ApJS*, 166, 298
Kirk, J. G. & Reville, B. 2010, *ApJ*, 710, L16
Kumar, P. & McMahon, E. 2008, *MNRAS*, 384, 33
Preece, R. D. *et al.* 2000, *ApJS*, 126, 19
Reville, B. & Kirk, J. G. 2010, ArXiv e-prints: astro-ph.HE/1010.0872; this conf. proc.
Toptygin, I. N. & Fleishman, G. D. 1987, *Astrophys. Space. Sci.*, 132, 213

Advances in Plasma Astrophysics
Proceedings IAU Symposium No. 274, 2010
A. Bonanno, E. de Gouveia Dal Pino & A.G. Kosovichev, eds.

© International Astronomical Union 2011
doi:10.1017/S1743921311007046

Magnetic collimation of relativistic jets: the role of the black hole spin

N. Globus, C. Sauty and V. Cayatte

LUTh, Observatoire de Paris, F-92190 Meudon, France
email: noemie.globus@obspm.fr

Abstract. An ideal engine for producing ultrarelativistic jets is a rapidly rotating black hole threaded by a magnetic field. Following the 3+1 decomposition of spacetime of Thorne *et al.* (1986), we use a local inertial frame of reference attached to an observer comoving with the frame-dragging of the Kerr black hole (ZAMO) to write the GRMHD equations. Assuming θ-self similarity, analytical solutions for jets can be found for which the streamline shape is calculated exactly. Calculating the total energy variation between a non polar streamline and the polar axis, we have extended to the Kerr metric the simple criterion for the magnetic collimation of jets developed by Sauty *et al.* (1999). We show that the black hole rotation induces a more efficient magnetic collimation of the jet.

Keywords. galaxies: jets, ISM: jets and outflows, black hole physics, MHD

1. Context of the study

Many mechanisms are invoked to collimate jets. We shall not consider here collimation due to the external medium or by energy dissipation (shocks, reconnection). In Active Galactic Nuclei, it is believed that the disk wind can confine the inner spine jet but this can also be done by magnetized flows. If the jet carries a net current, then it can be magnetically self-confined under the tension associated with the toroidal field lines. A more realistic approach must take into account the contribution of particle pressure as well as the dynamical effects of the Kerr metric. GRMHD probably provides the better description for combining those ingredients.

2. Meridionally self-similar models

The steady GRMHD equations consist of a set of coupled, nonlinear, partial differential equations expressing momentum, magnetic and mass flux conservation, together with Ohm's law for a perfect conductor. Under the assumption of axisymmetry, streamlines and magnetic fieldlines are roped on the same flux tube of constant mass and magnetic flux. Four free integrals exist that depends only on the magnetic flux function : the total specific angular momentum carried by the flow and the magnetic field, L, the corotation angular velocity of each streamline at the base of the flow, Ω, the ratio of the mass and magnetic fluxes, Ψ_A and the Bernoulli energy constant, $\mathcal{E}$. In terms of these integrals and the square of the poloidal Alfvén Mach number, we can express the components of the magnetic field and the bulk velocity. To construct classes of exact solutions, we assume a spherical shape for the Alfvén surface and a dipolar angular dependence for the poloidal velocity and magnetic fields. By means of a separation of the variables in several physical key quantities, the GRMHD equations reduce to a system of ordinary differential equations, suitable for a numerical integration. In meridionally self-similar models, the θ-dependance is prescribed a priori while the radial dependance is derived

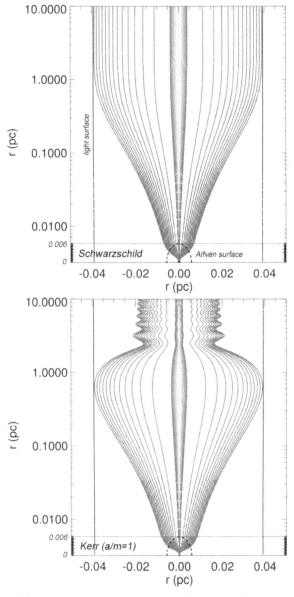

Figure 1. Topology of the solutions in the Schwarzschild metric (upper panel) and in the Kerr metric for a Kerr parameter $a/m = 1$ (lower panel).

from the equations. Such a treatment allows to study the physical properties of the outflow close to its rotational axis.

Figure 1 displays two solutions to the GRMHD equations, corresponding to the same physical parameters, except for the spin of the central engine. The lower jet is launched by a Schwarzschild black hole while the upper by a maximally rotating ($a/m = 1$) Kerr black hole. The jet solution in the Schwarzschild case is asymptotically cylindrical while in the Kerr case the solution recollimates at 1 pc, becoming an oscillatory solution with an asymptotic width about half the Schwarzschild one.

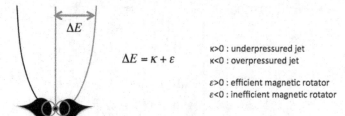

$$\Delta E = \kappa + \varepsilon$$

$\kappa > 0$: underpressured jet
$\kappa < 0$: overpressured jet

$\varepsilon > 0$: efficient magnetic rotator
$\varepsilon < 0$: inefficient magnetic rotator

A simple criterion for collimate the flow : $\Delta E > 0$

3. The magnetic collimation parameter

A general criterion for the jet collimation has been established. This criterion was developed successively in the non-relativistic case (Sauty *et al.*, 1999), in the Schwarzschild metric (Meliani *et al.* (2006)), in the Kerr metric (Globus *et al.*, in prep.). It is based on the variation of energy across the streamlines of the flow : *to allow self-confinement, there must be an excess of volumetric energy along a non polar streamline with respect to the axis.*

The two solutions have the same κ parameter, i.e. the same contribution to collimation of the pressure forces. This implies that the frame-dragging of the Kerr black hole have increased the contribution to collimation of the magnetic forces ϵ. Let express ϵ at the base of the flow (subscript $_o$),

$$\epsilon = \mathcal{E}_{\text{Rot},o} + \mathcal{E}_{\text{Poynt.},o} - \|\Delta \mathcal{E}_G^*\| + L\omega_o \,, \tag{3.1}$$

where $\mathcal{E}_{\text{Rot},o} + \mathcal{E}_{\text{Poynt.},o}$ is energy of the magnetic rotator, $\|\Delta \mathcal{E}_G^*\|$ is the gravitational potential which is not compensated by thermal driving (and thus must be supplied by magnetic means), ω_o is the frame dragging potential at the base of the flow. This last term represents the coupling between the orbital angular momentum of the fluid particle L and the frame dragging of the Kerr black hole.

Two new effects due to the Kerr metrics tend to increase ϵ (Globus *et al.*, in prep.). First $\|\Delta \mathcal{E}_G^*\|$ which is negative, decreases in absolute magnitude in the Kerr metric. Second, the new term $\mathcal{E}_{\text{drag}} = L\omega_o$ leads also to an increase of ϵ because it is positive.

4. Conclusion

Compared to the Schwarzschild case, the Kerr metric increases the efficiency of the magnetic rotator to collimate the outflow. Collimation occurs at large distances due to the building of the toroidal magnetic field in the super-Alfvénic regime. The rotation of the black hole acts on the collimation since it redistributes the energy of the magnetic rotator and not because it induces directly a strong toroidal magnetic field.

References

Sauty, C., Tsinganos, K. & Trussoni, E. 1999, *A&A*, 348, 327
Meliani, Z., Sauty, C., Vlahakis, N., Tsinganos, K. & Trussoni, E. 2006, *A&A*, 447, 797
Thorne, K. S., Price, R. H. & McDonald, D. A. 1986, *Black Holes: The Membrane Paradigm.*

Advances in Plasma Astrophysics
Proceedings IAU Symposium No. 274, 2010
A. Bonanno, E. de Gouveia Dal Pino & A.G. Kosovichev, eds.

© International Astronomical Union 2011
doi:10.1017/S1743921311007058

Current instabilities
in the pulsar magnetosphere

Axel Jessner[1], Harald Lesch[2] and Michael Kramer[1]

[1] Max Planck Institute for Radio Astronomy, Auf dem Hügel 69, D-53121, Bonn, Germany
email: jessner@mpifr-bonn.mpg.de

[2] Universitäts-Sternwarte München, Scheinerstr. 1, D-81679 München, Germany
email: lesch@usm.uni-muenchen.de

Abstract. Pulsars are rotating neutron stars with strong magnetic dipole fields ($B = 10^4 - 10^9$T), and high induced surface potentials (ca. 10^{14}V). A strong charged particle current is driven out of the polar cap. It returns along an equatorial current sheet. The total dissipated power of the current system is a significant fraction of the observed spin-down power of the pulsar. The Pierce instability occurs when particles are constrained to move in only one dimension and the field from the accumulated space charge exceeds the accelerating background field. Relativistic particle motion enhances the instability which forms narrow regions (cm) of high particle densities and low velocities separated by much longer but more tenuous relativistic flows. The calculated spectrum, power budget and time scales of the magnetospheric Pierce instabilities match the observed radio properties of pulsars.

Keywords. (stars:) pulsars: general, acceleration of particles, instabilities.

A strong charged particle current is driven out of the polar caps of a pulsar as a consequence of its rotation and its high surface magnetic field. Contopoulos *et al.* (1999) has found that the current will return along an equatorial current sheet. Observed pulsar braking indices are not compatible with energy losses arising from waves caused by magnetic dipole rotation. Some pulsars have their magnetic and spin axes very closely aligned. Their dipole spin-down losses would consequently be very small, compared to what is observed. Furthermore, observations of the difference in spin-down losses of PSR B1931+24, a pulsar with intermitting radio emission find their natural and very accurate explanation in an intermitting current with the co-rotational charge density dissipating the polar cap potential difference (Kramer *et al.*, 2006). For PSR B1931+24 the current is responsible for a third of the spin down losses.

Therefore a net non-neutral current is expected to flow across a net potential difference in order to provide for the electrical spin-down loss that occurs in addition to the dipole radiation loss. Komissarov (2006) has suggested, that losses in the current sheet may convert a significant amount of spin-down energy into optical and γ-ray emission, but the origin of the required resistivity, or the stability of the magnetospheric current system and its implication for pulsar radio emission have not been scrutinized. One of the basic assumptions of MHD simulations is the vanishing of $E_\parallel$ along the flow and that makes it difficult to derive other than ohmic losses.

Solving the equation of motion for particles along the magnetic field lines self-consistently under consideration of their losses by inverse Compton scattering and curvature radiation can provide estimates of particle energies and losses when the force-free constraints are violated (Jessner *et al.* 2001). This approach provides a natural explanation for the acceleration of particles to high energies, giving rise to X-ray and optical emissions and perhaps pair creation.

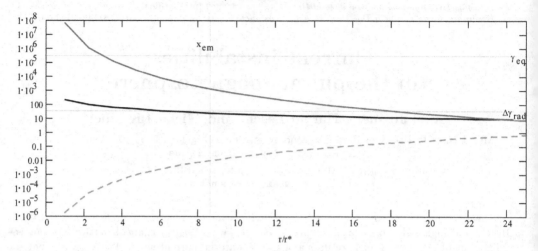

Figure 1. Characteristics of the Pierce instability on the critical field line of PSR B2021+51 as a function of distance from the neutron star. Black: Pierce length L_p [m], red: max. Lorentz factor γ_{max}, green dashed: thickness of virtual cathode layer L_{vc} [m]. The estimated location of the emission region x_{em} and the Lorentz factor corresponding to the radio emission losses $\Delta\gamma_{rad}$ are also indicated.

However, by omitting to solve Poisson's equation for the self-field and densities of the streaming particles at the same time, such modelling will fall short of being a true representation of the actual conditions in the pulsar magnetosphere.

Non-neutral currents can become unstable, especially when constrained to move in only one dimension by a strong magnetic field. This was noticed by Pierce (1949) in experiments with strong electron beams in a magnetic field. Davidson (2001) gives a comprehensive treatment of the space charge limited flow problem and provides solutions for different geometrical configurations and particle energies. The existence of a Pierce-type instability in pulsar magnetospheres had first been described by Mestel et al. (1985). Jessner et al. (2002) and Beloborodov (2008) discussed the existence of the Pierce instability in pulsar magnetospheres.

Particles start from the neutron star with thermal velocities where a cathode is formed in a boundary layer close to the surface when the charge density n_0 is sufficient to compensate the surface electric field. The particles will initially be non-relativistic ($v = \beta_0 c$) but the strong electric field will quickly accelerate them to relativistic energies and the *charge density* of the current will decrease to about $\beta_0 q_e n_0$ by virtue of current conservation. Hence a thermal emission current with a non-relativistic co-rotation surface density cannot compensate the strong parallel electric field as soon as charges become relativistic after leaving the cathode layer. The particles will continue to gain energy until their space charge is sufficient to compensate the background $E_\parallel$. From then on, their own field will repel them and they loose energy until they come to a stop and a new 'virtual' cathode is formed. The cycle will repeat, but as for a dipolar field $E_\parallel \propto r^{-4}$ and the current density $j \propto r^{-3}$ these oscillations will end at a certain distance from the neutron star and the flow will cease to be relativistic.

A virtual cathode will form whenever the space charge electric field grows faster than the external electric field. A homogeneous charged relativistic beam, having constant charge density and a linearly growing space-charge field forms a virtual cathode when the background electric field is only slowly changing. In that case the virtual cathode will form at a distance of $L_p = 4\frac{\epsilon_0 E_\parallel}{q_e n_e}$ and the maximum Lorentz factor will be given by $\gamma_{max} = 2\frac{\epsilon_0 E_\parallel^2}{n_e m_e c^2}$ as the equilibrium between the energy density of the background field and the current energy density $\gamma_{max} n_e m_e c^2$. Although L_p grows with $E_\parallel$, the width of the non-relativistic virtual cathode region $L_{vc} = \frac{m_e c^2}{q_e E_\parallel}$ decreases with $E_\parallel$ and in the non-relativistic regime L_p and L_{vc} become comparable.

The pulsar B2021+51 with period $P = 0.529$ s and surface magnetic field of $B_0 = 1.3 \times 10^8$ T serves as an illustrative example. For the pulse width of 10° at 1.4 GHz we obtain an emission height of $x_{em} = 8.7 \times r_{ns}$ and a background electric field $E_\parallel$ of 5.4×10^9 V/m together with a current density of 2.5×10^4 A/m^2. We compute numerical solutions for the set of coupled differential equations (Poisson's equation and the relativistic equation of motion)

$$\frac{dE}{dz} = -\frac{\kappa I_{tot}}{A_{em}\,\epsilon_0\,c|\beta|} \quad \text{and} \quad \frac{d}{dz}\beta\gamma = \frac{q_e}{m_e c^2}(E - (1-\kappa)E_\parallel)$$

assuming a charge density of 99% ($\kappa = 0.99$) of the full equilibrium value, which will shield the background $E_\parallel$ by a factor of $1 - \kappa$. Background $E_\parallel$ and current density $j = \frac{I_{tot} \cdot \kappa}{c A_{em}}$ vary slowly on the scales of the instability, justifying a 1-D cartesian coordinate approximation. The numerical results agree with the derived values for γ_{max}, L_p and L_{vc}. Fig. 1 shows the variation of γ_{max}, L_p and L_{vc} along the outermost open field line which still has an accelerating background field all the way out to the light cylinder. For the emission region we find $L_{vc} = 1$ cm, a repetition of the pattern after $L_p = 23$ m with a maximum Lorentz factor of $\gamma_{max} = 1234$ which is about fourty times the amount of $\Delta\gamma_{rad} = 33.6 = \frac{L_{rad}q_e}{I_{tot}m_e c^2}$ that needs to be dissipated by the beam particles to yield the observed radio luminosity of $L_{rad} = 5.8 \times 10^{18}$ W. The space charge electric field peaks in the narrow virtual cathode regions where the density is also at its maximum, giving rise to strong pulses of sub-nanosecond $\tau > L_{vc}/c = 10^{-11}$ s duration. Such Nano-shots are observed in giant pulses from the Crab pulsar. and there are indications of regular patterns of narrow emission regions separated on scales of a few m in Crab high-frequency giant pulses (Jessner *et al.* 2010). Loehmer *et al.* (2008) have shown that the superposition of sub-nanosecond emissions can also provide an explanation for the observed form of pulsar spectra. Our described solutions are stationary, but highly sensitive to the value of $\frac{1-\kappa}{\kappa}$ and we expect small variations of the current density to cause large fluctuations in the local radio emission.

References

Beloborodov, A. M. 2008, *ApJ*, 683, L41-44

Davidson, 2001, Physics of non-neutral beams, Imperial College Press London 2001

Komissarov, S. S. 2006, *MNRAS*, 367, 19

Kramer, M., Lyne, A. G., O'Brien, J. T., Jordan, C. A., & Lorimer, D. R. 2006, *Science*, 312, 549

Contopoulos, I., Kazanas, & D. Fendt, C. 1999, *ApJ*, 511, 351

Jessner, A., Lesch, H. & Kunzl, T. 2001, *ApJ*, 547, 959

Jessner, A., Lesch, H. & Kunzl, T. 2002, 270. WE-Heraeus Seminar on Neutron Stars, Pulsars and Supernova Remnants. (Eds.) W. Becker *et al.* MPE-Report No. 278, 209-214

Jessner, A., Popov, M. V., Kondratiev, V. I., Kovalev, Y. Y., Graham, D., Zensus, A., Soglasnov, V. A., Bilous, A. V. , & Moshkina, O. A. 2010, *A&* in press

Loehmer, O., Jessner, A., Kramer, M., Wielebinski, R. & Maron, O. 2008, *A&A*, 480, 623

Mestel, L., Robertson, J. A., Wang, Y. M. & Westfold, K. C. 1985, *MNRAS*, 217, 443

Pierce, J. R. 1949, The Theory and Design of Electron Beams, van Nostrand, (Toronto, New York, London 1949)

Advances in Plasma Astrophysics
Proceedings IAU Symposium No. 274, 2010
A. Bonanno, E. de Gouveia Dal Pino & A.G. Kosovichev, eds.

© International Astronomical Union 2011
doi:10.1017/S174392131100706X

Electromagnetic emission by subsequent processes L→L'+S and L+L'→T

Marian Karlický and Miroslav Bárta

Astronomical Institute of the Academy of Sciences of the Czech Republic,
CZ – 25165 Ondřejov, Czech Republic
email: karlicky@asu.cas.cz, barta@asu.cas.cz

Abstract. Using a 2.5-D electromagnetic particle-in-cell (PIC) model, very early stages of a generation of the electromagnetic emission produced by a monochromatic Langmuir wave are studied. It is found that the electromagnetic emission, which is dominant on the harmonic of the plasma frequency, starts to be generated in a very small region of k-vectors. Later on the k-vectors of this emission are scattered around a 'circle' (in our 2-D case), given by the relations for the L+L'→T process. Analytical analysis of two subsequent processes L→L'+S a L+L'→T confirms these results.

Keywords. waves, radiation mechanisms: general, Sun: radio radiation

1. Introduction

In astrophysical plasmas the Langmuir waves are easily generated (Benz 1993). But Langmuir waves are local waves and distant observers can obtain an information about them only after their transformation to electromagnetic ones. Several processes were proposed for this transformation, e.g.: $L \to T \pm S$ and $L + L' \to T$ (Melrose 1980). Details of these processes were studied by e.g. Robinson(1997), Bárta & Karlický (2000), and Willes *et al.* (1996). Furthermore, there are papers which studied this transformation numerically, usually after a generation of Langmuir waves by the two-stream instability (Sakai & Nagasuki 2007, Yu & Guangli 2008). In the present paper, using the PIC model, we study very early stages of this transformation.

2. Numerical modelling

We use a 2.5-D relativistic electromagnetic PIC code. The system size is $L_x = 1024\Delta$ and $L_y = 1024\Delta$, where Δ is a grid size. The electron-proton plasma with the proton-electron mass ratio m_p/m_e=1836 is considered. In each numerical cell we initiated 200 electrons and 200 protons. The electron thermal velocity was taken to be the same in the whole numerical box as $v_T = 0.06\ c$, where c is the speed of light. The Debye length is $\lambda_D = 0.6\ \Delta$. The initial magnetic field in the system is zero. Then we initiated a monochromatic Langmuir wave with the electric field oriented along the x-coordinate ($\boldsymbol{E} \equiv (E_x, 0, 0)$) with the k-vector corresponding to $k = 0.125$ (expressed in the ratio to $1/\lambda_D$), and with the electrostatic field energy $E_L/W_T = 0.029$, where W_T is the thermal plasma energy. Computations were made for 40 plasma periods. Here, we present only the results for the magnetic field component B_z, which represents the electromagnetic waves in the system. In Fig. 1 (left) the $k_x - \omega$ dispersion diagram of the B_z component, computed in the time interval 0-32.6 plasma periods, is shown. As seen here the electromagnetic emission is generated on the plasma frequency and its harmonic. The emission on the harmonic frequency is stronger, see Fig. 1 (right). For this reason we concentrate

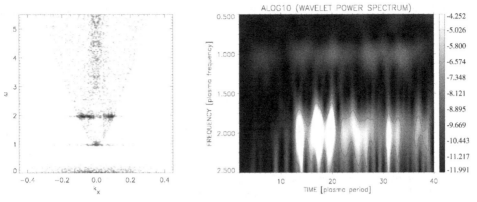

Figure 1. Left: The $k_x - \omega$ dispersion diagram obtained by the 2D Fourier transform of the magnetic field components B_z. The frequency is expressed in the ratio to the plasma frequency. Right: The wavelet spectrum of the B_z.

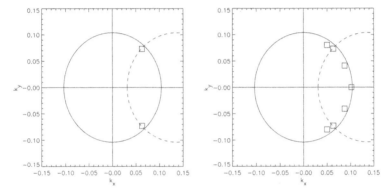

Figure 2. The $k_x - k_y$ diagram of the enhanced energy density (squares) of the B_z component at the time 12.7 (left part) and 14.3 (right part) plasma periods, respectively. Full circle corresponds to the relation (3.4) and dashed circle to the relations (3.5 and 3.6).

our attention only to a formation of this harmonic electromagnetic emission. We wanted to know how this emission is formed in the k-vector space. We recognized that the first k-vectors corresponding to the harmonic emission appeared at 12.7 plasma periods in the very small region around $k = (0.063, 0.073)$, see the square in Fig. 2 (left). Then the k-vectors of this emission started to be scattered around a 'circle', see the squares in Fig. 2 (right) at the time 14.3 plasma periods.

3. Analytical analysis of the L→L'+S and L+L'→T processes

Starting from resonant conditions and dispersion relations for the decay process, after some algebraic manipulations, we can write:

$$|K| = 2k_0 \cos(\phi) - 2/3v_s \tag{3.1}$$

$$K_x = |K| \cos(\phi), \tag{3.2}$$

$$K_y = |K| \sin(\phi), \tag{3.3}$$

where $k_0 = (k_0, 0)$ and K are the normalized k-vectors of the pumping Langmuir and ion-sound waves, and v_s is the constant proportional to the sound speed.

On the other hand, for the k-vector of the electromagnetic wave (T) (on the harmonic frequency), generated by the coalescence L+L'→T we derived:

$$(c/v_T)^2(k_{Tx}^2 + k_{Ty}^2) = 3 \tag{3.4}$$

which is the equation of circle with the center in the point (0,0) and the radius $v_T/c\sqrt{3}$. Now, assuming that the coalescence process includes the mother Langmuir and daughter waves from the parametric decay, then we can write:

$$k_{Tx}^{ML+DL} = 2k_0 - K_x, \tag{3.5}$$
$$k_{Ty}^{ML+DL} = -K_y, \tag{3.6}$$

where components of K are expressed by (3.2) and (3.3). ML and DL mean the mother and daughter Langmuir wave. Thus, when we consider the T-wave (on harmonic frequency) as a result of the coalescence of the mother Langmuir wave with daughter Langmuir ones (1-order in the cascading process) then the k-vectors of this T-wave have to correspond to the crossing point of curves given by the relations (3.5), (3.6), and (3.4), see the 'circles' in Fig. 2. Comparing the 'circles' with locations of the squares computed in PIC model, we can see a good agreement between numerical and analytical results. We also determined the angle between the k-vectors of the mother Langmuir wave and the electromagnetic one corresponding to the crossing point of the 'circles' in Figs. 2. In our case this angle is 49 degrees. It increases with the increase of the k_0 vector of the mother Langmuir wave.

4. Conclusions

Both the numerical modelling and analytical analysis show that the first k-vectors of the harmonic electromagnetic emission are generated in the vicinity of the crossing point of two 'circles' given by two subsequent resonant wave processes L→L'+S and L+L'→T. In the following times the k-vector of this emission is scattered around the full 'circle' given by the coalescence process L+L'→T. The angle of the k-vector of the first electromagnetic plasmon with that of the mother Langmuir wave increase with the v_{ph}/c decrease. It changes from 37 degrees for $v_{ph}/c = 0.7$ to 66 degrees for $v_{ph}/c = 0.1$. Thus, in the interval of beam velocities ($v_{beam} \approx v_{ph}$) considered for type III bursts in the corona, this angle deviates from that (~ 90 degree) considered in the so called head-on approximation of the coalescence process (Melrose 1980). It was found that in very early phases (for parameters considered here), the electromagnetic emission on the harmonic frequency is stronger than that on the fundamental frequency.

Acknowledgements

This research was supported by the grants 300030701 and 300030804 (GA AS CR), 205/07/1100 (GA CR) and the Centre for Theoretical Astrophysics, Prague.

References

Bárta, M. & Karlický, M. 2000, A&A, 353, 757
Benz, A. O. 1993, Plasma Astrophysics (Kluwer Acad. Publ., Dordrecht, The Netherlands)
Melrose, D. B. 1980, Plasma Astrophysics (Gordon and Breach Sci. Publ., New York)
Robinson, P. A. 1977, Reviews of Modern Physics, 69, 509
Sakai, J. I. & Nagasuki, Y. 2007, A&A, 474, L33
Willes, A. J., Robinson, P. A., & Melrose, D. B. 1996, Phys. Plasmas, 3(1), 149
Yu, H., & Guangli, H. 2008, Adv. Space Res., 41, 1202

Advances in Plasma Astrophysics
Proceedings IAU Symposium No. 274, 2010
A. Bonanno, E. de Gouveia Dal Pino & A.G. Kosovichev, eds.

© International Astronomical Union 2011
doi:10.1017/S1743921311007071

Quasi-periodic oscillations in solar X-ray sources

Hana Mészárosová[1], Marian Karlický[1], and František Fárník[1]

[1] Astronomical Institute of the Academy of Sciences,
CZ-25165 Ondřejov, Czech Republic
email: hana@asu.cas.cz

Abstract. We have searched for quasi-periodic oscillations in the hard X-ray emission of solar flares. We have selected 14 flare events which were divided into two groups: a) the events with the X-ray sources located at the flare loop footpoints and b) the events with the X-ray source above the solar limb, i.e. with the loop-top X-ray source. We found that while in the case with the footpoints X-ray sources the quasi-periods of the recorded oscillations were in the interval 2–380 s, in the events with loop-top sources only the quasi-periods longer than 50 s were recognized. These results are probably connected with the MHD oscillation modes of the flaring loop. While the long periods, which are dominant in loop-top sources, are produced by acoustic oscillations along the whole long loop, in the layers close to the loop footpoints also the MHD wave modes in shorter structures with shorter periods are generated.

Keywords. Sun: flares, Sun: X-rays, Sun: oscillations, methods: data analysis

1. Introduction

As presented in the papers by e.g. Aschwanden (2004), Mészárosová et al. (2006), Nakariakov & Melnikov (2009), Karlický *et al.* (2010), the quasi-periodic oscillations with periods from seconds to tens of minutes are often observed in solar flares, especially in the radio and X-ray bands. It is commonly believed that these oscillations are generated by various types of MHD waves or by periodic magnetic reconnection processes. Among various possibilities for long period oscillations, the most promising mode is the acoustic standing harmonic mode generated along the whole length of the loop. The estimated period is then $P/s = 6.7 \times (L/Mm)/(T/MK)^{1/2}$, where L is the length of the loop and T is the mean loop temperature (Nakariakov *et al.* 2004). The standing harmonic acoustic mode is connected with the plasma density variations, especially at the loop top, therefore this mode can be directly observable in the flare loop X-ray emissions (because the X-ray emission flux is proportional to the plasma density).

2. Observations and analysis

We have analyzed 14 solar hard X-ray flare emissions (Table 1) observed by the Czech-made Hard X-Ray Spectrometer (HXRS, launched onboard the U.S. MTI satellite) with the 0.2 s time resolution. For a comparison the HXT/Yohkoh (hard X-ray emissions, 0.5 s time resolution) of the same individual events were used. We have studied quasi-periodic oscillations in the solar X-ray sources located in the footpoints (Table 1, No. 1–7) and in the tops of flaring loops (Table 1, No. 8–14), where locations of the X-ray sources were derived from SXT/Yohkoh (soft X-ray) images. All the emissions quoted in Table 1 were analyzed in detail using the wavelet method (R. Sych, http://pwf.iszf.irk.ru/). The shortest detectable periods depend on the data time resolution (minimal period $P = 1$ s

H. Mészárosová, M. Karlický, & F. Fárník

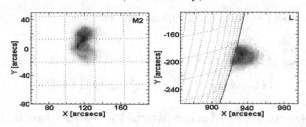

Figure 1. Examples of the SXT/Yohkoh soft X-ray image of the July 25, 2000 event (left panel) and of the July 26, 2000 event (right panel) with superimposed HXT/Yohkoh hard X-ray sources of the M2 and L energy bands located in the footpoints and the top of the flaring loop, respectively.

Table 1. Basic parameters and the characteristic periods P for HXRS/MTI emissions observed in: a) footpoints of flaring loop (No. 1–7, only P < 150 s are present) and b) top of flaring loop (No. 8–14).

No.	Event	Time interval [UT]	Energy band [keV]	Periods [s]
1	Mar 31, 2000	10:15:05–10:25:13	29.0–100.2	5, 7, 14, 32, 48, 75, 87
2	Jul 25, 2000	02:46:47–02:54:03	29.0–100.2	8, 11, 22, 45,66
3	Sep 30, 2000	23:16:50–23:30:46	29.0–100.2	11, 13,17, 27, 57, 81, 115
4	Oct 01, 2000	13:59:24–14:05:31	29.0–44.0	14, 28, 34, 52, 73
5	Nov 26, 2000	16:37:25–17:02:23	44.0–100.2	10, 14, 26, 32, 43, 50, 85, 110
6	Jun 05, 2001	04:44:25–04:49:25	12.6–67.2	17, 27, 38, 60, 76
7	Aug 31, 2001	10:39:00–10:42:21	29.0–100.2	2, 5, 12, 25, 40, 50, 68
8	Jul 09, 2000	07:19:44–07:28:18	12.6–29.0	60, 170
9	Jul 26, 2000	07:41:42–07:46:42	19.0–44.0	54, 55
10	Sep 30, 2000	20:12:34–20:17:34	12.6–29.0	55, 68, 100
11	Sep 30, 2000	23:16:50–23:30:46	12.6–29.0	98, 210, 280
12	Dec 27, 2000	15:35:39–15:45:13	12.6–29.0	90, 115, 145, 190
13	Jun 13, 2001	11:35:29–11:48:15	12.6–29.0	55, 65, 130, 155, 190, 260
14	Sep 24, 2001	10:07:02–10:44:59	12.6–29.0	130, 200, 270, 300, 450

here) and the longest ones on the length of the individual time series. As likelihood periods are considered periods out of the "cone of influence" (hatched regions in the magnitude wavelet spectra) effected by a finite length of time series. We have found that the hard X-ray emissions located at the footpoints of flaring loop usually show quasi-periodic oscillations with the characteristic periods in whole period range under study $P = 2$–380 s (Table 1, No. 1–7). But the emissions located in the top of loops show oscillations only with the period P longer than about 50 s ($P = 54$–450 s, Table 1, No. 8–14). The shorter periods ($P < 50$ s) are missing in these events.

Here, we present two examples showing both the cases. The left part of Figure 1 shows the SXT/Yohkoh image of the July 25, 2000 event with superimposed HXT/Yohkoh sources (M2 energy band) located in the footpoints of the flaring loop (where M2 = 32.7–52.7 keV). On the other hand, the left part of Figure 2 presents the corresponding hard X-ray time series (HXRS/MTI, upper panel) and magnitude wavelet spectra with the periods in the broad range, i.e. also with the periods below 50 s. Furthermore, the right part of Figure 1 shows the SXT/Yohkoh image of the July 26, 2000 event with super-imposed HXT/Yohkoh source (L energy band) located in the top of the flaring loop (where L = 13.9–22.7 keV). The corresponding time series (HXRS/MTI, upper panel) and wavelet spectra are shown in Figure 2 (right part, middle and bottom panels). These wavelet spectra show only the periods greater than 50 s. Any shorter periods are missing.

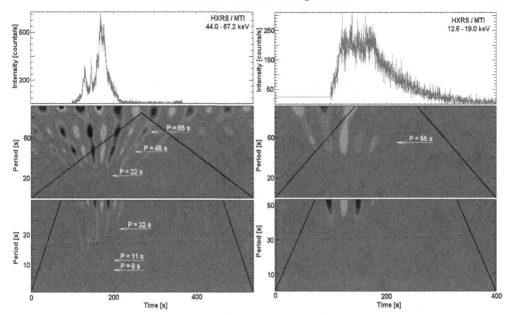

Figure 2. Example of the hard X-rays time series (HXRS/MTI, upper panels) and magnitude wavelet spectra (middle and bottom panels) with the characteristic periods for emissions: a) July 25, 2000 of the flaring loop footpoints (left panels). The energy band is 44.0–67.2 keV. ii) July 26, 2000 of the top of flaring loop (right panels). The energy band is 12.6–19.0 keV.

3. Conclusions

We have found that the oscillations in solar hard X-ray emissions from the flare loop-footpoints and from the loop-tops differ in the range of periods. The footpoint emissions revealed the quasi-periodic oscillations with characteristic periods $P < 50$ s in the wavelet spectra. Such periods were missing in the emissions from the loop-top X-ray sources. The results obtained for the HXRS/MTI emissions were confirmed by those obtained by the analysis of the HXT/Yohkoh data at similar energy bands. These results are probably connected with the MHD oscillation modes of the flaring loop. While the long periods, which are dominant in loop-top sources, are produced by acoustic oscillations along the whole long loop, in the layers close to the loop footpoints also the MHD waves modes in shorter structures with shorter periods are generated.

Acknowledgements

H. M. and M. K. acknowledge support from the Grant IAA300030701 of the Academy of Sciences of the Czech Republic and the research project AVOZ10030501 of the Astronomical Institute AS CR. The authors also thank Dr. R. Sych for the wavelet package used in this study.

References

Aschwanden, M. J. 2004, *Physics of the Solar Corona*, Berlin-Springer Praxis Books
Karlický, M., Zlobec, P., & Mészárosová, H. 2010, *Solar Phys.*, 261, 281
Mészárosová, H., Karlický, M., Rybák, J., Fárník, F., & Jiřička, K. 2006, *A&A*, 460, 865
Nakariakov, V. M. & Melnikov, V. F. 2009, *Space Sci. Revs*, 149, 119
Nakariakov, V. M., Tsiklauri, D., Kelly, A., Arber, T. D., & Aschwanden, M. J. 2004, *A&A*, 414, L25

Advances in Plasma Astrophysics
Proceedings IAU Symposium No. 274, 2010
A. Bonanno, E. de Gouveia Dal Pino & A.G. Kosovichev, eds.
© International Astronomical Union 2011
doi:10.1017/S1743921311007083

Two component relativistic acceleration and polarized radiation of the parsec-scale AGN jet

Oliver Porth

Max Planck Institut für Astronomie
69117 Heidelberg, Germany
email: porth@mpia.de

Abstract. We perform axisymmetric simulations of two-component jet acceleration using the special relativistic MHD code PLUTO (Mignone *et al.*, 2007). The inner, thermally driven component constitutes a dilute relativistic plasma originating in a high enthalpy central corona. The second component is a Poynting-dominated wind driven by a global current system. Once a near-stationary state is reached, we solve the polarized Synchrotron radiation transport incorporating self-absorption and (internal) Faraday rotation. With this approach we obtain high-resolution radio maps and spectra that can help in the interpretation of observational data from nearby active galactic nuclei by predicting spine-sheath polarization structures and Faraday rotation gradients.

Keywords. galaxies: active - galaxies: jets - ISM: jets and outflows - plasmas - polarization - radiation mechanisms: non-thermal - Radiative transfer - relativity

---

1. Introduction

Observations of core-dominated active galactic nuclei (AGN) hold the Synchrotron process responsible for the radio emission. The observed high linear polarization degrees (up to $\sim 30\%$, Marscher *et al.*, 2002) indicate that the emitting region is characterized by ordered magnetic fields. In current magneto hydrodynamical (MHD) models of jet formation, the magnetic fields form a large-scale helix twisted by the underlying accretion disk or rotating black hole (McKinney, 2006). The bulk of the energy is first carried in terms of Poynting-flux which is gradually converted to kinetic energy by the Lorentz force of the global current system (Vlahakis and Königl, 2004; Komissarov *et al.*, 2007). In order to efficiently transport energy to larger scales and eventually feed a radio lobe, the ordered field structure must survive against instabilities and dissipation to distances well beyond the parsec scale. Under the assumption that the relativistic electrons gyrate around this helical fields while emitting the corresponding Synchrotron radiation, it is interesting to ask how the geometry imprints on the observed radiation (Lyutikov *et al.* 2005). In the presence of relativistic motion, a kinematic jet model is needed to properly account for the transformation effects such as aberration, polarization swing (Blandford and König, 1979) and relativistic Faraday rotation (Broderick and Loeb, 2009).

In order to provide a global model of the outflow, we simulate the jet comprised of two components: a thermal *spine* (e.g. Sauty *et al.* 2004 and N. Globus in this volume) and an outer disk wind similar to Porth and Fendt (2010). Beyond the acceleration region, Meliani and Keppens (2009) showed that the interaction between the two components can give rise to a Raleigh-Taylor-type instability and ultimately cause jet disruption.

2. Jet acceleration

The slow conversion of Poynting-flux to kinetic energy asks for enormous spatial scales that are a challenge to dynamical codes. With a stretched grid, we are able to simulate several thousand gravitational radii and still avoid any causal interaction with the outgoing boundaries. The resulting collimated flow is thus a consequence of jet self-collimation alone. Simulations presented here cover 2555^2 grid cells in the (r, z)-plane (12000^2 gravitational radii,r_g) of which only 1417×2356 ($1200 \times 6000 \; r_g$) are used for the subsequent analysis.

As initial setup we assign a non-rotating corona threaded by a force-free poloidal field with $B_z(z=0) \propto r^{-1}$. The pressure gradient is balanced by a point-mass gravity situated $2r_g$ below the simulation domain. We assign boundary conditions for pressure, density and the current distribution and inject a slow-magnetosonic wind into the simulation domain. All remaining variables are treated as outflowing to properly account for the outgoing characteristics of the sub-Alfénic flow.

Specifically, the profiles read

$$\rho(R) = \rho_1 \left[(1 - \theta)R + \theta R^{-1.5} \right] \tag{2.1}$$

$$p(R) = p_1 \left[(1 - \theta)(1 - \rho_1 \ln R) + \theta R^{-2.5} \right] \tag{2.2}$$

$$B_\phi(r) = B_{\phi,1} \left[(1 - \theta)r + \theta r^{-s} \right] \tag{2.3}$$

$$v_p(r) = v_s(r) \tag{2.4}$$

with†

$$\theta = \begin{cases} 0 & ; \quad r < 1 \\ 1 & ; \quad r \geqslant 1 \end{cases}. \tag{2.5}$$

To treat the hot spine together with the disk-wind, we choose an equation of state satisfying the Taub1948 inequality discussed by Mignone *et al.*, 2005.

Figure 1 illustrates the near-stationary solution and acceleration of the inner and outer component. Helical magnetic fields are shown in the three dimensional rendering of the axisymmetric model. In the right-hand panels energy-conversion along selected field lines is plotted against cylindrical radius, clearly indicating the thermal driving in the spine and magnetic acceleration in the outer component. The total energy flux is only approximately conserved due to more efficient cooling in the central part of the inner component and absence of stationarity in the outer part of the solution. When the jet reaches the end of our simulation box, the outer component is still dominated by Poynting flux. We can see how the acceleration is coupled to the collimation: As the kinetic energy along a field line increases approximately linear with the cylindrical radius, enormous vertical scales are needed in order to follow the complete acceleration process for near-collimated flows (see also the contribution of A. Ferrari in this volume).

3. Radiation transport

Before the radiation transport can be conducted, we have to devise physical scales for the simulations. We do so by assuming a radial scale of $6r_g = 6GM_\bullet/c^2$ for the transition from the inner corona to the Poynting driven jet and by normalizing the energy flow to 10^{44}erg/s. The mock-observations of the parsec-scale AGN core assume a black-hole mass

---

† where R is the spherical radius to the origin below the simulation domain and r the cylindrical radius.

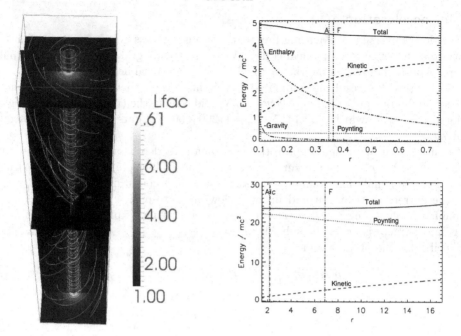

Figure 1. 3D rendering of jet model with asymptotically closed current, $s = 1.25$ *(left)*. The slices indicate bulk-flow Lorentz factor and the helical magnetic field lines are traced by the white lines. *Right:* Acceleration along selected field-lines against the cylindrical radius r showing thermal acceleration for a field line in the spine (footpoint $r_{\rm fp} = 0.1$, above) and magnetic acceleration in the jet ($r_{\rm fp} = 1.5$, below). Vertical lines indicate the crossing of the Alfvèn (A) and fast (F) critical point as well as the light cylinder (lc).

of $M_{\bullet} = 10^9 M_{\odot}$ and a photometric distance of $100\,{\rm Mpc}$. In this scaling, the active region of the radiation transport covers a physical volume of $0.12 \times 0.12 \times 0.6\,{\rm pc}^3$.

Ray-tracing linearly polarized radiation in the observers system, we solve the corresponding coupled linear equations

$$\frac{d\mathbf{I}}{ds} = \mathcal{E} - \mathbf{A}\,\mathbf{I} \tag{3.1}$$

taking into account relativistic beaming, boosting and swing of the polarization for the three Stokes parameters $\mathbf{I} = \{I^{(l)}, I^{(r)}, U^{(lr)}\}$. The comoving coefficients of the transfer equation are valid for power-law electron distributions $dn_e/dE_e = N_0 E_e^{-2\alpha - 1}$ as defined by Pacholczyk (1970). In the following we adopt $\alpha = 0.5$ and the lower (upper) cutoff $\gamma_{\rm l} = 100$ ($\gamma_{\rm u} = \infty$) to reproduce the empirical SED in the optically thin regime. Faraday rotation is calculated in accordance with Broderick & Loeb (2009) by utilizing the relativistic generalization for the rotation angle

$$\frac{d\chi_F}{ds} = \frac{e^3}{2\pi m_e^2 c^2} \frac{f(\gamma_t) n_e D^2}{\nu^2} (\hat{\mathbf{n}} - \beta) \cdot \mathbf{b} \tag{3.2}$$

in relation (3.1) allowing to resolve internal Faraday rotation†.

The largest uncertainties concerning the jet radiation in pure MHD models arise due to the lack of information about the location and mechanism of the particle acceleration

† here, e, m_e, n_e are the comoving electron -charge, -mass and -density, $\mathbf{b}$ is the comoving magnetic field vector and $\hat{\mathbf{n}}$ the (observer-frame) photon direction. The function $f(\gamma_e)$ interpolates between the limits for cold and relativistic electron temperatures γ_e according to Shcherbakov (2008).

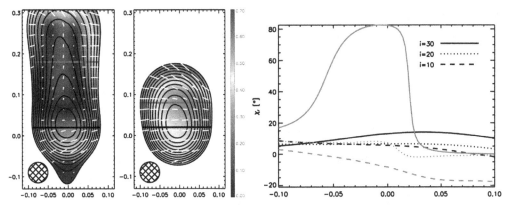

Figure 2. Polarization ê vectors *(white sticks)* for $i \in \{30°, 10°\}$ emitted from regions with co-moving pinches $B'_\phi / B'_p > 1$. The polarization degree $\Pi_{43\,\mathrm{GHz}}$ is indicated by the filling and $I_{43\,\mathrm{GHz}}$ contour lines are overplotted in solid black. The latter are spaced by a factor of 2 out to $\simeq 5 \cdot 10^{-4} I_{\nu,\mathrm{peak}}$ where the image is cropped. Spatial scale is given in milli arcseconds and a restoring beam with FWHM=0.05 mas was used. The right-hand panel shows polarization angles for cuts along core *(black)* and jet *(gray)*.

process. Here we have to resort to a simple recipe by assuming equipartition between the relativistic electrons and the large-scale magnetic field. This treatment is consistent with minimum-energy arguments inversely used to derive magnetic field-strength and electron density from the synchrotron emission Burbidge 1956.

Even in global VLBI experiments, the AGN core emission is almost certainly under-resolved and hence we need to investigate the beam-averaging effect on the observables. Figure 2 shows a mock polarization observation at $\nu = 43\,\mathrm{GHz}$ for two inclinations i (where $i = 0°$ would directly look into the jet) with a certainly optimistic beam-width of 0.05 mas. To demonstrate how the kinematic information of the jet model imprints on the polarization, Faraday rotation is neglected in this case. For moderate inclinations $i \geqslant 30°$ we observe a clear spine and sheath polarization structure tracing the lower pitch of the outer field lines. The polarization degree increases towards the edges and becomes beam-depolarized between perpendicular regions. When approaching the blazar case, polarizations become predominantly perpendicular to the jet direction.

4. Conclusions

Using large-scale axisymmetric RMHD simulations combined with rigorous ray-tracing, we demonstrate a way to model AGN core radio emission. Within the simulated domain extending into the parsec scale, our high-energy models accelerate from $\Gamma \simeq 1.1$ to $\Gamma \lesssim 8$ with acceleration still ongoing when the jet leaves the domain.

Depending on the position of the emitting region, the polarization shows characteristic swings by $\sim 90°$ that can appear as often observed "spine and sheath" or "jet and core" shift. Although numerous features can be reproduced, the helical field-models always have electric vectors perpendicular to the jet direction when seen "into" the jet and can thus not predict the majority of BL-Lac sources (Marscher *et al.* 2002).

An investigation that takes into account various particle acceleration recipes, resolution effects and Faraday rotation is in preparation by Porth & Fendt (2010b).

Acknowledgements

This work was carried out under the HPC-EUROPA project (RII3-CT-2003-506079), with the support of the European Community - Research Infrastructure Action under the FP6 "Structuring the European Research Area" Program.

References

Blandford, R. D. & Königl, A.: 1979, *ApJ* **232**, 34
Broderick, A. E. & Loeb, A.: 2009, *APjL* **703**, L104
Burbidge, G. R.: 1956, *ApJ* **124**, 416
Hawley, J. F. & Krolik, J. H.: 2006, *ApJ* **641**, 103
Komissarov, S. S., Barkov, M. V., Vlahakis, N., & Königl, A.: 2007, *MNRAS* **380**, 51
Lyutikov, M., Pariev, V. I., & Gabuzda, D. C.: 2005, *MNRAS* **360**, 869
Marscher, A. P., Jorstad, S. G., Mattox, J. R., & Wehrle, A. E.: 2002, *ApJ* **577**, 85
McKinney, J. C.: 2006, *MNRAS* **368**, 1561
Meliani, Z. & Keppens, R.: 2009, *ApJ* **705**, 1594
Mignone, A., Bodo, G., Massaglia, S., Matsakos, T., Tesileanu, O., Zanni, C., & Ferrari, A.: 2007, *Apjs* **170**, 228
Mignone, A., Plewa, T., & Bodo, G.: 2005, *Apjs* **160**, 199
Porth, O. & Fendt, C.: 2010, *ApJ* **709**, 1100
Sauty, C., Meliani, Z., Trussoni, E., & Tsinganos, K.: 2004, *Apss* **293**, 75
Shcherbakov, R. V.: 2008, *ApJ* **688**, 695
Taub, A. H.: 1948, *Physical Review* **74**, 328
Vlahakis, N. & Königl, A.: 2004, *ApJ* **605**, 656

Advances in Plasma Astrophysics
Proceedings IAU Symposium No. 274, 2010
A. Bonanno, E. de Gouveia Dal Pino & A.G. Kosovichev, eds.

© International Astronomical Union 2011
doi:10.1017/S1743921311007095

Particle acceleration in Blazars

Matthias Weidinger[1] and Felix Spanier[2]

Lehrstuhl für Astronomie, Universität Würzburg,
Am Hubland, 97074 Würzburg, Germany
[1] email: mweidinger@astro.uni-wuerzburg.de
[2] email: fspanier@astro.uni-wuerzburg.de

Abstract. Understanding the variable emission of blazars observed with gamma-ray telescopes and Fermi has become a major challenge for theoretical models of particle acceleration. Here, we introduce a novel time-dependent emission model in which the maximum energy of particles is determined from a balance between Fermi type I and II acceleration energy gains and radiative energy losses, allowing for an explanation of both the characteristic spectral energy distribution of blazars and their intrinsic sub-hour variability. Additionally, we can determine the physical condition of the emitting plasma concerning its turbulence and typical shock speeds.

Keywords. acceleration of particles, shock waves, radiation mechanisms: nonthermal, galaxies: active, galaxies: jets, BL Lacertae objects: individual (1 ES 1218+30.4, PKS 2155-304)

1. Introduction

Blazars, with all their subcategories from Flat Spectrum Radio Quasars (FSRQs) to Low-, Intermediate- and High Frequency Peaked BL-Lac Objects (LBLs, IBLs, HBLs), are believed to be Active Galactic Nuclei (AGN) where the highly relativistic, collimated outflow from the central black hole emerges under a small angle to the line of sight (Urry & Padovani 1995). The blazar flavours can be distinguished based on their spectral features. While all types show a more or less pronounced double humped spectral energy distribution (SED), FSRQs are the most luminous ones with the first peak occurring arround 10^{14} Hz, in LBLs and IBLs the total luminosity decreases while the peak frequency increases up to 10^{18} Hz when considering HBLs (e.g. Fossati *et al.* 1998). The jet as the origin of the characteristic very high energy (VHE) emission from blazars is beyond doubt, but the microphysical processes, the composition as well as typical physical conditions such as the size of the emitting region in the jet are still a matter of debate. To derive those features one has to understand the particle distributions leading to the observed SED in a microphysical way i.e. acceleration mechanisms will be essential. This extends models for the emission processes (e.g. Ghisellini 1988, Chiang & Böttcher 2002 or Mannheim 1993) where ad-hoc particle spectra are assumed. Fortunately blazars are non-static objects showing variability in their VHE emission down to timescales of minutes, hence providing information about acceleration and cooling processes. Secondly, if all types of blazars are essentially the same objects, the distinctions in their SEDs should as well arise from processes within the jet, this could mean variations in the acceleration mechanism itself or the dominant particle species.

With our "Code On Jetsystems Of Non-thermally Emitting Sources", a box-model considering acceleration due to diffusive shock acceleration as well as all the relevant radiation mechanisms self-consistently and time-dependently, we investigate those properties by comparing model SEDs and lightcurves during outbursts of blazars with multiwavelength observations provided by X-Ray satellites, Fermi-LAT and Air-Cerenkov telescopes. The parameters of the modelling can then be used to find boundaries for the properties of the underlying plasma. We apply our model on the low-state emission and variability data of the two HBLs PKS 2155-304 and 1 ES 1218+30.4 . The model gives predictions how these objects behave spectrally resolved during a flare, even though only one energy band has been observationally covered and gives hints for their plasma-physical properties.

2. The Model

Here we will give an overview of the main features of the model, for a complete description see Weidinger et al. (2010) and Weidinger & Spanier (2010b). Our model slightly follows the ansatz of Kirk et al. (1998) and extends it with stochastic acceleration. We solve the corresponding one dimensional Vlasov-equation in the diffusion approximation (see e.g. Schlickeiser 2002) for every particle species i considered in two spatially different zones. A setup is used where the upstream acceleration zone is nested within a bigger radiation zone. Both are assumed to be spherical symmetrical (radius R_z), containing isotropic particle distributions. The kinetic equation in the acceleration zone is

$$\partial_t n_i = \partial_\gamma \left[(\beta_{s,i}\gamma^2 - t_{\mathrm{acc},i}^{-1}\gamma) \cdot n_i \right] + \partial_\gamma \left[[(a+2)t_{\mathrm{acc},i}]^{-1}\gamma^2 \partial_\gamma n_i \right] + Q_{0,i} - \frac{n_i}{t_{\mathrm{esc},i}} \quad (2.1)$$

with Q_0 being a monoenergetic injection function for particles streaming into the considered region of the jet and β_s the corresponding synchrotron loss rate, t_{acc} and a can be derived from microphysics to

$$a = \frac{9}{4}\frac{v_s^2}{v_A^2} , \quad t_{\mathrm{acc}}^{-1} = \frac{v_s^2}{4K_{||}} + 2\frac{v_A^2}{9K_{||}} \quad (2.2)$$

for a parallel shock and Alfvén waves, here we made use of the hard-sphere approximation (e.g. Lerche & Schlickeiser 1985). $K_{||}$ is the spatial diffusion coefficient, which for relativistic particles can be estimated as $K_{||} = 1/3cl_i$ (Schlickeiser 2002) (l_i: mean free path). In the steady state this results in a power law, as expected from shock theory. After $t_{\mathrm{esc},i} = \eta R_{\mathrm{acc}}/c$ particles enter the radiation zone

$$\partial_t N_i = \partial_\gamma \left[(\beta_{s,i}\gamma^2 + P_{\mathrm{IC}}(\gamma)) \cdot N_i \right] - \frac{N_i}{t_{\mathrm{rad,esc}}} + b^3\frac{n_i}{t_{\mathrm{esc}}} \quad (2.3)$$

with $b = R_{\mathrm{acc}}/R_{\mathrm{rad}}$. The IC loss rate is calculated exploiting the full Klein-Nishina-Cross section (e.g. Rüger et al. 2010). Shifting the frame of reference to the Laboratory frame one finds the model SED from the arising photon distribution in the considered region.

3. Results and Conclusion

While FSRQs often require different particle species to be modelled, almost every HBL is reproduced using only electrons; i.e. our model is similar to a Synchrotron-Self Compton ansatz, but self-consistent and time-dependent. This model has been applied to the two nearby HBLs 1 ES 1218+30.4 and PKS 2155-304 (Weidinger & Spanier 2010a, Weidinger & Spanier 2010b). For both HBLs there is plenty of multiwavelength as well as archival data available. Both show variability in their VHE emission, though PKS 2155-304 with its remarkably short timescales is the more extreme one.

In the first step the low-state SED of an object is reproduced, which then is used to model the lightcurve of the considered flare. Here only the relevant results for the investigation at hand are shown, the complete results and implications are to be found in the individual publications. Fig. 1 shows the SEDs as well as the light curves in different

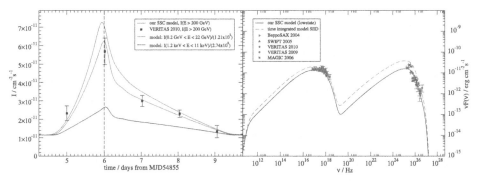

Figure 1. (from Weidinger & Spanier (2010a)) a) Model lightcurves in different energy bands and the VERITAS measurement from 2009 for 1 ES 1218+30.4. The flare is modeled injecting more e^- into the steady state emission region for a certain amount of time. b) Low-state model SED, the high-state (dashed line) is computed by averaging over the whole flare shown in a).

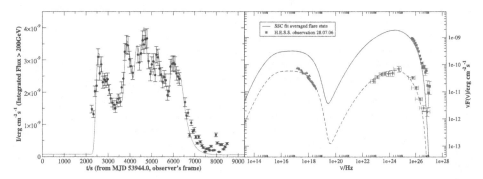

Figure 2. (from Weidinger & Spanier (2010b)) a) Lightcurve of the famous outburst of PKS 2155-304 measured by H.E.S.S. in 2006 and our model. b) The low-state model SED (dashed line) along with the multiwavelength data of the 2009 campaign with Fermi, this line also matches the archival data of H.E.S.S. from 2003 (Weidinger *et al.* 2010). The high-state reproducing the H.E.S.S. data is the time average over the whole outburst shown in a).

energy bands for the modelling process described above for all the available data from BeppoSAX, SWIFT, VERITAS and MAGIC (Donato *et al.* 2005, Tramacere *et al.* 2007, The VERITAS collaboration 2010, Albert *et al.* 2006) of 1 ES 1218+30.4 . Although the data is not simultaneous, it can be used for the modelling, for 1 ES 1218+30.4 is in a low-state most of the time, as the comparison of the VHE data yields. As one can see in Fig. 1a) we are able to explain the outburst recorded by VERITAS with our acceleration model. In Fig. 2 we show the results of applying our model to PKS 2155-304 and its famous outburst. To derive the low-state as a basis for the outburst we used the latest multiwavelength data of Fermi from 2009 (Aharonian *et al.* 2009). For the flare, density fluctuations as the blob travels down the jet axis, i.e. a varying injection function in eq. (2.1), was assumed. This reproduces the H.E.S.S. data of the event (Aharonian *et al.* 2007) and proofs the acceleration assumptions to be correct, hence the fit parameters can be used for further investigations (see Table 1).

We model the acceleration process self-consistently, hence we are able to compute upper boundaries for the parameters of the underlying plasma (Table 1). Since $l_i \ll R_{\mathrm{acc}}$ to ensure the diffusion approximation to remain valid one finds an upper boundary for $K_{\parallel}$ setting $l = R_{\mathrm{acc}}$. For the values of t_{acc} and a from the modelling of a blazar, this results in an upper boundary for v_s. In the non-relativistic case this can be used for further investigations because then the scaling for the acceleration-timescale is known

Table 1. Important parameters used to model the low-state of the two HBLs shown in Fig. 1 and Fig. 2 and the resulting plasma-physical boundaries for the assumed acceleration process using $l = R_{\mathrm{acc}}$.

	R_{acc}(cm)	t_{acc}(s)	a	α^1	$K_{\|}^{\mathrm{max}}$(cm)	v_s^{max}(cm s^{-1})	v_A^{max}(cm s^{-1})
1ES1218	$6 \cdot 10^{14}$	$2.22 \cdot 10^5$	10	1.9	$6 \cdot 10^{24}$	$9.9 \cdot 10^9$	$4.7 \cdot 10^9$
PKS2155	$1 \cdot 10^{13}$	$3.77 \cdot 10^3$	20	2.0	$1 \cdot 10^{23}$	$1.0 \cdot 10^{10}$	$3.4 \cdot 10^{10}$

[1] Powerlaw-index of the underlying accelerated electrons.

analytically, eq. (2.2). In highly relativistic cases one might use e.g. PIC simulations (e.g. Spitkovski 2008) for a deeper analysis of the correlation of the acceleration timescale with the energies involved. Nevertheless this gives us a first idea about shock speeds in jets and whether our assumptions were correct. As one can see from Table 1 even for such a large $K_\|$ the shock speed only reaches up to $0.3c$. For more realistic values for the spatial diffusion coefficient ($\ll 0.01 K_\|^{max}$) the shock seems to be non-relativistic. For the compression ratio we find $r = 3.7$ for both considered HBLs as one would expect for a strong non-relativistic shock (e.g. Schlickeiser 2002). Note that r is not to be calculated simply from α since Fermi-II processes together with the synchrotron cooling slightly hardens the spectrum. The v_A in Table 1 is calculated from v_s and a which directly comes out of our modelling process. In further analysis one might use e.g. MHD simulations to find typical v_A which then gives v_s (or at least a lower boundary) directly and as long as $v_s \ll c$ the diffusion coefficient as well as the compression ratio.

With our model we are able to reproduce the emission of different types of blazars due to different particle species being accelerated within an emitting region along the jet axis. This allows us to derive certain values for the diffusion coefficient and the speed of the accelerating shock, which for HBLs seems to be non-relativistic. This is in agreement with the compression ratio of $r = 3.7$ as calculated from the particle spectra. A systematic modelling along the blazar sequence should result in distinctive statements about the main distinctions in those objects.

References

Aharonian et al. 2007, *The Astrophysical Journal Letters*, 664 L71
Aharonian et al. 2009, *The Astrophysical Journal Letters*, 696, L150
Albert et al. 2010, *The Astrophysical Journal Letters*, 642, L119
Chiang, J. & Böttcher, M. 2002, *The Astrophysical Journal*, 564, 92
Donato, D. et al. 2005, *Astronomy & Astrophysics*, 433, 1163
Fossati, G. et al. 1998, *Bulletin of the American Astronomical Society*, 30, 768
Ghisellini, G. 1988, *Advances in Space Research*, 8, 579
Kirk et al. 1998, *Astronomy & Astrophysics*, 333, 452
Lerche, I. & Schlickeiser, R. 1985, *Astronomy & Astrophysics*, 151, 408
Mannheim, K. 1993, *Astronomy & Astrophysics*, 269, 67
Rüger, M., Spanier, F. & Mannheim, K. 2010, *Mon. Not. R. Astron. Soc.*, 401, 973
Schlickeiser, R. 2002, *Cosmic Ray Astrophysics*, Springer, Berlin
Spitkovsky, A. 2008, *The Astrophysical Journal Letters*, 682, L5
The VERITAS collaboration 2010, *The Astrophysical Journal Letters*, 709, L163
Tramacere, A. et al. 2007, *Astronomy & Astrophysics*, 467, 501
Urry, C. M. & Padovani, P. 1995, *Publ. Astron. Soc. Pac.*, 107, 803
Weidinger, M., Rüger, M. & Spanier, F. 2010, *Astrophysics and Space Sciences Trans.*, 6, 1
Weidinger, M & Spanier, F. 2010a, *Astronomy & Astrophysics*, 515, A18
Weidinger, M. & Spanier, F. 2010b, *Int. J. Mod. Phys. D*, 19, 887

Observational and modelling programs for plasma astrophysics

K.G. Strassmeier, M. Hanasz and S. Colgate

Advances in Plasma Astrophysics
Proceedings IAU Symposium No. 274, 2010
A.Bonanno, E. de Gouveia Dal Pino & A.G. Kosovichev, eds.
© International Astronomical Union 2011
doi:10.1017/S1743921311007101

An overview of the *Planck* mission

N. Mandolesi, C. Burigana, A. Gruppuso, P. Procopio, S. Ricciardi
on behalf of the *Planck* Collaboration

INAF - IASF Bologna, Via P. Gobetti 101, 40129 Bologna

Abstract. This paper provides an overview of the ESA *Planck* mission and its scientific promises. *Planck* is equipped with a 1.5–m effective aperture telescope with two actively-cooled instruments observing the sky in nine frequency channels from 30 GHz to 857 GHz: the Low Frequency Instrument (LFI) operating at 20 K with pseudo-correlation radiometers, and the High Frequency Instrument (HFI) with bolometers operating at 100 mK. After the successful launch in May 2009, *Planck* has already mapped the sky twice (at the time of writing this review) with the expected behavior and it is planned to complete at least two further all-sky surveys. The first scientific results, consisting of an Early Release Compact Source Catalog (ERCSC) and in about twenty papers on instrument performance in flight, data analysis pipeline, and main astrophysical results, will be released on January 2011. The first publications of the main cosmological implications are expected in 2012.

Keywords. Cosmology, cosmic microwave background, space missions

1. Introduction

In 1992, the Cosmic Background Explorer (*COBE*) team announced the discovery of intrinsic temperature fluctuations in the cosmic microwave background radiation (CMB) on angular scales greater than 7° and at a level of a few tens of μK (Smooth *et al.*, 1992). One year later two spaceborne CMB experiments were proposed to the European Space Agency (ESA) in the framework of the Horizon 2000 Scientific Programme: the Cosmic Background Radiation Anisotropy Satellite (COBRAS; Mandolesi *et al.*, 1994), an array of receivers based on High Electron Mobility Transistor (HEMT) amplifiers; and the SAtellite for Measurement of Background Anisotropies (SAMBA), an array of detectors based on bolometers (Tauber *et al.*, 1994). The two proposals were accepted for an assessment study with the recommendation to merge. In 1996, ESA selected a combined mission called COBRAS/SAMBA, subsequently renamed *Planck*, as the third Horizon 2000 Medium-Sized Mission. The *Planck* CMB anisotropy probe†, the first European and third generation mission after *COBE* and *WMAP* (Wilkinson Microwave Anisotropy Probe‡), represents the state-of-the-art in precision cosmology today (Tauber *et al.*, 2009; Bersanelli *et al.*, 2010; Lamarre *et al.*, 2010). The *Planck* payload (telescope instrument and cooling chain) is a single, highly integrated spaceborne CMB experiment. *Planck* is equipped with a 1.5–m effective aperture telescope with two actively-cooled instruments that will scan the sky in nine frequency channels from 30 GHz to 857 GHz: the

† *Planck* (http://www.esa.int/Planck) is a project of the European Space Agency - ESA - with instruments provided by two scientific Consortia funded by ESA member states (in particular the lead countries: France and Italy) with contributions from NASA (USA), and telescope reflectors provided in a collaboration between ESA and a scientific Consortium led and funded by Denmark. We acknowledge the support by the ASI/INAF Agreement I/072/09/0 for the *Planck* LFI Activity of Phase E2.

‡ http://lambda.gsfc.nasa.gov/

Low Frequency Instrument (LFI) operating at 20 K with pseudo-correlation radiometers, and the High Frequency Instrument (HFI; Lamarre *et al.*, 2010) with bolometers operating at 100 mK. The coordinated use of the two different instrument technologies and analyses of their output data will allow optimal control and suppression of systematic effects, including discrimination of astrophysical sources. All the LFI channels and four of the HFI channels will be sensitive to the linear polarisation of the CMB. A summary of the LFI and HFI performances is reported in Table 1. Note that *Planck* is sensitive to polarization up to the 353 GHz. The constraints on the thermal behaviour, required to minimize systematic effects, dictated a *Planck* cryogenic architecture that is one of the most complicated ever conceived for space. Moreover, the spacecraft has been designed to exploit the favourable thermal conditions of the L2 orbit. The thermal system is a combination of passive and active cooling: passive radiators are used as thermal shields and pre-cooling stages, active cryocoolers are used both for instrument cooling and pre-cooling (Collaudin & Passvogel, 1999). *Planck* is a spinning satellite. Thus, its receivers will observe the sky through a sequence of (almost great) circles following a scanning strategy (SS) aimed at minimizing systematic effects and achieving all-sky coverage for all receivers (Dupac & Tauber, 2005; Ashdown *et al.*, 2007).

The data analysis, its scientific exploitation, and the core cosmology programme of Planck are mostly carried out by two core teams (one for LFI and one for HFI), working in close connection with the Data Processing Centres (DPCs), and closely linked to the wider *Planck* scientific community, consisting of the LFI, HFI, and Telescope consortia, organized into various working groups. *Planck* is managed by the ESA *Planck* science team. *Planck* will open a new era in our understanding of the Universe and of its astrophysical structures (Planck Collaboration, 2006). To achieve these ambitious goals (Tauber *et al.*, 2010) an extremely accurate and efficient data analysis and a careful separation of CMB and astrophysical emissions is demanded.

Table 1. *Planck* performances. The average sensitivity, $\delta T/T$, per FWHM^2 resolution element (FWHM is reported in arcmin) is given in CMB temperature units (i.e. equivalent thermodynamic temperature) for 28 months of integration. The white noise (per frequency channel for LFI and per detector for HFI) in 1 sec of integration (NET, in $\mu K \cdot \sqrt{s}$) is also given in CMB temperature units. The other used acronyms are: DT = detector technology, N of R (or B) = number of radiometers (or bolometers), EB = effective bandwidth (in GHz). Adapted from Mandolesi *et al.*, 2010 and Lamarre *et al.*, 2010.

LFI

Frequency (GHz)	30	44	70
InP DT	MIC	MIC	MMIC
FWHM	33.34	26.81	13.03
N of R (or feeds)	4 (2)	6 (3)	12 (6)
EB	6	8.8	14
NET	159	197	158
$\delta T/T$ [$\mu K/K$] (in T)	2.48	3.82	6.30
$\delta T/T$ [$\mu K/K$] (in P)	3.51	5.40	8.91

Frequency (GHz)	217	353
FWHM in T (P)	4.6 (4.6)	4.7 (4.6)
N of B in T (P)	4 (8)	4 (8)
EB in T (P)	72 (63)	99 (102)
NET in T (P)	91 (132)	277 (404)
$\delta T/T$ [$\mu K/K$] in T (P)	3.4 (6.4)	14.1 (26.9)

HFI

Frequency (GHz)	100	143
FWHM in T (P)	(9.6)	7.1 (6.9)
N of B in T (P)	(8)	4 (8)
EB in T (P)	(33)	43 (46)
NET in T (P)	100 (100)	62 (82)
$\delta T/T$ [$\mu K/K$] in T (P)	2.1 (3.4)	1.6 (2.9)

Frequency (GHz)	545	857
FWHM in T	4.7	4.3
N of B in T	4	4
EB in T	169	257
NET in T	2000	91000
$\delta T/T$ [$\mu K/K$] in T	106	4243

2. From data analysis to cosmology and astrophysics

A key step of the *Planck* data analysis is the separation of astrophysical from cosmological components. For cosmological purposes, the most sensitive channels are between 70 GHz and 143 GHz. However, the other channels are essential for achieving the accurate separation of the CMB from astrophysical emissions, particularly for polarization, maximizing the effective sky area used in the analysis. The extraction of compact objects, is pursued with non-blind and blind codes (Lopez-Caniego *et al.*, 2006). To deal with the diffuse emission a variety of methods have been developed (Leach *et al.*, 2008) like: ILC (internal linear combination), ITF (internal template fitting), parametric methods. The goal for such tools is also that of propagating instrumental and foreground uncertainties to provide an accurate description of the separated components, a fundamental ingredient to carry on the further steps of the analysis. Since the CMB field is Gaussian to a large extent, assuming rotational invariance, most of the information is compressed in the two-point correlation function or equivalently in the angular power spectrum (APS), C_ℓ (Scott & Smoot, 1998). For an ideal experiment, the estimated APS could be directly compared to a Boltzmann code (http://camb.info/) prediction to constrain the cosmological parameters. In the case of incomplete sky coverage and realistic noise a more thorough analysis is necessary (Jewell *et al.*, 2004; Wandelt *et al.*, 2004; Hamimeche & Lewis, 2008; Rocha *et al.*, 2009; Efstathiou, 2004).

The quality of the recovered APS is a good predictor of the efficiency of extracting cosmological parameters by comparing the theoretical predictions of Boltzmann codes. Neglecting systematic effects (and correlated noise), the sensitivity of a CMB anisotropy experiment to C_ℓ, at each multipole ℓ, is summarized by the equation (Knox, 1995)

$$\delta C_\ell / C_\ell \simeq \sqrt{2/[f_{\rm sky}(2\ell+1)]}\left[1 + A\sigma^2/(NC_\ell W_\ell)\right]$$

where A is the size of the surveyed area, $f_{\rm sky} = A/4\pi$, σ is the rms noise per pixel, N is the total number of observed pixels, and W_ℓ is the beam window function. For a symmetric Gaussian beam, $W_\ell = \exp(-\ell(\ell+1)\sigma_{\rm B}^2)$, where $\sigma_{\rm B} = {\rm FWHM}/\sqrt{8\ln 2}$. Even for $\sigma = 0$, the accuracy in the APS is limited by the so-called cosmic and sampling variance, particularly relevant at low ℓ.

Higher sensitivity of *Planck* will permit to observe up to about the 7th or 8th peak of APS of CMB temperature anisotropies, i.e. 4 more peaks with respect to *WMAP*. As well as the temperature APS, *Planck* can measure polarisation anisotropies up to 353 GHz. Fig. 1 shows *Planck* sensitivity to the 'E' and 'B' polarisation modes. Note that the cosmic and sampling variance implies a dependence of the overall sensitivity at low multipoles on r (the green lines refer to the same r values as above), which is relevant to the parameter estimation; instrumental noise only determines the capability of detecting the B mode. We considered 28 months of integration (but for the case of a smoothing to the LFI 70 GHz the upper curve displays also the case of 15 months of integration). At frequencies close to *WMAP* V band and LFI 70 GHz channel the polarised foreground is minimal. The figure shows estimates of the residual contribution of unsubtracted extragalactic sources, $C_\ell^{\rm res,PS}$, and the corresponding uncertainty, $\delta C_\ell^{\rm res,PS}$ (plotted as thick and thin green dashes). The Galactic foreground dominates over them, the CMB B mode and also over the CMB E mode by up to multipoles of several tens. However, foreground subtraction at an accuracy of 5−10% of the map level is enough to reduce residual Galactic contamination to well below both the CMB E mode and the CMB B mode for a wide range of multipoles for $r = T/S \simeq 0.3$ (here r is defined in Fourier space). If we are able to model Galactic polarised foregrounds with an accuracy

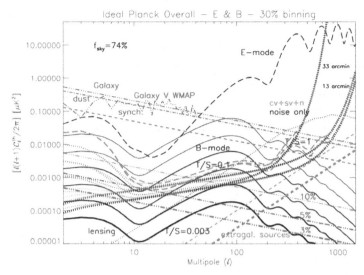

Figure 1. CMB E polarisation modes (black long dashes) compatible with *WMAP* data and CMB B polarisation modes (black solid lines) for different tensor-to-scalar ratios of primordial perturbations ($r \equiv T/S = 1, 0.3, 0.1, 0.03, 0.01, 0.003$, at increasing thickness) are compared to *Planck* overall sensitivity to the APS, assuming the noise expectation has been subtracted. The plots include cosmic and sampling variance plus instrumental noise (green dots for B modes, green long dashes for E modes, labeled with cv+sv+n; black thick dots, noise only) assuming a multipole binning of 30% and a smoothing to two different FWHM corresponding to LFI 30 and 70 GHz. The B mode induced by lensing (blue dots) is also shown for comparison. We display also the comparison with Galactic and extragalactic polarised foregrounds at a reference frequency of 70 GHz. Galactic synchrotron (purple dashes) and dust (purple dot-dashes) polarised emissions produce the overall Galactic foreground (purple three dot-dashes). *WMAP* 3-yr power-law fits for uncorrelated dust and synchrotron have been used. For comparison, *WMAP* 3-yr results derived directly from the foreground maps using the HEALPix package (Górski *et al.* (2005)) are shown: power-law fits provide (generous) upper limits to the power at low multipoles. Residual contamination levels by Galactic foregrounds (purple three dot-dashes) are shown for 10%, 5%, and 3% of the map level, at increasing thickness. See also the text.

at the several % level, then the main limitation will come from instrumental noise. As shown in Fig. 1, values of $r > 0.05$ are potentially achievable with *Planck*.

Planck will permit to significantly improve the determination of the cosmological parameters that are expected to be obtained with a relative error of the order of 1% or lower (see Fig. 5 of Mandolesi *et al.*, 2010 where a detailed forecast is provided), opening a new phase in our understanding of cosmology. Moreover, *Planck* will test the Gaussianity of the CMB anisotropies usually assumed by simple cosmological models. However, important information may come from mild deviations from Gaussianity (see e.g., Bartolo *et al.* (2004) for a review). *Planck* data will either provide the first true measurement of non-Gaussianity (NG) in the primordial curvature perturbations, or tighten the existing constraints (based on *WMAP* data) by almost an order of magnitude. In addition, *Planck* will provide independent investigations of the large-scale anomalies suspected in the WMAP temperature data that could be indicative of new (and fundamental) physics beyond the concordance model or simply the residuals of imperfectly removed astrophysical foreground (e.g. see Maris *et al.*, 2010) or systematic effects. *Planck* will also extend those investigations to polarization maps (see Mandolesi *et al.*, 2010; Tauber *et al.*, 2010 and references therein).

Planck will carry out an all-sky survey of the fluctuations in Galactic emission at its nine frequency bands. The HFI channels at $\nu \geqslant 100\,\mathrm{GHz}$ will provide the main improvement in the charcterization of the large-scale Galactic dust emission particularly in polarisation, while LFI will provide crucial information about the low frequency tail of this component. The LFI frequency channels, in particular at 30 and 44 GHz, will be relevant to the study of the diffuse, significantly polarised synchrotron emission and the almost unpolarised free-free emission. Results from *WMAP*'s lowest frequency channels inferred an additional contribution, probably correlated with dust (see Dobler *et al.*, 2009 and references therein). Several interpretations of microwave (Hildebrandt *et al.*, 2007; Bonaldi *et al.*, 2007) and radio (La Porta *et al.*, 2008) data, and in particular the AR-CADE 2 results (Kogut *et al.*, 2009), seem to support the identification of this anomalous component as spinning dust (Drain *et al.*, 1998; Lazarian & Finkbeiner, 2003). Another interesting component is the so-called "haze" emission in the inner Galactic region, possibly generated by synchrotron emission from relativistic electrons and positrons produced in the annihilations of dark matter particles (see e.g., Hooper *et al.*(2007), Cumberbatch *et al.* (2009), Hooper *et al.*(2008) and references therein). Furthermore, the full interpretation of the Galactic diffuse emissions in *Planck* maps will benefit from a joint analysis with both radio and far-IR data. The ultimate goal of these studies is the development of a consistent Galactic 3D model, which includes the various components of the ISM, and large and small scale magnetic fields (see e.g., Waelkens *et al.*, 2009), and turbulence phenomena (Cho & Lazarian, 2003). While having moderate resolution and being limited in flux to a few hundred mJy, *Planck* will also provide multifrequency, all-sky information about many classes of discrete Galactic sources, having a chance to observe some Galactic micro-blazars (such as e.g., Cygnus X-3) in a flare phase. Finally, *Planck* will provide unique information for modelling the emission from moving objects and diffuse interplanetary dust in the Solar System (Cremonese *et al.*, 2003, Maris & Burigana, 2009, Maris *et al.*, 2006b).

The higher sensitivity, angular resolution, and frequency coverage of *Planck* will allow us to obtain very rich samples of extragalactic sources at mm and sub-mm wavelengths (Herranz *et al.*, 2009) and a good statistics for different subpopulations of sources, some of which are not (or only poorly) represented in the *WMAP* sample. Also, interesting will be the synergy with high energy astrophysics observations, e.g. with *Fermi Gamma-ray Space Telescope* (Abdo *et al.*, 2009, Fermi/LAT Collaboration (2009)). Another noteworthy example is the *Planck* contribution to the astrophysics of clusters. *Planck* will detect $\approx 10^3$ galaxy clusters out to redshifts of order unity by means of their thermal Sunyaev-Zel'dovich effect (Leach *et al.*, 2008, Bartlett *et al.*, 2008). This sample will be extremely important for understanding both the formation of large-scale structure and the physics of the intracluster medium. *Planck*, supplemented by ground-based, follow-up observations planned by the *Planck* team, will allow, in particular, accurate correction for the contamination by sources (Lin *et al.*, 2009).

3. Conclusion

After the successful launch in May 2009, *Planck* has already mapped the sky twice (at the time of writing this review) with the expected behavior and it is planned to complete at least two further all-sky surveys. The release of the first scientific results is expected for January 2011 and will consists of an Early Release Compact Source Catalog (ERCSC) and in about twenty papers on instrument performance in flight, data analysis pipeline, and main astrophysical results. The first publications of the main cosmological implications are expected in 2012.

References

Abdo, A. A., *et al.* 2009, *ApJ*, 700, 597

Ashdown, M. A. J., *et al.* 2007, *A&A*, 467, 761

Bartlett, J. G., *et al.* 2008, *Astronomische Nachrichten*, 329, 147

Bartolo, N., *et al.* 2004, *Phys. Rep.*, 402, 103

Bersanelli, M., *et al.* 2010, *A&A*, 520, A4

Boesgaard, A. M., *et al.* 1998, *ApJ*, 493, 206

Bonaldi, A., *et al.* 2007, *MNRAS*, 382, 1791

Burigana, C., *et al.* 1997, Int. Rep. TeSRE/CNR 198/1997, arXiv: astro-ph/9906360

Burigana, C., *et al.* 2001, *Experimental Astronomy*, 12, 87

Cho, J. & Lazarian, A., 2003, *New Astronomy Review*, 47, 1143

Collaudin, B., Passvogel, T. 1999, *Cryogenics*, 39, 157

Cremonese, G., *et al.* 2003, *New Astronomy*, 7, 483

Cumberbatch, D. T. *et al.* 2009, ArXiv e-prints 0902.0039

Dobler, G., *et al.* 2009, *ApJ*, 699, 1374

Draine, B. T. & Lazarian, A., 1998, *ApJL* 494, L19

Dupac, X., Tauber, J. 2005, *A&A*, 430, 363

Efstathiou, G. 2004, *MNRAS*, 349,603

Fermi/LAT Collaboration, 2009, *ApJ.*, 697, 1071

Górski, K. M. *et al.* 2005, *ApJ*, 622, 759

Hamimeche, S., Lewis, A. 2008, *Phys. Rev.* D77, 103013

Herranz, D. *et al.* 2009, *MNRAS*, 394, 510

Hildebrandt, S. R., *et al.* 2007, *MNRAS*, 382, 594

Hooper, D. *et al.* 2007, *Phys. Rev. D* 76, 083012

Hooper, D. *et al.* 2008, *Phys. Rev. D* 77, 043511

Janssen, M. A., Gulkis, S. 1992, in NATO ASIC Proc. 359: The Infrared and Submillimetre Sky after COBE, ed. M. Signore and C. Dupraz, 391

Jewell, J., *et al.* 2004, *ApJ*, 609, 1

Knox, L. 1995, *Phys. Rev. D* 52, 4307

Kogut, A., *et al.* 2009, arXiv:0901.0562

Lamarre, J. M., *et al.* 2010, *A&A*, 520, A9

La Porta, L., *et al.* 2008, *A&A*, 479, 641

Lazarian, A. & Finkbeiner, D., 2003, *New Astronomy Review*, 47, 1107

Leach, S. M., *et al.* 2008, *A&A*, 491. 597

Leahy, P., *et al.* 2010, *A&A*, 520, A8

Leach, S. M., *et al.* 2008, *A&A*, 491, 597

Lin, Y.-T., *et al.* 2009, *ApJ*, 694, 992

Lopez-Caniego, M., *et al.* 2006, *MNRAS*, 370, 2047

Maino, D., *et al.* 1999, *A&AS*, 140, 383

Mandolesi, N., *et al.* 1994, in Lecture Notes in Physics, Berlin Springer Verlag, Vol. 429, Present and Future of the Cosmic Microwave Background, ed. J. L. Sanz, *et al.*, 228

Mandolesi, N. *et al.* 2010, *A&A*, 520, A3

Maris, M., *et al.* 2006b, *A&A*, 452, 685

Maris, M. & Burigana, C., 2009, *Earth, Moon, and Planets*, 105, 2, 81

Maris, M., *et al.* 2010, arXiv:1010.0830

Rocha, G., *et al.* 2009, arXiv:1008.4948

Scott, D. & Smoot, G. 1998, *Phys. Lett. B*, 667, 246

Smoot, G. F., *et al.* 1992, *ApJ*, 396, L1

Tauber, J., *et al.* 1994, *ESA Journal*, 18, 239

Tauber, J., *et al.* 2010, *A&A*, 520, A1

The Planck Collaboration, 2006, astro-ph/0604069

Waelkens, A. *et al.* 2009, *A&A* 495, 697

Wandelt, B. D., *et al.* 2004, *Phys. Rev. D* 70, 083511

Wright, E. L., *et al.* 1996, *ApJ*, 458, L53

Advances in Plasma Astrophysics
Proceedings IAU Symposium No. 274, 2010
A. Bonanno, E. de Gouveia Dal Pino & A.G. Kosovichev, eds.

© International Astronomical Union 2011
doi:10.1017/S1743921311007113

Towards observational MHD.
Advances in spectropolarimetry and the
prospects for the E-ELT

Klaus G. Strassmeier

Astrophysical Institute Potsdam, An der Sternwarte 16, D-14482 Potsdam, Germany
email: kstrassmeier@aip.de

Abstract. Polarization and wavelength are the bits of information attached to every photon that reveal the most about its formation and subsequent history. The E-ELT will, for the foreseeable future, be the most powerful optical light-collecting machine ever built. The strength of its combination, spectropolarimetry with the E-ELT, is the anchorage in physics of astronomical observations. I present a strawman design of a spectropolarimeter for its intermediate focus.

Keywords. Magnetic fields – stars: magnetic fields – instrumentation: polarimeters – techniques: polarimetric – techniques: spectroscopic

1. Size matters

Per wavelength resolution element, polarization detectable at the 0.1% level translates into a few million photons. For many classes of objects, even 8-m telescopes quickly run out of power when confronted with such a task. Therefore, as soon as the scientific objectives under consideration reach beyond the mere detection of an object, there is no stronger justification for increased telescope diameters than from spectropolarimetry. In fact, since modern detectors will reach negligible read noise levels, equipment that can measure polarization to better than 0.1% can also detect very dim sources. In polarimetry, the E-ELT will not do 8-m class science with shorter exposure times. Many of the science cases discussed below† will only be enabled by the E-ELT (see also Baade *et al.* 2006). This is further compounded by the variability of many types of sources which, in the environment of compact objects, tends to be rapid.

In order to explain complex spectroscopic and photometric observations of accretion processes, collimated jets, stellar rotational braking, coronal heating, non-spherical geometries, and many others, magnetic fields are frequently invoked. But only spectropolarimetry can measure and map magnetic fields and so substitute facts for speculation‡. Figure 1 is a state-of-the-art spectropolarimetric observation of a 11.2-mag star with a 3.6m telescope. The polarimetric signal of this star is ten times smaller than the noise, yet it can be reconstructed from time series of such spectra (e.g. Carroll *et al.* 2011) and used to infer the stellar surface magnetic field. Magnetism is one of the four fundamental forces in nature. Understanding the Universe is impossible without understanding cosmic magnetism, which plays an important role in the formation of celestial bodies and their evolution even during the early stages of the Universe (e.g. Gaenslera *et al.* 2004) as well as for our own existence on this planet.

† See the full community proposal to ESO at www.eso.org/sci/facilities/eelt/science/doc/
‡ As an often-heard joke puts it: "To understand the Universe, we examine galaxies and stars for radiation, small- and large-scale motions, temperatures, chemical composition, and much more. Anything we can't explain after that, we attribute to magnetic fields."

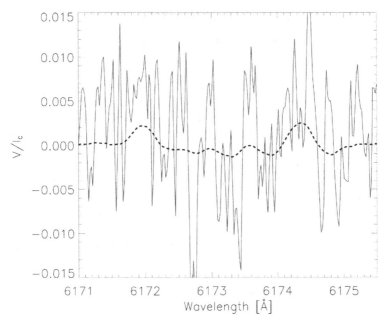

Figure 1. A state-of-the-art spectropolarimetric observation with a 3.6m telescope in Stokes V (full line) compared with the signal level that one would expect from a solar-like surface magnetic field (dashed line). Only a small wavelength section around the Fe I-6173 line is shown. The star is V410 Tauri, a weak-lined 11.2-mag T Tauri star, observed with CFHT and the ESPADONS spectropolarimeter at $R = 60,000$ and an integration time of 600 s. The magnetic-field signal is approximately ten times smaller than the observational noise.

2. Some selected science cases

2.1. *The Vegetation Red Edge and other terrestrial-life markers*

Vegetation has a fivefold higher reflectivity in the NIR than in the optical domain. The steep drop around 700 nm is called the Vegetation Red Edge (VRE). It has been detected in Earthshine observations off the face of the Moon (e.g. Arnold *et al.* 2002, Hamdani *et al.* 2006) and could be a prime target for E-ELT broad-band observations in integrated light. However, the spectral region of the VRE is contaminated by unknown amounts of O_3 absorption (plus O_2 bands). A possible way out is to observe in polarized light and use the known albedo-polarization relation for planetary surfaces (e.g. Stam 2008). Polarization degrees of up to 20% are expected from planets in short-period orbits.

The development of terrestrial life may have been helped through seeding by comets. Many molecules are biologically relevant in only one of their two chemically equivalent chiral forms. The origin of this peculiarity is not known. For instance, a chiral imbalance may have arisen from irradiation by circularly polarized starlight, in which case the chirality of terrestrial life would possibly be a mere coincidence. A recently analyzed meteorite suggests that chirality was, in fact, a property of very early solar-system material (Pizzarello *et al.* 2008). For the more pertinent gaseous matter, circular spectropolarimetry is the analysis method of choice and needs to be applied to cometary tails, the interstellar medium, and even exoplanets. Circular polarization was not detected in the spectrum of the Earthshine yet, which could have been interpreted as a signature of biomatter (chlorophyll in particular) on the Earth's surface (Sterzik & Bagnulo 2009; see also Boehnhardt *et al.* 2009).

2.2. *Protostars: the link to the star-formation process*

Magnetic fields are important ingredients of the star-formation process (McKee & Ostriker 2007). Models of magnetically driven accretion and outflows successfully reproduce many observational properties of low-mass pre-main sequence stars. Indirect observational evidence for the presence of magnetic fields in these stars manifests itself in strong X-ray, FUV, and UV emission (e.g., Feigelson & Montmerle 1999).

The first detections of magnetic fields in protostars of class I and II were obtained using NIR spectrographs and revealed kG fields (e.g., Johns-Krull *et al.* 2009). A VLT surface map in Stokes I of a young star in the Lupus star-forming region even resolved the regions of impacting circumstellar matter and predicted a 3 kG polar field (Strassmeier *et al.* 2005). The accreted matter is presumably funneled along magnetic field lines but lack of spectropolarimetry prevented the full empirical verification of this model. The first magnetic-field maps of T Tauri stars show some systems that have complex fields while some have much simpler dipolar/octupolar fields (Donati *et al.* 2008). Accretion models based on these maps demonstrate the strong dependence of accretion efficiency on both the size and geometry of the star's magnetic field. Fields have also been detected in half a dozen Herbig Ae/Be stars (Hubrig *et al.* 2009). The magnetic field strength and the X-ray emission of Herbig Ae/Be stars show hints for a decline with age in the range of ≈2–14 Myr supporting a dynamo mechanism that decays with age.

Very few systems can be studied even with 8m-class telescopes; high-resolution spectropolarimetry at NIR and MIR wavelengths would enable a large step forward in the understanding of star formation. Complex theories have been developed based on indirect indicators of magnetospheric accretion; high-resolution spectropolarimetry in all Stokes parameters is essential in testing them.

2.3. *Interstellar magnetic fields*

Optical and NIR observations can separate dust and magnetic effects and may also be applicable to high-resolution data of the Lα forest in distant quasars. But their potential scope was recently strongly extended by the so-called Atomic Magnetic Realignment effect (Yan & Lazarian 2008, and references therein). In its basic form, it has higher sensitivity to weak magnetic fields than the Zeeman effect. Its application to emission lines permits the exploration of relatively high-temperature regions. But it mandates extremely large telescope apertures.

Beyond 2-3 kpc from the Sun, the magnetic field of our Galaxy is not well explored (see Feinstein *et al.* 2008). Magnetic fields in the disk-halo interface regions are complex and only few observations are available. The E-ELT can contribute to many of these questions and generate synergy effects with upcoming radio telescopes, which will dominate in the area of extended sources. Note that optical/NIR measurements can provide the origin of the polarization direction while, in the radio domain, the Faraday effect induces a rotation of the polarization direction. Thus, covering both the radio and the optical/NIR range for such measurements is of high value (see Beck 2010).

2.4. *Type Ia supernovae*

Virtually every supernova that has been properly observed has displayed significant polarization, and hence some significant degree of asymmetry. This has provided new challenges to theory and has shaped the conceptual development of the field.

A particularly stunning reminder of polarimetric diagnostics unparalleled by more limited conventional observing techniques proved to be the observations by Wang *et al.* (2006) of SN 2004dt. Ca II 3968, Si II 3859/6355, Mg II 4481, and O I 7774 all had about the same velocity profile. Therefore, on a purely spectroscopic basis, one would have concluded that all these species have about the same radial distribution in the ejecta. Only polarimetry revealed this as a mistake because, contrary to the other species named, O I 7774 was virtually unpolarized. This is an important finding since of the four elements only oxygen is expected to have been present in major quantities before the explosion. Significant constraints on explosion models arise, which require ELT follow-up, especially with a view towards different metallicities and likely ages inferred from the timing of star bursts.

3. A strawman design for the E-ELT

3.1. *Constraints*

The current concept is to provide a simultaneous polarimetric fibre feed to an optical spectrograph like CODEX, or similar, and a NIR spectrograph like SIMPLE, or similar. The applicable wavelength range will be set on the short wavelength side by the fiber throughput as it is rather poor for wavelengths shorter than 380nm. On the long wavelength side it is set by our intention not to cool the polarimeter optics and thus stay shorter than the thermal IR, say, 1.8 μm. Currently, the polarimeter exit is laid out to feed both an optical as well as a NIR spectrograph simultaneously although the option to either feed an optical or a NIR instrument alone is retained. A further option is to consider a dedicated, fibre-fed and bench mounted spectrograph of moderately high resolution and fixed format with three arms (Blue, Visual-Red, NIR), somewhat similar to X-Shooter on the VLT but bench mounted.

Such a polarimeter is not foreseen to be operated with laser guide stars, only with natural guide stars. The deformable M4 mirror is only needed in a stiff position. Note that no image derotation is required because all polarimetric targets are observed on the optical axis. However, the polarimeter concept must enable that the acquisition, guiding, and wave-front sensing for M1 is still done through the guide probes in a ring-like FOV with diameter 5–10 arcmin in one of the two Nasmyth foci (which one is tbd). We proceed with a science FOV for the polarimeter of 3 arcsec and try to fit the instrument diameter, its length, and the distance to the focus into the free inner FOV. This comes down to designing a compact collimator that matches these dimensions.

3.2. *Collimator*

The need for a collimator comes from the requirement that the incidence angles into the (plane) polarimetric optical components must be the same across the FOV and remain perpendicular to their respective optical surfaces. Our current design foresees a 15mm diameter for this parallel beam. The larger the beam the more problematic is the production of homogeneous crystals with birefringent behavior. The smaller the beam the higher are the requirements on the optical surface quality and homogeneity of the birefringent material. Our overall design is an inverted Schwarzschild system (a microscope in principle) solely based on reflections (Fig. 2). It has the big advantage that cross talk due to birefringency introduced by thermal and mechanical stresses during transmission does not exist. Another advantage is that it is free of chromatic problems, and thus useable for optical and NIR spectrographs at the same time. The disadvantage is that the collimator entrance angles become very large, up to 10 degrees in the present f/4.15 case, and the available FOV very small (3 arc sec in our case).

3.3. *Polarimetric components*

The first component of the polarimeter is the linear polarizer and beam splitter, in our case we propose a non-standard Foster prism following the PEPSI design (Strassmeier

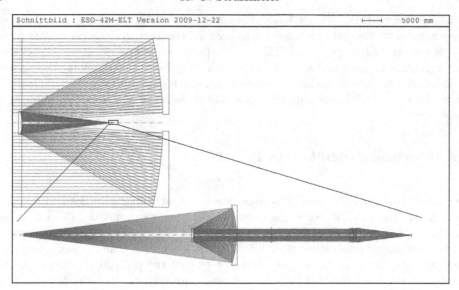

Figure 2. Optical design of a collimator for the intermediate f/4.15 focus of the E-ELT. The top drawing shows the telescope together with the collimator (in the box as indicated). The bottom drawing is a blow-up of the collimator design alone. The polarimetric components are located in the parallel beam before the camera (only one beam and one camera is shown).

et al. 2008). Currently, the calcite for such a prism is principally – but not easily – available and may require additional R&D with the optical industry (e.g. Spanó *et al.* 2006). The entire unit consists of three prisms, two are made of calcite, one is N-Bk7. The latter is needed to make the output ordinary (o)- and extra-ordinary (eo)-beams parallel to each other. This enables a compact and slim mechanical design in order to minimize diffraction of the telescope beam. In linear polarizing mode, the Foster unit will be rotated and thus must be turnable with respect to the sky by 270°.

The second component of the polarimeter is the elliptical retarder in front of the Foster subunit. It must be additionally rotateable with respect to the Foster unit. We envision to use a similar retarder type that we implemented in the PEPSI spectropolarimeter for the LBT as well as for the calibration unit for the GREGOR 1.5m solar telescope on Tenerife. These retarders are based on plastic sheets (polymethylmetacrylat) and we envision a single, elliptical retarder for the entire wavelength range from 380–1800nm. This will require some R&D with the optical industry but appears reasonable.

3.4. *Dichroic beam splitters*

For our conceptual design, we decided to proceed with two simultaneous wavelength regions, optical and NIR, i.e. feeding proposed instruments like CODEX and SIMPLE at the same time. The wavelength cut is set at 1000nm with wings extending ±30nm. The dichroic beam splitter would be situated behind the Foster subunit and behind a common ADC for both beams (o and eo) and follow the design for a multi-channel focal reducer by Laux *et al.* (2009).

4. Conclusions

The special and unique sensitivity of polarization to the interaction between light and matter gives polarimetry diagnostic qualities not available with other observing techniques. However, amongst the recent and near-future infrared space missions (Spitzer,

Herschel, JWST), there is no polarimetric option. In space there is an obvious imperative to keep mechanical systems as simple and trouble-free as possible, and in wavelength regions with useful atmospheric transmission, polarimetry can be achieved more effectively from the ground. Yet, the 2007 proposal for the construction for the US Thirty Meter Telescope ignores polarimetry, except in the context of the search for extra-solar planets. Similarly, the Science Case from 2006 for the Giant Magellan Telescope only mentions supernovae polarization explicitly. However, large-area polarization surveys are foreseen with the LSST (Clemens *et al.* 2009), or are even ongoing at IRSF (Kandori *et al.* 2006), and will provide a wealth of polarized targets for the E-ELT.

I conclude that a megastructure like the E-ELT with its 1380 m^2 light-collecting area should be in the position to recover the full wavefront information.

Acknowledgements

I would like to thank Alfio Bonanno and his team for a great meeting. Also thanks to my AIP colleagues Thoroton Carroll, Michil Weber and Thomas Granzer for continuing discussions and Uwe Laux for his efforts in the optical design of the collimator. The E-ELT science case for a spectropolarimeter was a joint effort of 80 colleagues across Europe with particular input from Dietrich Baade at ESO, whom I thank for this. I acknowledge grant STR645-1 from the Deutsche Forschungsgemeinschaft (DFG) and the support from the Bundeministerium für Bildung & Forschung (BMBF) through the Verbundforschung grant 05A09IPA.

References

Arnold, L., Gillet, S., Lardiere, O., Riaud, & P., Schneider, & J. 2002, *A&A* 392, 231

Baade, D., Wang, L., Hubrig, S., & Patat, F. 2006, in IAU Symp. 232, CUP, p.248

Beck, R. 2010, in "Astronomy with Megastructures. Joint Science with E-ELT and SKA", arXiv:1008.3806

Boehnhardt, H., Tozzi, G. P. Sterzik, M. *et al.* 2009, *Earth, Moon & Planets*, 105, 95

Carroll, T. A., Strassmeier, K. G., Ilyin, I., & Rice, J. B. 2011, in IAU Symp. 273, CUP, in press

Clemens, D. P., Pinnick, A., Pavel, M. *et al.* 2009, *BAAS* 41, 459

Donati, J.-F. *et al.* 2008, *MNRAS* 386, 1234

Feigelson, E. & Montmerle, T. 1999, *ARA&A* 37, 363

Feinstein, C., Vergne, M. M., Martinez, R., & Orsatti, A. M. 2008, *MNRAS* 391, 447

Gaenslera, B. M., Beck, R., & Feretti, L. 2004, *New Astron. Reviews* 48, 1003

Hamdani, S., Arnold, L., Foellmi, C. *et al.* 2006, *A&A* 460, 617

Hubrig, S., Grady, C., Schöller, M. *et al.* 2009, in IAU Symp. 259, CUP, p.395

Johns-Krull, C., Greene, T. P., Doppmann, G. W., & Covey, K. R. 2009, *ApJ* 700, 1440

Kandori, R., Kusakabe, N., Tamura, M. *et al.* 2006, *SPIE* 6269, 159

Laux, U., Klose, S., & Greiner, J. 2009, poster at the fall meeting of the AG 2009, Potsdam

McKee, C. F. & Ostriker, E. C. 2007, *ARA&A* 45, 565

Pizzarello, S. *et al.* 2008, *PNAS* 105, 3010

Spanó, P., Zerbi, F. M., Norrie, C. J. *et al.* 2006, *AN* 327, 649

Stam, D. M. 2008, *A&A* 482, 989

Sterzik, M. F. & Bagnulo, S. 2009, in Bioastronomy 2007, *ASPC* 420, p.371

Strassmeier, K. G., Rice, J. B., Ritter, A. *et al.* 2005, *A&A* 440, 1105

Strassmeier, K. G., Woche, M., Ilyin, I. *et al.* 2008, *SPIE* 7014, 21

Wang, L., Baade, D., Höflich, P. *et al.* 2006, *ApJ* 653, 490

Yan, H. & Lazarian, A. 2008, *ApJ* 677, 1401

Advances in Plasma Astrophysics
Proceedings IAU Symposium No. 274, 2010
A. Bonanno, E. de Gouveia Dal Pino & A.G Kosovichev, eds.
© International Astronomical Union 2011
doi:10.1017/S1743921311007125

New interactive solar flare modeling and advanced radio diagnostics tools

Gregory D. Fleishman[1,2], Gelu M. Nita[1] and Dale E. Gary[1]

[1] New Jersey Institute of Technology, Newark, NJ 07102, USA

[2] Ioffe Institute, St. Petersburg 194021, Russia
email: gfleishm@njit.edu

Abstract. The coming years will see routine use of solar data of unprecedented spatial and spectral resolution, time cadence, and completeness in the wavelength domain. To capitalize on the soon to be available radio facilities such as the expanded OVSA, SSRT and FASR, and the challenges they present in the visualization and synthesis of the multi-frequency datasets, we propose that realistic, sophisticated 3D active region and flare modeling is timely now and will be a forefront of coronal studies over the coming years. Here we summarize our 3D modeling efforts, aimed at forward fitting of imaging spectroscopy data, and describe currently available 3D modeling tools. We also discuss plans for future generalization of our modeling tools.

Keywords. Sun: corona, Sun: magnetic fields, radiation mechanisms: non-thermal, methods: numerical, Sun: radio radiation, Sun: flares, stars: flares

1. Introduction

Solar activity, although energetically driven by subphotospheric processes, depends critically on coronal magnetism, which, broadly speaking, includes magnetic field generation, evolution, and transformation into kinetic, thermal, and nonthermal energies in the corona. Reliable tools for doing direct diagnostics have been lacking, although the situation is currently changing. Indeed, new space- and ground-based solar optical telescopes are already capable of precise measurements of the photospheric magnetic field with sub-arcsecond angular resolution and high temporal resolution. When combined with modern extrapolation algorithms, these data offer important clues on the coronal magnetic field structure and evolution. However, given the finite angular resolution, sensitivity, observational errors, and even theoretical limitations, such extrapolations are not unique, so the extrapolations require independent verification. An opportunity for quantitative verification will be available when the new generation of high-resolution solar-dedicated radio instruments (expanded OVSA, SSRT, and FASR) become operational. Microwave radiation is produced by the gyrosynchrotron (GS) mechanism as accelerated fast electrons gyrate in the coronal magnetic field. As has been recently proven using simulated microwave data, the coronal magnetic field can indeed in principle be reliably recovered at the flare dynamic time scales from the radio data, along with the key parameters of the thermal plasma and accelerated electrons (Fleishman *et al.* 2009). The ability to detect the magnetic field and its changes on dynamic time scales is a critically needed element to uncover the fundamental physics driving solar flares, eruptions, and activity.

2. Modeling Methods and Tools

Direct Modeling. 3D models of the solar flares are not yet numerous. Those available models (Preka-Papadema & Alissandrakis 1992, Kucera *et al.* 1993, Bastian *et al.* 1998,

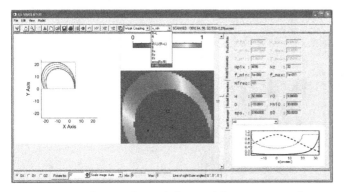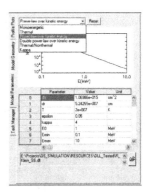

Figure 1. Left: interface of the model geometry and non-uniform parameter input—model magnetic loop (left window); nonuniform distribution of fast electrons given in a color code (middle window); and numeric input and graph display of the parameters involved. Right: selection of the fast electron energy distribution from a number of pre-defined analytical functions and parameters of this distribution.

Simões & Costa 2006, 2010, Tzatzakis *et al.* 2008, Fleishman *et al.* 2009) are built on an idealized (e.g., dipole) magnetic loop, rather than a realistic magnetic geometry.

Our currently available modeling tool, **GS Simulator**, is also built based on an analytical (dipole) magnetic field model. A flaring loop model (i.e., user specified dipole flux tube, Figure 1, left) is produced with a newly developed interactive IDL widget application intended to provide a flexible tool that allows the user to generate spatially resolved GS spectra. To do so, the user populates the loop with thermal plasma by selecting plasma temperature and density in the adjustable parameter list, Figure 1, right, and specifies a fast electron population by selecting one of a few pre-defined distributions of the electrons over energy and pitch-angle and choosing numeric parameters for these distribution (Figure 1, right). The spatial distribution of the electrons is also specified along the loop, as shown in the color coded image (Figure 1, left). After selecting the angle from which to view the loop with the mouse, the tool then calculates the physical parameters along each line of sight needed to solve the radiation transfer equation, which is performed by external callable computing blocks. The default codes generating the GS (and free-free) emission based on the input geometrical line-of-sight model data were written in FORTRAN and C++ based on fast GS codes newly developed by ?, and compiled as a DLL (or SO in case of Linux) callable by IDL.

The object-based architecture of this application provides the user with full 3D interaction with a predefined, but adjustable, magnetic loop geometry, as well as with any user defined analytical geometrical model that would inherit the basic properties of the generic "gs_model" IDL object defined in this package. To allow for more realistic 3D flare modeling, in place of drawing a loop "by hand", the tool must be further developed to include (i) a numerical magnetic field structure, such as would be obtained from a photospheric extrapolation (Figure 2) or a full MHD model based on vector photospheric measurements of the magnetic and velocity fields, along with the thermal plasma distribution, (ii) a realistic electron acceleration and transport model, and (iii) the ability to quickly compute emission in various wavelength regimes. The tool will enable the user to identify, from the model or through comparison with observations, the subset of field lines involved in flaring, consider fast electron transport in this realistic magnetic structure, calculate radiation from this evolving volume, and so simulate a solar flare.

Forward Fitting. The coronal magnetic field is a key parameter controlling most solar flaring activity, particle acceleration and transport. It has been understood, and

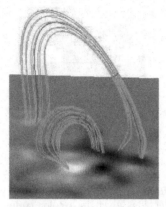 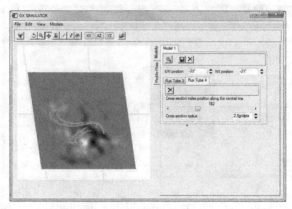

Figure 2. Left: Two flux tube models. They can be defined independently, and may represent interacting loops. Additional controls are being prepared for populating the loops with particles and ambient medium. Even dynamics of the particles is planned. Right: The loops now viewed from a specified location on the solar disk (22 E 21 S in this example). The tabs labeled Flux Tube 3 and Flux Tube 4 represent the two loops.

often proposed, that the coronal magnetic field along with fast electron distribution can in principle be evaluated from the microwave GS radiation, which is indeed sensitive to the instantaneous magnetic field strength and orientation relative to the line of sight and to the fast electron spectrum.

Diagnostics, understood as the determination of physical parameters of a system under study from arrays of observed parameters, is a key outstanding problem in Solar Physics. In some (basically linear) cases regularized true inversions can work well (e.g., Kontar *et al.* 2004). In most of the cases, however, such true inversions fail because of the highly nonlinear nature of physical systems. In such cases, the forward fitting, i.e., finding a number of free parameters of a physically motivated model of the system from fitting the model to observations, can often be used successfully in place of true inversions.

Anticipating a large breakthrough in the radio imaging spectroscopy observations, which will become possible soon due to the next generation of radio instruments, we have developed a practical forward fitting method, based on the SIMPLEX algorithm with shaking, that allows reliable derivation of the magnetic field and other parameters along a solar flaring loop using microwave imaging spectroscopy of GS emission, which is calculated with newly developed fast GS codes (Fleishman & Kuznetsov 2010). We illustrate the method using a model loop with spatially varying magnetic field, filled with uniform ambient density and an evenly distributed fast electron population with an isotropic, power-law energy distribution (Fleishman *et al.* 2009).

From the flare radio model (Figure 3, left) described above we have a sequence of spatially resolved microwave spectra (in a general case, both intensity and polarization data, one spectrum per pixel). Then, we fit the data to a model microwave spectrum pixel by pixel (Figure 3, middle) to derive physical parameters of the source (e.g., the magnetic field, Figure 3, right). Although the exact GS formulae are very computationally expensive, much faster codes giving the same accuracy have recently been developed by Fleishmann and Kuznetsov (2010), which are used in practice as the forward fitting input. Then, having a fitting procedure resulting in fast and reliable finding of the true source parameters is exceedingly important. The problem here is that most of the minimization algorithms often find a *local* minimum of the normalized residual (or of the reduced chi-square), while the ultimate goal of the fitting is to identify the *global* minimum. So far,

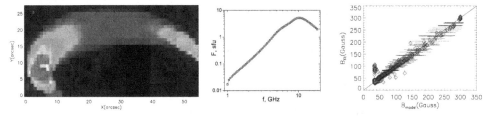

Figure 3. Simulated image of the radio emitting loop source at 4 GHz as observed by an ideal radio heliograph with ~1" pixel size resolution (left). Example of the model (symbols) and fit (solid curve) spectra corresponding to one particular pixel of the image displayed in the left panel (middle). Model-to-fit comparison of the magnetic field (right).

we determined that the simplex algorithm is very efficient in finding a local minimum. Then, it needs to be 'shaken' for the simplex solution to overcome any local minima and continue downhill towards the global minimum (a version of the stimulated annealing approach). Even when the algorithm performance is overall good, there is a non-zero probability that the algorithm fails to find the true solution in some pixels. We use post-processing to identify and flag/remove those pixels.

3. Conclusions

Modeling flare geometry, its full 3D visualization, and interactive adjustments to the user-specific needs are highly complicated tasks in themselves. Our modeling tools offer a united solution for these problems, which are widely applicable to various external data cube inputs and so offer a convenient framework for diverse studies of coronal magnetism, including flares and the active region magnetosphere. The modeling tools, computational libraries, and documentation are available via author's web page, see http://web.njit.edu/~gfleishm/.

The outlined modeling efforts can only bring fundamental knowledge about flare/active region physics if used in conjunction with modern, high-resolution observations. Key observations of the coronal plasma parameters can only be made by radio instruments that combine high sensitivity, temporal, spatial, and spectral resolution, which are unavailable now. A small part of the required science will be possible soon with the expanded OVSA instrument (anticipated operation of the upgraded instrument begins in fall, 2013) and the upgraded multi-wavelength SSRT. However, the full required capability has to wait until the full FASR (Gary 2003) has been built.

Acknowledgments

This work was supported in part by NSF grants ATM-0707319, AST-0908344, and AGS-0961867 and NASA grant NNX10AF27G to New Jersey Institute of Technology, and by the RFBR grants No. 08-02-92228, 09-02-00226, 09-02-00624.

Advances in Plasma Astrophysics
Proceedings IAU Symposium No. 274, 2010
A. Bonanno, E. de Gouveia Dal Pino & A.G. Kosovichev, eds.
© International Astronomical Union 2011
doi:10.1017/S1743921311007137

ALMA and solar research

Marian Karlický and Miroslav Bárta

Astronomical Institute of the Academy of Sciences of the Czech Republic,
CZ – 25165 Ondřejov, Czech Republic
email: karlicky@asu.cas.cz, barta@asu.cas.cz

Abstract. The ALMA (Atacama Large Millimeter/sub-millimeter Array) is the large inter-ferometer that will consist up to 64 high-precision antennas operating in the 31.3 – 950 GHz frequency range. In this range unique observations in cosmology, cold universe, galaxies, stars and their formations, and so on are expected. Among these objectives there is a unique possibil-ity to observe the Sun and to address outstanding issues of solar physics. The ALMA is shortly described and then the new ESO-ALMA European node (ARC) built at Ondřejov Observatory is presented. The new ARC is the only one in Europe oriented to solar physics. The requirements and limitations for ALMA solar observations, as well as some examples of possible solar-oriented ALMA projects, are mentioned.

Keywords. instrumentation: high angular resolution, interferometers, Sun: radio radiation

1. Introduction

The Atacama Large Millimeter/sub-millimeter Array (www.almaobservatory.org) is a worldwide project to construct a large interferometer (up to 64 high-precision 12 m and 7 m antennas, operating in the 31.3–950 GHz frequency range) in Chajnantor plain of the Chilean Andes at the altitude 5000 m above sea level. The telescope will be operated from the nearby operational center located at the elevation 2900 m. Four main centers are built for the preparation of proposals for observations, data distri-bution, data analysis and ALMA science and technical supports. Besides the center in Santiago de Chile, there is main European ESO-ALMA center in Garching, Germany (www.eso.org/sci/facilities/alma/arc/), North American ALMA Science Center (NAASC) in Charlottesville, Virginia, USA (science.nrao.edu/alma/index.shtml), and East-Asian ALMA Regional Center in Mitaka, Japan (alma.mtk.nao.ac.jp/EA-ARC/). The European ALMA Regional Center (ARC) is, besides the main center in Garching, formed by a distributed structure of seven additional nodes (ARC nodes) which cover various branches of expected ALMA science and provide better geographical accessibility for ALMA users in Europe.

The science program of ALMA is very broad. It includes molecular clouds in the cold universe, cosmology, the formation of galaxies and clusters of galaxies, formation of stars and planets, astrochemistry, comets and also the solar research. The solar physics was included into this ambitious project primarily thanks to A. O. Benz, T. Bastian, S. White, M. Kundu and others. There are several papers advertising ALMA as a future observing instrument for solar research, e.g Bastian (2002) and Loukitcheva *et al.* (2008).

In this paper, we want to inform about the present status of the ALMA project, about the new ESO-ALMA node at Ondřejov Observatory which is oriented mainly to solar physics, and about a unique potential of ALMA for the solar physics research.

Table 1. ALMA parameters. λ_{mm} is the wavelength in mm.

Antennas	64
	(25 μm rms, 0.6″ pointing)
Collecting Area	> 7000 m^2
Receivers	10 bands in 31.3–950 GHz
	0.3–9.6 mm
Field of view	21″ × λ_{mm}
Spatial resolution	0.02″ × λ_{mm} × (10 km/baseline)
Number of baselines	up to 2016

2. ALMA

ALMA construction started in 2002 and the project will be completed in 2012. The telescope is located in the Atacama desert in Chile, at the elevation of 5000 m, where superior transmission properties of the atmosphere at mm/sub-mm wavelengths are found. The first call for proposals of observations is expected to be released at the beginning of 2011.

ALMA is a Fourier synthesis telescope. It will consist up to 64, 12 m and 7 m antennas (for ALMA parameters, see Table 1). Each pair of antennas measures one Fourier component of the brightness distribution of the radio source. An image of the source is then computed by the inverse Fourier transform of all $N(N-1)/2$ components, where N is the number of antennas in the array, i.e. 2016 in ALMA case. The instantaneous field of view of the instrument is given by the antenna size (12 m) and scales linearly with the wavelength (see Table 1). The spatial resolution is determined by the maximum separation of antennas (up to 14 km) and for its wavelength scaling see Table 1. ALMA is also designed to perform very sensitive spectral line observations in the 31.3–950 GHz frequency range, which will be divided into 10 frequency bands. For more details about ALMA, see www.almaobservatory.org.

Proposals for ALMA observations will be prepared by the ALMA Observing Tool software. This will be a two-step procedure, where in the first step the scientific objectives and goals are considered, and if approved by the scientific committee then, in the second step, detailed technical specifications are checked. On the other hand, observed data will be reduced and analyzed by the Common Astronomy Software Applications (CASA) (casa.nrao.edu). Now, both these software packages are in a final testing phase.

3. ESO-ALMA node (ARC) in Ondřejov

In November 2008, we have prepared and presented to ESO a proposal to build a new ESO-ALMA node (ARC) in Ondřejov. In December 2009 the ESO-ALMA node at the Astronomical Institute of the Academy of Sciences was formally approved by the ESO Director. The main purpose of the European ARC nodes is to provide on-line as well as face-to-face support to the ALMA users in matters of proposal preparation, observation planning, and data reduction. The newly formed node in Ondřejov will provide scientific and technical support mainly in the field of solar physics, but also in galactic/extragalactic and relativistic astrophysics, and laboratory measurements and quantum-physics modelling of molecular spectral lines (in cooperation with the Institute of Chemical Technology in Prague). Since it is a unique node providing support in solar research with ALMA in Europe it is, according to the ARC strategy, open to all European ALMA users requiring assistance with accomplishing their solar-oriented ALMA project. Formation of the new node in Ondřejov also improved geographical distribution of European ARC structure. Thus, namely for topics which are covered also by other nodes (e.g. galactic physics), it represents a natural regional center for support of users from Central and Eastern Europe. For details about the new ESO-ALMA node, see http://www.asu.cas.cz/alma. Present activities of the team of this node are: a) preparation of the infrastructure of the ARC node (computers, data servers, fast internet connections), b) participation in official tests of CASA and ALMA Observing Tool softwares, c) communication with other European ESO-ALMA nodes through telephone conferences and face-to-face meetings and d) presentation of lectures about ALMA for students.

4. ALMA and solar research

In the following the limitations of ALMA and requirements for solar observations are summarized:

• Field of view (FOV) of ALMA is rather small (Table 1). To increase the FOV, an observing technique called "on-the-fly" has to be used (Bastian, 2002).

• Comparing with other astrophysical radio sources the solar radio flux is very strong. Therefore, an appropriate attenuation of the signal is necessary.

• It will be good to scatter visible/IR part of the spectrum by the milling of the surface of antennas.

• An advanced calibration technique is necessary.

• For transient phenomena such as solar flares, a flexible communication between scientists and observing staff regarding observed targets is necessary.

Examples of possible ALMA solar-physics studies follow:
• Study of the quiet chromosphere
• Study of prominences and filaments
• Study of solar radio recombination lines
• Study of chromospheric oscillations and waves
• Study of microjets in sunspot penumbrae
• Study of solar flares

5. Concluding remarks

ALMA is designed to observe many objects in the Universe. Since the observing capabilities of ALMA are advanced, an involvement of the solar community in ALMA is highly desirable. To encourage potential solar observers, the new ESO-ALMA node (ARC) at Ondřejov Observatory is being built under the supervision of ESO. Although we are aware of problems which are specific for solar observations (small field of view, strong radio flux, calibration and so on), we hope that the advanced techniques will overcome these difficulties.

Acknowledgements

The authors are indebted to Prof. T. Wilson and Dr. P. Andreani from ESO for their great support in building the Ondřejov ALMA node. This paper was supported by Grant 300030701 of the Grant Agency of the Academy of Sciences of the Czech Republic.

References

Bastian, T. S. 2002, *Astron. Nachr.*, 323, 271
Loukitcheva, M. A., Solanki, S. K., & White, S. 2008, *Astrophys. Space Sci.*, 313, 197

Advances in Plasma Astrophysics
Proceedings IAU Symposium No. 274, 2010
A. Bonanno, E. de Gouveia Dal Pino & A.G. Kosovichev, eds.
© International Astronomical Union 2011
doi:10.1017/S1743921311007149

Investigations of solar plasma in the interior and corona from Solar Dynamics Observatory

A. G. Kosovichev

W. W. Hansen Experimental Physics Laboratory, Stanford University,
Stanford, CA 94305, AKosovichev@solar.stanford.edu

Abstract. The Sun is a plasma laboratory for astrophysics, which allows us to investigate many important phenomena in turbulent magnetized plasma in detail. Solar Dynamics Observatory (SDO) launched in February 2010 provides unique information about plasma processes from the interior to the corona. The primary processes of magnetic field generation and formation of magnetic structures are hidden beneath the visible surface. Helioseismic diagnostics, based on observations and analysis of solar oscillations and waves, give insights into the physical processes in the solar interior and mechanisms of solar magnetic activity. In addition, simultaneous high-resolution multi-wavelength observations of the solar corona provide opportunity to investigate in unprecedented detail the coronal dynamics and links to the interior processes. These capabilities are illustrated by initial results on the large-scale dynamics of the Sun, the subsurface structure and dynamics of a sunspot and observations of a X-class solar flare.

Keywords. Plasmas, Sun: helioseismology, magnetic fields, sunspots, photosphere, flares

1. Introduction

It is well accepted that the Sun represents a unique laboratory for plasma astrophysics. The most fundamental plasma processes in the Universe, including generation of magnetic fields by dynamo, magnetic self-organization in turbulent plasma, magnetic energy release, particle acceleration, MHD waves and shocks, are observed on the Sun in significantly more detail than in other astrophysical objects. Thus, investigation of these processes on the Sun can shed light in other cosmic astrophysics phenomena.

2. SDO science goals and instrumentation

The primary goal of the SDO mission is to study the origin of solar magnetism and variability, to characterize and understand the Sun's interior and how the physical processes inside the Sun are related to surface and coronal magnetic fields and activity. It carries three instruments, which observe the Sun uninterruptedly (except short eclipse periods in March and September) with high spatial and temporal resolution.

The basic characteristics of the instruments are the following:
- Helioseismic and Magnetic Imager (HMI) provides spectro-polarimetric observations in photospheric Fe 6173A line (Schou *et al.*, 2010):
  - 4096x4096 images every 45 sec; spatial resolution ~ 0.5 arcsec/pixel (~ 350 km);
  - Doppler velocity;
  - Line-of-sight magnetograms;
  - Vector magnetograms and plasma parameters every 12 minutes.
- Atmospheric Imaging Assembly (AIA) has 4 telescopes to obtain EUV images of the solar atmosphere and corona in 8 channels (Golub, 2006):

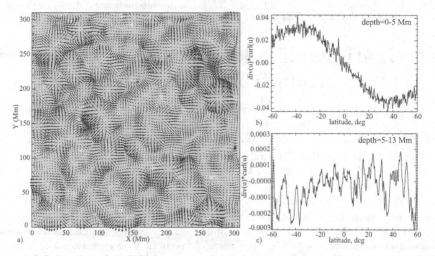

Figure 1. a) Subsurface flow field beneath a quiet-Sun region in the depth range of 1–3 Mm, obtained from the SDO/HMI data using the time-distance helioseismology pipeline (Zhao *et al.*, 2011); the proxy of kinetic helicity, $(\nabla \cdot \mathbf{u})(\nabla \times \mathbf{u})_z$ as a function of latitude in two subsurface layers: b) 0–5 Mm and c) 5–13 Mm deep. The grayscale background in panel a) is the corresponding photospheric magnetogram showing concentrations of small-scale magnetic structures at the boundaries of supergranular convective cells.

- o 4096x4096 images every 3 sec; spatial resolution ~ 0.5 arcsec/pixel;
- o 7 EUV channels in a sequence of iron lines and He II 304Å;
- o One UV Channel with 1600Å, 1700Å, white light filters;
- o Plasma temperature and differential emission measure of the solar transition region and corona are determined in the range of 10^5-10^7 K.
- • EUV Variability Experiment (EVE) makes high-spectral resolution observations of UV irradiance using 2 main systems with 6 spectral channels (Woods *et al.*, 2010):
  - o Multiple EUV Grating Spectrometer: 0.1-37 nm, 10 sec cadence;
  - o EUV Spectrophotometer: 0.1-38 nm, 0.25 sec cadence.

The HMI and AIA collect ~ 4 TB of data per day. The data products are available from the Joint Science Operations Center at Stanford (**http://jsoc.stanford.edu**).

3. Investigation of solar dynamo

Investigation of the solar dynamo is one of the primary objectives of the SDO mission. The current paradigm is that the poloidal magnetic of the Sun is generated by turbulent helicity (so-called α-effect) in the convection zone, primarily at the boundaries, in a subsurface shear layer and in the tachocline (a thin layer at the bottom of the convection zone), and that the toroidal magnetic field is generated in the tachocline by the differential rotation (Ω-effect). It is also assumed that the toroidal magnetic emerged at the surface at low latitudes in the form of bipolar magnetic active regions is transported by the meridional circulation to the polar regions, where it reverses the polarity of the dipolar poloidal field. Therefore, it is critical to obtain the information about the subsurface convection flows, variations of the differential rotation and meridional circulation. Such measurements are obtained through a helioseismology data analysis pipeline, which includes measurements of frequencies of normal oscillation modes of the Sun and inferences of the 2D sound-speed and rotation (radial and latitudinal) profiles (so-called global helioseismology), and also 3D maps of subsurface sound-speed perturbations and detailed

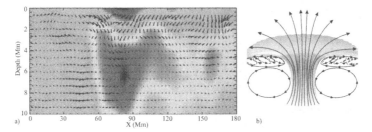

Figure 2. a) A vertical cut through a sunspot structure, obtained by inversion of acoustic travel-time variations from the SDO/HMI data analysis pipeline. The background shows variations of the wave speed, which are negative in a shallow subsurface layer and positive in the deeper interior. Arrows show plasma velocity with a typical speed 0.5–1 km/s, forming a converging flow pattern. b) a schematic illustration of a possible structure of subsurface flows: a thin layer of near-surface shear (Evershed) flows formed by magnetoconvection in inclined magnetic field, and deeper converging downflow, which maintains the compact magnetic structure.

flow in the upper convection zone, about 30 Mm deep (local helioseismology) (see e.g. Kosovichev & HMI Science Team, 2007; Zhao *et al.*, 2011). Figure 1 shows a sample of the subsurface flow field, and estimates of a proxy of the kinetic helicity, $(\boldsymbol{\nabla} \cdot \mathbf{u})(\boldsymbol{\nabla} \times \mathbf{u})_z$, in two subsurface layers. It seems that the latitudinal dependence of the helicity parameter changes with depth. Initial results have been also obtained for the meridional circulation (Zhao *et al.*, 2011). For investigations of the solar cycle it is important to monitor the large-scale solar dynamics over a significant portion of the solar cycle, and the mission plan is to observe at least the raising phase of the solar cycle and the maximum.

4. Magnetic self-organization in solar plasma

The solar magnetic field after emergence on the surface becomes quickly organized in stable structures, sunspots, which are stable for much longer than the convective turnover time. The local helioseismology and magnetic field observations with SDO/HMI provide unique data for studying the processes of formation and stability of the magnetic structures. Figure 2a illustrates initial results of local helioseismology imaging of the subsurface structure and plasma flows beneath a stable sunspot. These results confirm the previous evidence from analysis of the SOHO/MDI and Hinode data (Zhao *et al.*, 2010) that the magnetic self-organized structures are formed and maintained by converging flows and downdrafts in subsurface layers as illustrated in Fig. 2b, and supported by recent numerical simulations (Kitiashvili *et al.*, 2010).

5. Solar flares and magnetic energy release

The understanding of magnetic energy release in solar flares is one of the outstanding problems of plasma astrophysics. The AIA instrument provides high-resolution high-cadence images and diagnostics of temperature and differential emission measure of plasma in the solar corona. The HMI measures detailed configuration of the photospheric magnetic fields and their variations in flaring regions. The EVE monitors the UV flux and has made an important discovery of a secondary peak of the EUV emission in solar flares, which has about four times more energy than the EUV energy during the time of the X-ray flare peak. Together with the high-resolution Hinode data and X- and γ-ray measurements from the RHESSI and FERMI missions we have now a unique collection of data for understanding the mechanism of solar flares and their effects in the solar interior and atmosphere. Figure 3 illustrates the photospheric and helioseismic effects of the first X-class flare, observed by SDO (Kosovichev, 2011). The flare occurred in a

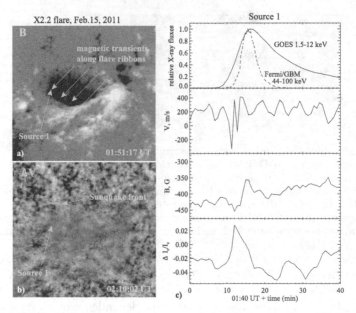

Figure 3. SDO/HMI observations of the impact of the X2.2 solar flare (15 Feb 2011) on the photospheric plasma: a) magnetogram showing the magnetic transients on along both sides of the magnetic neutral line during the impulsive phase; b) the Dopplergram difference taken about 20 minutes after the impact showing a circular helioseismic wave ('sunquake') originating from the impact place (Source 1); c) variations of the soft and hard X-ray emissions measured by GOES-15 and Fermi/GBM instruments, and variations of velocity, magnetic field and continuum intensity in Source 1 from HMI (Kosovichev, 2011).

δ-type sunspot near the magnetic neutral line. The HMI magnetograms, Dopplergrams, and continuum intensity images reveal strong impacts in the photosphere during the impulsive phase. The flare impacts, which are believed to be caused by precipitation of high-energy particles, are organized in two ribbons rapidly expanding from the neutral line (Fig. 3a). In one place (marked 'Source 1'), the flare impact initiated a strong helioseismic wave ('sunquake'), observed as expanding circular-shaped ripples on the surface (Fig. 3b). Comparison of the photospheric impact with the flare X-ray emission observed with the GOES-15 and FERMI/GBM instruments shows that the photospheric impact correlates with the initial soft X-ray increase, and not with the following hard X-ray impulse as thought before. These observations challenge the standard model of solar flares.

Acknowledgements

I thank the participants of the ISSI team on solar magnetism for stimulating discussions, and the ISSI (Bern) for support.

References

Golub, L. 2006, *Sp. Sci. Rev.*, 124, 23
Kitiashvili, I. N., Kosovichev, A. G., Wray, A. A., & Mansour, N. N. 2010, *ApJ*, 719, 307.
Kosovichev, A. G., & HMI Science Team 2007, *Astron. Nach.*, 328, 339
Kosovichev, A. G. 2011, arXiv:1102.3954
Schou, J., *et al.* 2010, *Sol. Phys*, in press.
Woods, T. N., *et al.* 2010, *Sol. Phys*, in press
Zhao, J., Kosovichev, A. G., & Sekii, T. 2010, *ApJ*, 708, 304
Zhao, J., *et al.* 2011, *Sol, Phys.*, submitted.

Advances in Plasma Astrophysics
Proceedings IAU Symposium No. 274, 2010   © International Astronomical Union 2011
A. Bonanno, E. de Gouveia Dal Pino & A.G. Kosovichev, eds.   doi:10.1017/S1743921311007150

Inclusion of velocity gradients in the Unno solution for magnetic field diagnostic from spectropolarimetric data

Guillaume Molodij[1] and Véronique Bommier[2]

Observatoire de Paris-Meudon, 5 place J. Janssen, 92195 Meudon principal

[1] LESIA, UMR 8109 CNRS
email: `Guillaume.Molodij@obspm.fr`

[2] LERMA UMR 8112 CNRS
email: `Veronique.Bommier@obspm.fr`

Abstract. We present an extension of the Unno-Rachkovsky solution that provides the theoretical profiles coming out of a Milne-Eddington atmosphere imbedded in a magnetic field, to the additional taking into account of a vertical velocity gradient. Thus, the theoretical profiles may display asymmetries as do the observed profiles, which facilitates the inversion based on the Unno-Rachkovsky theory, and leads to the additional determination of the vertical velocity gradient. We present UNNOFIT inversion on spectropolarimetric data performed on an active region of the Sun with the french-italian telescope THEMIS operated by CNRS and CNR on the island of Tenerife.

Keywords. Sun: magnetic fields, polarization, line: profiles, radiative transfer, methods: numerical.

1. Introduction

Many observations of solar Stokes profiles show asymmetries that can be well explained by depth gradients in the line-of-sight velocity (Auer & Heasley, 1978 and Landolfi & Landi Degl'Innocenti, 1996). Fourier Transform Spectrometer (FTS) observations at disk center (Stenflo *et al.*, 1984) have shown that Stokes V profiles have larger blue lobes than red, indicating the presence of such gradients, if we exclude other explanation related to non-LTE effects (Solanki, 1986; Pantellini *et al.*, 1988). Sanchez, Almeida & Lites (1992) showed that the observations can be reproduced by postulating sufficiently large vertical velocity gradients. We present an extension of the Unno-Rachkovsky solution (Landi Degl'Innocenti & Landolfi, 2004) that provides the theoretical Stokes profiles taking into account a vertical velocity gradient. Thus, the theoretical profiles may display asymmetries as do the observed profiles, which facilitates the inversion based on the Unno theory, and leads to the additional determination of the vertical velocity gradient.

2. The Unno theory modified for velocity gradients

We present a modelling involving a flow inside a magnetic element with a gradient along the line of sight to reproduce the observed asymmetry. The comparison between the modelling and observations leads us to call in question again the hypothesis of a stationary flow inside the magnetic element. The first reason is that the stationary flow doesn't describe the behavior of the mean slope bisector that is proportional to the velocity gradients. Secondly, Ribes *et al.* (1985) as well as Solanki & Pahkle (1988) show that the calculated profiles fail completely to match the observations especially in the

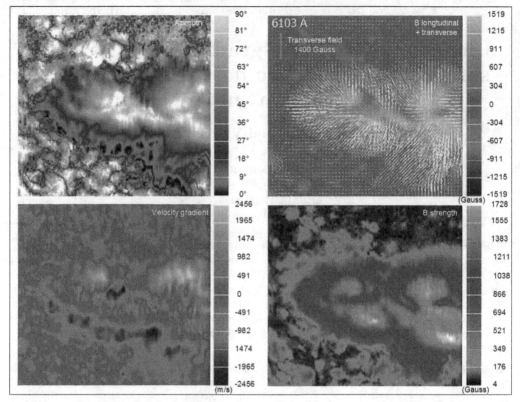

Figure 1. UNNOFIT inversion code applied on sunspot observed the 20th, August 2008 at THEMIS for ion Ca I 6103 Å. Top left: horizontality of the field vector; Top right: The longitudinal magnetic field component displayed in color and the tranverse magnetic field in dashed lines (Gauss). Bottom left: The additional determination of the vertical velocity gradients is displayed and expressed in ms^{-1}. Bottom right: The magnetic field derived in terms of local average field strength (Gauss).

comparaison of lines with different stengths and excitation potentials. A systematical comparison between the modelling and observations leads us to allow to the velocity gradient a wavelength distribution function proportional to the profile itself in order to keep the linear behavior of the observed mean bisectors as noticed by J. Rayrole (Molodij & Rayrole, 2006). We have implemented this extension in the UNNOFIT Milne-Eddington inversion code. We propose the following modification of the theory.

We modified the absorption coefficient entering the Unno-Rachkovsky formalism, Unno (1956), Rachkovsky (1961). To generalize the transfer equations to account for the magnetic field splitting in the presence of a velocity field gradient, we propose the following modification of the quantities $\eta_{p,l,r}$ that denote the ratio between the line absorption and the continuous absorption for Zeeman triplets.

Let be v_r the radial velocity at the line center formation depth and v_h the Zeeman shift expressed both in Doppler width unit to match to the different observed line profiles:

$$\begin{cases} v_r = \alpha \dfrac{v_s}{\xi} \\ v_h = \dfrac{4,67.10^{-2}\lambda^2 \bar{g} H}{\xi} \end{cases} \qquad (2.1)$$

where v_s is the velocity expressed in ms^{-1}, $\bar{g}$ is the effective Landé factor, α a constant to convert the velocity in Doppler width unity ξ. Introducing the velocity gradient δ_V

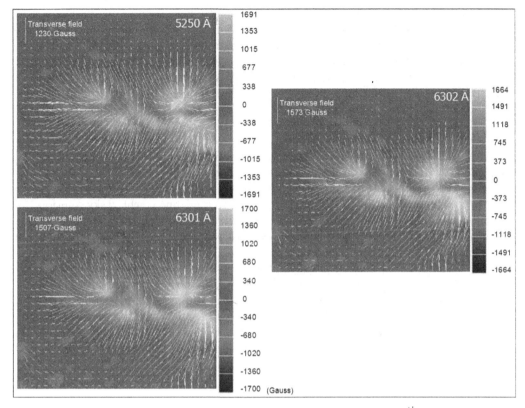

Figure 2. UNNOFIT inversion code applied on sunspot observed the 20th, August 2008 at THEMIS for ions Fe I 6301 Å, Fe I 6302 Å and 5250 Åsimultaneously observed. We obtain a tomography of the 3D shape of the magnetic field depending on the height of formation of the different lines of the solar atmosphere.

(indeed, velocity difference between the line center and far wings formation depths), one obtains the absorption coefficients $\eta_{p,l,r}$ respectively for each of the π, σ_+ and σ_- components:

$$
\left\{
\begin{aligned}
\eta_p &= \eta_0 \, e^{-\left(\frac{\lambda-\lambda_0}{\xi}+v_r+\delta V_p\right)^2} \\
\eta_l &= \eta_0 \, e^{-\left(\frac{\lambda-\lambda_0}{\xi}-v_h+v_r+\delta V_l\right)^2} \\
\eta_r &= \eta_0 \, e^{-\left(\frac{\lambda-\lambda_0}{\xi}+v_h+v_r+\delta V_r\right)^2}
\end{aligned}
\right.
\tag{2.2}
$$

with:

$$
\left\{
\begin{aligned}
\delta V_p &= \frac{\delta_V}{\xi} \, e^{-\left(\frac{\lambda-\lambda_0}{\xi}+v_r\right)^2} \\
\delta V_l &= \frac{\delta_V}{\xi} \, e^{-\left(\frac{\lambda-\lambda_0}{\xi}-v_h+v_r\right)^2} \\
\delta V_r &= \frac{\delta_V}{\xi} \, e^{-\left(\frac{\lambda-\lambda_0}{\xi}+v_h+v_r\right)^2}
\end{aligned}
\right.
\tag{2.3}
$$

3. UNNOFIT inversion icluding velocity gradients

We inverted a spectropolarimetric scan of a sun spot region achieved with THEMIS on 20, August 2008 in the line Ca I 6103 Å. We use the UNNOFIT code of Bommier *et al.* (2007) improved by introducing the velocity gradient parameter. The tests run show that the inversion is faster and reproduce successfully asymmetries modelled with the velocity gradients assumption. Figure 1 displays the magnetic field solution of the UNNOFIT procedure. The field vector is drawn in terms of longitudinal (in colors) and transverse (in dashes) components. These components are expressed in the line-of-sight and plane of the sky coordinates. The magnetic field can be derived in terms of local average field strength and horizontality of the field vector. We show an additional determination of the vertical velocity gradients.

4. Conclusion

We have performed UNNOFIT inversion on spectropolarimetric data obtained for CaI 6103 Å on a sunspot. UNNOFIT is an inversion code (Landolfi *et al.*, 1984) that includes the magneto-optical and damping effects (Landolfi & Landi Degl'Innocenti, 1982) and that is based on the Marquardt algorithm applied to the Unno-Rachkowsky solution and modified to take into account the velocity gradient for the Stokes parameters emerging from a Milne-Eddington atmosphere. UNNOFIT was complemented by introducing a two-component atmosphere, having a magnetic and a non-magnetic component (Bommier *et al.*, 2007). We modified the absorption coefficient entering the Unno-Rachkovsky formalism in order to derive the theoretical profiles. The theoretical profiles display asymmetries as do the observed profiles, which facilitates the inversion based on the Unno-Rachkovsky theory, and leads to the additional determination of the vertical velocity gradient. An interest of the present work us to provide 3D plot of the local average magnetic field for simultaneously observed lines as displayed in figure 2.

References

Auer,L. & Heasley, J., 1978, *A & A*, 64,67

Bommier, V., Landi Degl'Innocenti, E., Landolfi, M., & Molodij, G., 2007, *A & A*, 464, 323

Landi Degl'Innocenti, E., & Landolfi, M. 2004, Polarization in Spectral Lines (Kluwer Academ. Publ., Dordrecht)

Landolfi, M. & Landi Degl'Innocenti, E., 1982, *Sol.Phys.*, 78, 355

Landolfi, M., Landi Degl'Innocenti, E., & Arena, P. 1984, *Sol.Phys.*, 93, 269

Landolfi, M. & Landi Degl'Innocenti, E., 1996, *Sol.Phys.*, 164, 191

Molodij, G. & Rayrole, J., 2006, (in R. Casini and B.W. Lites eds.), ASPW4, 358, 132

Rachkovsky D. N., 1961, *Izv. Crim. Astrphys. Obs.*, 25, 277.

Ribes E., Rees, D. E., & Fang Ch; 1985, *Astrophy. J.*, 296, 268.

Sánchez Almeida, J. & Lites, B. W., 1992, *ApJ*, 398, 359

Solanki S. K., 1986, *A & A*, 168, 311

Solanki, S. W. & Pahkle, K. D., 1988, *A & A*, 201, 143

Stenflo,J.O., Harvey, J. W., Brault, J. W., Solanki, S. K.,1984, *A & A*, 131,333

Pantellini, F. G. E., Solanki, S. K., & Stenflo, J. O., 1988, *A & A*, 189,263

Unno, W., 1956, *Publ. Astr. Soc. Japan*, 8, 108.

Advances in Plasma Astrophysics
Proceedings IAU Symposium No. 274, 2010
A. Bonanno, E. de Gouveia Dal Pino, & A. G. Kosovichev, eds.
© International Astronomical Union 2011
doi:10.1017/S1743921311007162

Solar wind turbulence: Advances in observations and theory

J. J. Podesta[1]

[1]Los Alamos National Laboratory, Los Alamos, NM, 87545, USA
email: jpodesta@solar.stanford.edu

Abstract. Observations of plasma and magnetic field fluctuations in the solar wind provide a valuable source of information for the study of turbulence in collisionless astrophysical plasmas. Scientific data collected by various spacecraft over the last few decades has fueled steady progress in this field. Theoretical models, numerical simulations, and comparisons between theory and experiment have also contributed greatly to these advances. This review highlights some recent advances on the observational side including measurements of the anisotropy of inertial range fluctuations as revealed by the different scaling laws parallel and perpendicular to the mean magnetic field, measurements of the normalized cross-helicity spanning the entire inertial range which demonstrate that this quantity is scale invariant, and improved measurements of the spectrum of magnetic field fluctuations in the dissipation range that show a spectral break near the lengthscale of the electron gyro-radius. The theoretical implications of these results and comparisons between theory and observations are briefly summarized.

Keywords. Solar wind, turbulence, interplanetary medium, waves

1. Introduction

Turbulence is found in a wide variety of laboratory and astrophysical plasmas and MHD turbulence with high kinetic and magnetic Reynolds numbers is believed to be common in many astrophysical systems. The solar wind is one of the only such systems in which the turbulence can be measured in exquisite detail by means of spacecraft instrumentation and such measurements provide a solid foundation for developing an understanding of this fundamental physical process. The study of solar wind turbulence, both in theory and observations, has advanced steadily over the past few decades. In this review, I shall briefly summarize recent advances in three specific areas in which I have played a role including measurements of the normalized cross-helicity spectrum spanning the entire inertial range, measurements of the anisotropy of power spectra parallel and perpendicular to the mean magnetic field, and measurements of the spectrum of magnetic field fluctuations in the dissipation range.

2. Energy spectrum and cross-helicity spectrum

Some researchers believe that the energy spectrum of driven, steady state, homogeneous, incompressible MHD turbulence cannot be characterized by a *universal* power-law (Lee *et al.* 2010), contrary to hydrodynamic turbulence in non-conducting fluids which experimental measurements indicate has a universal scaling exponent of 5/3 (Sreenivasan & Dhruva 1998; Pope 2000). For MHD turbulence there is another school of thought that suggests there *is* a universal scaling with a perpendicular energy spectrum that scales like $k_\perp^{-3/2}$ when the mean magnetic field is strong, that is, when $(\delta B/B_0)^2 \ll 1$, where δB is the rms amplitude of the fluctuations at the outer scale or the largest inertial range scale and B_0 is the amplitude of the ambient magnetic field (Müller *et al.* 2003; Boldyrev 2005, 2006; Mason *et al.* 2006, 2008; Müller 2009; Perez & Boldyrev 2010). In a recent study of solar wind turbulence at 1 AU, Podesta & Borovsky (2010) have shown that the inequality $(\delta B/B_0)^2 \ll 1$ is usually well satisfied and that the scaling law for the total energy (kinetic plus magnetic) is usually closer to 3/2 than 5/3, in agreement with these theoretical predictions. In particular, for highly Alfvénic wind, wind with high normalized cross-helicity, the spectral index for the total energy is found to be 1.540 ± 0.033 (Podesta & Borovsky 2010). Highly Alfvénic wind provides the best example of MHD turbulence because it has undergone relatively little dissipation during its transit time from the sun so that, in the absence of continuous *in situ* forcing, its characteristics more closely resemble the turbulence generated closer to the sun. The analysis of corotating interaction regions (CIRs) has shown that large scale velocity gradients (shear) do not provide *in situ* forcing of solar wind turbulence near 1 AU as was previously thought because the turbulence amplitude is not observed to rise in the vicinity of the shear zone (Borovsky & Denton 2010).

The study by Podesta & Borovsky (2010) is the first large statistical study of energy and cross-helicity spectra spanning the entire inertial range at 1 AU. The measurements show that the total energy spectrum and the cross-helicity spectrum typically have the same spectral index and that the normalized cross-helicity σ_c, the ratio of the cross-helicity spectrum to the total energy spectrum, is approximately constant throughout the inertial range. That is, σ_c is independent of wavenumber. Compared to previous measurements of σ_c by Tu *et al.* (1989) and Marsch & Tu (1990), the measurements by Podesta & Borovsky (2010) increase the wavenumber range of the observations by more than one decade and show that the scale invariance of σ_c covers the entire inertial range at 1 AU. This result is of significant importance for theories of solar wind turbulence and turbulence in astrophysical plasmas. It implies that the ratio of the two Elsasser spectra, the Elsasser ratio, is a constant, independent of wavenumber. This, combined with Kolmogorov's equations expressing the constancy of energy flux in k-space, implies that the ratio of the energy cascade times of the two Elsasser energies τ^+/τ^- is also constant, independent of wavenumber. These ideas are often incorporated into theoretical models and now, for the first time, they have direct observational support.

3. Anisotropy of the magnetic field spectrum

It is well known that fluctuations in turbulent magnetized plasmas are anisotropic with respect to the direction of the local magnetic field B_0: the fluctuations vary much more rapidly in the direction perpendicular to B_0 than in the direction parallel to B_0. This has been observed in laboratory plasma experiments since the 1960s and in computer simulations since the 1970s or 1980s. However, it took many years before a practical turbulence phenomenology emerged that can account for this anisotropy in a simple way. The most influential phenomenological theory of this kind is the theory of incompressible MHD turbulence developed by Goldreich & Sridhar (1995, 1997) which has been reviewed by Goldreich (2001) and Sridhar (2010). The theory of Goldreich & Sridhar assumes that the eddy turnover time is approximately equal to the Alfvén wave period so that

$$\frac{1}{k_\perp \delta v_\perp} \simeq \frac{1}{k_\parallel v_A} = \frac{1}{\omega_A}. \tag{3.1}$$

In the inertial range, the energy flux in $k_\perp$-space is constant, $(\delta v_\perp)^2/\tau = \varepsilon$, and using (3.1) this implies $(\delta v_\perp)^2 = \varepsilon^{2/3} k_\perp^{-2/3}$. Therefore, the perpendicular energy spectrum

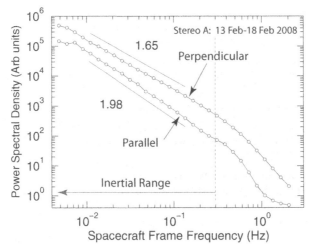

Figure 1. Example of power spectra measured perpendicular and parallel to the local mean magnetic field using wavelets (Podesta 2009). The best fit lines, in red, are offset for easier viewing and the measured inertial range slopes are 1.65 and 1.98.

defined by $k_\perp E(k_\perp) = (\delta v_\perp)^2$ takes the form

$$E(k_\perp) = C_0 \varepsilon^{2/3} k_\perp^{-5/3}, \tag{3.2}$$

where C_0 is a dimensionless constant of order unity. For wavenumbers $k_\perp$ and $k_\parallel$ related by (3.1), one may define a parallel energy spectrum such that $k_\parallel E(k_\parallel) = k_\perp E(k_\perp)$. This yields

$$E(k_\parallel) \propto k_\parallel^{-2}. \tag{3.3}$$

Thus, in the Goldreich & Sridhar theory the energy spectrum is proportional to $k_\perp^{-5/3}$ in the direction perpendicular to the mean field and to $k_\perp^{-2}$ in the direction parallel to the mean field. A different theory developed by Boldyrev (2005, 2006) predicts spectra proportional to $k_\perp^{-3/2}$ and $k_\parallel^{-2}$ in the directions perpendicular and parallel to B_0, respectively,

Motivated by these theoretical results, it is of interest to examine solar wind fluctuations to see if the observed energy spectrum exhibits different power-law behaviors in the directions perpendicular and parallel to the mean field. The first successful measurements of this kind were performed by Horbury *et al.* (2008) and Podesta (2009) who found that the spectral index of magnetic field fluctuations changes from roughly 5/3 for measurements perpendicular to the local mean magnetic field B_0 to approximately 2 for measurements parallel to B_0 (see also Podesta 2010). Horbury *et al.* (2008) analyzed high speed solar wind data from the *Ulysses* spacecraft taken above the poles of the sun at heliocentric distances near 1.4 AU and Podesta (2009) analyzed high speed streams in the ecliptic plane at 1 AU using data from the two *Stereo* spacecraft. These studies have been confirmed and expanded by Luo & Wu (2010) and by Wicks *et al.* (2010). An example of the anisotropy measurements is shown in Figure 1.

While the results in Figure-1 appear to agree with the scaling laws in the Goldreich & Sridhar theory, it is premature to conclude that the Goldreich & Sridhar scaling provides the best agreement with solar wind data. The solar wind measurements performed so far have only considered the magnetic field spectrum because of the availability of high cadence magnetic field data needed to resolve the smaller scale inertial range fluctuations. However, theories are generally based on the total energy spectrum—kinetic plus magnetic—which has typical spectral slopes between 1.5 and 1.6 in the solar wind and slopes closer to 3/2 for highly Alfvénic wind (Podesta & Borovsky 2010). Therefore, it is expected that measurements of the scaling laws for the total energy perpendicular and parallel to B_0 will yield results in better agreement with the Boldyrev scaling. These measurements will be performed in the near future.

It is important to emphasize that the Goldreich & Sridhar theory and the Boldyrev theory only apply to turbulence with vanishing cross-helicity and, therefore, these theories must be generalized to turbulence with non-vanishing cross-helicity before they can properly be applied to solar wind turbulence.

How do we measure the scaling laws in solar wind turbulence and what exactly is being measured? A single spacecraft records spatial variations of the fields along a line parallel to the average solar wind velocity. Consequently, the idea is to make measurements when the mean magnetic field B_0 is directed either parallel or perpendicular to the flow.

The measured quantity is the mean square magnetic field $\langle |B(x+r) - B(x)|^2 \rangle$, where the displacement vector r is always parallel to the solar wind flow velocity. Therefore, when B_0 is perpendicular to the mean flow, the displacement r is perpendicular to B_0 and one measures the energy in the plane perpendicular to B_0; when B_0 is parallel to the mean flow, the displacement r is parallel to B_0 and one measures the energy in the direction parallel to B_0. The technique just described can also be used to make measurements at any angle θ with respect to B_0, not just $\theta = 0$ (parallel) and $\theta = \pi/2$ (perpendicular). Note that the studies by Horbury et al. (2008) and Podesta (2009) employed wavelet analysis to effect the decomposition in time and scale, however, the analysis can also be performed using second order structure functions, as just described (Luo & Wu 2010).

To be able to measure differences in the scaling exponents for fluctuations parallel and perpendicular to B_0 it is important to use the *local* mean magnetic field (short time average) rather than the *global* mean magnetic field (long time average) and also to use sufficiently high cadence magnetic field data to obtain good statistics. The use of the *local* mean field is crucial because the dynamics of turbulent eddies with a given lengthscale $\lambda_\perp$ are most sensitive to the local mean magnetic field at around the same lengthscale. The necessity of using the local rather than the global mean field was shown by Cho & Vishniac (2000) who analyzed the anisotropy of turbulent eddies in simulations of three-dimensional incompressible MHD turbulence.

Soon after the solar wind studies by Horbury et al. (2008) and Podesta (2009), an independent study based on a different analysis technique was performed by Tessein et al. (2009) who found no change in the spectral indices parallel and perpendicular to B_0. Unfortunately, this negative result cast doubt on the previously obtained results. This negative result is partly a consequence of the fact that the study by Tessein et al. (2009) used a 1 hr average for the mean magnetic field B_0 (long time average) rather than the *local* mean magnetic field. It has since been shown that if the same analysis technique employed by Tessein et al. (2009) is applied to shorter time intervals, 7.5 min instead of 1 hr, and if higher cadence data is used, 1 sec instead of 64 sec, then the change in the spectral index from approximately 5/3 in the perpendicular direction to 2 in the parallel direction is also seen using their technique (unpublished work by various groups).

Unfortunately, many other important studies cannot be discussed here due to space restrictions. For example, the work of Hnat & Chapman (2007) and of Narita et al. (2010a, 2010b).

4. Spectral break and Dissipation range

The spectrum of solar wind magnetic field fluctuations exhibits a spectral break, a steepening of the spectral slope, that marks the transition from the inertial range at large MHD scales to the dissipation range at kinetic scales (Leamon *et al.* 1998, 1999, Bale *et al.* 2005). The spectral break usually occurs around $1/2$ Hz in the spacecraft frame which, by Taylor's hypothesis, is equivalent to wavenumbers of order $k_\perp \rho_i \sim 1$ or $k_\parallel d_i \sim 1$, where ρ_i is the proton Larmor radius and $d_i = c/\omega_{pi}$ is the proton inertial length. Therefore, it should more precisely be referred to as the *proton spectral break* to distinguish it from the *electron spectral break* at higher wavenumbers. The dissipation range, sometimes called the kinetic regime or kinetic range, begins near the proton spectral break where a wide range of kinetic process come into play. These kinetic processes are believed to dissipate the turbulent energy cascade, heat the plasma, and regulate the particle distribution functions although details of the heating process are not completely understood.

Measurements of the magnetic field spectrum extending from the proton spectral break $\sim 1/2$ Hz through the dissipation range to ~ 100 Hz have recently been performed using the search coil magnetometers on board the four *Cluster* spacecraft (Saharoui *et al.* 2009, 2010; Alexandrova *et al.* 2009; Kiyani *et al.* 2009). With a 450 Hz sampling rate in burst mode, these measurements provide unprecedented time and frequency resolution for the investigation of kinetic processes and dissipation range physics. Unfortunately, however, *Cluster* is principally a magnetospheric mission and the four spacecraft spend relatively little time in the solar wind. Moreover, during the brief periods near apogee at 19 R_e when solar wind measurements are possible the spacecraft are close enough to the bow shock that many observed plasma kinetic effects are often caused by the bow shock and have nothing to do with processes inherent to the unobstructed solar wind (Balogh *et al.* 2005, Eastwood *et al.* 2005, Burgess *et al.* 2005). Whether particular intervals of "solar wind" data are devoid of bow shock or foreshock effects is difficult to determine with certainty and this is a serious concern that should be kept in mind when the data are used for solar wind science.

An example of the magnetic energy spectrum (trace spectrum) obtained from the burst mode data is shown in Figure-2. The data show that the spectrum falls off rapidly at the proton spectral break with a typical spectral slope around 3 or 4, as already known. At higher frequencies the spectrum flattens out and for approximately one decade or more, from roughly 3 Hz to 30 Hz, the spectrum is often well fit by a straight line on a log-log plot indicating a power-law behavior in this range. The spectral slope in this range is near $7/3$, the predicted value for a kinetic Alfvén wave (KAW) cascade and also for a whistler mode cascade, although large variations in this slope from ~ 1.7 to ~ 3 are also seen in the data—possibly caused by electron foreshock effects. Around 40 Hz the spectrum shows clear evidence of a high frequency spectral break first reported by Saharoui *et al.* (2009) and attributed to the onset of collisionless damping at wavenumbers near the electron gyroradius $k_\perp \rho_e \sim 1$ and the electron inertial length $k_\parallel d_e \sim 1$. Since electron physics dominates at these scales, the term *electron spectral break* seems appropriate.

The physical interpretation of these interesting observations is fundamental for understanding solar wind physics. Saharoui *et al.* (2009) interpreted the transition from a spectral slope near $5/3$ in the inertial range to a spectral slope near $7/3$ in the kinetic range as a transition from an Alfvén wave cascade to a KAW cascade at $k_\perp \rho_i \sim 1$ and they attributed the electron spectral break to Landau damping of the KAW cascade at electron scales $k_\perp \rho_e \sim 1$. This interpretation, which seemed to fit the theoretical picture elaborated by Howes *et al.* (2008) and Schekochihin *et al.* (2009), was sharply criticized by Podesta *et al.* (2010) who demonstrated that for typical high-speed solar wind at 1 AU

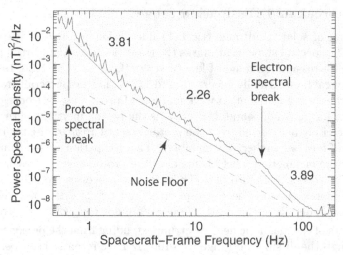

Figure 2. Example of the *Cluster* search coil spectrum, trace spectrum, for a 145.6 s interval on 19 March 2006. The line segments are the linear least squares fits obtained over different frequency ranges and have been displaced downward for easier viewing. The spectral slopes obtained from the fits are also shown.

the damping of KAWs becomes significant for $1 < k_\perp \rho_i < 10$ and, therefore it is likely that the KAW cascade will damp before reaching electron scales where $k_\perp \rho_i \gg 10$. This was also realized by Leamon *et al.* (1999). Thus, the high frequency part of the search coil spectrum, beyond approximately 5 Hz, must be supported by some other types of wave modes—the most natural candidate being electron whistler waves as first suggested by Beinroth & Neubauer (1981) and Denskat *et al.* (1983). The study by Podesta *et al.* (2010) indicates that the damping of KAWs should occur in a narrow range near the proton spectral break at frequencies roughly between 1/2 Hz and 4 Hz in the spacecraft frame, where the most rapid decrease in the observed spectrum occurs. That is where the dominant dissipation of solar wind turbulence is believed to occur. Leamon *et al.* (1999) reached similar conclusions using a different theoretical model.

In light of the study by Podesta *et al.* (2010), Saharoui *et al.* (2010) in their second *Physical Review Letter* changed the physical interpretation given in Saharoui *et al.* (2009) by emphasizing the significant Landau damping of KAWs that is expected to occur in a narrow range near the proton spectral break $k_\perp \rho_i \sim 1$ and by showing that this coincides with the steep drop in the spectrum at that point. Saharoui *et al.* (2010) also made a significant contribution to the subject by using the wave telescope technique to determine the observed dispersion relation of the waves as a function of the propagation angle in the wavenumber range $0.1 < k_\perp \rho_i < 2$. The observations were shown to be in reasonable agreement with the linear Vlasov-Maxwell dispersion relation for KAWs. The analysis also showed that the propagation direction of the waves was nearly perpendicular to the mean magnetic field B_0 as predicted by anisotropic turbulence phenomenologies. Thus, Saharoui *et al.* (2010) provided new evidence to corroborate that KAWs are the energetically dominant wave mode in the vicinity of the proton spectral break and that the damping of these waves is the dominant dissipation mechanism for solar wind turbulence.

References

Alexandrova, O., Saur, J., Lacombe, C., Mangeney, A., Mitchell, J., Schwartz, S. J., & Robert, P. 2009, *Phys. Rev. Lett.*, 103, 165003

Bale, S. D., Kellogg, P. J., Mozer, F. S., Horbury, T. S., & Reme, H. 2005, *Phys. Rev. Lett.*, 94, 215002

Balogh, A., *et al.* 2005, *Space Sci. Rev.*, 118, 155

Beinroth, H. J., & Neubauer, F. M. 1981, *J. Geophys. Res.*, 86, 7755

Boldyrev, S. 2005, *Astrophys. J.*, 626, L37

—. 2006, *Phys. Rev. Lett.*, 96, 115002

Borovsky, J. E., & Denton, M. H. 2010, *J. Geophys. Res.* (Space Physics), 115, A10101

Burgess, D., *et al.* 2005, *Space Sci. Rev.*, 118, 205

Cho, J., & Vishniac, E. T. 2000, *Astrophys. J.*, 539, 273

Denskat, K. U., Beinroth, H. J., & Neubauer, F. M. 1983, *J. Geophysics Zeitschrift Geophysik*, 54, 60

Eastwood, J. P., Lucek, E. A., Mazelle, C., Meziane, K., Narita, Y., Pickett, J., & Treumann, R. A. 2005, *Space Sci. Rev.*, 118, 41

Goldreich, P. 2001, *Astrophys. Space Sci.*, 278, 17

Goldreich, P., & Sridhar, S. 1995, *Astrophys. J.*, 438, 763

—. 1997, *Astrophys. J.*, 485, 680

Horbury, T. S., Forman, M., & Oughton, S. 2008, *Phys. Rev. Lett.*, 101, 175005

Howes, G. G., Cowley, S. C., Dorland, W., Hammett, G. W., Quataert, E., & Schekochihin, A. A. 2008, *J. Geophys. Res.*, 113, 5103

Kiyani, K. H., Chapman, S. C., Khotyaintsev, Y. V., Dunlop, M. W., & Sahraoui, F. 2009, *Phys. Rev. Lett.*, 103, 075006

Leamon, R. J., Smith, C. W., Ness, N. F., Matthaeus, W. H., & Wong, H. K. 1998, *J. Geophys. Res.*, 103, 4775

Leamon, R. J., Smith, C. W., Ness, N. F., & Wong, H. K. 1999, *J. Geophys. Res.*, 104, 22331

Lee, E., Brachet, M. E., Pouquet, A., Mininni, P. D., & Rosenberg, D. 2010, *Phys. Rev. E*, 81, 016318

Luo, Q. Y., & Wu, D. J. 2010, *Astrophys. J.*, 714, L138

Mason, J., Cattaneo, F., & Boldyrev, S. 2006, *Phys. Rev. Lett.*, 97, 255002

—. 2008, *Phys. Rev. E.*, 77, 036403

Müller, W. 2009, in Lecture Notes in Physics, Berlin Springer Verlag, Vol. 756, Interdisciplinary Aspects of Turbulence, ed. W. Hillebrandt & F. Kupka, 223–254

Müller, W.-C., Biskamp, D., & Grappin, R. 2003, *Phys. Rev. E.*, 67, 066302

Narita, Y., Glassmeier, K., Sahraoui, F., & Goldstein, M. L. 2010a, *Phys. Rev. Lett.*, 104, 171101

Narita, Y., Sahraoui, F., Goldstein, M. L., & Glassmeier, K. 2010b, *J. Geophys. Res. A*, 115, 4101

Perez, J. C. & Boldyrev, S. 2010a, *Astrophys. J.*, 710, L63

—. 2010b, *Phys. Plasmas*, 17, 055903

Podesta, J. J. 2009, *Astrophys. J.*, 698, 986

Podesta, J. J. 2010, in AIP Conference Series, Vol. 1216, Twelfth International Solar Wind Conference, ed. M. Maksimovic, K. Issautier, N. Meyer-Vernet, M. Moncuquet & F. Pantellini, 128–131

Podesta, J. J. & Borovsky, J. E. 2010, *Phys. Plasmas* (in press)

Podesta, J. J., Borovsky, J. E., & Gary, S. P. 2010, *Astrophys. J.*, 712, 685

Pope, S. B. 2000, Turbulent Flows (Cambridge University Press)

Sahraoui, F., Goldstein, M. L., Belmont, G., Canu, P., & Rezeau, L. 2010, *Phys. Rev. Lett.*, 105, 131101

Sahraoui, F., Goldstein, M. L., Robert, P., & Khotyaintsev, Y. V. 2009, *Phys. Rev. Lett.*, 102, 231102

Schekochihin, A. A., Cowley, S. C., Dorland, W., Hammett, G. W., Howes, G. G., Quataert, E., & Tatsuno, T. 2009, *Astrophys. J. Suppl.*, 182, 310

Sreenivasan, K. R., & Dhruva, B. 1998, *Progress of Theoretical Physics Supplement*, 130, 103

Tessein, J. A., Smith, C. W., MacBride, B. T., Matthaeus, W. H., Forman, M. A., & Borovsky, J. E. 2009, *Astrophys. J.*, 692, 684

Wicks, R. T., Horbury, T. S., Chen, C. H. K., & Schekochihin, A. A. 2010, *Mon. Not. R. Astron. Soc.*, 407, L31

Advances in Plasma Astrophysics
Proceedings IAU Symposium No. 274, 2010
A. Bonanno, E. de Gouveia Dal Pino & A.G. Kosovichev, eds.
© International Astronomical Union 2011
doi:10.1017/S1743921311007174

"Ambipolar diffusion" and magnetic reconnection

Yuriy T. Tsap[1,2] and Alexander V. Stepanov[2]

[1] Crimean Astrophysical Observatory,
Nauchny, Crimea, 98409, Ukraine
email: yur@crao.crimea.ua

[2] Central (Pulkovo) Astronomical Observatory,
St.Petersburg, 196140, Russia
email: stepanov@gao.spb.ru

Abstract. Based on the three-fluid approximation the influence of the neutral component of hydrogen plasma on Joule dissipation of electric currents are considered. As distinguished from Mestel & Spitzer (1956) and Parker (1963) it has been shown that the magnetic flux may be not conserved in the case of the "ambipolar diffusion" due to collisions between ions and neutrals. This is explained by the ion acceleration under the action of Ampere's force. Joule dissipation is determined by electron and ion collisions in a partially ionized plasma. Plasma evacuation from current sheets is the effective mechanism of its cooling. Thickness of a current sheet can achieve up to hundreds of kilometers in the solar chromosphere. The origin of the solar chromospheric jets observed with the Hinode satellite are discussed.

Keywords. plasmas, magnetic fields, diffusion, stars: formation, Sun: chromosphere

---

1. Introduction

Magnetic field diffusion and dissipation in a partially ionized plasma were considered in detail more than a half of century ago by Piddington (1954), Mestel & Spitzer (1956), and Cowling (1957). These processes play an important role in many space phenomena. Since different theoretical approaches are used up to now (e.g., Zaitsev & Stepanov 1992, Zweibel & Brandenburg 1997), this problem should be studied additionally.

Mestel & Spitzer (1956) studied star formation from the magnetic dust cloud for the first time. It was pointed out that the cloud cannot break up into fragments of mass less than $500M_\odot$ due to the magnetic flux conservation. However, this difficulty can be remedied if we take into account the freezing of magnetic field lines into the charged particles. According to the proposed mechanism named the "ambipolar diffusion" (e.g., Parker 1963; Spitzer 1978; Zweibel & Brandenburg 1997), the force of magnetic tension, arising due to the gravitational contraction, can drive the ionized matter and frozen–in magnetic field lines through a "gas" of neutrals thereby decoupling the field from the neutral matter. This suggests that the "ambipolar diffusion" does not alter the total magnetic flux, it simply redistributed the flux within the plasma. Consequently, the proposed mechanism cannot directly result in magnetic reconnection.

In contrast to Mestel–Spitzer's approach, Cowling (1957) turned to the generalized Ohm's law. He showed that the electrical conductivity (Cowling conductivity) of the non–stationary plasma can be significantly decreased owing to the ion acceleration by Ampere's force. Collisions between ions and neutral particles become very effective in view of the high ion velocities. As a result, the magnetic flux is not conserved and the rate of magnetic reconnection might be considerably increased because of Joule (Ohmic) dissipation.

The influence of Mestel–Spitzer's "diffusion" on magnetic reconnection was considered by many authors (e.g., Parker 1963; Zweibel 1989; Vishniac & Lazarian 1999). On the other hand, Tsap (1994) and Ni *et al.* (2007) proceeded from the generalized Ohm's law but they did not take into account the plasma evacuation in the energy balance of a current sheet.

2. Joule dissipation in a partially ionized plasma

Using the standard notation, the simplified momentum equations for electrons (e), ions (i), and neutrals (n) of hydrogen plasma can be written as

$$n_e m \frac{d\mathbf{V}_e}{dt} = -en_e\mathbf{E} - \frac{en}{c}\mathbf{V}_e \times \mathbf{B} + n_e m\nu_{ei}(\mathbf{V}_i - \mathbf{V}_e) + n_e m\nu_{en}(\mathbf{V}_n - \mathbf{V}_e); \quad (2.1)$$

$$n_i M \frac{d\mathbf{V}_i}{dt} = en_i\mathbf{E} + \frac{en_i}{c}\mathbf{V}_i \times \mathbf{B} + n_i M\nu_{in}(\mathbf{V}_n - \mathbf{V}_i) + n_i M\nu_{ie}(\mathbf{V}_e - \mathbf{V}_i); \quad (2.2)$$

$$n_n M \frac{d\mathbf{V}_n}{dt} = n_n M\nu_{ni}(\mathbf{V}_i - \mathbf{V}_n) + n_n M\nu_{ne}(\mathbf{V}_e - \mathbf{V}_n). \quad (2.3)$$

It will be noted that adding termwise (2.1)–(2.3) and taking into account that $n_i = n_e = n$, introducing the velocity of plasma with density $\rho = M(n + n_n) = \rho_i + \rho_n$ as whole

$$\mathbf{v} = \frac{n\mathbf{V}_i + n_n\mathbf{V}_n}{n + n_n}, \quad \mathbf{v_i} = \mathbf{V_i} - \mathbf{v}, \quad \mathbf{v_n} = \mathbf{V_n} - \mathbf{v}, \quad (2.4)$$

at $|\mathbf{v}| \gg |\mathbf{v}_i|, |\mathbf{v}_n|$ we get the MHD momentum equation

$$\rho \frac{d\mathbf{v}}{dt} = \frac{\mathbf{j} \times \mathbf{B}}{c}. \quad (2.5)$$

In the case of weakly ionized plasma, when the number density of neutrals $n_n \gg n$, neglecting by collisions with electrons and inertial terms, equations (2.1) and (2.2) yield

$$\mathbf{V}_i = \mathbf{V}_n + \frac{\mathbf{j} \times \mathbf{B}}{nM\nu_{in}c}, \quad (2.6)$$

where the electric current $\mathbf{j} = en(\mathbf{V}_i - \mathbf{V}_e)$. Equation (2.6) describes the ion drift under action of Ampere's force, i.e., the "ambipolar diffusion". This term becomes widely used probably due to Parker (1963) (see also Nakano *et al.* 2002). However, in spite of the charge separation, the discussed process has not any relation to the classical (real) ambipolar *diffusion*, which are caused by the density gradient in the collisional inhomogeneous plasma.

Mestel & Spitzer (1956) suggested that magnetic field lines are frozen into electrons and ions and the magnetic flux is conserved. Therefore the "ambipolar diffusion" did not result in Joule dissipation (see also Parker 1963; Shu *et al.* 1987; Zweibel & Brandenburg 1997). In our opinion, such approach is not quite correct because of the following reasons.

Multiplying equations (2.1), (2.2), and (2.3) by $\mathbf{V}_e$, $\mathbf{V}_i$, and $\mathbf{V}_n$, respectively, using equation (2.4), after some algebra, we find that the electric energy is

$$\mathbf{jE} = \rho \frac{d\mathbf{v}^2}{2dt} + nM\nu_{in}(\mathbf{V}_n - \mathbf{V}_i)^2 + nm\nu_{ei}(\mathbf{V}_i - \mathbf{V}_e)^2 + n_n m\nu_{ni}(\mathbf{V}_n - \mathbf{V}_e)^2. \quad (2.7)$$

As is easily seen from equation (2.7), the energy of magnetic field is transformed into the kinetic and thermal ones associated with electron and ion collisions. Substituting (2.5) into (2.7), we get

$$\mathbf{j}\mathbf{E} = (\mathbf{j} \times \mathbf{B})\mathbf{v}/c + Q,$$

where Joule dissipation is

$$Q = \mathbf{E}^*\mathbf{j} = nM\nu_{in}(\mathbf{V}_n - \mathbf{V}_i)^2 + nm\nu_{ei}(\mathbf{V}_i - \mathbf{V}_e)^2 + n_n m\nu_{ni}(\mathbf{V}_n - \mathbf{V}_e)^2, \qquad (2.8)$$

and $\mathbf{E}^* = \mathbf{E} + \mathbf{v} \times \mathbf{B}/c$. As follows from (2.8), Joule dissipation is a work of the electric field on the electric current without the mechanical energy and it equals the sum of terms describing collisions between particles. Meanwhile, Mestel & Spitzer (1956) (see also Parker 1963) judged that Joule dissipation is only caused by electron collisions and did not take into consideration the ion–neutral ones.

Following Cowling (1957) (see also Zaitsev & Stepanov 1992), introducing the neutral density fraction $F = n_n/(n+n_n)$, it is easy to show from (2.1)–(2.3) that the generalized Ohm's law at $\mathbf{j} \perp \mathbf{B}$ can be represent in terms of the Cowling (σ_C) and Spitzer (σ_S) conductivities in the form (Tsap 1994)

$$\left(\mathbf{E} + \frac{1}{c}\mathbf{v} \times \mathbf{B}\right)_\perp = \left(\frac{1}{\sigma_C} + \frac{1}{\sigma_S}\right)\mathbf{j}_\perp, \quad \sigma_C = \frac{c^2 n_n M\nu_{ni}}{F^2 B^2}, \quad \sigma_S = \frac{ne^2}{m(\nu_{ei} + \nu_{en})}. \qquad (2.9)$$

Note that the value of F is arbitrary and the Cowling conductivity σ_C describes the energy release caused by the ion–neutral collisions.

3. Sweet–Parker model, Cowling conductivity, and solar chromospheric jets

Scaling laws, following from equations of continuity, equilibrium, and magnetic diffusion, at $\sigma_C \ll \sigma_S$ can be written as

$$\rho L v = \rho_o l v_o; \quad \frac{1}{2}\rho_o v_o^2 = p_o - p; \quad \frac{B^2}{8\pi} + p = p_o; \quad vB = \eta_C \frac{B}{l}; \qquad (3.1)$$

where the lower index (o) denotes parameters inside a current sheet, L and l are the half width and half thickness of a current sheet, respectively, v and v_0 are the inflow and outflow velocities, $\eta_C = c^2/(4\pi\sigma_C)$ is the diffusion coefficient.

Combining equations (3.1), we obtain the rate of magnetic reconnection

$$v \approx \left(\frac{c^2 v_A}{4\pi\sigma_C L}\right)^{1/2},$$

which is in agreement with results of Vishniac & Lazarian (1999), Tsap (1994), and Ni et al. (2007).

As follows from scaling laws (3.1) and Ampere's law the Joule dissipation rate is

$$Q = \frac{j^2}{\sigma_c} \approx \frac{B^2}{4\pi}\frac{v_A}{L}\frac{\rho_0}{\rho}. \qquad (3.2)$$

Since $n_0 kT_0 \approx B^2/8\pi$, the characteristic Ohmic heating time

$$\tau_h \approx \frac{3n_0 kT_0}{2Q} \approx \frac{3B^2}{16\pi Q} \approx \frac{3}{4}\frac{\rho}{\rho_0}\tau_A. \qquad (3.3)$$

In turn, the characteristic time of the dynamical cooling caused by plasma evacuation is $\tau_e \approx L/v_A$. Joule heating will not ionize plasma within a current sheet until $\tau_e \lesssim \tau_h$ or, as is easy to see from (3.3), $\rho_0 \lesssim \rho$. It should be stressed that the energy loss due to the dynamical cooling significantly exceeds energy losses caused by the plasma radiation and thermal conductivity in the solar chromosphere.

Effective collision frequencies ν_{ei}, ν_{ea}, and ν_{ia} under the condition of the solar atmosphere are

$$\nu_{ei} \approx \frac{60n}{T^{3/2}} \; [s^{-1}], \quad \nu_{en} \approx 5 \cdot 10^{-10} n_n \sqrt{T} \; [s^{-1}], \quad \nu_{in} \approx 10^{-10} n_n \sqrt{T} \; [s^{-1}]. \tag{3.4}$$

Adopting $B = 30$ G, $n_n = 10^{11}$ cm^{-3}, $n = 10^{10}$ cm^{-3}, $T = 10^4$ K, equations (2.9) and (3.4) give $\sigma_C \approx 10^7$ s^{-1} and $\sigma_S \approx 10^{13}$ s^{-1}, i.e. the Cowling conductivity is 6 orders of magnitude less than the Spitzer one. As follows from (3.1), the characteristic half thickness of a current sheet is

$$l \sim \sqrt{\eta_C L / v_A}. \tag{3.5}$$

Taking $L = 10^8$ cm, $\sigma_C = 10^7$ s^{-1}, $v_A = 10^7$ cm/s, from (3.5) we obtain $l \sim 100$ km ($l \sim 100$ m for the Spitzer conductivity). Consequently, the thickness of a current sheet l can achieve hundreds of kilometers in the chromosphere.

Recently thin spicules and chromospheric jets were revealed with the SOT/Hinode telescope in the line Ca II H (De Pontieu et al. 2007; Shibata *et al.* 2007). These magnetic features with the characteristic widthes 100–300 km and $\lesssim 200$ km, respectively are observed outside of solar spots and active regions. To our view, the origin of these phenomena can be connected with the Sweet–Parker magnetic reconnection.

4. Conclusions

(a) There are two types of the ambipolar diffusions: real and formal.

(b) Joule dissipation in a partially ionized plasma is determined by collisions of neutral particles not only with electrons but with ions too.

(c) The "ambipolar diffusion" and the Cowling conductivity describe the same phenomena in the collisional partially ionized plasma.

(d) Plasma evacuation is an effective mechanism of its cooling in a current sheet.

(e) Sweet–Parker reconnection can give rice to the formation of thick (~ 100 km) current sheets in the solar chromosphere.

References

Cowling, T. G. 1957, *Magnetohydrodynamics*, Interscience, New York
De Pontieu, B., McIntosh, S., Hansteen, V. H. et al. 2007, *PASJ*, 59, 655
Mestel, L., & Spitzer, L.Jr. 1956, *MNRAS*, 116, 503
Nakano, T., Nishi, R., & Umebayashi, T. 2002, *ApJ*, 573, 199
Ni, L., Yang, Zh., & Wang, H. 2007, *ApSS*, 312, 139
Parker, E. N. 1963, *ApJS*, 8, 177
Piddington, J. H. 1954, *MNRAS*, 114, 551
Shibata, K., Nakamura, T., Matsumoto, T. et al. 2007, *Science*, 318, 1591
Shu, F. H., Adams, F. C., & Lizano, S., 1987, *ARA&A*, 25, 23
Spitzer, L. Jr. 1978, *Physical Processes in the Interstellar Medium*, J. Wiley & Sons, New York
Tsap, Y. T. 1994, *Astron. Lett.*, 20, 127
Vishniac, E. T., & Lazarian, A. 1999, *ApJ*, 511, 193
Zaitsev, V. V., & Stepanov A. V. 1992, *Solar Phys.*, 139, 343
Zweibel, E. G. 1989, *ApJ*, 340, 550
Zweibel, E. G., & Brandenburg, A. 1997, *ApJ*, 478, 563

Advances in Plasma Astrophysics
Proceedings IAU Symposium No. 274, 2010
A. Bonanno, E. de Gouveia Dal Pino & A.G. Kosovichev, eds.
© International Astronomical Union 2011
doi:10.1017/S1743921311007186

Plasmoid ejections driven by dynamo action underneath a spherical surface

Jörn Warnecke[1,2], Axel Brandenburg[1,2] and Dhrubaditya Mitra[1]

[1]Nordita, AlbaNova University Center,
Roslagstullsbacken 23, SE-10691 Stockholm, Sweden
email: joern@nordita.org

[2]Department of Astronomy, AlbaNova University Center,
Stockholm University, SE 10691 Stockholm, Sweden

Abstract. We present a unified three-dimensional model of the convection zone and upper atmosphere of the Sun in spherical geometry. In this model, magnetic fields, generated by a helically forced dynamo in the convection zone, emerge without the assistance of magnetic buoyancy. We use an isothermal equation of state with gravity and density stratification. Recurrent plasmoid ejections, which rise through the outer atmosphere, is observed. In addition, the current helicity of the small–scale field is transported outwards and form large structures like magnetic clouds.

Keywords. MHD, Sun: magnetic fields, Sun: coronal mass ejections (CMEs), turbulence

1. Introduction

Usually, magnetic phenomena in the atmosphere of the Sun, e.g., the formation of active regions and sunspots, emergence of magnetic fields, and coronal mass ejections, are described in terms of magnetic flux tubes. These flux tubes are supposed to form at the tachocline, rise through the convection zone almost undeformed by the effects of magnetic buoyancy and reach into the photosphere by creating bipolar regions such as sunspots. Above the photosphere twisted magnetic fields are observed to form arch-like structure. However, there is no clear evidence for the existence of magnetic flux tubes inside the deep convection zone. Direct numerical simulations of large-scale dynamos suggest that flux tubes are primarily a feature of the kinematic regime, but tend to be less pronounced in the nonlinear stage (Käpylä *et al.* 2008). In addition, magnetic buoyancy, which is usually believed to be the driver of flux tube emergence, can be compensated sufficiently by downward pumping resulting from the stratification of turbulence intensity in the solar convection zone (Nordlund *et al.* 1992; Tobias *et al.* 1998).

In an earlier work (Warnecke & Brandenburg 2010) we have suggested an alternative approach. This is a two-layered approach, where the simulation of a turbulent large-scale dynamo in the convection zone is coupled to a simplified solar atmosphere model. Magnetic buoyancy does not play an important role in this model. Such a simplified model, in Cartesian coordinates, was able to capture some of the observed qualitative features. In this work we extend this model further, by going to a spherical coordinate system, in the presence of density stratification and gravity. Here we present preliminary results from such a study. We find that helical fields, generated by the large-scale dynamo below, emerge above the solar surface. We expect such fields to drive flares and coronal mass ejections via Lorentz force.

2. The Model

Similar to Warnecke & Brandenburg (2010), a two-layer system is used. We model the convection zone starting at $r = 0.7\,R_\odot$ to the solar corona till $r = 2\,R_\odot$, where $R_\odot$ is the solar radius, used from here on as our unit length. Our simulation domain is a spherical shell extending in the θ (colatitude) direction from $\pi/3$ to $2\pi/3$ and in the ϕ direction from 0 to 0.3. This means, the surface in the photosphere, which will be described by this simulation, would $600 \times 200\,\mathrm{Mm}^2$ large. A helical random force drives the velocity in the lower layer. For our model the momentum equation can be written as followed:

$$\frac{\mathrm{D}\boldsymbol{U}}{\mathrm{D}t} = \theta_w(r)\boldsymbol{f} + \nabla h + \boldsymbol{g} + \boldsymbol{J} \times \boldsymbol{B}\rho + \boldsymbol{F}_{\mathrm{visc}}, \tag{2.1}$$

where $\theta_w(r) = \frac{1}{2}\left(1 - \mathrm{erf}\frac{r}{w}\right)$, a profile function the connect the two layers, where w is the width of the transition. and $\boldsymbol{F}_{\mathrm{visc}} = \rho^{-1}\nabla \cdot (2\rho\nu\mathsf{S})$ is the viscous force, $\mathsf{S}_{ij} = \frac{1}{2}(U_{i;j} + U_{j;i}) - \frac{1}{3}\delta_{ij}\nabla\cdot\boldsymbol{U}$ is the traceless rate-of-strain tensor, semi-colons denote covariant differentiation, $\boldsymbol{g} = -\frac{GM}{r^2}\hat{\boldsymbol{r}}$ the gravitational acceleration, $h = c_{\mathrm{s}}^2 \ln\rho$ is the specific pseudo-enthalpy, $c_{\mathrm{s}} = \mathrm{const}$ is the isothermal sound speed, and $\boldsymbol{f}$ is a forcing function that drives turbulence in the interior. The pseudo-enthalpy term emerges from the fact that for an isothermal equation of state the pressure is given by $p = c_{\mathrm{s}}^2\rho$, so the pressure gradient force is given by $\rho^{-1}\nabla p = c_{\mathrm{s}}^2 \nabla\ln\rho = \nabla h$. The continuity equation be written in terms of h

$$\frac{\mathrm{D}h}{\mathrm{D}t} = -c_{\mathrm{s}}^2 \nabla \cdot \boldsymbol{U}. \tag{2.2}$$

Equations (2.1) and (2.2) are solved together with the induction equation. In order to preserve $\nabla \cdot \boldsymbol{B} = 0$, we write $\boldsymbol{B} = \nabla \times \boldsymbol{A}$ in terms of the vector potential $\boldsymbol{A}$ and solve the induction equation in the form

$$\frac{\partial \boldsymbol{A}}{\partial t} = \boldsymbol{U} \times \boldsymbol{B} + \eta\nabla^2\boldsymbol{A}, \tag{2.3}$$

For the density we use an initial distribution, where $\rho \approx 1/r^2$.

The simulation domain is periodic in the azimuthal direction. For the velocity we use the stress-free conditions all other boundaries. For the magnetic field we adopt vertical field conditions for the $r = 2$ boundary and perfect conductor conditions for the $r = 0.7$ and both θ boundaries. Time is measured in non-dimensional units $\tau = tu_{\mathrm{rms}}k_{\mathrm{f}}$, which is the time normalized to the eddy turnover time of the turbulence. We use the PENCIL CODE†, which uses a sixth order centered finite-difference in space and a third-order Runge-Kutta scheme in time. See Mitra, Tavakol, Brandenburg & Moss (2009) for extension of the PENCIL CODE to spherical coordinates.

3. Results

The forcing gives rise to an α^2 dynamo in the turbulence zone. After a short phase of exponential growth, the magnetic field shows opposite polarities in the two hemispheres with oscillations and equatorward migration (Mitra *et al.* 2010). The maximum magnetic field at each hemisphere is about 63% of the equipartition value. This is a typical behavior of an efficient large-scale dynamo. The magnetic fields emerge through the surface and create field line concentrations, which reconnect, separate and rise to the outer boundary of the domain. This dynamical evolution is clearly seen in a sequence of field line images in Fig. 1, where the field lines of $\langle\boldsymbol{B}\rangle_\phi$ are shown as contours of $r\sin\theta\langle A_\phi\rangle_\phi$ and the color

† http://pencil-code.googlecode.com

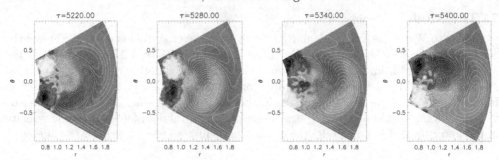

Figure 1. Time series of formation of a plasmoid ejection in spherical coordinates. Contours of $r\sin\theta\langle A_\phi\rangle_\phi$ are shown together with a color-scale representation of $\langle B_\phi\rangle_\phi$; dark blue stands for negative and red for positive values. The contours of $r\sin\theta\langle A_\phi\rangle_\phi$ correspond to field lines of $\langle B\rangle_\phi$ in the $r\theta$ plane. The dotted horizontal lines show the location of the surface at $r = 1\,R_\odot$.

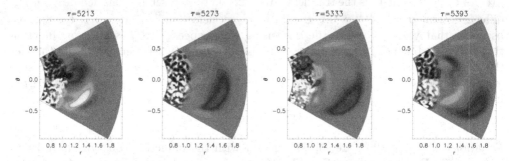

Figure 2. Times series of what looks a bit like coronal ejections in spherical coordinates. The normalized current helicity $\langle J \cdot B\rangle_\phi/\langle B^2\rangle_\phi$ is shown in a color-scale representation for different times; dark blue stands for negative and red for positive values. The dotted horizontal lines show the location of the surface at $r = 1\,R_\odot$.

representation stands for $\langle B_\phi\rangle_\phi$. Investigating the current helicity, we find a surprising result. In the turbulence zone the $\langle J \cdot B\rangle_\phi/\langle B^2\rangle_\phi$ is negative in the northern hemisphere and positive in south, i.e., has the same sign as the helicity of the external force. Above the surface, for each hemisphere, current helicity with sign opposite to the turbulent layer is ejected in large patches (Fig. 2). These structures correlate with reconnection events of strong magnetic fields. In Rust (1994) such phenomena have been described as *magnetic clouds*. In order to demonstrate that plasmoid ejection is a recurrent phenomenon, we plot the evolution of the ratio $\langle J \cdot B\rangle_\phi/\langle B^2\rangle_\phi$ versus τ and r in Fig. 3. We further find that the typical speed of plasmoid ejection is about 0.13 times the rms velocity of the turbulence in the interior region, which corresponds to 0.08 of the Alfvén speed. The time interval between successive ejections is about 100 τ. Note that the ejection of magnetic field and associated reconnection events are fairly regular, but the formation structures such as `magnetic clouds` are less regular. Note further that the sign of azimuthally averaged current helicity in the outer layer is always opposite to that of the turbulence zone. This is demonstrated in Fig. 4 where we plot time-series of azimuthally averaged current helicity at $r = 1.5\,R_\odot$. For the northern hemisphere the current helicity (solid black line) and the accumulated mean (solid red line) show positive values and for the southern hemisphere (dotted lines) negative values. This suggests that, even though the plasmoids observed in our simulations are shedding small-scale current helicity of opposite sign to that of the large-scale current helicity inside the turbulence zone, outside the turbulence zone the sign of large-scale current helicity has reversed and is now the same as that of the

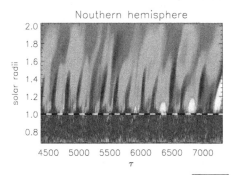
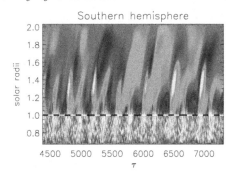

Nouthern hemisphere Southern hemisphere

Figure 3. Dependence of $\langle \boldsymbol{J} \cdot \boldsymbol{B}_\phi\, \theta \rangle / \overline{\langle \boldsymbol{B}^2 \rangle_{\phi\theta}}$ versus time τ and radius r in terms of the solar radius $R_\odot$. The left penal show a thin band in θ in the northern hemisphere, the right one a thin band in θ in the southern hemisphere, both are averaged over $20°$–$28°$ latitude. Dark blue stands for negative and red for positive values. The dotted horizontal lines show the location of the surface at $r = 1\,R_\odot$.

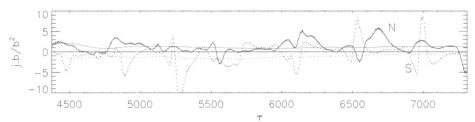

Figure 4. Dependence of $\langle \boldsymbol{J} \cdot \boldsymbol{B} \rangle_\phi / \overline{\langle \boldsymbol{B}^2 \rangle_\phi}$ versus time τ for a radius of $r = 1.5\,R_\odot$ at $\pm 28°$ latitude in arbitrary units. The solid lines stand for the northern hemisphere and the dotted for the southern hemisphere. The red colored lines represent accumulated means for each hemisphere.

small-scale helicity. This may be explained by the action of turbulent magnetic diffusion on the magnetic field; see e.g., Brandenburg, Candelaresi, & Chatterjee (2009).

In summary, it turns out that twisted magnetic fields generated by the helical dynamo beneath a spherical surface are able to produce flux emergence in ways that are reminiscent of that found in the Sun. We find phenomena that can be interpreted as recurrent plasmoid ejections, which then lead to magnetic clouds further out. A promising extension of this work would be to include a Parker–like wind that turns into a supersonic flow at sufficiently large radii.

References

Brandenburg, A., Candelaresi, S., & Chatterjee, P. 2009, *MNRAS*, 398, 1414
Käpylä, P. J., Korpi, M. J., & Brandenburg, A. 2008, *A&A*, 491, 353
Mitra, D., Tavakol, R., Käpylä, P. J., & Brandenburg, A. 2010, *ApJ*, 719, L1-L4.
Mitra, D., Tavakol, R., Brandenburg, A., & Moss, D. 2009, *ApJ*, 697, 923
Nordlund, Å., Brandenburg, A., Jennings, R. L., *et al.* 1992, ApJ, 392, 647
Rust, D. M. 1994, *Geophys. Res. Lett.*,21, 241
Tobias, S. M., Brummell, N. H., Clune, T. L., & Toomre, J. 1998, *ApJ*, 502 L177
Warnecke, J. & Brandenburg, A. 2010, *A&A*, 523, A19

Advances in Plasma Astrophysics
Proceedings IAU Symposium No. 274, 2010
A. Bonanno, E. de Gouveia Dal Pino & A. G. Kosovichev

© International Astronomical Union 2011
doi:10.1017/S1743921311007198

The EST project

Francesca Zuccarello[1,2] and the EST team

[1]Dipartimento di Fisica e Astronomia – Universitá di Catania, Via S. Sofia 78,
I-95123 Catania, Italy
[2]Istituto Nazionale di Astrofisica – Osservatorio Astrofisico di Catania, Via S. Sofia 78, I-95123
Catania, Italy
email: `fzu@oact.inaf.it`

Abstract. EST European Solar Telescope is a pan-european project, presently in its Conceptual Design Study financed by the European Commission in the framework of FP7, involving 29 partners, from 14 different countries. The EST project is aimed at the realization of a 4-m class telescope, characterized by an optical design and a set of instruments optimized for extremely high resolution imaging and spectropolarimetric observations from near UV to NIR. EST will be four times larger than any existing high resolution solar telescope and it is designated with the highest priority among the ground-based, medium term (2016-2020) new projects in the ASTRONET Roadmap (Panel C). The EST instruments will measure fundamental astrophysical processes at their intrinsic scales in the Sun's atmosphere to establish the mechanism of magnetic field generation and removal, and of energy transfer from the surface to the upper solar atmosphere and eventually to the whole heliosphere. The conceptual Design Study started on February 2008 and will finish during 2011. EST will be operational at the same time as major ESA and NASA space missions aimed at studying solar activity.

Keywords. Instrumentation: high angular resolution, Sun: photosphere, Sun: chromosphere, Sun: magnetic fields

---

1. Introduction

The comprehension of physical processes occurring in solar phenomena has been greatly improved in the past decades thanks to space-borne satellites (YOHKOH, SOHO, TRACE, RHESSI, HINODE) and to high resolution ground-based telescopes (THEMIS, VTT, SST, DOT, DST). Nevertheless, despite the important progresses achieved, there are still several aspects that need to be further investigated by means of higher temporal, spatial and spectral resolution data (Title 2010).

Moreover, if we take into account that some mechanisms occurring on the Sun are often considered to take place in other astrophysical contexts, it is becoming increasingly clear that the necessity of new-generation solar telescopes, of the 4-m class, can be shared, besides than by solar physicists, also by researchers involved in the study of plasma physics, Earth-magnetosphere, stellar activity, binary systems (see, e.g., Zhang 2007), accretion disks around black holes (Yuan *et al.* 2009), and other topics.

On the basis of these needs, new-generation solar telescopes have been promoted in the last years both in the USA and in Europe: the Advanced Technology Solar Telescope (ATST) is now at the beginning of the construction phase, while the European Solar Telescope (EST) is currently in the conclusive phase of the Design Phase.

In particular, the EST project is promoted by the european solar physics community by means of EAST (European Association for Solar telescopes), a consortium formed by research organizations from 15 European countries (Austria, Croatia, Czech Republic, France, Germany, Great Britain, Hungary, Italy, The Netherlands, Norway, Poland,

Slovakia, Spain, Sweden and Switzerland), aimed at the development, realization and management of the telescope.

The Conceptual Design Phase of EST, financed by the European Commission in the framework of FP7 – Capacities Specific Programme – Research Infrastructures, is lead by the Institute de Astrofísica de Canarias, and is developed by 29 european partners (14 research institutions and 15 companies).

2. The Current Design

The current baseline design of EST is based on a 4-meter class solar telescope characterized by an on-axis Gregorian configuration, which will allow to achieve a good polarimetric performance. Three different types of instruments will be located in a controlled laboratory at the Coudé focus: a Broad Band Imager, a Narrow-Band Tunable Filter Spectropolarimeter and a Grating Spectropolarimeter (Collados *et al.* 2010). Each instrument will have several channels in order to observe simultaneously at various wavelengths.

The telescope mechanical configuration is alt-azimuthal in order to obtain, compared to the equatorial configuration, a more compact system and a better air flushing on the primary mirror, besides than allowing to achieve a polarimetric compensated optical design with less optical surfaces.

The telescope design includes adaptive optics (AO) and multi-conjugate adaptive optics (MCAO), consisting of deformable mirrors at conjugated heights, able to correct for the diurnal variation of the Sun's distance to the turbulence layers. The MCAO are integrated in the telescope optical path, allowing to maximize the telescope throughput and to provide simultaneously a corrected image at the Coudé focus for the three instruments. It is also foreseen that the MCAO train would be by-passable.

In order to allow the user to find the desired target on the solar disk, an auxiliary full-disc telescope (AFDT, Klvana *et al.*, 2008) is foreseen. The AFDT has an aperture of 150 mm and can operate simultaneously in three spectral regions: Ca II K (394 nm), Hα (656.3 nm), and white light (450 - 460 nm).

Two different options have been considered for the enclosure: a completely retractable enclosure and a conventional dome. To date, the analysis carried out indicates that the foldable enclosure can provide better local seeing conditions and would allow the use of a reflecting heat rejecter at the Gregorian focus, but on the other hand it can produce higher wind effect on image quality. The conventional dome would allow a lower effect of wind on image quality, but the local seeing degradation might be larger and it would be necessary to control the environmental conditions inside the dome.

2.1. *The Telescope Optical Design*

The current baseline optical layout consists of 14 polarimetrically compensated mirrors. The elevation axis (line joining M4 and M5, see Fig. 1 (a)) is placed 1.5 m below the vertex of the primary mirror: this configuration facilitates M1 air flushing and allows space enough for the M1 cell and for an adequate placement of the transfer optics train vertically from the telescope to the Coudé focus. Moreover, the elevation and azimuth (line joining M7 and M8) axes are decentered with respect to the telescope optical axis so that the optical path is fold in an asymmetric way to produce a polarimetric compensated layout. In the F1 focus generated by the primary mirror M1, a heat-stop will be located, in order to remove most of the solar light and to allow a field-of-view of (2 - 3) × (2 - 3) arcmin^2. In focus F2, generated by the secondary mirror, M2, the calibration optics will be installed to analyze the polarimetric performance of the rest of the telescope, and

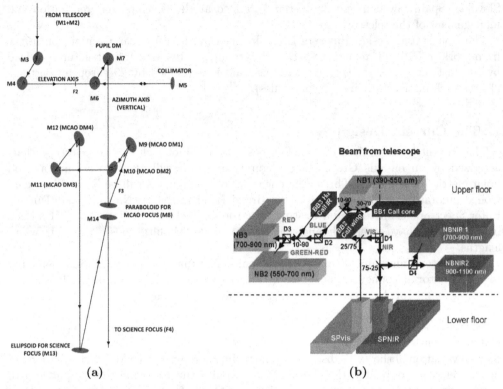

Figure 1. (a) Schematic drawing of the optical elements of the telescope. Mirrors M3 and M4 (as well as M6 and M7) are polarimetrically compensated. Mirrors M9 to M12 are the MCAO mirrors. (b) Schematic drawing of the instrument set-up designed for the EST telescope (from Cavaller *et al.* (2010)).

possibly also polarimetric modulators for polarization measurements. M5 is a collimator that produces an image of the pupil where a deformable mirror, DM (indicated by M7 in Fig. 1 (a)), is located to correct for the ground-layer turbulence. The flat mirror M6 will have tip-tilt correction capabilities to avoid image motion.

The MCAO mirrors are located after F3, the focus generated by M8: the present design has four MCAO DMs, conjugated at 5, 9, 15 and 30 km, which operate at 45 degrees in a polarimetric compensated configuration. M13 generates the science focal plane, F4.

The present optical design adds three additional flat mirrors to this basic layout. The first two define the elevation axis of the telescope, and are placed in such a way that their incidence-reflection planes are perpendicular one to the other (mirrors M3 and M4 in Fig. 1 (a)), so that their instrumental polarization is cancelled. Furthermore, M6 and M7 (tip-tilt mirror and pupil DM, respectively), arranged in the same geometrical configuration as M3 and M4, are also polarimetrically compensated. As the elevation axis is defined by the line joining M4 and M5 and the azimuth axis by the line joining M7 and M8, the polarization of the incoming light is not modified, independently of the pointing of the telescope to any direction on the sky. If we consider the transfer optics, which includes the MCAO system, the four high-altitude DMs are also distributed to compensate their polarization properties. M14 is introduced to send the light to the Coudé room.

The transfer optics can be used as an optical de-rotator if the input optical axis (line joining the centers of M7 and M8) and the output optical axis (line joining M14 and F4) coincide (see Fig. 1 (a)). This avoids the inclusion of any derotating device at instrument level, such as a rotating platform, so that instruments can be kept fixed.

2.2. *The Instruments*

The instruments, composed of different channels at the Coudé focus, will be placed in a laboratory with a controlled environment, distributed in two floors (see Fig. 1 (b)). A system composed of dichroic and intensity beam-splitters will be used to feed the different instruments channels, in order to allow simultaneous observations.

In the current design the light is first split by a dichroic beam-splitter D1 in a visible and a near-infrared beam (see Fig. 1 (b)). Dichroic beam-splitters D2 and D3 will split further the visible beam in three narrower band-passes (390 - 550 nm, 550 - 700 nm, and 700 - 900 nm), where the corresponding broad-band and narrowband channels are placed. Intensity beam-splitters (inclined lines crossing the beam in Fig. 1 (b)), are then used to feed all the instrument channels. The same conceptual design is used for the near-infrared branch. The Broad-Band Imager has two different operational modes: (a) high resolution mode (1×1 arcmin^2, 0.015 arcsec/pix, 4k $\times$ 4k detectors); (b) large field of view mode (2×2 arcmin^2, 0.03 arcsec/pix, 4k $\times$ 4k detectors).

The Narrow-Band, tunable filter imaging system, is based on Fabry-Pérot interferometers. The etalons could be mounted within the optical light beam either in the collimated configuration, or in the telecentric configuration.

Four different alternatives have been evaluated for the Grating Spectropolarimeters (each having a visible and a near-IR configuration): i) Long-Slit Standard Spectrograph; ii) Multi-Slit Multi-Wavelength Spectrograph; iii) Tunable Universal Narrow-band Imaging Spectrograph (TUNIS; López Ariste *et al.*, 2010); iv) New generation Multi-channel Subtractive Double Pass (MSDP).

3. Conclusions

The EST Design phase will finish during the year 2011: at this time several decisions will be taken on the different alternatives that have been proposed so far. The site seeing campaign will also allow to take a decision on the site in the Canary Islands where the telescope will be built. EST construction is expected to start in 2014 and first light is foreseen in 2019.

Acknowledgements

The conceptual design of EST is partially supported by the European Commission through the collaborative project Nr. 212482 *EST: the large aperture European Solar Telescope* under framework programme FP7.

References

Cavaller L., *et al.* 2010, "EST: European Solar Telescope. Design study phase II report", EST project documentation, report RPT-GTC

Collados, M.; Bettonvil, F.; Cavaller, L.; Ermolli, I.; Gelly, B.; Pérez, A.; Socas-Navarro, H.; Soltau, D.; Volkmer, R.; EST team 2010, *AN*, 331, 615

Klvana, M., Sobotka, M., Svanda, M. 2008, in: 12th European Solar Physics Meeting, Online at http://espm.kis.unifreiburg. de/, p. 2.73

López Ariste, A., LeMen, C., Gelly, B., Asensio Ramos, A. 2010, *AN* 331, 658

Title, A. 2010, *AN*, 331, 599

Yuan, F., Lin, J., Wu, K., & Ho, L. C. 2009, *Mon. Not. R. Astron. Soc.*, 395, 2183

Zhang, S. N. 2006, *Highlights of Astronomy*, 14, 119

Advances in Plasma Astrophysics
Proceedings IAU Symposium No. 274, 2010
A. Bonanno, E. de Gouveia Dal Pino & A.G. Kosovichev, eds.
© International Astronomical Union 2011
doi:10.1017/S1743921311007204

Optimized gyrosynchrotron algorithms and fast codes

Alexey A. Kuznetsov[1,2] and Gregory D. Fleishman[3,4]

[1] Armagh Observatory, Armagh BT61 9DG, Northern Ireland, UK
email: aku@arm.ac.uk

[2] Institute of Solar-Terrestrial Physics, Irkutsk 664033, Russia

[3] New Jersey Institute of Technology, Newark, NJ 07102, USA
email: gfleishm@njit.edu

[4] Ioffe Physico-Technical Institute, St. Petersburg 194021, Russia

Abstract. Gyrosynchrotron (GS) emission of charged particles spiraling in magnetic fields plays an exceptionally important role in astrophysics. In particular, this mechanism makes a dominant contribution to the continuum solar and stellar radio emissions. However, the available exact equations describing the emission process are extremely slow computationally, thus limiting the diagnostic capabilities of radio observations. In this work, we present approximate GS codes capable of fast calculating the emission from anisotropic electron distributions. The computation time is reduced by several orders of magnitude compared with the exact formulae, while the computation error remains within a few percent. The codes are implemented as the executable modules callable from IDL; they are made available for users via web sites.

Keywords. radiation mechanisms: nonthermal, methods: numerical, Sun: flares, Sun: radio radiation, stars: flare, radio continuum: general

---

Gyrosynchrotron (GS) radiation (incoherent radiation of mildly relativistic electrons spiraling in a magnetic field) makes a dominant contribution into the radio and microwave emission of solar and stellar flares. Therefore, radio observations potentially can be used to diagnose the parameters of the emission sources (such as energy and pitch-angle distributions of the energetic electrons, magnetic field, and plasma density). The diagnostics can be made, e.g., by using the forward-fitting methods (e.g., Fleishman *et al.* 2009), which requires fast and accurate methods of calculating the GS emission.

The exact formulae for the gyromagnetic radiation are known for several decades (Eidman 1958, 1959; Melrose 1968; Ramaty 1969). However, these formulae are computationally slow, especially when high harmonics of the cyclotron frequency are involved (i.e., in a relatively weak magnetic field). A number of simplified approaches has been proposed (e.g., Petrosian 1981; Dulk & Marsh 1982; Klein 1987), but those approximations were developed for the isotropic electron distributions only, while real distributions in the solar and stellar flares are often expected to be highly anisotropic. Thus, there is a need to develop algorithms and computer codes that would be fast, accurate, and applicable to both isotropic and anisotropic electron distributions.

In developing such algorithms, we follow the approximation proposed by Petrosian (1981) and Klein (1987), which has been proved to provide a reasonable accuracy in a high-frequency range in the case of the isotropic or weakly anisotropic electron distributions. We have found that, after a few improvements (involving a wider use of numerical methods), this approximation can be extended to the strongly anisotropic distributions, provided that the pitch-angle distribution function is continuous together with its first and second derivatives. The key element of this new algorithm (which we call "continu-

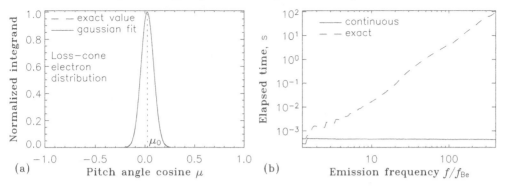

Figure 1. (a) Example of the angular integrand (dashed line) and its gaussian fit (solid line). (b) Time required to calculate the intensity and polarization of the GS emission at a given frequency (a 2 GHz Intel Pentium processor was used for the calculations).

ous") is a possibility to approximate the angular integrand of the exact GS equations by a Gaussian fit, see Fig. 1a, which is then integrated analytically saving the computation time significantly (see Fig. 1b). The comprehensive description of the new approximation is given in the recently published article of Fleishman & Kuznetsov (2010). Besides a wider applicability range, the new approximation improves the accuracy for the isotropic distributions in comparison with the original Petrosian-Klein algorithm.

Figure 2 demonstrates an example of the GS emission calculated using the exact formulae and new continuous algorithm. The energetic electrons are assumed to have a power-law distribution over energy and a loss-cone distribution over pitch-angle. One can see that the contunuous algorithm provides very high accuracy, especially at high frequencies. The degree of polarization is reproduced with high accuracy. Spectral index of the emission in the optically thin range is reproduced very well, too. At low frequencies, the continuous algorithm is unable to reproduce the harmonic structure of the GS emission; however, the mean level of the spectrum is reproduced well. In Figs. 2a-2c, an additional line shows the corresponding parameters calculated for the isotropic electron distribution. One can see that the considered anisotropy affects significantly the emission intensity, polarization, and spectral index. Nevertheless, the continuous algorithm works excellently even for this (strongly anisotropic) distribution.

Being that accurate, the continuous algorithm is much faster than the exact expressions. As one can notice from Fig. 1b, computation time for the exact formulae grows exponentially with the harmonic number, while for the continuous algorithm this time is nearly constant. As a result, at high frequencies (i.e., where the continuous algorithm is very accurate), the computation time can be reduced by several orders of magnitude in comparison with the exact formulae. In addition to this continuous algorithm, Fleishman & Kuznetsov (2010) developed a number of codes, which can be gradually tuned to optimize either computation time or accuracy including recovery of the low-frequency harmonic structure of the GS radiation, remaining much faster than the exact codes.

The new algorithms are implemented as executable modules (Windows dynamic link libraries and Linux shared objects) callable from IDL. The codes are called using the IDL `call_external` function. This approach combines high computation speed with the IDL visualization capabilities. A factorized form of the electron distribution function is adopted: $F(E, \alpha) = u(E)g(\alpha)$, where E and α are the electron energy and pitch-angle, respectively. Currently, the code contains 9 built-in energy distributions (including thermal, power-law, kappa, etc.) and 5 pitch-angle distributions (including isotropic, loss-cone, and beam-like); any combination of the energy and pitch-angle distribution

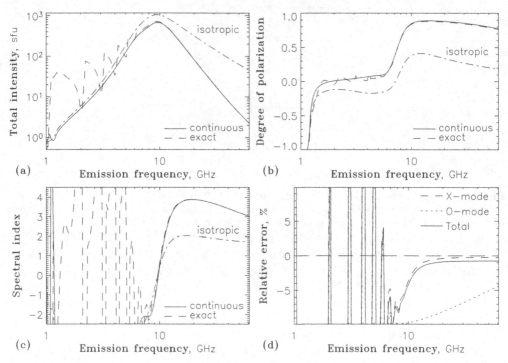

Figure 2. Calculated parameters of the GS emission for the electron distribution of loss-cone type. In the panels (a-c), the dash-dotted line shows the corresponding parameters for the isotropic distribution. Panel (d) shows the relative difference between the emission intensities calculated using the exact and continuous codes.

is possible. The code returns the intensities of the ordinary and extraordinary modes from a homogeneous source located at the Sun (and observed at the Earth), as well as the corresponding gyroabsorption factors which can directly be used for solving the equation of radiation transfer in an inhomogeneous case. Fast GS codes together with the comprehensive documentation are available as the online supplement to the article of Fleishman & Kuznetsov (2010), and at the web site: star.arm.ac.uk/~aku/gs/

Acknowledgements

A.K. thanks the Leverhulme Trust for financial support. This work was supported in part by NSF grants ATM-0707319, AST-0908344, and AGS-0961867 and NASA grant NNX10AF27G to New Jersey Institute of Technology, and by the RFBR grants 08-02-92204, 08-02-92228, 09-02-00226, and 09-02-00624.

References

Dulk, G. A., & Marsh, K. A. 1982, *ApJ*, 259, 350
Eidman, V. Y. 1958, *Sov. Phys. — JETP*, 7, 91
Eidman, V. Y. 1959, *Sov. Phys. — JETP*, 9, 147
Fleishman, G. D., Nita, G. M., & Gary, D. E. 2009, *ApJ*, 698, L183
Fleishman, G. D., & Kuznetsov, A. A. 2010, *ApJ*, 721, 1127
Klein, K.-L. 1987, *A&A*, 183, 341
Melrose, D. B. 1968, *Ap&SS*, 2, 171
Petrosian, V. 1981, *ApJ*, 251, 727
Ramaty, R. 1969, *ApJ*, 158, 753

Plasmas in galaxies and galaxy clusters

Luigina Feretti

Advances in Plasma Astrophysics
Proceedings IAU Symposium No. 274, 2010
A. Bonanno, E. de Gouveia Dal Pino & A.G. Kosovichev, eds.
© International Astronomical Union 2011
doi:10.1017/S1743921311007216

Development of the theory of instabilities of differentially rotating plasma with astrophysical applications

J. G. Lominadze[1,2]

[1]Georgian National Astrophysical Observatory, The Ilia State University, Tbilisi 0160, Georgia

[2]M. Nodia Institute of Geophysics, Tbilisi 0193, Georgia
email: contact@gsa.gov.ge

Abstract. Instabilities of nonuniform flows is a fundamental problem in dynamics of fluids and plasmas. This presentation outlines atypical dynamics of instabilities for unmagnetized and magnetized astrophysical differentially rotating flows, including, our efforts in the development of general theory of magneto rotation instability (MRI) that takes into account plasma compressibility, pressure anisotropy, dissipative and kinetic effects. Presented analysis of instability (transient growth) processes in unmagnetized/hydrodynamic astrophysical disks is based on the breakthrough of the hydrodynamic community in the 1990s in the understanding of shear flow non-normality induced dynamics. This analysis strongly suggests that the so-called bypass concept of turbulence, which has been developed by the hydrodynamic community for spectrally stable shear flows, can also be applied to Keplerian disks. It is also concluded that the vertical stratification of the disks is an important ingredient of dynamical processes resulting onset of turbulence.

Keywords. Astrophysical disks, transient growth, MRI, turbulence.

1. Introduction

Nonuniform flows are ubiquitous in astrophysical rotating (stars, protoplanetary disks, galaxies, etc.) and plane flows (e.g. solar wind). Consequently, the appearance of complex dynamics of these systems is often a manifestation of nonuniform kinematics. For instance, the structure and dynamical appearance (such as turbulence) of astrophysical disks are largely defined by the differential character of the disk matter rotation. This concerns to both, unmagnetized and magnetized astrophysical disks – objects of our interest.

The consequent development in understanding the physics of turbulence in astrophysical disks has been irregular and has taken considerable time. Substantial progress has been achieved in the nineties with the discovery of a linear instability in magnetized disks, so called magneto rotation instability (MRI) (Balbus & Hawley, 1991; Balbus & Hawley, 1992; Hawley & Balbus, 1991; Hawley & Balbus, 1992; Hawley *et al.*, 1995; Stone *et al.*, 1996). Behind the frameworks of this theory were the papers Velikhov (1959) and Chandrasekhar (1960), showing that the nondissipative Couette flow, i.e., the flow of an ideally conducting fluid between rotating cylinders can be destabilized by an axial magnetic field. As a whole, MRI and related instabilities became a focus point of astrophysical studies dealing with accretion disks around a compact object in binary systems. These studies were summarized in the review Balbus & Hawley (1998). These studies were mostly concerned with hydrodynamic regimes when standard MHD is applicable. On other hand, there are many astrophysical objects where regimes different from standard MHD occur. In a series of recent papers, the MRI theory has been extended taking into account finite plasma compressibility, pressure anisotropy, dissipative and kinetic effects thereby allowing applications to various of astrophysical objects. We aim to present these recent developments.

In contrast to the magnetized disks, the solution of the turbulence problem in the unmagnetized/hydrodynamic case has not yet reached sufficient maturity. Moreover, the very occurrence of turbulence in hydrodynamic disks has been questioned by Balbus *et al.* (1996) and Balbus & Hawley (1998). However, there is irrefutable observational evidence that such disks have to be turbulent. Due to this apparent contradiction, disk turbulence is often considered as some sort of mystery. The reason for this situation is that cylindrical flows with Keplerian profile belong to the class of smooth shear flows, i.e. which present no inflection point; it is well known that these flows are spectrally stable, although they may become turbulent in the laboratory. This dilemma, that existed also in laboratory/engineering flows, has been solved by the hydrodynamic community in the 90s of the last century, where a breakthrough was accomplished in the comprehension of turbulence in spectrally/asymptotically stable shear flows (e.g. in the plane Couette flow). We also aim to outline the breakthrough of the hydrodynamic community and its application to unmagnetized/hydrodynamic astrophysical disks.

2. Generalization of MRI

The incompressible case studied in Balbus & Hawley (1991) corresponds to the high-β plasma (β- ratio of the plasma and the magnetic field pressures) and can be called the simplest MRI in the simplest astrophysical situation. In Mikhailovskii *et al.*, (2008), it was suggested that the MRI can be treated as a plasmaphysical phenomenon. The electrodynamic approach to the MRI problem has been formulated within a framework of an appropriate plasma permittivity tensor. One more step in the MHD theory of MRI has been done in Kim & Ostriker (2000) and Blaes & Socrates (2001) taking into account compressibility. Thereby, the analysis of Balbus & Hawley (1991) has been generalized to the case of an arbitrary-β plasma. It was also pointed out that, in addition to one-fluid MRI theory, a kinetic theory is needed for a collisionless plasma with anisotropic pressure, and the axisymmetric modes in the simplest astrophysical configuration were analyzed in Quataert *et al.* (2002) and Sharma *et al.* (2003) where the basic kinetic theory of the MRI in the isotropic plasma and the related instabilities in the presence of pressure anisotropy were studied. In Mikhailovskii *et al.*(2008a) a family of hybrid instabilities due to the differential plasma rotation was found: the so-called rotational-firehose and rotational-mirror instabilities.

Electrodynamic theory for a larger part of the phenomena was developed in Mikhailovskii *et al.*, (2008b). The dynamics can be described by either the one-fluid MHD or the kinetic theory. We have shown that the local dispersion relation for all these situations and for both approaches in terms of the permittivity tensor has the universal form. We have developed four versions of electrodynamic theory: one-fluid theory for the simplest astrophysical plasma, kinetic theory for the simplest astrophysical plasma, one-fluid theory for the case with both the gravitation force and the pressure gradient present and kinetic theory for this case. These versions differ from each other by the permittivity tensor components. These equations show that the nonaxisymmetric modes are less dangerous than the axisymmetric being, first, stronger stabilized by the magnetoacoustic effect, and, second, because of the overstable effect. The dispersion relation describes two instabilities: the MRI and the Convective Instability. The latter induced by the pressure gradient together with the density gradient.

The kinetic treatment reveals the pressure anisotropy effects on the instabilities. It is known that the plasma pressure anisotropy may be a drive of the collisionless plasma instabilities (Mikhailovskii, 1975). Three varieties of the pressure anisotropy-driven instabilities in the nonrotating plasma (Kitsenko & Stepanov, 1960; Rudakov & Sagdeev, 1958; Vedenov & Sagdeev, 1958; Chandrasekhar, Kaufman & Watson, 1958; Parker, 1958) are known: the mirror instability (Kitsenko & Stepanov, 1960; Rudakov & Sagdeev, 1958; Vedenov & Sagdeev, 1958) and two kinds of firehose instabilities, one related to the Alfven oscillation branches (Kitsenko & Stepanov, 1960; Chandrasekhar, Kaufman & Watson, 1958; Parker, 1958), other to the magnetoacoustic ones (Kitsenko & Stepanov, 1960; Rudakov & Sagdeev, 1958; Vedenov & Sagdeev, 1958). Initially, the rotation effect has been included into firehose instability theory in Sharma *et al.*, (2006) and Ferriere (2006). The rotational-firehose and rotational-mirror instabilities have been found in Mikhailovskii *et al.* (2008a) for the simplest astrophysical plasma with axisymmetric perturbations. The axisymmetry is a restriction of Mikhailovskii *et al.* (2008a), Sharma *et al.* (2006) and Ferriere (2006). In general, the technique developed in Mikhailovskii *et al.* (2008b) allows to obtain a broader view of the rotation effect on the pressure-anisotropy-driven instability. Comparison shows that, in contrast to the one-fluid approach, the axisymmetric MRI in the collisionless laboratory plasma is not affected by the Velikhov or the plasma density gradient effects. Predictions of the one-fluid MHD and the kinetics are different since the MHD implies a coupling of the perpendicular and parallel plasma motions. Behavior of the axisymmetric and nonaxisymmetric modes in the kinetic laboratory plasma model is similar. The main difference between them comes from the overstable effect for nonaxisymmetric modes. In addition to the local dispersion relations, in Mikhailovskii *et al.* (2008b) is derived a series of electrodynamic mode equations. These can be used for the development of the electrodynamic theory of nonlocal instabilities complementing the MHD theory of such instabilities (Mikhailovskii *et al.*, 2008c).

Our electrodynamic theory is developed for the pure plasma, similar to MHD theory. It can be generalized to the case of the dusty plasma, which can be an alternative to the MHD theory of instabilities in the rotating dusty plasma (Mikhailovskii *et al.*, 2008b; Mikhailovskii *et al.*, 2008d). Both the one-fluid and kinetic regimes considered here concern the magnetized plasma with the ion cyclotron frequency larger than the oscillation frequency and the plasma rotation frequency. A weak magnetization implies the so-called Hall regime which was broadly analyzed in astrophysics (Wardle, 1999; Urpin & Rudiger, 2005; see also Mikhailovskii *et al.*, 2007; Mikhailovskii *et al.*, 2008e), with both the MHD and electrodynamic approaches. Then the axisymmetric modes only have been studied. It seems that our technique can be applied for the electrodynamic theory of nonaxisymmetric modes. According to Krolik & Zweibel (2006), in some cases the electron inertia should be allowed for in astrophysics. The electrodynamic theory of axisymmetric modes with the electron inertia has been developed in Mikhailovskii *et al.* (2007a) and Mikhailovskii *et al.* (2008e). With our technique, this can be extended to the nonaxisymmetric modes. An important area of application of our technique is the electromagnetic instabilities in the rotating plasma in the approximation of unmagnetized electrons. A first step in this direction has been made in Mikhailovskii *et al.* (2007a) in the framework of electron hydrodynamics allowing for the collisionless parallel viscosity. However, the effects of electron pressure anisotropy leading to the Weibel-type instabilities (Weibel, 1959; Shukla & Shukla, 2007) were problematic in this topic. It seems that electrodynamic treatment can allow one to fill the gap in the theory. We have restricted ourselves to the linear approximation. It seems that inclusion of the three-wave interaction, as in Lindgren, Larsson & Stenflo (1982), and nonlinear zonal flow generation, as in Mikhailovskii *et al.*, (2007b), may be important generalization of our theory.

3. Hydrodynamic Keplerian disk flows

The breakthrough of the hydrodynamic community (see Chagelishvili *et al.* (2003), for details). Traditional stability theory followed the approach of Rayleigh (1880) where the instability is determined by the presence of exponentially growing modes that are solutions of the linearized dynamic equations. Only recently has one become aware that operators involved in the modal analysis of plane shear flows are not normal, hence that the corresponding eigenfunctions are non-orthogonal and would strongly interfere (Reddy *et al.*, 1993). For this reason, the emphasis was shifted in the 90s from the analysis of long time asymptotic flow stability to the study of short time behavior. It was established that asymptotically/Rayleigh stable flows allow for linear transient growth of vortex and/or wave mode perturbations (cf. Gustavsson, 1991; Butler & Farrell, 1992; Reddy & Henningson, 1993; Trefethen *et al.*, 1993). This fact incited a number of fluid dynamists to examine the possibility of a subcritical transition to turbulence, with the linear stable flow finding a way to bypass the usual route to turbulence (via linear classical/exponential instability). On closer examination, the perturbations reveal rich and complex behavior in the early transient phase, which leads to the expectation that they may become self-sustaining when there is nonlinear positive feedback.

Based on the interplay of linear transient growth and nonlinear positive feedback, a new concept emerged in the hydrodynamic community for the onset of turbulence in spectrally stable shear flows and was named bypass transition (cf. Boberg & Brosa (1988), Butler & Farrell (1992), Farrell & Ioannou (1993), Reddy & Henningson (1993), Gebhardt & Grossmann (1994), Henningson & Reddy (1994), Baggett *et al.* (1995), Grossmann (2000), Reshotko (2001), Chagelishvili *et al.* (2002), Chapman (2002)). The bypass scenario differs fundamentally from the classical scenario of turbulence. In the classical model, exponentially growing perturbations permanently supply energy to the turbulence and they do not need any nonlinear feedback for their self-sustenance, so the role of nonlinear interaction is just to reduce the scale of perturbations to that of viscous dissipation. In the bypass model, nonlinearity plays a key role. The nonlinear processes are conservative, but in the case of positive feedback, they ensure the repopulation of perturbations that are able to extract energy transiently from the mean flow. The self-sustenance of turbulence is then the result of a subtle and balanced interplay of linear transient growth and nonlinear positive feedback. Consequently, thorough examination of the nonlinear interaction between perturbations is a problem of primary importance, and the first step is to search and to describe the linear perturbation modes that will participate in the nonlinear interactions.

Transient dynamics of hydrodynamic disk flows. Described above linear transient growth is also at work in rotating hydrodynamic disk flows; however, the Coriolis force causes a quantitative reduction of the growth rate there which delays the onset of turbulence. Keplerian flows are therefore expected to become turbulent for Reynolds numbers a few orders of magnitude higher than for plane subcritical flows (see: Longaretti (2002), Tevzadze *et al.* (2003)). The possibility of an alternate route to turbulence gave new impetus to the research on the dynamics of astrophysical disks (Lominadze *et al.* (1988), Richard & Zahn 1999, Richard (2001), Ioannou & Kakouris (2001), Tagger (2001), Longaretti (2002), Chagelishvili *et al.* (2003), Tevzadze *et al.* (2003), Klahr & Bodenheimer (2003), Yecko (2004) 2004, Afshordi *et al.* (2004), Umurhan & Regev (2004), Umurhan & Shaviv (2005), Klahr (2004) 2004; Bodo *et al.* (2005), Mukhopadhyay *et al.*

(2005), Barraco & Marcus (2005), Johnson & Gammie (2005), Umurhan (2006)). By adapting the progress of the hydrodynamic community to the disks flow, this research is promising for solving the disks' hydrodynamic turbulence problem.

But it remains to be seen whether this route to turbulence actually applies to astrophysical disks. Compared to plane shear flows, these possess two additional properties: differential rotation and vertical stratification. Separate studies of these factors show that each exerts a stabilizing effect on the flow: these include numerical calculation of the stability of unstratified flows by Shen et al. (2006), experiments on Keplerian rotation without stratification by Ji et al. (2006), estimates of the growth rates with stratification by Brandenburg & Dintrans (2006). However, it appears that the combined action of differential rotation and stratification introduces a new degree of freedom that may influence the flow stability and lead to turbulence at a high enough Reynolds number. The study of the linear perturbations in strato-rotational flow in the local limit can be found in Tevzadze et al. (2003) and Tevzadze et al. (2008) where it is shown the following:

The combined action of vertical gravity and Coriolis forces in 3D case results the conservation of the potential vorticity that indicates the existence of a vortex/apperiodic mode in the perturbation spectrum of the system. This vortex mode is the primary in the transient amplification phenomenon. In the absence of any one of these (vertical gravity or Coriolis) forces vortex mode degenerates into the trivial solution of the system – it disappears.

In 3D hydrodynamic disks with vertical gravity the density-spiral wave mode (which is internal gravity wave modified by the disk rotation) does not extract the energy of differential rotation efficiently, but is linearly coupled with the vortex mode. This coupling indicates the importance of the density-spiral waves along with the vortices in the overall dynamics of the system. The linear dynamics of small scale perturbations can be analyzed by a single spatial Fourier harmonic: a leading Fourier harmonic of 3D vortex mode gains background shear flow energy and are transiently amplified by several orders of magnitude. Reaching the point of maximal amplification and switching to the trailing, it gives rise to the corresponding harmonic of the density-spiral wave due to the shear flow induced coupling. The generated wave maintains the energy of perturbations. In fact, overall energetic dynamics is similar to that occurred in the plane parallel constant shear flow (see Fig. 2 of Tevzadze et al. (2003)). So, these investigations strongly suggest that the linear dynamics in vertically stratified 3D hydrodynamic Keplerian disks matches requirements of the bypass concept developed for the plane-parallel flows. This conjecture may be confirmed by appropriate numerical simulations that take the vertical stratification and consequent mode coupling into account in the high Reynolds number regime.

Acknowledgements

This work was supported by Georgian National Science Foundation grant GNSF/ST08/4-420. It is a pleasure to acknowledge helpful discussions with Drs. A. I. Smolyakov, G. D. Chagelishvili and A. G. Tevzadze.

References

Afshordi, N., Mukhopadhyay, B., & Narayan, R. 2005, *ApJ*, 629, 373
Baggett, J. S., Driscoll, T. A., & Trefethen, L. N. 1995, *Physics of Fluids*, 7, 833
Balbus, S. A. & Hawley, J. F. 1991, *ApJ*, 376, 214
Balbus, S. A. & Hawley, J. F. 1992, *ApJ*, 400, 610
Balbus, S. A., Hawley, J. F. & Stone, J. M. 1996, *ApJ*, 467, 76

Balbus, S. A. & Hawley, J. F. 2006, *Review of Modern Physics*, 70, 1

Balbus, S. A. & Hawley, J. F. 2006, *ApJ*, 652, 1020

Barranco, J. A., & Marcus, P. S. 2005, *ApJ*, 623, 1157

Blaes, O. M. & Socrates, A. 2001, *ApJ*, 553, 987

Broberg, L. & Brosa, U. 1988, *Z. Naturforschung*, 43a, 697

Bodo, G., Chagelishvili, G., Murante, *et al.* 2005, *A&A*, 437, 9

Brandenburg, A. & Dintrans, B. 2006, *A&A*, 450, 437

Butler, K. M. & Farrell, B. F. 1992, *Phys. Fluids*, 4, 1637

Chagelishvili, G. D., Chanishvili, R. G., Hristov, T. S., & Lominadze, J. G. 2002, *Sov. Phys. JETP*, 94, 434

Chagelishvili, G. D., Zahn, J.-P., Tevzadze, A. G., & Lominadze, J. G. 2003, *A&A*, 402, 401

Chandrasekhar, S., Kaufman, A. N., & Watson, K. M. 1958, *Proc. R. Soc. London A*, 245, 435

Chapman, S. J. 2002, *J. Fluid Mech.*, 451, 35

Chandrasekhar, S. 1960, *Proc. Nat. Acad. Sci. U.S.A.* 46, 253

Farrell, B. F. & Ioanou, P. J. 1993, *Phys. Fluids A*, 5, 1390

Ferriere, K. 2006, *Space Sci. Rev.*, 122, 247

Gebhardt, T. & Grossmann, S. 1994, *Phys. Rev. E*, 50, 3705

Grossmann, S. 2000, *Rev. Mod. Phys.*, 72, 603

Gustavsson, L. H. 1991, *J. Fluid Mech.*, 224, 241

Hawley, J. F. & Balbus, S. A. 1991, *ApJ*, 376, 223

Hawley, J. F. & Balbus, S. A. 1992, *ApJ*, 400, 595

Hawley, J. F., Gammie, C. F., & Balbus, S. A. 1995, *ApJ*, 440, 742

Quataert, E., Dorland, W., & Hammett, G. W. 2002, *ApJ*, 577, 524

Henningson, D. S. & Reddy, S. C. 1994, *Phys. Fluids*, 6, 1396

Ioannou, P. J. & Kakouris, A. 2001, *ApJ*, 550, 931

Ji, H., Burin, M., Schartman, E. & Goodman, J. 2006, *Nature*, 444, 343

Johnson, B. M. & Gammie, C. F. 2005, *ApJ*, 626, 978

Kim, W. T. & Ostriker, E. C. 2000, *ApJ*, 540, 372

Klahr, H. & Bodenheimer, P. 2003, *ApJ*, 582, 869

Klahr, H. 2004, *ApJ*, 606, 1070

Kitsenko, A. B. & Stepanov K. N. 1960, *Sov. Phys. JETP*, 432, 31

Krolik, J. H. & Zweibel, E. G. 2006, *ApJ*, 644, 651

Lindgren, T., Larsson, J. & Stenflo, L. 1958, *Phys. Plasmas*, 24, 1177

Lominadze, J. G., Chagelishvili, G. D., & Chanishvili, R. G. 1988, *Sov. Astr. Lett.*, 14, 364

Longaretti, P.-Y. 2002, *ApJ*, 576, 587

Mikhailovskii, A. B. 1975, in Leontovich M. A. (ed) *Reviews of Plasma Physics*, 6, p. 77

Mikhailovskii, A. B., Lominadze, J. G., *et al.* 2007, *Phys. Plasmas*, 14, 112108

Mikhailovskii, A. B., Lominadze, J. G., *et al.* 2007a, *Phys. Lett. A*, 372, 49

Mikhailovskii, A. B., Lominadze, J. G., *et al.* 2007b, *Phys. Plasmas*, 14, 082302

Mikhailovskii, A. B., Lominadze, J. G., *et al.* 2008, *Sov. Phys. JETP*, 106, 154

Mikhailovskii, A. B., Lominadze, J. G., *et al.* 2008a, *Sov. Phys. JETP*, 106, 371

Mikhailovskii, A. B., Lominadze, J. G., *et al.* 2008b, *Plasma Phy. Control. Fusion*, 50, 085012

Mikhailovskii, A. B., Lominadze, J. G., *et al.* 2008c, *Phys. Plasmas*, 15, 052109

Mikhailovskii, A. B., Lominadze, J. G., *et al.* 2008d, *Phys. Plasmas*, 15, 014504

Mikhailovskii, A. B., Lominadze, J. G., *et al.* 2008e, *Plasma Phys. Rep.*, 34, 052109

Mukhopadhyay, B., Afshordi, N., & Narayan, R. 2005, *ApJ*, 629, 383

Parker, E. N. 1958, *Phys. Rev.*, 109, 1874

Rayleigh, Lord 1880, *Scientific Papers (Cambridge Univ. press)*, 1, 474

Reddy, S. C., Schmid, P. J., & Hennigson, D. S. 1993, *SIAM J. Appl. Math.*, 53, 15

Reddy, S. C. & Henningson, D. S. 1993, *J. Fluid Mech.*, 252, 209

Reshotko, E. 2001, *Phys. Fluids*, 13, 1067

Richard, D. & Zahn, J.-P. 1999, *A&A*, 347, 734

Richard, D. 2001, *Ph.D. Thesis, Universite Paris 7*

Rudakov, L. I. & Sagdeev, R. Z. 1958, *Physics and the Problem of Controlled Thermonuclear Reactions*, (New York: Pergamon), 3, P. 321

Sharma, P., Hammett, G. W., & Quataert, E. 2003, *ApJ*, 596, 1121

Sharma, P., Hammett, G. W., & Quataert, E. 2006, *ApJ*, 637, 952

Shen, Y., Stone, J. M., & Gardiner, T. A. 2006, *ApJ*, 653, 513

Stone, J. M., Hawley J. F., Gammie, C. F., & Balbus, S. A. 1996, *ApJ*, 463, 656

Shukla, N. & Shukla, P. K. 2007, *Phys. Lett. A*, 362, 221

Tagger, M. 2001, *A&A*, 380, 750

Tevzadze, A. G., Chagelishvili, G. D., Zahn, J.-P., Chanishvili, R. G., & Lominadze, J. G. 2003, *A&A*, 407, 779

Tevzadze, A. G., Chagelishvili, G. D., & Zahn J.-P. 2008, *A&A*, 478, 9

Trefethen, L. N., Trefethen, A. E., Reddy, S. C., & Discoll, T. A. 1993, *Science*, 261, 578

Vedenov, A. A. & Sagdeev, R. Z. 1958, *Physics and the Problem of Controlled Thermonuclear Reactions*, (New York: Pergamon), 3, P. 332

Velikhov, E. P. 1959, *Sov. Phys. JETP* 9, 995

Umurhan, O. M. 2006, *MNRAS*, 368, 85

Umurhan, O. M. & Regev, O. 2004, *A&A*, 427, 855

Umurhan, O. M. & Shaviv, G. 2005, *A&A*, 432, 31

Urpin, V. & Rudiger, G 2005, *A&A*, 437, 23

Wardle, M. 1999, *A&A*, 307, 849

Weibel E. S. 1959, *Phys. Rev. Lett.*, 2, 83

Yecko, P. A. 2004, *A&A*, 425, 385

Advances in Plasma Astrophysics
Proceedings IAU Symposium No. 274, 2010
A. Bonanno, E. de Gouveia Dal Pino & A.G. Kosovichev, eds.

© International Astronomical Union 2011
doi:10.1017/S1743921311007228

Magnetism in galaxies –
Observational overview and next generation radio telescopes

Rainer Beck

Max-Planck-Institut für Radioastronomie, Auf dem Hügel 69, 53121 Bonn, Germany
email: rbeck@mpifr-bonn.mpg.de

Abstract. The strength and structure of cosmic magnetic fields is best studied by observations of radio continuum emission, its polarization and its Faraday rotation. Fields with a well-ordered spiral structure exist in many types of galaxies. Total field strengths in spiral arms and bars are 20–30 μG and dynamically important. Strong fields in central regions can drive gas inflows towards an active nucleus. The strongest *regular fields* (10–15 μG) are found in interarm regions, sometimes forming "magnetic spiral arms" between the optical arms. The typical degree of polarization is a few % in spiral arms, but high (up to 50%) in interarm regions. The detailed field structures suggest interaction with gas flows. *Faraday rotation measures* of the polarization vectors reveals large-scale patterns in several spiral galaxies which are regarded as signatures of large-scale (coherent) fields generated by dynamos. – Polarization observations with the forthcoming large radio telescopes will open a new era in the observation of magnetic fields and should help to understand their origin. Low-frequency radio synchrotron emission traces low-energy cosmic ray electrons which can propagate further away from their origin. LOFAR (30–240 MHz) will allow us to map the structure of weak magnetic fields in the outer regions and halos of galaxies, in galaxy clusters and in the Milky Way. Polarization at higher frequencies (1–10 GHz), to be observed with the EVLA, MeerKAT, APERTIF and the SKA, will trace magnetic fields in the disks and central regions of galaxies in unprecedented detail. All-sky surveys of Faraday rotation measures towards a dense grid of polarized background sources with ASKAP and the SKA are dedicated to measure magnetic fields in distant intervening galaxies and clusters, and will be used to model the overall structure and strength of the magnetic field in the Milky Way.

Keywords. Instrumentation: polarimeters, techniques: polarimetric, ISM: magnetic fields, galaxies: magnetic fields, galaxies: spiral, radio continuum: galaxies, radio continuum: ISM

1. Introduction

The amplification of cosmic fields from weak seed fields to the present-day level, their strength in intergalactic space, and their dynamical importance for galaxy evolution are important but unanswered astrophysical questions. Magnetic fields are believed to be a major agent in the ISM. They control the density and distribution of cosmic rays. Cosmic rays accelerated in supernova remnants can provide the pressure to drive galactic outflows and buoyant loops of magnetic fields via the Parker instability. Outflows from starburst galaxies in the early Universe may have magnetized the intergalactic medium.

2. Origin of galactic magnetic fields

Primordial seed fields of $10^{-20} - 10^{-14}$ G could originate from phase transitions in the early Universe (Widrow 2002, Caprini *et al.* 2009), from the time of cosmological structure formation by the Weibel instability (Lazar *et al.* 2009) or from injection by the first stars or jets generated by the first black holes (Rees 2005). High-energy γ-ray observations with HESS and FERMI indicate that the secondary particles are deflected by intergalactic fields of at least 10^{-16} G strength with a filling factor larger than 60% (Dolag et al. 2010).

The most promising mechanism to sustain magnetic fields in the interstellar medium of galaxies is the dynamo (Beck et al. 1996). A small-scale dynamo in protogalaxies may have amplified seed fields to the energy density level of turbulence within less than 10^8 yr (Schleicher et al. 2010). To explain the generation of large-scale fields in galaxies, the mean-field dynamo has been developed. It is based on turbulence, differential rotation and helical gas flows (α-effect), generated by supernova explosions (Gressel et al. 2008, and this volume) and cosmic rays (Hanasz et al. 2009, and this volume). The mean-field dynamo in galaxy disks predicts that within a few 10^9 yr large-scale regular fields are excited from the turbulent fields (Arshakian et al. 2009), forming spiral patterns (modes) with different azimuthal symmetries in the disk and vertical symmetries in the halo (Section 6).

The mean-field dynamo generates large-scale helicity with a non-zero mean in each hemisphere. As total helicity is a conserved quantity, the dynamo is quenched by the small-scale fields with opposite helicity unless these are removed from the system (Shukurov et al. 2006). Outflows are probably essential for effective mean-field dynamo action.

3. Measuring magnetic fields in galaxies

Magnetic fields need illumination to be detectable. Most of what we know about interstellar magnetic fields comes through the detection of radio waves. *Zeeman splitting* of radio spectral lines directly measures the field strength in gas clouds of the Milky Way (Heiles & Troland 2005) and in starburst galaxies (Robishaw et al. 2008). The intensity of *synchrotron emission* is a measure of the number density of cosmic-ray electrons in the relevant energy range and of the strength of the total magnetic field component in the sky plane. Assuming energy equipartition between these two components allows us to calculate the total magnetic field strength from the synchrotron intensity (Section 4).

Polarized emission emerges from ordered fields in the sky plane. As polarization "vectors" are ambiguous by $180°$, they cannot distinguish *regular (coherent) fields*, defined to have a constant direction within the telescope beam, from *anisotropic fields*, which are generated from turbulent fields by compressing or shearing gas flows and frequently reverse their direction within the telescope beam. Unpolarized synchrotron emission indicates *turbulent (random) fields* which have random directions in 3-D and have been amplified and tangled by turbulent gas flows.

The intrinsic degree of linear polarization of synchrotron emission is about 75%. The observed degree of polarization is smaller due to the contribution of unpolarized thermal emission, which may dominate in star-forming regions, by *Faraday depolarization* along the line of sight and across the beam (Sokoloff et al. 1998), and by geometrical depolarization due to variations of the field orientation within the beam.

At long radio wavelengths, the polarization vector is rotated in a magnetized thermal plasma by *Faraday rotation*. If Faraday rotation is small (in galaxies typically at wavelengths shorter than a few centimeters), the observed B-vector gives the intrinsic field orientation in the sky plane, so that the magnetic pattern can be mapped directly. The rotation angle is proportional to the square of the wavelength λ^2 and to the *Rotation Measure (RM)*, defined as the line-of-sight integral over the product of the plasma density and the strength of the field component along the line of sight. As the rotation angle is sensitive to the sign of the field direction, only regular fields give rise to Faraday rotation, while anisotropic and random fields do not. Measurements of the Faraday rotation

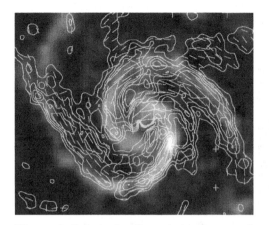

Figure 1. Polarized radio emission (contours) and B–vectors of M 51, combined from observations at 6 cm wavelength with the VLA and Effelsberg telescopes at to 8" resolution (Fletcher *et al.* 2010), overlaid onto an image of the CO(1-0) line emission by Helfer *et al.* (2003) (Copyright: MPIfR Bonn).

Figure 2. Polarized radio emission (contours) and B–vectors of M 83, combined from observations at 6 cm wavelength with the VLA and Effelsberg telescopes and smoothed to 15" resolution (Beck, unpublished), overlaid onto an optical image from Dave Malin (Anglo Australian Observatory).

from multi-wavelength observations allow to determine the strength and direction of the regular field component along the line of sight. Dynamo modes of regular fields can be identified from the pattern of polarization angles and of RMs of the diffuse polarized emission of galaxy disks (Section 6).

Distinct emitting & rotating regions located along the line of sight generate several RM components, and the observed average RM is a function of wavelength. In such cases, multi-channel spectro-polarimetric radio data are needed that can be Fourier-transformed into Faraday space, called *RM Synthesis* (Brentjens & de Bruyn 2005). If the medium has a relatively simple structure, the 3-D structure of the magnetized interstellar medium can be determined (*Faraday tomography*).

A grid of RM measurements of polarized background sources is another powerful tool to study magnetic field patterns in galaxies (Stepanov *et al.* 2008). A large number of background sources is required to recognize the field patterns, to separate the Galactic foreground contribution and to account for intrinsic RMs of the extragalactic sources.

4. Total galactic magnetic fields

The typical average *equipartition strength* of the total magnetic field (Beck & Krause 2005) in spiral galaxies is about 10 μG, assuming energy equipartition between cosmic rays and magnetic fields. Radio-faint galaxies like M 31 and M 33, our Milky Way's neighbors, have weaker total magnetic fields (about 5 μG), while gas-rich spiral galaxies with high star-formation rates, like M 51 (Fig. 1), M 83 (Fig. 2) and NGC 6946 (Fig. 3), have total field strengths of 20–30 μG in their spiral arms. The strongest total fields of 50–100 μG are found in starburst galaxies, like M 82 (Klein *et al.* 1988) and the "Antennae" NGC 4038/9 (Chyży & Beck 2004), and in nuclear starburst regions, like in the center of NGC 1097 (Fig. 4) and of other barred galaxies (Beck *et al.* 2005).

If energy losses of cosmic-ray electrons are significant, especially in starburst regions or massive spiral arms, the equipartition values are lower limits (Beck & Krause 2005). In starburst galaxies it is probably underestimated by a factor of a few (Thompson *et al.*

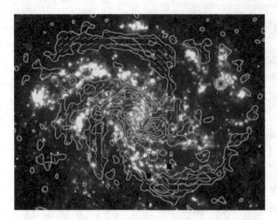

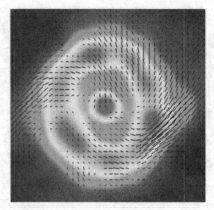

Figure 3. Polarized radio emission (contours) and B–vectors of NGC 6946, combined from observations at 6 cm wavelength with the VLA and Effelsberg telescopes and smoothed to 15" resolution (Beck 2007), overlaid onto an Hα image by Ferguson *et al.* (1998) (Copyright: MPIfR Bonn; graphics: *Sterne und Weltraum*).

Figure 4. Total radio emission (contours) and B–vectors of the circum-nuclear ring of the barred galaxy NGC 1097, observed at 3 cm wavelength with the VLA and smoothed to 3" resolution (Beck *et al.* 2005) (Copyright: MPIfR Bonn).

2006). Field strengths of 0.5–18 mG were detected in starburst galaxies by the Zeeman effect in the OH megamaser emission line at 18 cm (Robishaw *et al.* 2008).

The relative importance of various competing forces in the interstellar medium can be estimated by comparing the corresponding *energy densities*. The mean energy densities of the total (mostly turbulent) magnetic field and the cosmic rays in NGC 6946 and M 33 are $\simeq 10^{-11}$ erg cm^{-3} and $\simeq 10^{-12}$ erg cm^{-3}, respectively (Beck 2007, Tabatabaei *et al.* 2008), similar to that of the turbulent gas motions across the whole star-forming disk, but about 10 times larger than that of the ionized gas (*low-beta plasma*). Magnetic fields are dynamically important.

The integrated luminosity of the total radio continuum emission at centimeter wavelengths (frequencies of a few GHz), which is mostly of nonthermal synchrotron origin, and the far-infrared (FIR) luminosity of star-forming galaxies are tightly correlated. This correlation is one of the tightest correlations known in astronomy. It extends over five orders of magnitude (Bell 2003) and is valid in starburst galaxies to redshifts of at least 3 (Murphy 2009). Hence the total radio emission serves as a tracer of magnetic fields and of star formation out to large distances. The correlation requires that total (mostly turbulent) magnetic fields and star formation are connected, but the tightness needs multiple feedback mechanisms which are not yet understood (Lacki *et al.* 2010).

5. Structure of ordered galactic magnetic fields

Ordered (regular and/or anisotropic) field traced by the polarized synchrotron emission form *spiral patterns* in almost every galaxy, even in ring galaxies (Chyży & Buta 2008), in flocculent galaxies without massive spiral arms (Soida *et al.* 2002), and in the central regions of galaxies and circum-nuclear gas rings (Fig. 4). Ordered fields are generally strongest (10–15 μG) in the regions *between* the optical spiral arms and oriented parallel to the adjacent spiral arms. In some galaxies *magnetic arms* are formed, e.g. in IC 342 (Krause *et al.* 1993) and NGC 6946 (Fig. 3), with exceptionally high degrees of polarization (up to 50%), possibly the result of higher dynamo modes (Section 6). In

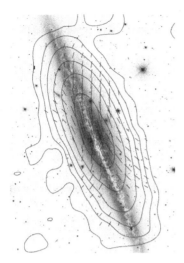

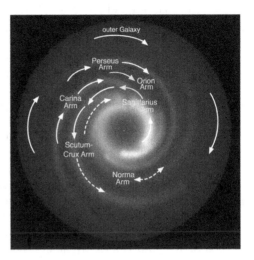

Figure 5. Total radio emission (84" resolution) and B–vectors of the edge-on galaxy NGC 891, a galaxy similar to the Milky Way, observed at 8.4 MHz with the Effelsberg telescope (Krause 2009). The background optical image is from the CFHT. Copyright: MPIfR Bonn and CFHT/Coelum.

Figure 6. Model of the magnetic field in the Milky Way, derived from Faraday rotation measures of pulsars and extragalactic sources. Generally accepted results are indicated by yellow vectors, while white vectors are not fully confirmed (from Brown, priv. comm.).

galaxies with strong density waves like M 51 (Fig. 1) and M 83 (Fig. 2) enhanced ordered (anisotropic) fields occur at the inner edges of the inner optical arms, in the interarm regions and in the outer optical arms.

In galaxies with massive *bars* the field lines follow the gas flow. As the gas rotates faster than the bar pattern of a galaxy, a shock occurs in the cold gas which has a small sound speed, while the flow of warm, diffuse gas is only slightly compressed but sheared. The ordered field is also hardly compressed. It is probably coupled to the diffuse gas and strong enough to affect its flow (Beck *et al.* 2005). The polarization pattern in barred galaxies can be used as a tracer of shearing gas flows in the sky plane and hence complements spectroscopic measurements of radial velocities.

The *central regions* of galaxies are often sites of ongoing intense star formation and strong magnetic fields. NGC 1097 hosts a bright circum-nuclear ring with about 1.5 kpc diameter (Fig. 4) and an active nucleus in its center. The ordered field in the ring has a spiral pattern and extends towards the nucleus. Magnetic stress in the circumnuclear ring due to the strong total magnetic field (about 50 μG) can drive gas inflow at a rate of several $M_\circ$/yr, which is sufficient to fuel the activity of the nucleus (Beck *et al.* 2005).

Nearby galaxies seen "edge-on" generally show a disk-parallel field near the disk plane. Edge-on galaxies like NGC 891 (Fig. 5) and NGC 253 (Heesen *et al.* 2009) reveal vertical field components in the halo forming an X-shaped pattern. The field is probably transported into the halo by an outflow emerging from the disk.

6. Faraday rotation

Spiral fields can be generated by compression at the inner edge of spiral arms, by shear in interarm regions, or by dynamo action (Section 2). Large-scale patterns of Faraday rotation measures (RM) are signatures of the mean-field dynamo and can be identified from the diffuse polarized emission of the galaxy disks (Krause 1990) or from RM data of polarized background sources (Stepanov *et al.* 2008). If several dynamo modes are superimposed, Fourier analysis of the RM variation is needed. The resolution of present-day observations is sufficient to identify not more than 2–3 modes.

The disks of a few spiral galaxies reveal large-scale RM patterns. The Andromeda galaxy M 31 is the prototype of a dynamo-generated axisymmetric spiral disk field (Fletcher *et al.* 2004). Other candidates for a dominating axisymmetric disk field (dynamo mode $m = 0$) are the nearby spiral IC 342 (Krause *et al.* 1989) and the irregular Large Magellanic Cloud (LMC) (Gaensler *et al.* 2005). Dominating bisymmetric spiral fields (dynamo mode $m = 1$) are rare, which is also predicted by dynamo models. Faraday rotation in NGC 6946 and in other similar galaxies with magnetic arms can be described by a superposition of two azimuthal dynamo modes ($m = 0$ and $m = 2$) (Beck 2007).

However, the spiral pattern of magnetic fields cannot be solely the result of mean-field dynamo action. If the beautiful spiral pattern of M 51 seen in radio polarization (Fig. 1) were only due to a regular field, its line-of sight component should generate a large-scale pattern in Faraday rotation, which is not observed. This means that a large amount of the ordered field is anisotropic and probably generated by compression and shear of the non-axisymmetric gas flows in the density-wave potential. The anisotropic field is strongest at the positions of the prominent dust lanes on the inner edge of the inner gas spiral arms, due to compression of turbulent fields in the density-wave shock. Regular fields (dynamo modes $m = 0$ and $m = 1$) also exist in the disk of M 51 but are much weaker than the anisotropic field (Fletcher *et al.* 2010).

In many other observed galaxy disks no clear patterns of Faraday rotation were found. Either several dynamo modes are superimposed and cannot be distinguished with the limited sensitivity and resolution of present-day telescopes, or the timescale for the generation of large-scale modes is longer than the galaxy's lifetime (Arshakian *et al.* 2009).

While the azimuthal symmetry of the magnetic field is known for many galaxies, the vertical symmetry (even or odd) is much harder to determine. The RM patterns of even and odd modes are similar in mildly inclined galaxies. The field of odd modes reverses its sign above and below the galactic plane. The symmetry type becomes only visible in strongly inclined galaxies, from the RM sign above and below the plane (Heesen *et al.* 2009), but the data are not yet conclusive.

7. Magnetic fields in the Milky Way

The detection of ultrahigh-energy cosmic rays (UHECRs) with the AUGER observatory and the anisotropic distribution of their arrival directions (Abreu *et al.* 2010) calls for a proper model of particle propagation. As UHECR particles are deflected by large-scale regular fields and scattered by turbulent fields, the structure and extent of the fields in the disk and halo of the Milky Way need to be known.

The all-sky maps of polarized synchrotron emission at 1.4 GHz from the Milky Way from DRAO and Villa Elisa and at 22.8 GHz from WMAP and the Effelsberg RM survey of polarized extragalactic sources were used to model the regular Galactic field (Sun *et al.* 2008). One large-scale *field reversal* is required at about 1–2 kpc from the Sun towards the Milky Way's center (Fig. 6), which is also supported by the detailed study of RMs from extragalactic sources near the Galactic plane (Van Eck *et al.* 2010). More large-scale reversals possibly exist (Han *et al.* 2006).

In mildly inclined spiral galaxies, no large-scale field reversals at certain radial distances from a galaxy's center have yet been detected, although high-resolution RM maps are available for many galaxies. The reversals in the Milky Way may be of limited azimuthal extent, and then are difficult to observe in external galaxies with the resolution of present-day telescopes. Alternatively, the reversals in the Milky Way may be part of a disturbed field structure, e.g. due to interaction with the Magellanic clouds.

The signs of RMs of extragalactic sources and of pulsars at Galactic longitudes l=90°–270° are the same above and below the plane (Taylor *et al.* 2009): the local magnetic field is symmetric, while the RM signs towards the inner Galaxy (l=270°–90°) are *opposite* above and below the plane. This can be assigned to an antisymmetric halo field (Sun *et al.* 2008) or to deviations of the local field (Wolleben *et al.* 2010). In conclusion, the overall structure of the regular field in the disk of the Milky Way is not known yet - its structure cannot be described by a simple large-scale pattern (Noutsos 2009). A larger sample of pulsar and extragalactic RM data is needed.

Little is known about the halo field in the Milky Way. The synchrotron scale height of about 1.5 kpc indicates a scale height of the total field of at least 6 kpc. The local regular Galactic field, according to RM data from extragalactic sources, has no significant vertical component towards the northern Galactic pole and only a weak vertical component of $B_z \simeq 0.3~\mu$G towards the south (Mao *et al.* 2010).

8. Outlook

High-resolution, high-sensitivity observations at high frequencies with the planned *Square Kilometre Array* (SKA) will directly map the detailed field structure and the interaction with the gas. Synchrotron emission, signature of total magnetic fields, can be detected with the SKA out to very large redshifts for starburst galaxies, depending on luminosity and magnetic field strength (Murphy 2009). If the emission from galaxies is too weak to be detected, the method of *RM grids* towards background QSOs can still be applied. Here, the distance limit is given by the polarized flux of the background QSO which can be much higher than that of the intervening galaxy. Regular fields of several μG strength were already detected in distant galaxies (Bernet *et al.* 2008). Mean-field dynamo theory predicts RMs from evolving regular fields with increasing coherence scale at $z \leqslant 3$ (Arshakian *et al.* 2009).

The SKA "Magnetism" Key Science Project plans to observe an all-sky survey (at least 10^4 deg^2) around 1 GHz which will measure at least 1500 RMs per square degree, in total at least 2×10^7 RMs from compact polarized extragalactic sources at a mean spacing of $\simeq 90''$ and at least 20 000 RMs from pulsars (Gaensler *et al.* 2004). This will allow the detailed reconstruction of the 3-D field structure in the Milky Way and in many nearby galaxies, while simple patterns of regular fields can be recognized out to distances of about 100 Mpc (Stepanov 2008). All-sky and deep-field RM surveys are also planned with the SKA precursor telescopes, ASKAP (Gaensler et al. 2010) and MeerKAT, and with the pathfinder telescope APERTIF.

The recently completed low-frequency radio telescopes *Low Frequency Array* (LOFAR) and the *Murchison Widefield Array* (MWA) (under construction) are suitable instruments to search for extended synchrotron radiation at the lowest possible levels in outer galaxy disks and halos and the transition to intergalactic space. Low frequencies are also ideal to search for small Faraday rotation measures from weak interstellar and intergalactic fields (Beck 2009) and in galaxy clusters (see Feretti and Dolag, this volume).

References

Abreu, P. & The Pierre Auger Collaboration 2010, *Astroparticle Physics*, 34, 314

Arshakian, T. G., Beck, R., Krause, M., & Sokoloff, D. 2009, *A&A*, 494, 21

Beck, R. 2007, *A&A*, 470, 539

Beck, R. 2009, *Rev. Mex. AyA*, 36, 1

Beck, R. & Hoernes, P. 1996, *Nature*, 379, 47

Beck, R. & Krause, M. 2005, *AN*, 326, 414

Beck, R., Brandenburg, A., Moss, D., Shukurov, A., & Sokoloff, D. 1996, *ARAA*, 34, 155

Beck, R., Fletcher, A., Shukurov, A., *et al.* 2005, *A&A*, 444, 739

Bell, E. F. 2003, *ApJ*, 586, 794

Bernet, M. L., Miniati, F., Lilly, S. J., Kronberg, P. P., & Dessauges-Zavadsky, M. 2008, *Nature*, 454, 302

Brentjens, M. A. & de Bruyn, A. G. 2005, *A&A*, 441, 1217

Caprini, C., Durrer, R., & Fenu, E. 2009, *J. Cosmology & Astroparticle Physics*, 11(2009)001

Chyży, K. T. 2008, *A&A*, 482, 755

Chyży, K. T. & Beck, R. 2004, *A&A*, 417, 541

Chyży, K. T. & Buta, R. J. 2008, *ApJ*, 677, L17

Dolag, K., Kachelriess, M., Ostapchenko, S., & Tomas, R. 2010, arXiv:1009.1782

Ferguson, A. M. N., Wyse, R. F. G., Gallagher, J. S., & Hunter, D. A. 1998, *ApJ*, 506, L19

Fletcher, A., Berkhuijsen, E. M., Beck, R., & Shukurov, A. 2004, *A&A*, 414, 53

Fletcher, A., Beck, R., Shukurov, A., Berkhuijsen, E. M., & Horellou, C. 2010, *MNRAS*, in press

Gaensler, B. M., Beck, R., & Feretti, L. 2004, *New Astr. Revs*, 48, 1003

Gaensler, B. M., Haverkorn, M., Staveley-Smith, L., *et al.* 2005, *Science*, 307, 1610

Gaensler, B. M., Landecker, T. L., & Taylor, A. R. 2010, *BAAS*, 42, 470

Gressel, O., Elstner, D., Ziegler, U., & Rüdiger, G. 2008, *A&A*, 486, L35

Han, J. L., Manchester, R. N., Lyne, A. G., Qiao, G. J., & van Straten, W. 2006, *ApJ*, 624, 868

Hanasz, M., Wóltański, D., & Kowalik, K. 2009, *ApJ*, 706, L155

Heesen, V., Krause, M., Beck, R., & Dettmar, R.-J. 2009, *A&A*, 506, 1123

Heiles, C. & Troland, T. H. 2005, *ApJ*, 624, 773

Helfer, T. T., Thornley, M. D., Regan, M. W., *et al.* 2003, *ApJS*, 145, 259

Klein, U., Wielebinski, R., & Morsi, H. W. 1988, *A&A*, 190, 41

Krause, M. 1990, in: R. Beck *et al.* (eds.), *Galactic and Intergalactic Magnetic Fields* (Dordrecht: Kluwer), p. 187

Krause, M. 1993, in: F. Krause *et al.* (eds.), *The Cosmic Dynamo* (Dordrecht: Kluwer), p. 303

Krause, M. 2009, *Rev. Mex. AyA*, 36, 25

Krause, M., Hummel, E., & Beck, R. 1989, *A&A*, 217, 4

Lacki, B. C., Thompson, T. A., & Quataert, E. 2010, *ApJ*, 717, 1

Lazar, M., Schlickeiser, R., Wielebinski, R., & Poedts, S. 2009, *ApJ*, 693, 1133

Mao, S. A., Gaensler, B. M., Haverkorn, M., *et al.* 2010, *ApJ*, 714, 1170

Murphy, E. 2009, *ApJ*, 706, 482

Noutsos, A. 2009, in: K. G. Strassmeier *et al.* (eds.), *Cosmic Magnetic Fields: From Planets, to Stars and Galaxies* (Cambridge: Cambridge Univ. Press), p. 15

Rees, M. J. 2005, in: R. Wielebinski & R. Beck (eds.), *Cosmic Magnetic Fields* (Berlin: Springer), p. 1

Robishaw, T., Quataert, E., & Heiles, C. 2008, *ApJ*, 680, 981

Schleicher, D. R. G., Banerjee, R., Sur, S., *et al.* 2010, *A&A*, 522, A115

Shukurov, A., Sokoloff, D., Subramanian, K., & Brandenburg, A. 2006, *A&A*, 448, L33

Soida, M., Beck, R., Urbanik, M., & Braine, J. 2002, *A&A*, 394, 47

Sokoloff, D. D., Bykov, A. A., Shukurov, A., *et al.* 1998, *MNRAS*, 299, 189, and Erratum in *MNRAS*, 303, 207

Stepanov, R., Arshakian, T. G., Beck, R., Frick, P., & Krause, M. 2008, *A&A*, 480, 45

Sun, X. H., Reich, W., Waelkens, A., & Enßlin, T.A. 2008, *A&A*, 477, 573

Tabatabaei, F., Krause, M., Fletcher, A., & Beck, R. 2008, *A&A*, 490, 1005

Taylor, A. R., Stil, J. M., & Sunstrum, C. 2009, *ApJ*, 702, 1230

Thompson, T. A., Quataert, E., Waxman, E., Murray, N., & Martin, C. L. 2006, *ApJ*, 645, 186

Van Eck, C. L., Brown, J. C., Stil, J. M., *et al.* 2010, *ApJ*, in press

Widrow, L. M. 2002, *Rev. Mod. Phys.*, 74, 775

Wolleben, M., Fletcher, A., Landecker, T. L., *et al.* 2010, *ApJ*, 724, L48

Advances in Plasma Astrophysics
Proceedings IAU Symposium No. 274, 2010
A. Bonanno, E. de Gouveia Dal Pino & A. G. Kosovichev, eds.

© International Astronomical Union 2011
doi:10.1017/S174392131100723X

MHD turbulence-Star Formation Connection: from pc to kpc scales

E. M. de Gouveia Dal Pino[1], R. Santos-Lima[1], A. Lazarian[2], M. R. M. Leão[1], D. Falceta-Gonçalves[3], and G. Kowal[1]

[1]IAG, Universidade de São Paulo, Rua do Matão 1226, São Paulo 05508-090, Brazil
email: dalpino@astro.iag.usp.br

[2]Astronomy Department, University of Wisconsin, Madison, WI, USA

[3]NAC, Universidade Cruzeiro do Sul, Rua Galvão Bueno 868, São Paulo 01506-000, Brazil

Abstract. The transport of magnetic flux to outside of collapsing molecular clouds is a required step to allow the formation of stars. Although ambipolar diffusion is often regarded as a key mechanism for that, it has been recently argued that it may not be efficient enough. In this review, we discuss the role that MHD turbulence plays in the transport of magnetic flux in star forming flows. In particular, based on recent advances in the theory of fast magnetic reconnection in turbulent flows, we will show results of three-dimensional numerical simulations that indicate that the diffusion of magnetic field induced by turbulent reconnection can be a very efficient mechanism, especially in the early stages of cloud collapse and star formation. To conclude, we will also briefly discuss the turbulence-star formation connection and feedback in different astrophysical environments: from galactic to cluster of galaxy scales.

Keywords. MHD, turbulence, plasmas, stars: formation

1. Introduction

It is generally believed that stars form within molecular clouds. This is a natural consequence of their low temperatures and high densities which help the gravitational force to overcome the internal pressures that act "outward" to prevent collapse.

The internal motions in the molecular clouds are governed by turbulence in a cold, magnetized gas. Broad line widths ranging from a few to more than 10 times the sound speed are observed in the molecular clouds and the interstellar magnetic field strength is a few microgauss, in rough equipartition with the kinetic energy in the interstellar medium. For this reason the turbulent motions are regarded as mainly supersonic and trans-Alfvénic (Elmegreen & Scalo 2004; Heiles & Troland 2005). The role of this turbulence in the interstellar dynamics and star formation, although amply discussed in the literature (see, e.g., the reviews by Elmegreen & Scalo 2004; McKee & Ostriker 2007), is still poorly understood.

A vital question that frequently permeates these debates is the diffusion of the magnetic field through a collapsing cloud. If the magnetic field were perfectly frozen into the interstellar gas during the gravitational collapse, the magnetic field strength of a typical star like our Sun would be more than 10 orders of magnitude larger than we observe today. Thus, there must be some mechanisms that are effective in removing the excess of magnetic flux during the star formation process. To address this problem, researchers usually appeal to the ambipolar diffusion (AD) of the magnetic field through the neutral component of the plasma (Mestel & Spitzer 1956; Shu 1983; Tassis & Mouschovias 2005).

However, it is unclear yet if AD can be high enough to provide the magnetic flux transport in collapsing fluids (Shu *et al.* 2006; Krasnopolsky *et al.* 2010).

Here, we discuss an alternative mechanism based on turbulent magnetic reconnection that seems to be able to provide an efficient magnetic flux removal, from the initial stages of accumulating interstellar gas to the final stages of the accretion to form protostars.

2. Magnetic Flux transport by turbulent reconnection

The idea that, due to turbulence, the magnetic field lines can reconnect and then lead to removal of magnetic flux from collapsing clouds was first discussed by Lazarian (2005) and more recently explored numerically by Santos-Lima *et al.* (2010). This was based on the magnetic reconnection model of Lazarian & Vishniac (1999) which was successfully confirmed by three-dimensional numerical simulations in Kowal *et al.* (2009). This model establishes that in the presence of *weak* turbulence, reconnection of the field lines is *fast* since many reconnection events occur simultaneously at smaller and smaller scales (even with Ohmic resistivity only). As a consequence, a robust removal of magnetic flux can be expected both in partially and fully ionized plasma.

3. Three-Dimensional simulations of magnetic flux transport by turbulent reconnection in gravitating clouds

We have performed 3-D MHD simulations employing a modified version of the Godunov-based MHD code developed by Kowal *et al.* (2007). We first considered *cubic* cloud systems as shown in Figure 1, subject to a gravitational potential with cylindrical geometry, an isothermal equation of state, and immersed in an initial vertical magnetic field. We work with dimensionless variables defined in terms of the background sound speed and the dimensions of the cloud. The boundaries were assumed to be periodic and the simulations were started with the cloud either in magneto-hydrostatic equilibrium (Figure 1, left) or already collapsing under the effect of the gravitational potential. In all cases, transonic, sub-Alfénic, non-helical turbulence with an rms velocity (V_{rms}) around unity (normalized by the sound speed) was injected in the system which was then let to evolve (see Santos-Lima *et al.* 2010 for details).

As an example, Figure 1 (right) shows the ratio between the magnetic field and the density (both averaged over the z direction) as a function of the cloud radius for a collapsing system with and without turbulence. For a cylindrical symmetry, this ratio gives a measure of the magnetic flux to mass ratio of the system. The results clearly show that in the presence of weak turbulence there is an efficient removal of magnetic flux from the central regions (where gravity is stronger and the density is higher) to the outer regions of the system. The fact that in the absence of turbulence there is no change in the magnetic-flux-to-mass ratio is a clear evidence that in the case with turbulence, the transport is due to turbulent reconnection diffusion. The same effect was detected for clouds starting in magneto-hydrostatic equilibrium. This is an indication that the process of turbulent magnetic field removal should be applicable both to quasi-static subcritical molecular clouds and to collapsing cores. These results were also found to be insensitive to the numerical resolution. Tests made with resolutions of 128^3, and 512^3 gave essentially the same results as those of 256^3 (shown in the figures) thus confirming the robustness of the results above.

We have also examined an extensive parameter space considering different strengths of the gravitational potential, the turbulent velocity and the magnetic field. The increase of the gravitational potential as well as the magnetization of the gas (decrease of β)

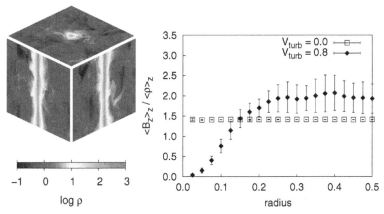

Figure 1. Left: Logarithm of the density field in $t = 3$ c.u. (where 1 time code unit is given by the box size L divided by c_s (with $c_s = 1$ c.u. being the sound speed) for a system initially in magneto-hydrostatic equilibrium, in a cylindrically symmetric gravitational potential, with a constant ratio between the thermal and the magnetic pressure ($\beta = 1$), and $v_A/c_s = 1.4$ c.u., where v_A is the Alfvén speed. The gravitational potential is $\Psi \sim A/R$, where R is the radial distance to the center of the cloud, and A is a gravitational parameter given in units of $c_s^2 L$. In this case, $A = 0.9$ c.u. The central xy, xz, and yz slices of the system were projected on the respective walls of the cubic computational domain. Right: Ratio between the magnetic field and the density (averaged over the z direction) normalized by the initial value, as a function of the cloud radius for a collapsing system (starting out of the magneto-hydrostatic equilibrium) both with and without turbulence, in $t = 8$ c.u. The cloud has the same initial conditions as those of the left system. The magnetic field is given in units of $B = c_s \sqrt{4\pi\rho}$. Error bars show the standard deviation. The numerical resolution is 256^3. Extracted from Santos-Lima *et al.* (2010).

increases the efficiency of the transport of magnetic flux to the periphery of the system. This is expected, as turbulence brings the system to a state of minimal energy. The effect of varying magnetization in some sense is analogous to the effect of varying gravity. The physics is simple, the lighter fluid (magnetic field) is segregated from the heavier fluid (gas), supporting the notion that the reconnection-enabled diffusivity relaxes the magnetic field + gas system in the gravitational field to its minimal energy state.

When the turbulent velocity is increased, there is at first a trend to remove more magnetic flux from the center allowing a more efficient infall of matter to the center. However, if we keep increasing the turbulent velocity, there is a threshold above which cloud collapse fails. Gravity becomes negligible compared with the turbulent kinetic energy and matter is removed from the center along with the magnetic flux and this causes cloud fragmentation rather than its collapse to form stars. Indeed, this effect may be very frequent in the ISM of our Galaxy since the star formation efficiency is known to be very small.

An enhanced Ohmic resistivity of unknown origin has been invoked in the literature as a way to remove magnetic flux from cores and accretion disks (e.g., by Shu *et al.* 2006). We have then also performed numerical simulations of systems without turbulence but with enhanced Ohmic diffusivity. The comparison of these models with those with turbulent have shown that turbulent reconnection diffusivity can mimic the effects of an enhanced Ohmic resistivity in gravitating clouds.

Recently, we have been testing more realistic cloud systems, considering spherically symmetric potentials. The effects of self-gravity, which become more significant at the late stages of the collapse, have been also introduced. These assumptions are consistent with observed evidence that the large, star-forming clouds are confined by their own

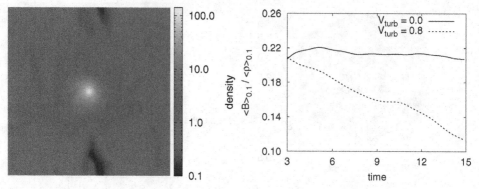

Figure 2. Left panel: density map of the central slice of a collapsing cloud with a spherically symmetric potential (central mass M = 50 M$_\odot$) in t = 10 c.u. = 100 Myr. The initial magnetic field is in the vertical direction of the image and has an intensity of 0.1μG. The initial density in the medium is 1 c.u. = 100 cm^{-3}, and β = 3. Right panel: time evolution of the magnetic field-to-density ratio, averaged within the core region of radius R = 0.1 c.u. = 0.3 pc. The curves were smoothed for making the visualization clearer. (Leão *et al.* 2011, in prep.)

gravity (like stars, planets, and galaxies) rather than by external pressure (like clouds in the sky). This evidence comes from the fact that the "turbulent" velocities inferred from CO line-widths scale in the same manner as the orbital velocity (like in virial systems). Figure 2 shows a snapshot of one of these preliminary tests and the temporal evolution of the magnetic field-to-density ratio averaged within a region of radius R = 0.1 c.u., where L = 1 c.u. = 3 pc is the size of the box. Again, it indicates an efficient removal of magnetic flux in comparison to the case when no turbulence is injected.

4. The role of turbulent reconnection on the formation of rotationally supported circumstellar disks

In the late stages of star formation, former studies showed that the observed magnetic fields in molecular cloud cores (which imply magnetic flux-to-mass ratios of the order of a few times unity; Crutcher 2005) are high enough to inhibit the formation of rationally supported disks during the main protostellar accretion phase of low mass stars (see e.g., Krasnopolsky, Li, and Shang 2010 and references therein).

For realistic levels of core magnetization and ionization, recent work has demonstrated that AD is not sufficient to weaken the magnetic braking. Further, Krasnopolsky, Li, and Shang (2010) have found that in order to enable the formation of rotationally supported disks during the protostellar mass accretion phase, a resistivity a few orders of magnitude larger than the classic microscopic resistivity values ($\eta \sim 10^{19}$ cm^{2}s^{-1}) would be needed.

We have also explored the effects of turbulent reconnection diffusion on removing magnetic flux during the formation of these disks. We found that turbulent magnetic reconnection diffusivity causes an efficient transport of magnetic flux to the outskirts of the disk at time scales compatible with the accretion time scales, allowing the formation of a rotationally supported protostelar disk with nearly Keplerian profile (see Figure 3, from Santos-Lima, de Gouveia Dal Pino, Lazarian 2011, in prep.).

5. Star formation turbulence feedback into the environment

So far, we have focused on the effects of turbulence on star formation (SF). Now, to conclude, we would like to address very briefly the SF feedback into ISM and environment

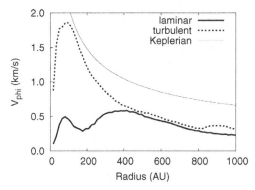

Figure 3. Radial profile of the rotational velocity of a protostellar disk with and without turbulence in $t = 1.5 \times 10^{12}$ s. The Keplerian profile is shown for comparison. The progenitor cloud has initially uniform density $\rho = 1.4 \times 10^{-19}$ g cm^{-3} and uniform magnetic field parallel to the rotation axis with intensity $B = 100\mu$G. The sound speed is $c_s = 2 \times 10^4$ cm s^{-1}. The central proto-star has a mass of 0.5 M$_\odot$. Extracted from Santos-Lima, de Gouveia Dal Pino, Lazarian 2011 (in prep.).

at galactic scales. This is a fundamental issue in the evolution of galaxies. SF in galaxies depends on both the gas content and the energy budget of the ISM. Since the most efficient stellar energy power is exerted by supernovae (SNe) and, particularly by the explosions of shortly living massive stars as type II SNe, their feedback is of fundamental relevance. There has been extensive investigation about the regulation of the ISM by the SF feedback through SN bubbles (see de Avillez 2000; de Avilez & Breitschwerdt 2005; Melioli & de Gouveia Dal Pino 2004; Melioli *et al.* 2005; Hensler 2010 and references therein).

Likewise, galactic winds and outflows emerging from star forming galaxies are also believed to be driven mainly by SN turbulence (Strickland & Stevens 2000; Tenorio-Tagle *et al.* 2003; Cooper, Bicknell & Sutherland 2008; Melioli *et al.* 2008, 2009).

Recently, SF/SN driven turbulence has been assessed even at the scales of galaxy cluster cores. For instance, Perseus, the brightest galaxy cluster observed in X-ray, presents a rich structure of cold filaments and loops seen in stellar continuum, Hα, and [NII] line emission, which are immersed in the hot gas around the central galaxy of the cluster (NGC1275) extending for up to 50 kpc (Fabian *et al.* 2008). The origin of the filaments is still unknown. We performed 2.5 and 3-dimensional MHD simulations of the central region of the cluster in which turbulent energy, triggered by star formation and supernovae (SNe) explosions was introduced. The simulations have revealed that the turbulence injected by massive stars could be responsible for the nearly isotropic distribution of filaments and loops that drag magnetic fields upward as indicated by the observations. Weak shell-like shock fronts propagating into the intra-cluster medium with velocities of $100 - 500$ km/s were found, also resembling the observations (Falceta-Gonçalves *et al.* 2010a).

As the turbulence is subsonic over most of the simulated volume, the turbulent kinetic energy is not efficiently converted into heat and additional heating is required to suppress the cooling flow at the core of the cluster. The central galaxy of the Perseus cluster (NGC1275) is the host of gigantic hot bipolar bubbles which are inflated by the AGN jets observed in the radio. It is generally believed that this is the main agent to suppress the cooling flow (see also Feretti et al. in this volume). Simulations combining the MHD SN driven turbulence with the AGN outflow were able to reproduce the temperature radial profile observed around NGC1275. While the AGN is the main heating source, the supernova turbulence seems to be an important mechanism to isotropize the energy distribution, besides being able to explain the rich filamentary and weak shock structures (Falceta-Gonçalves *et al.* 2010a, 2010b).

6. Summary

We first addressed the effects of turbulence on star formation (SF) and discussed a new possible mechanism for magnetic field flux removal from collapsing clouds and cores based on MHD turbulent reconnection, which seems to be more efficient than ambipolar diffusion. 3D numerical modeling spanning an extensive parameter space has revealed that turbulent reconnection can play an important role in the removal of magnetic field flux in different stages of SF and make molecular clouds which are initially subcritical to become supercritical and therefore, ready to collapse. Besides, this mechanism can also help the transport of magnetic flux to the outer regions of protostellar disks allowing the formation of rotationally supported disks.

We have also briefly addressed the star formation and supernovae-driven turbulence feedback into the environment at galactic scales. 3D MHD simulations have revealed that this can be a mechanism very efficient to generate outflows, and cold, magnetized filaments and clouds even in cores of galaxy clusters.

References

Cooper, J. L., Bicknell, G. V., Sutherland, R. S., & Bland-Hawthorn, J. 2008, *ApJ*, 674, 157
Crutcher, R. M. 2005, in: R. Cesaroni, M. Felli, E. Churchwell, & M. Walmsley (eds.), *Massive Star Birth: A Crossroads of Astrophysics*, Proc. IAU Symposium No. 227, p. 98
de Avillez, M. A. & Breitschwerdt, D. 2005, *A&A*, 436, 585
de Avillez, M. A. 2000, *MNRAS*, 315, 479
Elmegreen, B. G. & Scalo, J. 2004, *ARAA*, 42, 211
Fabian, A. C., Johnstone, R. M., Sanders, J. S., Conselice, C. J., Crawford, C. S., Gallagher, J. S., III, & Zweibel, E. 2008, *Nature*, 454, 968
Falceta-Gonçalves, D., Caproni, A., Abraham, Z., Teixeira, D. M., & de Gouveia Dal Pino, E. M. 2010, *ApJ* (Letters), 713, L74
Falceta-Gonçalves, D., de Gouveia Dal Pino, E. M., Gallagher, J. S., & Lazarian, A. 2010, *ApJ* (Letters), 708, L57
Heiles, C. & Troland, T. H. 2005, *ApJ*, 624, 773
Hensler, G. 2010, in: J. Alves, B. Elmegreen, & V. Trimble (eds.), *Computational Star Formation*, Procs. IAU Symposium No. 270, in press
Kowal, G., Lazarian, A., & Beresnyak, A. 2007, *ApJ*, 658, 423
Kowal, G., Lazarian, A., Vishniac, E. T., & Otmianowska-Mazur, K. 2009, *ApJ*, 700, 63
Krasnopolsky, R., Li, Z.-Y., & Shang, H. 2010, *ApJ*, 716, 1541
Lazarian, A. 2005, in: E. M. de Gouveia dal Pino, G. Lugones, & A. Lazarian (eds.), *Magnetic Fields in the Universe: From Laboratory and Stars to Primordial Structures.*, American Institute of Physics Conference Series No. 784, p. 42
Lazarian, A. & Vishniac, E. T. 1999, *ApJ*, 517, 700
Leão, M. R. M., Santos-Lima, R., de Gouveia Dal Pino, E. M., Lazarian, A. 2011, in prep.
McKee, C. F. & Ostriker, E. C. 2007, *ARAA*, 45, 565
Melioli, C., & de Gouveia Dal Pino, E. M. 2004, *A&A*, 424, 817
Melioli, C., de Gouveia dal Pino, E. M., & Raga, A. 2005, *A&A*, 443, 495
Melioli, C., Brighenti, F., D'Ercole, A., & de Gouveia Dal Pino, E. M. 2008, *MNRAS*, 388, 573
Melioli, C., Brighenti, F., D'Ercole, A., & de Gouveia Dal Pino, E. M. 2009, *MNRAS*, 399, 1089
Mestel, L., & Spitzer, L., Jr. 1956, *MNRAS*, 116, 503
Santos-Lima, R., Lazarian, A., de Gouveia Dal Pino, E. M., & Cho, J. 2010, *ApJ*, 714, 442

Santos-Lima, R., A., de Gouveia Dal Pino, E. M., Lazarian 2011, in prep.
Shu, F. H. 1983, *ApJ*, 273, 202
Shu, F. H., Galli, D., Lizano, S., & Cai, M. 2006, *ApJ*, 647, 382
Strickland, D. K., & Stevens, I. R. 2000, *MNRAS*, 314, 511
Tassis, K. & Mouschovias, T. C. 2005, *ApJ*, 618, 769
Tenorio-Tagle, G., Silich, S., & Muñoz-Tuñón, C. 2003, *ApJ*, 597, 279

Advances in Plasma Astrophysics
Proceedings IAU Symposium No. 274, 2010
A. Bonanno, E. de Gouveia Dal Pino & A.G. Kosovichev, eds.
© International Astronomical Union 2011
doi:10.1017/S1743921311007241

Relativistic plasma and ICM/radio source interaction

Luigina Feretti[1], Gabriele Giovannini[1,2], Federica Govoni[3], and Matteo Murgia[3]

[1] INAF Istituto di Radioastronomia,
Via P. Gobetti n. 101, 40129 Bologna, Italy
email: lferetti@ira.inaf.it

[2] Dipartimento di Astronomia, Universitá di Bologna,
via Ranzani n.1, 40127 Bologna, Italy

[3] INAF Osservatorio Astronomico di Cagliari,
Strada 54, Loc. Poggio dei Pini, 09012 Capoterra, Italy

Abstract. The first detection of a diffuse radio source in a cluster of galaxies, dates back to the 1959 (Coma Cluster, Large *et al.* 1959). Since then, synchrotron radiating radio sources have been found in several clusters, and represent an important cluster component which is linked to the thermal gas. Such sources indicate the existence of large scale magnetic fields and of a population of relativistic electrons in the cluster volume. The observational results provide evidence that these phenomena are related to turbulence and shock-structures in the intergalactic medium, thus playing a major role in the evolution of the large scale structure in the Universe. The interaction between radio sources and cluster gas is well established in particular at the center of cooling core clusters, where feedback from AGN is a necessary ingredient to adequately describe the formation and evolution of galaxies and host clusters.

Keywords. galaxies: clusters: general, cooling flows, intergalactic medium, magnetic fields, radio continuum: general, radiation mechanisms: nonthermal

1. Introduction

Clusters of galaxies are the largest gravitationally bound systems in the Universe. Most of the gravitating matter in any cluster is in the form of dark matter ($\sim 80\%$). Some of the luminous matter is in galaxies (~ 3-5%), the rest is in diffuse hot gas (~ 15-17%), detected in X-ray through its thermal bremsstrahlung emission. This thermal plasma, consisting of particles of energies of several keV, is commonly referred to as Intracluster Medium (ICM). In recent years it has become clear that the ICM can also contain highly relativistic particles, whose number density is of the order of 10^{-10} cm^{-3}. Although the relativistic plasma has an energy density of $<1\%$ than that of the thermal gas, it is nevertheless very important in the cluster formation and evolution.

Clusters are formed by hierarchical structure formation processes. In this scenario, smaller units formed first and merged to larger and larger units in the course of time. The merger activity appears to be continuing at the present time, and explains the relative abundance of substructure and temperature gradients detected in Abell clusters by optical and X-ray observations. At the end of their evolution, clusters reach a relaxed state, with a giant galaxy at the center, and enhanced X-ray surface brightness peak in the cores. The hot gas in the centre has a radiative cooling time shorter than the expected cluster age, therefore energy losses due to X-ray emission are important and lead to a temperature drop towards the centre (Fabian 1994). The relaxed clusters are thus referred to as cooling core clusters.

From the radio point of view, clusters can host diffuse radio emission, which has been now revealed in several conditions (merging and relaxed clusters), at different cluster locations (center, periphery, intermediate distance), and on very different size scales (100 kpc to >Mpc),(see Fig. 1 for several examples). All diffuse radio sources have in common the short lifetimes of the radiating particles, which therefore need to be reaccelerated. The properties of the radio emission are linked to those of the host cluster, therefore the connection between the thermal and relativstic plasma in clusters of galaxies is important for the cluster formation and evolution. The understanding of magnetic field and relativistic particle properties is important for a comprehensive physical description of the intracluster medium in galaxy clusters.

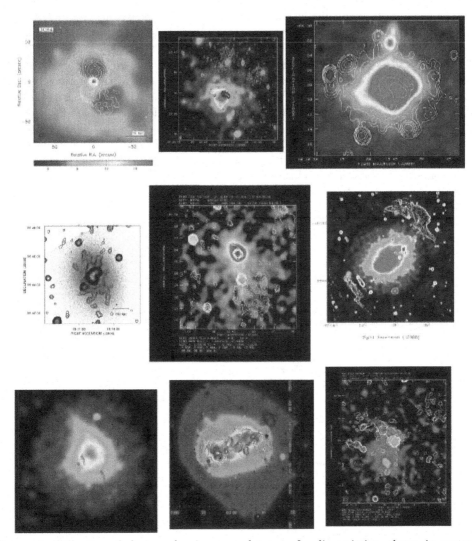

Figure 1. Collection of clusters showing several types of radio emission, shown in contours, overlaid onto the X-ray emission, shown in colors. Clusters are (from left to right and from top to bottom) Perseus (mini-halo and X-ray cavity), A 548b (relic), A 2163 (halo), A 2029 (mini-halo), A 1664 (relic), A 3667 (double relics), A 119 (radio galaxies), A 754 (halo plus relic), A 115 (relic).

2. Diffuse cluster radio sources in merging clusters

The most spectacular example of diffuse cluster radio sources is represented by giant radio halos. They are associated with clusters undergoing merging processes, and believed to be energized by the turbulence produced during the cluster mergers (see Feretti & Giovannini 2008 for a review). Radio halos (Fig. 2) can reach a size of 1 - 2 Mpc and more, although smaller size halos have also been found. New halos have recently been detected in A 851, A 1213, A 1351, A 1995, A 2034 and A 2294 (Giovannini *et al.* 2009, also Giacintucci *et al.* 2009 for A 1351). The most powerful radio halo known so far is found in the distant cluster MACS J0717.5 +3745 (Bonafede *et al.* 2009b, van Weeren *et al.* 2009) at z = 0.55. A peculiar example of a double radio halo in a close pair of galaxy clusters is represented by A 399 and A 401 (Murgia *et al.* 2010), where the two radio halos could be either originated by previous merger histories of the two clusters, or due to the currently ongoing interaction.

The radio power of both small and giant halos correlates with the cluster X-ray luminosity, i.e. gas temperature and total mass (Cassano *et al.* 2006, Giovannini *et al.* 2009), in the sense that highly luminous X-ray clusters host the most powerful radio halos. The radio halos are generally associated with clusters with X-ray luminosity in the 0.1-2.4 keV range $>> 10^{44}$ erg s^{-1}. Recently, however, radio halos have been found also in clusters with X-ray luminosity around 10^{43} erg s^{-1}, which are typical of low density environments: the first example is the cluster A 1213 (Giovannini *et al.* 2009), other cases are presented by Brown & Rudnick 2009.

Another important link between the relativistic and thermal plasma is represented by the connection between the cluster temperature and the radio halo spectral index, first suggested by Feretti *et al.* 2004, and now confirmed by Giovannini *et al.* 2009.

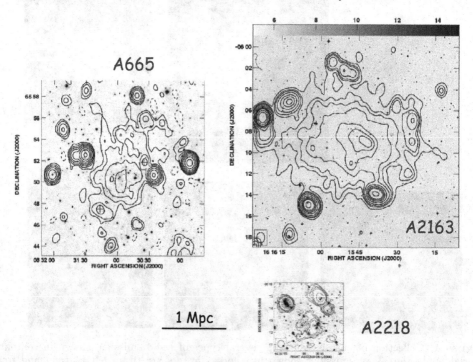

Figure 2. Images of the clusters A 665, A 2163 and A 2218, hosting radio halos: radio emission is represented by contours, which are overlaid onto the optical image. The maps are all scaled to the same linear scale.

It is found that clusters at higher temperature tend to host halos with flatter spectra (see Fig. 3). This correlation reinforces the connection between radio emission and cluster mergers, since hot clusters are more massive and may derive from more energetic merging processes, supplying more energy to the radiating electrons.

Other diffuse radio sources associated with cluster mergers are relics, located in cluster peripheral regions, and characterized by high polarized emission (see Giovannini & Feretti 2004 for a review and references). Relics can be also detected in clusters containing radio halos. A very narrow giant relic, which displays highly aligned magnetic field, has been recently detected by van Weeren *et al.* 2010. Remarkable are the giant double relics, located on opposite sides with respect to the cluster center, whose prototype is A 3667 (Röttgering *et al.* 1997). Several other cases have been found more recently: RXCJ1314.4-2515 (Feretti *et al.* 2005), A 3376 (Bagchi *et al.* 2006), A 1240 and A 2345 (Bonafede *et al.* 2009a). Radio relics are likely energized by shock waves occurring during the cluster mergers, as confirmed by observational results obtained by e.g. Solovyeva *et al.* 2008, and Finoguenov *et al.* 2010.

Diffuse radio emission is also detected beyond clusters, on very large scales. An example is represented by the filament of galaxies ZwCl 2341.1+0000 at z ~ 0.3, first detected by Bagchi *et al.* 2002, and fecently confirmed by Giovannini *et al.* 2010, who also detect polarized emission. Noticeable are the complex radio structures detected on large scale at more than 2 Mpc from the center of A 2255 (Pizzo *et al.* 2008), wich have been suggested to be connected with large scale structure formation shocks.

3. Radio - X-Ray interaction in cooling core clusters

The diffuse radio sources which may be detected at the center of cooling core clusters are classified as mini-halos, since they are morphologically similar to giant halos associated with merging clusters, but they are smaller in size (a few hundreds kpc). Although these sources are generally surrounding a powerful central radio galaxy, it has been argued that the energetics necessary to their maintenance is not supplied by the radio galaxy itself, but the electrons are reaccelerated by MHD turbulence in the cooling core region (Gitti *et al.* 2002). A peculiar case is represented by the cluster RXJ 1347.5-1145, one of the highest X-ray luminous clusters known so far. It is a relaxed clusters hosting a mini-halo, but it is also characterized by minor mergers in the cluster periphery. The

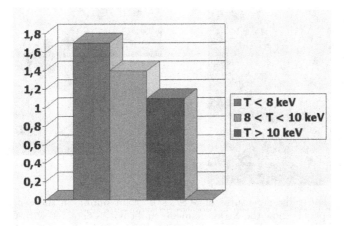

Figure 3. Average spectral index of radio halos for clusters in different ranges of temperature, showing that hotter clusters tend to host halos with flatter spectra.

mini-halo, detected by Gitti *et al.* 2007, shows an asymmetrical structure, with elongation coincident with a X-ray subclump, suggesting that additional energy for the electron re-acceleration might be provided by the sub-merger event.

In some cooling core clusters, as A 13, A 85, A 133, A 4038, diffuse radio sources offset from the cluster center have been detected (Slee *et al.* 2001). They can be classified as mini relics, because of their small-intermediate size. They are characterized by very steep spectra ($\alpha \gtrsim 2$), strong polarization, and filamentary structure when observed with sufficient resolution. The phenomenology of diffuse mini halos and mini relics, may be related to the radio source/ISM interaction in the central regions of cooling core clusters.

A spectacular example of the interaction between radio sources and the hot intra-cluster medium is represented by cavities in the X-ray gas distribution, filled with radio plasma. The first case has been detected by Böhringer *et al.* 1993 in the ROSAT image of NGC 1275 (3C 84) in the Perseus cluster. Here the thermal plasma is displaced by the inner parts of the radio lobes, causing a significant decrease of the X-ray surface brightness in those regions. The high spatial resolution of the Chandra X-ray Observatory has allowed the detection of X-ray cavities in the inner region of many cooling core clusters. These systems have been extensively studied in recent years, and a clear correspondence between regions of radio emission and deficits in the X-ray has been found in several cases (see Blanton *et al.* 2010 and references therein).

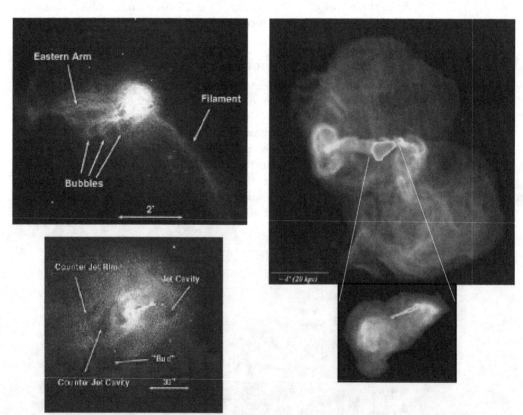

Figure 4. X-ray and radio structures of M 87, a buoyant bubble in the center of a cool core cluster. The 2 left panels show the X-ray filaments and cavities detected with Chandra (Forman *et al.* 2007). The right panels show the large scale radio bubbles, with the radio jets in the inset (Owen *et al.* 2000). The angular scale is indicated within each image by a bar, which corresponds to 2′ in the top-left panel, to 30″ in the bottom-left panel, and 4′ in the top-right panel.

The interaction between the AGN jets and the ICM is believed to be the primary feedback mechanism between the AGN driven by supermassive black holes and its environment. Cooling of the hot intracluster medium in cluster centers can feed the supermassive black holes found in the nuclei of the dominant cluster galaxies, leading to AGN outbursts, which can reheat the gas. The relativistic radio jets of the associated radio sources are creating cavities, through one strong episode or several small episodes of energy release. AGN heating can come in the form of shocks, and buoyantly rising bubbles that have been inflated by radio lobes (Best *et al.* 2007, Birzan *et al.* 2008).

Recently, Falceta-Gonçalvez *et al.* 2010 used hydrodynamical simulations to study the structure and formation of the inflated cavities of NGC 1275, in the Perseus cluster, as due to the interaction between precessing jets and the warm intracluster plasma. They show that precession can create multiple pairs of bubbles, which rise and move away from the cluster core, and may therefore be related to the extended diffuse radio emission.

Another well studied radio source is M 87 (Owen *et al.* 2000) in the nearby Virgo cluster (Fig. 4), where Forman *et al.* 2007 show X-ray structure and filaments which are related to the radio emission. Radio bubbles, blown into the cluster gas, rise buoyantly, expand and give rise to complex structures. Radio jets are confined within cavities at early times, but at later times the cosmic rays diffuse into the surrounding hot gas to form radio lobes with sizes much larger than the cavities. According to Mathews & Brighenti 2008, the bubbles may eventually expand to the cluster outskirts, where the cosmic rays would impact into the surrounding medium, giving rise to mini relics.

4. Magnetic fields in clusters

The presence of diffuse radio sources in clusters demonstrates the existence of magnetic fields in the intracluster medium. An indipendent way to obtain information on the cluster magnetic field is from the analysis of Faraday Rotation. This kind of studies has allowed a breakthrough in the knowledge of the strength and structure of magnetic fields in clusters of galaxies, by analyzing the Rotation Measure of sources seen through the magnetized cluster medium (see e.g. Govoni & Feretti 2004). Most recent studies are those on the Coma Cluster (Bonafede *et al.* 2010a), A 665 (Vacca *et al.* 2010) and the radio galaxy 3C 449 (Guidetti *et al.* 2010). The results obtained so far can be summarized as follows (see also Bonafede *et al.* 2010b): (i) magnetic fields are present in all clusters; (ii) at the center of clusters undergoing merger activity the field strenght is around 1 μG, whereas at the center of relaxed cooling core clusters the intensity is much higher (~ 10 μG); (iii) a model involving a single magnetic field coherence scale is not suitable to describe the observational data, because of different scales of field ordering and tangling.

Assuming a magnetic field power spectrum: $|B_\kappa|^2 \propto \kappa^{-n}$ (Murgia *et al.* 2004), the range of spatial scales is found between $30 - 500$ kpc and the spectral index n is in the range $2 - 4$. In A 2255, Govoni *et al.* 2006, from the comparison between the observations and the simulation of both the radio halo and the polarization of radio galaxies, obtain a flatter power spectrum at the center, and a steeper power spectrum at the periphery. This could originate from the different turbulence development in central and peripheral cluster regions.

The cluster magnetic field intensity shows a radial decline linked to the thermal gas density n_e as $B \propto n_e^x$. A trend with $x - 1/2$ is expected if the B field energy scales as the thermal energy, while $x = 2/3$ if the B field results from a frozen-in field during the cluster collapse. The values of x derived so far are in the range $0.5 - 1$, thus not conclusive.

5. Summary and future prospects

The presence of relativistic plasma in clusters of galaxies is demonstrated by the existence of halos and relics detected in merging clusters, and the mini-halos and mini relics in cooling core clusters. These features, which were detected in the past only in a few clusters, are now becoming more and more common, in particular, it is remarkable that diffuse structures, on size scales of hundreds kpc are detected in low density environments. Diffuse radio emission is also detected in filaments and in the large scale structure. X-ray cavities filled by radio bubbles are found at the center of cooling core clusters and represent the extreme example of interaction between the thermal and relativistic plasma.

Future prospects for the study of new radio sources of the different classes, in particular at low powers and at large redshifts, and study of the polarization come from the new generation radio telescopes, like the eVLA, LOFAR, the SKA precursors ASKAP and MeerKAT, and finally SKA. The study of relics and large scale diffuse sources is promising to trace the large scale structure formation and the cosmic web. Observations in the X-ray and optical domain are required to establish the cluster conditions, the merger evolutionary stage, the presence and properties of shocks, and the signatures of cluster turbulence. The comparison between the observational results and the theoretical expectations will be crucial to understand the physical link between the relativstic and thermal plasma.

Acknowledgements

L.F. wishes to thank the conference organizers for the invitation to this very stimulating meeting.

References

Bagchi, J., Enßlin, T. A., Miniati, F., Stalin, C. S., Singh, M., Raychaudhury, S., & Humeshkar, N. B., 2002, *NewA* 7, 249

Bagchi, J., Durret, F., Lima Neto, G. B., & Paul, S., 2006, *Sci* 314, 791

Best, P. N., von der Linden, A., Kauffmann, G., Heckman, T. M., & Kaiser, C.R., 2007, *MNRAS* 379, 894

Birzan, L., McNamara, B. R., Nulsen, P. E. J., Carilli, C.,L., & Wise, M. W., 2008, *ApJ* 686, 859

Blanton, E. L., Clarke, T. E., Sarazin, C. L., Randall, S. W., & McNamara, B. R., 2010, Publications of the National Academy of Sciences, Vol. 107, p. 7174

Böhringer, H., Voges, W., Fabian, A. C., Edge, A. C., & Neumann, D. M., 1993, *MNRAS* 264, L25

Bonafede, A., Giovannini, G., Feretti, L., Govoni, F., & Murgia, M., 2009a, *A&A* 494, 429

Bonafede, A., Feretti, L., Giovannini, G., Govoni, F., Murgia, M., Taylor, G. B., Ebeling, H., Allen, S., Gentile, G., & Pihlström, Y., 2009b, *A&A* 503, 707

Bonafede, A., Feretti, L., Murgia, M., Govoni, F., Giovannini, G., Dallacasa, D., Dolag, K., & Taylor, G. B., 2010a, *A&A* 513, 30

Bonafede, A., Feretti, L., Murgia, M., Govoni, F., Giovannini, G., & Vacca, V., 2010b, *A&A* in press, arXiv1009.1233

Brown, S. & Rudnick, L., 2009, *AJ* 137, 3158

Cassano R., Brunetti G., & Setti G., 2006 *MNRAS* 369, 1577

Fabian, A. C., 1994 *ARA&A* 32, 277

Falceta-Gonçalves, D., Caproni, A., Abraham, Z., Teixeira, D. M., & de Gouveia Dal Pino, E. M., 2010 *ApJ* 713, L74

Feretti, L., Brunetti, G., Giovannini, G., Kassim, N., Orrú, E., & Setti, G., 2004, *JKAS* 37, 315

Feretti, L., Schücker, P., Böhringer, H., Govoni, F., & Giovannini, G., 2005, *A&A* 444, 157

Feretti, L. & Giovannini, G., 2008, in A Pan-Chromatic View of Clusters of Galaxies and the Large-Scale Structure, M. Plionis, O. Lopez-Cruz & D. Hughes Eds., Lecture Notes in Physics, Vol. 740, p. 474

Finoguenov, A., Sarazin, C. L., Nakazawa, K., Wik, D. R., & Clarke, T. E., 2010, *ApJ* 715, 1143

Forman, W., Jones, C., Churazov, E., Markevitch, M., Nulsen, P., Vikhlinin, A., Begelman, M., Böhringer, H., Eilek, J., Heinz, S., Kraft, R., Owen, F., & Pahre, M., 2007, *ApJ*, 665, 1057

Giacintucci, S., Venturi, T., Cassano, R., Dallacasa, D., & Brunetti, G., 2009, *ApJ* 704, L54

Giovannini, G. & Feretti, L., 2004, *JKAS* 37, 323

Giovannini, G., Bonafede, A., Feretti, L., Govoni, F., Murgia, M., Ferrari, F., & Monti, G., 2009, *A&A* 507, 1257

Giovannini, G., Bonafede, A., Feretti, L., Govoni, F., & Murgia, M., 2010, *A&A* 511, L5

Gitti, M., Brunetti, G., & Setti, G., 2002, *A&A* 386, 456

Gitti, M., Ferrari, C., Domainko, W., Feretti, L., & Schindler, S., 2007, *A&A* 470, L25

Govoni, F. & Feretti, L., 2004, *Int. J. Mod. Phys. D*, Vol., 13, 1549

Govoni, F., Murgia, M., Feretti, L., Giovannini, G., Dolag, K., & Taylor, G. B., 2006, *A&A* 460, 425

Guidetti, D., Laing, R. A., Murgia, M., Govoni, F., Gregorini, L., & Parma, P., 2010, *A&A* 514, 50

Large, M. I., Mathewson, D. S., & Haslam, C. G. T., 1959, *Nat* 183, 1663

Mathews, W. G. & Brighenti, F., 2008, *ApJ* 676, 880

Murgia, M., Govoni, F., & Feretti, L. , 2004, *A&A* 424, 429

Murgia, M., Govoni, F., Feretti, L., & Giovannini, G., 2010, *A&A* 509, 86

Owen, F. N., Eilek, J. A., & Kassim, N. E., 2000, *ApJ*, 543, 611

Pizzo, R. F., de Bruyn, A. G., Feretti, L., & Govoni, F., 2008, *A&A* 481, L91

Röttgering, H. J. A., Wieringa, M. H., Hunstead, R. W., & Ekers, R. D., 1977, *MNRAS* 290, 577

Slee, O. B., Roy, A. L., Murgia, M., Andernach, H., & Ehle, M., 2001, *AJ* 122, 1172

Solovyeva, L., Anokhin, S., Feretti, L., Sauvageot, J. L., Teyssier, R., Giovannini, G., Govoni, F., & Neumann, D., 2008, *A&A* 484, 621

Vacca, V., Murgia, M., Govoni, F., Feretti, L., Giovannini, G., Orrú E., & Bonafede, A., 2010, *A&A* 514, 71

van Weeren, R. J., Röttgering, H. J. A., Brüggen, M., & Cohen, A., 2009, *A&A* 505, 991

van Weeren, R. J., Röttgering, H. J. A., Brüggen, M., & Hoeft, M., 2010, *Sci* 330, 347

Advances in Plasma Astrophysics
Proceedings IAU Symposium No. 274, 2010
A. Bonanno, E. de Gouveia Dal Pino & A.G. Kosovichev, eds.

© International Astronomical Union 2011
doi:10.1017/S1743921311007253

Supernova-driven interstellar turbulence and the galactic dynamo

Oliver Gressel[1], Detlef Elstner[2] & Günther Rüdiger[2]

[1] Astronomy Unit, Queen Mary, University of London, Mile End Road, London E1 4NS, UK
[2] Astrophysikalisches Institut Potsdam, An der Sternwarte 16, 14482 Potsdam, Germany
email: o.gressel@qmul.ac.uk, elstner@aip.de, gruediger@aip.de

Abstract. The fractal shape and multi-component nature of the interstellar medium together with its vast range of dynamical scales provides one of the great challenges in theoretical and numerical astrophysics. Here we will review recent progress in the direct modelling of interstellar hydromagnetic turbulence, focusing on the role of energy injection by supernova explosions. The implications for dynamo theory will be discussed in the context of the mean-field approach.

Results obtained with the test field-method are confronted with analytical predictions and estimates from quasilinear theory. The simulation results enforce the classical understanding of a turbulent Galactic dynamo and, more importantly, yield new quantitative insights. The derived scaling relations enable confident global mean-field modelling.

Keywords. turbulence – ISM: supernova remnants, dynamics, magnetic fields

---

1. Interstellar turbulence

Apart from stars, the baryonic matter within the Galaxy is in the form of an extremely dilute, turbulent plasma known as the interstellar medium (ISM). The multitude of physical processes within the ISM entails a rich heterogeneous structure (Spitzer, 1978).

Approximating radiative processes by a simplified cooling prescription, and restricting the computational domain to a local patch, the turbulent ISM is now routinely modelled by means of three-dimensional fluid simulations (Korpi *et al.*, 1999; Melioli & de Gouveia Dal Pino, 2004; Slyz *et al.*, 2005; Joung & Mac Low, 2006; Dib *et al.*, 2006). One main focus of these simulations has been to obtain filling factors of the different ISM phases and compare them to the classical predictions as well as observations (Dettmar, 1992). Further topics of interest include turbulent mixing (Balsara & Kim, 2005), thermodynamic distribution functions (Mac Low *et al.*, 2005), and line-of-sight integrated column densities (de Avillez & Breitschwerdt, 2005b).

1.1. *The small-scale dynamo*

While various simulations (Korpi *et al.*, 1999b; de Avillez & Breitschwerdt, 2005a; Mac Low *et al.*, 2005) discuss the influence of magnetic fields on the ISM morphology, little is said about the actual mechanism of field amplification. Balsara *et al.* (2004) have addressed this question by means of unstratified simulations of SNe turbulence. The authors relate the growth of small-scale magnetic fields to vorticity production in supernova shocks (Balsara *et al.*, 2001), and chaotic field line-stretching (Balsara & Kim, 2005).

The fact that vorticity production by colliding shells is almost inevitable in a clumpy and highly structured ISM has first been pointed out by Korpi *et al.* (1999b). The issue has then been investigated for the simplified case of driven expansion waves by Mee & Brandenburg (2006) and, more recently, by Del Sordo & Brandenburg (this volume). Considering turbulence driven by non-helical transverse waves, Haugen *et al.*, (2004) have shown that the small-scale dynamo becomes harder to excite in the super-sonic regime, albeit the critical Reynolds number for the onset of dynamo action only seems to depend weakly on the Mach number.

Because the eddy turnover time is short at small scales, a dynamo based on chaotic field line stretching will be fast. This is in-line with observations see and this volume, which exhibit dominant turbulent fields. Open issues remain with respect to the mechanism governing the saturation of the small-scale dynamo. Therefore, it is currently unclear whether equipartition field strengths can be obtained by a non-helical dynamo alone. Alternatively, the turbulent field might be explained as a "shredded" coherent field, i.e., as the by-product of a helical mean-field dynamo.

2. The large-scale galactic dynamo

Notwithstanding the above, the presence of the observed coherent fields on scales larger than the outer scale of the interstellar turbulence ($\sim$ 50 pc, see Fletcher *et al.*, 2010) clearly requires the presence of a coherent dynamo. The favoured candidates for such a dynamo are, in no particular order: (i) the kinetic driving by SNe (Korpi *et al.*, 1999; Gressel *et al.*, 2008b; Gissinger *et al.*, 2009) (ii) the buoyant cosmic ray-supported Parker instability (Parker, 1992; Hanasz *et al.*, 2004), (iii) the magneto-rotational instability (Sellwood & Balbus, 1999; Dziourkevitch *et al.*, 2004; Nishikori *et al.*, 2006; Piontek & Ostriker, 2007), and (iv) gravitational interactions (Kotarba *et al.*, 2010).†

The last of these effects is certainly dominant at the early stages of galaxy formation and will provide a seed field for subsequent processes. It remains to be shown, however, how important external interactions are in the presence of a realistic feedback from scales currently unresolved in cosmological simulations. Moreover, as Hanasz *et al.*, (2009) have shown, the cosmic ray (CR) dynamo critically relies on the anisotropy of the CR diffusion coefficient. Combined simulations, including both SNe and CRs, will prove whether this anisotropy remains effective for strongly tangled turbulent fields.

The MRI will be important (at least) in the outer regions of galaxies, where the star formation activity is low (Korpi *et al.* 2010) – even under moderate turbulence, it may operate efficiently. The stability criterion for a global isothermal disk of thickness $2H$ leads to the relation

$$8\sqrt{\mathrm{Pm_t}} < C_\Omega \equiv \frac{\Omega H^2}{\eta_t} \qquad (2.1)$$

for instability (Kitchatinov & Rüdiger, 2004). Depending on the local rotation frequency, Ω, turbulent diffusivity η_t, and turbulent magnetic Prandtl number $\mathrm{Pm_t}$ – which is generally expected to be of order unity (Fromang & Stone, 2009) – this value may be lower than the critical dynamo number for the supernova-driven dynamo. It should be worthwhile to address this question within state-of-the-art simulations of the ISM, with an Alfvén velocity of the external vertical field of the order $1\,\mathrm{km\,s^{-1}}$.

2.1. *The supernova-driven dynamo*

The energy input through SNe into the ISM is tremendous. The corresponding amplitude of the expected dynamo effect has first been estimated by Sokoloff & Shukurov (1990). Assuming hydrostatic equilibrium and applying quasi-linear theory (Rüdiger & Kitchatinov, 1993), these estimates have subsequently been refined by (Fröhlich & Schultz, 1996). A shortcoming of the approach was the neglect of a possible galactic wind. In general,

---

† Also see the respective reviews by Hanasz, Otmianowska-Mazur, and Lesch (this volume).

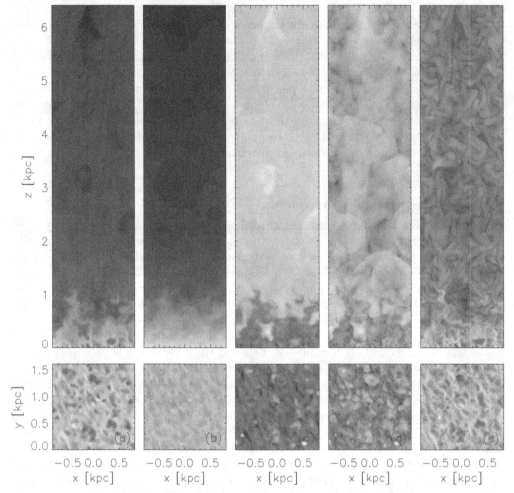

Figure 1. Snapshots at time $t = 72$ Myr of the top half of the now larger simulation box (*upper panels*), and the disc midplane (*lower panels*). The variables shown are: (a) number density [cm^{-3}], (b) column density [cm^{-2}], (c) temperature [K], (d) velocity dispersion [km s^{-1}], and (e) magnetic field strength [μG]. The logarithmic grey scales cover ranges $[-4.76, 1.01]$, $[17.56, 21.83]$, $[2.13, 7.03]$, $[-0.61, 2.64]$, and $[-5.98, -1.18]$, respectively.

there was a controversy as to whether the turbulence created by the SNe was too vigorous to warrant dynamo action – either because of a too strong wind (carrying the field away), or contrary to this, because of a too strong downward pumping (enhancing turbulent dissipation near the midplane).

In a series of papers, (Ferrière, 1998) analytically derived the electromotive force stemming from isolated remnants; the line of work was also later supported by simulations of single remnants (Kaisig *et al.*, 1993; Ziegler *et al.* 1996). To obtain the net α effect, a convolution with an assumed vertical SN distribution was applied. However, the approach suffered from a too weak dynamo and highly dominant (upward) pumping. Only when considering the stratified nature of the galactic disc (Ferrière, 1998), the issue was somewhat alleviated, although still predicting a strong upward pumping.

Pioneering semi-global simulations based on "first principles" where performed by Korpi *et al.*, (1999), and it was only for the fact of a too low dynamo number that no field amplification was observed in their simulations. The first direct simulations exhibiting

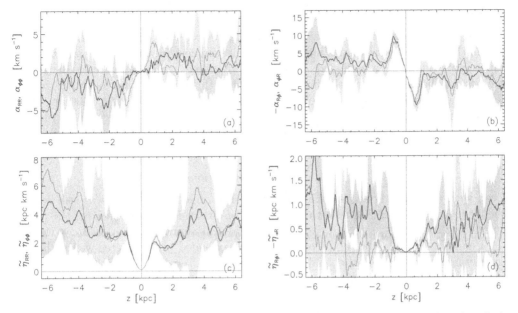

Figure 2. Dynamo coefficients as a function of the height z. The Variables are plotted in dark $(\alpha_{RR}, \ldots)$ or light $(\alpha_{\phi\phi}, \ldots)$ colours, respectively. Shaded areas indicate 1σ-fluctuations. In the region for $|z| < 2\,\mathrm{kpc}$, the results agree well with a previous run applying a much smaller box.

dynamo action were reported almost a decade later (Gressel *et al.*, 2008b). The general morphology of such simulations is illustrated in Figure 1; note the apparent correlation between the density (a) and magnetic field amplitude (e) near the midplane, indicating field amplification via compression. In contrast, away from the midplane (i.e., in the diffuse medium) the field shows more folded structures.

The fast growth of the dynamo can be understood rigorously in terms of mean-field theory, i.e. via a Reynolds-averaged induction equation

$$\partial_t \overline{\mathbf{B}} = \nabla \times \overline{\mathcal{E}} + \nabla \times \left[\overline{\mathbf{u}} \times \overline{\mathbf{B}} - \eta \nabla \times \overline{\mathbf{B}} \right], \tag{2.2}$$

and assuming a standard parametrisation

$$\overline{\mathcal{E}}_i = \alpha_{ij} \bar{B}_j - \tilde{\eta}_{ij} \varepsilon_{jkl} \partial_k \bar{B}_l, \quad i, j \in \{R, \phi\}, k = z, \tag{2.3}$$

with tensorial coefficients α_{ij} and $\tilde{\eta}_{ij}$. The determination of these closure parameters (Gressel *et al.*, 2008a) hugely benefited from the development of the so-called test field method (Schrinner *et al.*, 2005). For kinematically forced turbulence, this method allows to determine unambiguously all eight tensor components – and, in fact, none of them can be ignored (Gressel, 2009). Notably, we find a significant positive Rädler effect in the off-diagonal elements of the diffusivity tensor (see panel 'd' in Figure 2).

The measured α effect and turbulent diffusion are of the expected sign and magnitude (see Figure 2, panels 'a' and 'c'). The key finding is that of a moderate downward pumping (panel 'b') as predicted by Rüdiger & Kitchatinov (1993). This inward pumping has profound implications in compensating the effect of the equally strong upward mean flow (not shown), thus yielding ideal conditions for the dynamo. Moreover, a wind may aid the shedding of small-scale magnetic helicity as discussed in Sur *et al.*, (2007).

3. Recent advances

One major concern with the local box approximation was the issue of its limited dimension in the horizontal direction. In the low-pressure ambient medium of the galactic halo, supernova remnants can easily expand to several hundred parsec in diameter. Because of the periodic boundaries and the finite domain size, this leads to (spuriously) correlated self-interactions. The simulations of Korpi *et al.*, (1999) used $L_x = L_y = 500$ pc, which was justified for the limited vertical extent of their model. For our standard runs we applied $L_x = L_y = 800$ pc together with a vertical box size of ± 2 kpc.

3.1. *Large box simulations*

Even with near kpc horizontal box size, the issue of self-interaction was still seen in the far halo region. To eliminate the possibility of an artificial dependence of the results on the horizontal box size, we carried out a fiducial simulation run at $L_x = 1.6$ kpc (cf. Figures 1 & 2). In the region of overlap, the extended profiles agree well with their counterparts derived from the smaller boxes (cf. Fig. 4.5, model 'F4' in Gressel, 2009).

Beyond ± 2 kpc, the velocity dispersion increases significantly, enhancing the turbulent diffusion and resulting in noisy profiles for the other coefficients. Note, however, the pronounced systematic α effect in the region of strong vertical gradients near the midplane (panels 'a' and 'b' in Figure 2). This clearly supports the paradigm of helicity production being the direct result of inhomogeneous turbulence (Rüdiger & Kitchatinov, 1993).

3.2. *Scaling relations*

Beyond the purpose of mere diagnostics, the profiles shown in Figure 2 serve as a foundation for global large eddy-simulations solving (2.2). We believe that the semi-global approach of vertically stratified local boxes captures the essential physics behind the turbulent α effect in the galactic disc. At the same time, we are aware that the local geometry implies certain restrictions on the permitted dynamo modes. Identifying global symmetries is however necessary for a direct comparison with observations.

To warrant global mean-field modelling of the $\alpha\Omega$ dynamo, the scaling of the measured closure coefficients with the relevant input parameters has to be obtained. This is important because the angular velocity Ω, the local shear rate q, the supernova rate σ, and the midplane density ρ_c are functions of the galactocentric distance (Ferrière, 2001). Gressel *et al.*, (2009) already made a first step in this direction by inferring the dependence on the supernova rate. Corresponding mean-field models have been reported in (Elstner *et al.*, 2009). Further scaling relations with respect to Ω and ρ_c will be reported elsewhere.

First estimates (subject to small number statistics) for a range of 0.1–1 times the galactic star formation rate lead to the relations

$$C_\alpha \sim \Omega\, H, \quad \text{and} \quad C_\Omega \sim \Omega^{1.5}\, H^2\, \sigma^{-0.5}. \tag{3.1}$$

This has the consequence that the dynamo number $D \equiv C_\alpha C_\Omega \sim \Omega^{2.5} H^3 \sigma^{-0.5}$ scales inversely with the star formation rate. Therefore we would not expect a stronger amplification of the large-scale field for stronger star formation activity. Lastly, the pitch angle $p \simeq \sqrt{C_\alpha/C_\Omega} \sim \Omega^{-0.25} H^{-0.5} \sigma^{0.25}$, in the kinematic regime, only seems to depend weakly on the studied parameters. The strongest dependence here is, however, expected from the local shear rate $q \equiv \mathrm{d}\ln\Omega/\mathrm{d}\ln R$, suggesting a dedicated parameter survey.

4. Conclusions and prospects

Because current numerical simulations are limited to very moderate Reynolds numbers, it is important to understand how efficient the observed mechanisms remain under realistic conditions. To achieve this, it has proven fruitful to study simplified scenarios and run multiple parameter sets (see, for a comprehensive review Brandenburg & Subramanian, 2005). Fortunately, the emerging physical effects are dominated by the outer scale of the turbulence, i.e., as soon as a rudimentary scale separation is achieved, the turbulent quantities should become independent of the actual micro scale.

The growing complexity of models challenges the distribution of computing time: increasing physical realism leaves little margin for the variation of key parameters, let alone convergence checks or running multiple representations of a single parameter set. Dedicated studies remain mandatory to segregate artificial trends from genuinely physical ones (Hanasz *et al.*, 2009; Gressel *et al.*, 2009).

The dilemma becomes even more apparent when looking at the recent trend to performing "resolved" global simulations. For these, convergence checks are a rare exception. While an "enhanced" diffusivity may be sufficiently approximated by the numerical truncation error on the grid scale, the diamagnetic pumping term certainly is not. Yet vertical transport has profound implications on the emerging dynamo modes and growth rates (Bardou *et al.*, 2001).

In conclusion, we advocate a strategy that has been applied with great success in the design of aircraft, namely the concept of large eddy (or mean-field) simulations. Global fluid simulations currently cannot guarantee scale separation for all relevant physical scales. To obtain quantitatively correct results, we therefore believe that a sub-grid scale model is inevitable. Present local box simulations are a valuable means to provide a rigorous framework for the calibration of such a model.

Acknowledgements

This work used the NIRVANA code version 3.3 developed by Udo Ziegler at the Astrophysical Institute Potsdam. Computations were performed at the AIP `babel` cluster.

References

Balsara D., Benjamin R. A., & Cox D. P., 2001, *The Astrophysical Journal*, 563, 800

Balsara D. S. & Kim J., 2005, *The Astrophysical Journal*, 634, 390

Balsara D. S., Kim J., Mac Low M.-M., & Mathews G. J., 2004, *The Astrophysical Journal*, 617, 339

Bardou A., von Rekowski B., Dobler W., Brandenburg A., & Shukurov A., 2001, *Astronomy & Astrophysics*, 370, 635

Beck R., Brandenburg A., Moss D., Shukurov A., & Sokoloff D., 1996, *Ann. Rev. Astron. & Astroph.*, 34, 155

Brandenburg A. & Subramanian K., 2005, *Phys. Rep.*, 417, 1

de Avillez M. A. & Breitschwerdt D., 2005a, *Astronomy & Astrophysics*, 436, 585

de Avillez M. A. & Breitschwerdt D., 2005b, *The Astrophysical Journal*, 634, L65

Dettmar R. J., 1992, Fundamentals of Cosmic Physics, 15, 143

Dib S., Bell E., & Burkert A., 2006, *The Astrophysical Journal*, 638, 797

Dziourkevitch N., Elstner D., & Rüdiger G., 2004, *Astronomy & Astrophysics*, 423, L29

Elstner D., Gressel O., & Rüdiger G., 2009, in IAU Symposium Vol. 259, pp 467–478

Ferrière K., 1998, *Astronomy & Astrophysics*, 335, 488

Ferrière K. M., 2001, RMP, 73, 1031

Fletcher A., Beck R., Shukurov A., Berkhuijsen E. M., & Horellou C., 2010, (astro-ph:1001.5230)

Fröhlich H.-E. & Schultz M., 1996, *Astronomy & Astrophysics*, 311, 451

Fromang S. & Stone J. M., 2009, *Astronomy & Astrophysics*, 507, 19

Gissinger C., Fromang S., & Dormy E., 2009, *Monthly Notices of the Royal Astronomical Society*, 394, L84

Gressel O., 2009, PhD thesis, University of Potsdam (astro-ph:1001.5187)

Gressel O., Ziegler U., Elstner D., & Rüdiger G., 2008, AN, 329, 619

Gressel O., Elstner D., Ziegler U., & Rüdiger G., 2008, *Astronomy & Astrophysics*, 486, L35

Gressel O., Ziegler U., Elstner D., & Rüdiger G., 2009, in IAU Symposium Vol. 259, pp 81–86

Hanasz M., Kowal G., Otmianowska-Mazur K., & Lesch H., 2004, *The Astrophysical Journal*, 605, L33

Hanasz M., Otmianowska-Mazur K., Kowal G., & Lesch H., 2009, *Astronomy & Astrophysics*, 498, 335

Haugen N. E. L., Brandenburg A., & Mee A. J., 2004, *Monthly Notices of the Royal Astronomical Society*, 353, 947

Joung M. K. R. & Mac Low M.-M., 2006, *The Astrophysical Journal*, 653, 1266

Kaisig M., Rüdiger G., & Yorke H. W., 1993, *Astronomy & Astrophysics*, 274, 757

Kitchatinov L. L. & Rüdiger G., 2004, *Astronomy & Astrophysics*, 424, 565

Korpi M. J., Brandenburg A., Shukurov A., & Tuominen I., 1999b, *Astronomy & Astrophysics*, 350, 230

Korpi M. J., Brandenburg A., Shukurov A., Tuominen I., & Nordlund Å., 1999, *The Astrophysical Journal*, 514, L99

Korpi M. J., Käpylä P. J., & Väisälä M. S., 2010, Astronomische Nachrichten, 331, 34

Kotarba H., Karl S. J., Naab T., Johansson P. H., Dolag K., Lesch H., & Stasyszyn F. A., 2010, *The Astrophysical Journal*, 716, 1438

Mac Low M.-M., Balsara D. S., Kim J., & de Avillez M. A., 2005, *The Astrophysical Journal*, 626, 864

Mee A. J. & Brandenburg A., 2006, *Monthly Notices of the Royal Astronomical Society*, 370, 415

Melioli C. & de Gouveia Dal Pino E. M., 2004, *Astronomy & Astrophysics*, 424, 817

Nishikori H., Machida M. & Matsumoto R., 2006, *The Astrophysical Journal*, 641, 862

Parker E. N., 1992, *The Astrophysical Journal*, 401, 137

Piontek R. A. & Ostriker E. C., 2007, *The Astrophysical Journal*, 663, 183

Rüdiger G. & Kitchatinov L. L., 1993, *Astronomy & Astrophysics*, 269, 581

Schrinner M., Rädler K.-H., Schmitt D., Rheinhardt M., & Christensen U., 2005, AN, 326, 245

Sellwood J. A. & Balbus S. A., 1999, *The Astrophysical Journal*, 511, 660

Slyz A. D., Devriendt J. E. G., Bryan G., & Silk J., 2005, *Monthly Notices of the Royal Astronomical Society*, 356, 737

Sokoloff D. & Shukurov A., 1990, Nat., 347, 51

Spitzer L., 1978, Physical processes in the ISM. New York Wiley-Interscience, 1978. 333 p.

Sur S., Shukurov A. & Subramanian K., 2007, *Monthly Notices of the Royal Astronomical Society*, 377, 874

Ziegler U., Yorke H. W., & Kaisig M., 1996, *Astronomy & Astrophysics*, 305, 114

Advances in Plasma Astrophysics
Proceedings IAU Symposium No. 274, 2010
A. Bonanno, E. de Gouveia Dal Pino & A.G. Kosovichev, eds.

© International Astronomical Union 2011
doi:10.1017/S1743921311007265

Cosmic-ray driven dynamo in galaxies

M. Hanasz[1], D. Wóltanski[1], K. Kowalik[1] and H. Kotarba[2]

[1] Centre for Astronomy, Nicolaus Copernicus University, ul. Gagarina 11, PL-87-100 Torun,
Poland

[2] University Observatory Munich, Scheinerstr. 1, D-81679 Munich, Germany

Abstract. We present recent developments of global galactic-scale numerical models of the Cosmic Ray (CR) driven dynamo, which was originally proposed by Parker (1992). We conduct a series of direct CR+MHD numerical simulations of the dynamics of the interstellar medium (ISM), composed of gas, magnetic fields and CR components. We take into account CRs accelerated in randomly distributed supernova (SN) remnants, and assume that SNe deposit small-scale, randomly oriented, dipolar magnetic fields into the ISM. The amplification timescale of the large-scale magnetic field resulting from the CR-driven dynamo is comparable to the galactic rotation period. The process efficiently converts small-scale magnetic fields of SN-remnants into galactic-scale magnetic fields. The resulting magnetic field structure resembles the X-shaped magnetic fields observed in edge-on galaxies.

Keywords. Galaxies: ISM, Magnetic Fields; ISM: Cosmic Rays, Magnetic Fields; MHD: Dynamos

1. Introduction

The dynamical role of CRs was first recognized by Parker, (1966), who noticed that a vertically stratified ISM which consists of thermal gas, magnetic fields and CRs is unstable due to buoyancy of the weightless components, i.e. the magnetic fields and the CRs. According to diffusive shock acceleration models CRs are continuously supplied to the ISM by SN remnants. Therefore, the buoyancy effects caused by CRs are expected in all star forming galaxies. Theories of diffusive shock acceleration predict that about 10 % of the $\sim 10^{51}$ erg of the SN II explosion energy is converted to CR energy. Observational data indicate that gas, magnetic fields and CRs appear in approximate energetic equipartition, which means that all three components are dynamically coupled. In order to incorporate the CR propagation in MHD considerations we use the diffusion–advection equation (e.g. Schlickeiser & Lerche, 1985) and take into account the CR pressure gradient in the gas equation of motion (see e.g. Berezinski *et al.*, 1990).

The CR-driven dynamo was originally proposed by Parker, 1992. Our model of the CR-driven dynamo involves the following elements (Hanasz *et al.*, 2004; Hanasz *et al.*, 2006; Hanasz *et al.*, 2009b; Hanasz *et al.*, 2009c): (1) The CR nuclear component described by the diffusion–advection transport equation, supplemented to the standard set of resistive MHD equations (Hanasz & Lesch, 2003). (2) CRs supplied in SN remnants. The CR input of individual SNe is assumed to be 10% of the typical SN kinetic energy output (= 10^{51} erg), while the thermal energy output from supernovae is neglected. (3) Anisotropic CR diffusion along magnetic field lines (Giacalone & Jokipii, 1999), (4) Finite resistivity of the ISM in order to permit topological evolution of the galactic magnetic fields via anomalous resistivity processes (Hanasz *et al.*, 2002), and/or via turbulent reconnection (Kowal *et al.*, 2009) on small spatial scales, which are unresolved in our simulations. (5) An initial gas distribution in the disk which follows the model of the ISM in the Milky

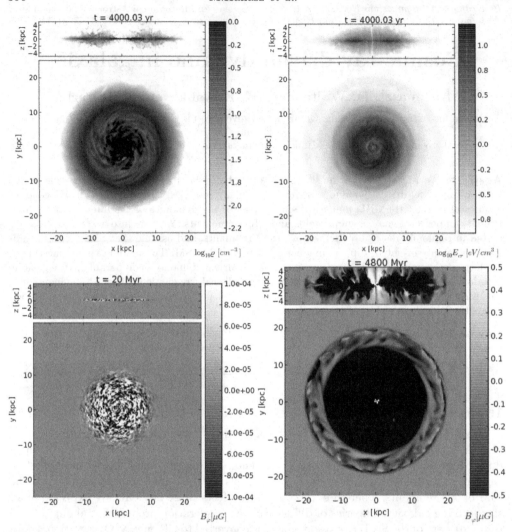

Figure 1. Top panels: logarithm of gas number density (left) and cosmic ray energy density (right) at $t = 4\,\mathrm{Gyr}$. Bottom panels: the distribution of toroidal magnetic field at $t = 20\,\mathrm{Myr}$ (left) and $t = 4.8\,\mathrm{Gyr}$ (right). Unmagnetized regions of the volume are grey, while positive and negative toroidal magnetic fields are marked lighter and darker, respectively. Note that the grey-scale scale in magnetic field maps is saturated to enhance weaker magnetic field structures in disk peripheries.

Way by Ferrière (1998). (6) Differential rotation of the interstellar gas, which currently follows an assumed form of a galactic gravitational potential.

We briefly mention that various properties of the shearing-box models of the CR-driven dynamo were discussed in a series of papers. Computations of the dynamo coefficients in Parker unstable disks with CRs and shear are described by Kowal *et al.*, (2006) and by Otmianowska-Mazur *et al.*, (2007). Synthetic radio-maps of a global galactic disk based on local CR-driven dynamo models exhibiting X-type structures were presented by Otmianowska-Mazur *et al.*(2009). More recently, Siejkowski *et al.*, (2010) (see also Siejkowski *et al.*, 2010, this volume) demonstrated that the CR-driven dynamo can also work given the physical conditions of irregular galaxies, characterized by a relatively weak rotation and shearing rate.

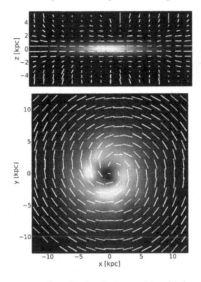

Figure 2. Synthetic radio maps of polarized intensity (PI) of synchrotron emission, together with polarization vectors are shown for the edge-on and face-on views of the galaxy at $t = 4.8$ Myr. Vectors direction resembles electric vectors rotated by $90°$, and their lengths are proportional to the degree of polarization.

2. Global CR-driven dynamo simulations

Our first realization of a global CR+MHD galactic disk model relies on the following assumptions: First of all, we adopt analytical formulae for the gravitational potential corresponding to a system consisting of a galactic halo, bulge and disk Allen & Santillan, 1991. We assume no magnetic field at $t = 0$, and that weak ($10^{-4}\,\mu$G), dipolar, small scale ($r \sim 50$ pc), randomly oriented magnetic fields are supplied locally in 10% of the SN remnants for $t \leqslant 1$ Gyr. We assume also that the SN rate is proportional to the star formation rate (SFR) which, on the other hand, is proportional to the initial gas column density.

We use the PIERNIK MHD code (Hanasz *et al.*, 2010a; Hanasz *et al.*, 2010b; Hanasz *et al.*, 2008; Hanasz *et al.*, 2009a) which is a grid-MHD code based on the Relaxing TVD (RTVD) scheme (Jin & Xin, 1995; Pen *et al.*, 2003). PIERNIK is parallelized by means of block decomposition with the aid of the MPI library. The original scheme was extended to deal with dynamically independent but interacting fluids, i.e. thermal gas and a diffusive CR gas, which is described within the fluid approximation (Hanasz *et al.*, 2008).

We find magnetic field amplification originating from the small-scale, randomly oriented dipolar magnetic fields, which is apparent through the exponential growth by several orders of magnitude of both the magnetic flux and the magnetic energy (details see Hanasz *et al.*, 2009). The growth phase of the magnetic field starts at the beginning of the simulation. The growth of the magnetic field strength saturates at about $t = 4$ Gyr, reaching values of $3-5\,\mu$G in the disk. During the amplification phase, magnetic flux and total magnetic energy grow by about 6 and 10 orders of magnitude, respectively. The average e-folding time of magnetic flux amplification is approximately equal to 270 Myr, corresponding to the rotation at the galactocentric radius (≈ 10 kpc). The magnetic field is initially entirely random ($t = 20$ Myr), since it originates from randomly oriented magnetic dipoles. Later on, the toroidal magnetic field component forms a spiral structure revealing reversals in the plane of the disk. The magnetic field structure evolves gradually towards larger and lager scales. The toroidal magnetic field component becomes

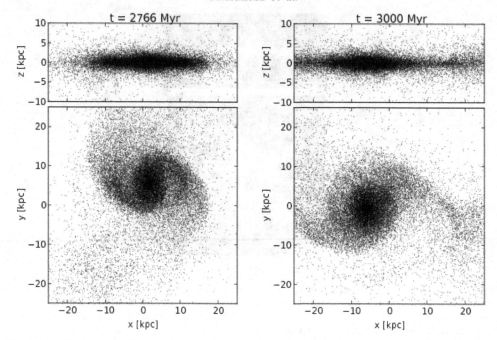

Figure 3. Two states of the N-body galactic disk prior to the galactic merger. The companion galaxy is apparent in the right panels.

almost uniform inside the disk at $t = 2.5\,\mathrm{Gyr}$. The volume occupied by the well-ordered magnetic field expands continuously until the end of the simulation.

In order to visualize the magnetic field structure in a manner resembling radio observations of external galaxies, we construct synthetic radio maps of the synchrotron radio-emission, assuming that energy density of CR electrons equals 1% energy density of CR nucleons. We apply standard procedures of line-of-sight integration of the stokes parameters I, Q, and U for the polarized synchrotron emissivity. We neglect the effects of of Faraday rotation.

In Figure 2, we show the polarized intensity of synchrotron emission (grey-scale maps), together with polarization vectors. Electric vectors, computed on the basis of integrated Stokes parameters, are rotated by 90° to reproduce the magnetic field direction averaged along the line-of-sight, assuming vanishing Faraday rotation effects. The polarization vectors, indicating the mean magnetic field direction, reveal a regular spiral structure in the face-on view, and the so-called *X-shaped structure* in the edge-on view. A particular similarity can be noticed between our edge-on synthetic radio map and the radio maps of observed edge-on galaxies such as NGC 891 (Krause, 2009). In the present global model, the X-shaped configuration is an intrinsic property of the magnetic field structure, since it corresponds closely to the flaring radial distribution of magnetic field in the disk and its neighborhood, as shown in Figure 1.

The face-on synchrotron radio map reveals a spiral structure of the magnetic field, however, due to the assumed axisymmetric gravitational potential no features resembling spiral magnetic arms are present. To make the model more realistic, we incorporate non-axisymmetry in the gravitational potential, applying two different approaches. The first approach relies on the addition of an analytical elliptical perturbation to the axisymmetric gravitational potential. The results of this approach, which are presented by Kulpa-Dybel *et al.*, (2010, this volume), indicate that the CR-driven dynamo model

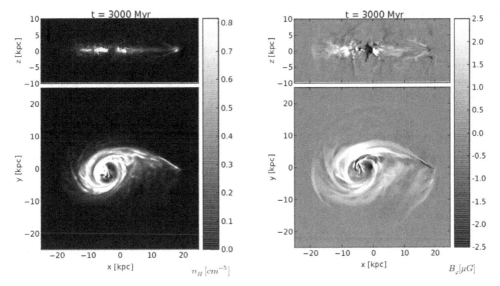

Figure 4. Gas density (left panel) and magnetic field (right panel) corresponding to the gravitational potential of the N-body system at the right plot of Fig.3

reveals new properties, such as the presence of a ring-like structure as well as a shift of the magnetic arms with respect to the crests of spiral density waves.

As a further step towards more realistic galactic magnetic field models we perform N-body simulations of a disk-bulge-halo system (Wóltański *et al.*, 2011), and interpolate the resulting gravitational potential onto the computational grid. We use this potential to compute the gravitational acceleration acting on the fluid components. In order to excite density waves in the galactic disk we add a small satellite galaxy, which ultimately merges with the main galaxy at $t = 3.2\,\text{Gyr}$. Two snapshots of the N-body disk simulation are displayed in Fig. 3. The N-body part of the computation is performed with the VINE code (Wetzstein *et al.*, 2009), and the CR+MHD part with the PIERNIK code. Fig. 4 shows the gas density and the toroidal magnetic field component. Similar to the case of a axisymmetric gravitational potential we observe efficient magnetic field amplification by the CR-driven dynamo in the presence of density waves in the galactic disk. In the presence of spiral arms in the stellar and gas components the magnetic field also reveals a spiral structure. Moreover, we notice that both polarities of the azimuthal magnetic field are present at an advanced stage of magnetic field evolution. It appears that during the merger phase and afterwards the magnetic field structure becomes even more disordered, providing a possible explanation for the less regular magnetic field structures observed in interacting galaxies such as M51 (Fletcher *et al.*, 2010).

3. Conclusions

We have shown that the contribution of CRs to the dynamics of the ISM on a global galactic scale, studied by means of CR+MHD simulations, leads to a very efficient magnetic field amplification on the timescale of galactic rotation. The model applying a fixed analytical gravitational potential reveals a large scale regular magnetic field with apparent spiral structure in the face-on view and a X-shaped structure in the edge-on view. In the presence of spiral perturbations excited in the stellar component, the magnetic field structure follows these perturbations in the stellar and gaseous components. The magnetic field structure becomes less regular compared to the axisymmetric case. Dynamical magnetic field structures with opposite polarities develop within the disk and are present even at the saturation phase of the dynamo. Moreover, during the coalescence phase of the two galaxies the magnetic field structure becomes irregular as observed in M51. An important part of the CR-driven dynamo is the galactic wind which reaches velocities of a few hundred km/s at galactic altitudes of a few kpc. The mass of gas transported out of the disk is about $1M_\odot$/yr for the star formation rate of the Milky Way.

Acknowledgements

This work was supported by Polish Ministry of Science and Higher Education through the grants 92/N–ASTROSIM/2008/0 and N N203 511038.

References

Allen, C. & Santillan, A.: 1991, *Revista Mexicana de Astronomia y Astrofisica* **22**, 255

Berezinskii, V. S., Bulanov, S. V., Dogiel, V. A., & Ptuskin, V. S.: 1990, *Astrophysics of cosmic rays*, Amsterdam: North-Holland, 1990, edited by Ginzburg, V.L.

Ferriere, K.: 1998, *ApJ* **497**, 759

Fletcher, A., Beck, R., Shukurov, A., Berkhuijsen, E. M., & Horellou, C.: 2010, *ArXiv e-prints*

Giacalone, J. & Jokipii, J. R.: 1999, *ApJ* **520**, 204

Hanasz, M., Kowal, G., Otmianowska-Mazur, K., & Lesch, H.: 2004, *ApJL* **605**, L33

Hanasz, M., Kowalik, K., Wóltański, D., & Pawłaszek, R.: 2008, *ArXiv e-prints*

Hanasz, M., Kowalik, K., Wóltański, D., & Pawłaszek, R.: 2009a, *ArXiv e-prints*

Hanasz, M., Kowalik, K., Wóltański, D., & Pawłaszek, R.: 2010a, in K. Goździewski, A. Niedzielski, & J. Schneider (ed.), *EAS Publications Series*, Vol. 42 of *EAS Publications Series*, pp 275–280

Hanasz, M., Kowalik, K., Wóltański, D., Pawłaszek, R., & Kornet, K.: 2010b, in K. Goździewski, A. Niedzielski, & J. Schneider (ed.), *EAS Publications Series*, Vol. 42 of *EAS Publications Series*, pp 281–285

Hanasz, M. & Lesch, H.: 2003, *A&A* **412**, 331

Hanasz, M., Otmianowska-Mazur, K., Kowal, G., & Lesch, H.: 2006, *Astronomische Nachrichten* **327**, 469

Hanasz, M., Otmianowska-Mazur, K., Kowal, G., & Lesch, H.: 2009b, *A&A* **498**, 335

Hanasz, M., Otmianowska-Mazur, K., & Lesch, H.: 2002, *A&A* **386**, 347

Hanasz, M., Wóltański, D., & Kowalik, K.: 2009c, *ApJL* **706**, L155

Jin, S. & Xin, Z.: 1995, *Comm. Pure Appl. Math.* **48**, 235

Kowal, G., Lazarian, A., Vishniac, E. T., & Otmianowska-Mazur, K.: 2009, *ApJ* **700**, 63

Kowal, G., Otmianowska-Mazur, K., & Hanasz, M.: 2006, *A&A* **445**, 915

Krause, M.: 2009, in *Revista Mexicana de Astronomia y Astrofisica Conference Series*, Vol. 36 of *Revista Mexicana de Astronomia y Astrofisica, vol. 27*, pp 25–29

Otmianowska-Mazur, K., Kowal, G., & Hanasz, M.: 2007, *ApJ* **668**, 110

Otmianowska-Mazur, K., Soida, M., Kulesza-Żydzik, B., Hanasz, M., & Kowal, G.: 2009, *ApJ* **693**, 1

Parker, E. N.: 1966, *ApJ* **145**, 811

Parker, E. N.: 1992, *ApJ* **401**, 137

Pen, U.-L., Arras, P., & Wong, S.: 2003, *ApJS* **149**, 447

Schlickeiser, R. & Lerche, I.: 1985, *A&A* **151**, 151

Siejkowski, H., Soida, M., Otmianowska-Mazur, K., Hanasz, M., & Bomans, D. J.: 2010, *A&A* **510**, A97+

Wetzstein, M., Nelson, A. F., Naab, T., & Burkert, A.: 2009, *ApJS* **184**, 298

Wóltański, D., Hanasz, M., Kowalik, K., & Kotarba, H.: 2011, *ApJ(in prep)*

Advances in Plasma Astrophysics
Proceedings IAU Symposium No. 274, 2010
A. Bonanno, E. de Gouveia Dal Pino & A.G. Kosovichev, eds.

© International Astronomical Union 2011
doi:10.1017/S1743921311007277

On the solution of the Kompaneets equation in cosmological context: a numerical code to predict the CMB spectrum under general conditions

C. Burigana, P. Procopio, and A. De Rosa

INAF - IASF Bologna, Via P. Gobetti 101, 40129 Bologna

Abstract. Interpretation of current and future data calls for a continuous improvement in the theoretical modeling of CMB spectrum. We describe the new version of a numerical code, KYPRIX, specifically written to solve the Kompaneets equation in a cosmological context under general assumptions. We report on the equation formalism, and structure and computational aspects of the code. New physical options have been introduced in the current code version: the cosmological constant in the terms controlling the general expansion of the Universe, the relevant chemical abundances, and the ionization history, from recombination to cosmological reionization. We present some of fundamental tests we carried out to verify the accuracy, reliability, and performance of the code. All the tests demonstrate the reliability and versatility of the new code version and its accuracy and applicability to the scientific analysis of current CMB spectrum data and of much more precise measurements that will be available in the future.

Keywords. Cosmology, cosmic microwave background, radiative processes, numerical methods

1. Introduction

Ground-based, balloon-born, and space experiments, and, in particular, the COBE/FIRAS results (Fixsen *et al.*, 1996), confirm that only very small deviations from a Planckian shape can be present in the CMB spectrum. Therefore, current and future CMB absolute temperature experiments aim at discovering very small distortions that are predicted to be associated with the cosmological reionization process or that could be generated by different kinds of earlier processes. Interpretation of future data (Fixsen & Mather, 2002; Burigana & Salvaterra, 2003, Mather, 2009) calls for a continuous improvement in the theoretical modeling of CMB spectrum. We describe the new version (Procopio & Burigana, 2009) of a numerical code, KYPRIX, specifically written to solve the Kompaneets equation in a cosmological context and first implemented in the years 1989–1991, to accurately compute the CMB spectral distortions under general assumptions. After the presentation of the necessary formalism and of the main structure of the code, we describe here the new physical options introduced in the current version.

2. The Kompaneets equation in cosmological context

Physical processes occurring at redshifts $z < z_{therm}$ may leave imprints on the CMB spectrum. Thus, it carries crucial informations on physical processes occurring during early cosmic epochs (see e.g. Danese & Burigana, 1993 and references therein) and the comparison between models of CMB spectral distortions and CMB absolute temperature measures can constrain their physical parameters (Burigana *et al.*, 1991b).

The time evolution of the photon occupation number, $\eta(\nu, t)$, under the effect of Compton scattering and of photon production processes, radiative Compton (RC) (Gould, 1984), bremsstrahlung (B) (Karzas & Latter, 1961; Rybicki & Lightman, 1979), plus other possible photon emission/absorption contributions (EM)†, is well described by the complete Kompaneets equation (Kompaneets, 1956; Burigana *et al.*, 1995):

$$\frac{\partial \eta}{\partial t} = \frac{1}{\phi}\frac{1}{t_C}\frac{1}{x^2}\frac{\partial}{\partial x}\left[x^4\left[\phi\frac{\partial \eta}{\partial x} + \eta(1+\eta)\right]\right] + \left[\frac{\partial \eta}{\partial t}\right]_{RC} + \left[\frac{\partial \eta}{\partial t}\right]_{B} + \left[\frac{\partial \eta}{\partial t}\right]_{EM}, \qquad (2.1)$$

where $t_C = t_{\gamma e}m_e c^2/(kT_e) \simeq 4.5\times10^{28}\,(T_0/2.7\,K)^{-1}\phi^{-1}\widehat{\Omega}^{-1}(1+z)^{-4}\,\sec$, $t_{\gamma e} = 1/(n_e\sigma_T c)$ is the photon–electron collision time, $\phi = (T_e/T_r)$, T_e and $T_r = T_0(1 + z)$ being respectively the electron and the CMB radiation temperature; and $x = h\nu_0/kT_0 = h\nu_0(1 + z)/kT_0(1 + z)$ is a dimensionless, redshift independent, frequency (ν_0 being the present frequency), being $\epsilon_{r0} = aT_0^4$ the radiation energy density today. This equation is coupled to the time differential equation governing the electron temperature evolution for an arbitrary radiation spectrum in the presence of the above processes plus adiabatic cooling and possible external heating sources, $q = a^{-3}(dQ/dt)$,

$$\frac{dT_e}{dt} = \frac{T_{eq,C} - T_e}{(27/28)t_{e\gamma}} + \left[\frac{dT_e}{dt}\right]_{RC,B,EM} - \frac{2T_e}{t_{exp}} + \frac{(32/27)q}{3n_e k}; \qquad (2.2)$$

here $T_{eq,C} = [h\int\eta(1+\eta)\nu^4 d\nu]/[4k\int\eta\nu^3 d\nu]$ is the Compton equilibrium electron temperature (Peyraud, 1968, Zeldovich & Levich, 1970), $t_{e\gamma} = 3m_e c/(4\sigma_T\epsilon_r)$, $\epsilon_r \simeq \epsilon_{r0}(1 + z)^4$.

3. The numerical code KYPRIX

The numerical code KYPRIX was written to overcome the limited applicability of analytical solutions and to get a precise computation of the evolution of the photon distribution function for a wide range of cosmic epochs and for many cases of cosmological interest (Burigana *et al.*, 1991a). KYPRIX makes use of the NAG libraries (NAG Ltd, 2009) and of a lot of numerical algorithms, available to the scientific community or written specifically for code dedicated tasks. The D03PCF routine of the current version of the NAG release has been used to reduce the Kompaneets equation into a system of ordinary differential equations (Dew & Walsh, 1981; Berzins *et al.*, 1989; Skeel & Berzins, 1990). The numerical framework is the same of the D03PGF routine, no longer available, used in the first versions of KYPRIX, but they come from different technical implementations and present remarkable differences.

The code is divided in several sections. 1) Main program, for the defiinition of: choice of the physical processes, choice of the cosmological parameters, initial conditions, characteristics of the numerical integration (accuracy, number of points of the grid), time interval of interest, choice of the boundary conditions, chemical abundances, ionization history. 2) Subroutine PDEDEF, the subprogram where the problem is numerically defined. This subroutine is also divided in subsections to allow modifications in a simple and practical way. 3) Subroutine BNDARY, where the boundary conditions are numerically specified. 4) Subroutines and auxiliary functions to perform specific calculations.

During the numerical integration, some subprograms use the distribution function

---

† A process that in principle should be included is the cyclotron emission. On the other hand, for realistic values of cosmic magnetic field and CMB realistic distorted spectra, the cyclotron process never plays an important role for (global) CMB spectral distortions when ordinary and stimulated emission and absorption are properly taken into account (Zizzo & Burigana, 2005).

calculated at that time to compute ϕ. The integrals to be computed are those that we find in the expression for $\phi_{eq,C} = T_{eq,C}/T_r$. In this calculation, the integration range is obviously the interval considered for the problem: $A \leqslant X \leqslant B$ (that, in terms of mesh ordering, corresponds to the range between 1 and $NPTS$ or $NPTS - 1$; the adopted grid uses $X = \log x$). For computing these integrals, all the points of the grid are used. The integration is based on the NAG D01GAF routine, suitable for tabulated functions. Thanks to the opportunity of having the correct value of ϕ for each time step, the update of the boundary conditions can be physically motivated. We implemented in the code the possibility to adopt a particular case of Neumann boundary conditions: the requirement that the current density, in the frequency space, is null at the boundaries of the integration range (Chang & Cooper, 1970): $[\phi \partial \eta / \partial x + \eta(1 + \eta)]_{x = x_{min}, x_{max}} = 0$.

4. Physical options of the code

The code KYPRIX has been written to solve the Kompaneets equation in many kinds of situations. The physical processes that can be considered are: Compton scattering, bremsstrahlung, radiative Compton scattering, sources of photons, energy injections without photon production, energy exchanges (heating or cooling processes) associated to $\phi \neq 1$ at low redshifts, radiative decays of massive particles, and so on (see e.g. Danese & Burigana, 1993 for some applications). This code could be easily implemented to consider other kinds of physical processes. Various kinds of initial conditions for the problem can be considered and many of them have been already implemented in KYPRIX. The first obvious case is a pure Planckian spectrum. Several ways to model an instantaneuos heating implying deviations from the Planckian spectrum have been introduced: a pure Bose-Einstein (BE) spectrum or a BE spectrum modified to become Planckian at low frequencies; a grey-body spectrum; a superposition of blackbodies.

We have then updated the numerical integration code KYPRIX to include the cosmological constant and the curvature in the terms controlling the general expansion of the Universe, thus making the code suitable to be accurately applied at late ages. Remarkable examples are spectral distortions associated to the reionization of the Universe (Burigana *et al.*, 2008) that, in typical astrophysical scenarios, starts $z < 10 - 20$.

In this new version of the code it is possible to choose the primordial abundances of hydrogen, H, and helium, He, and to consistently compute n_e. Denoting with f_H the fraction in mass of primordial H, since $n_e = n_H + 2n_{He}$, we have $n_e^{free} = n_e^{tot} = [(1 + f_H)/2](\rho_b/m_b)$. This obviously impacts the physical processes involved in the code. Note that Compton scattering, radiative Compton, and bremsstrahlung depend linearly on n_e^{free} (for bremsstrahlung, the dependence on the densities of nuclei is now explicit).

We have then included the accurate modeling of the ionization history in the current implementation of KYPRIX. In particular, the fraction of each state of ionization of the relevant elements (H and He) has been implemented. Given the electron ionization fraction, χ_e, from the charge conservation law we have a constraint on the number of the free ions in the considered plasma. The simplest way to take count of them in the code is to assume an equal fraction of ionization for H and He. A more accurate treatment of the physics of reionization/recombination processes implemented in the code is based on the Saha equation. The code can also ingest a table with the desired evolution of χ_e and, as necessary for a physical modeling of cosmological reionization, of the electron temperature. Obviously, we have implemented the best way to perform the exact calculation of the rates of considered processes in scenarios involving reionization/recombination which is that of using a co-running code, coupled to KYPRIX, able to supply the ionization fraction for all the species. For the recombination process, we developed an interface that allows to call an external program, in our case RECFAST (Seager et al., 1999), and run it with the same cosmological parameters selected for KYPRIX.

5. Code porting, tests, and conclusion

We successfully finalized the porting of the code KYPRIX through different platforms and environments, i.e. VAX machines (original platform, disused), DEC machines, IBM Power5/Power6 and Intel based machines. All the working machines in which the code was run on have a 64 bit OS. We carried out fundamental tests to verify the accuracy, reliability, and performance of the code: energy conservation, time behavior of electron temperature, comparison between the results obtained with both the update and the original version of the code, and properties of the free-free distortions relevant at long wavelengths. All the tests demonstrate the reliability and versatility of the new code version and its accuracy and applicability to the scientific analysis of current CMB spectrum data and of much more precise measurements that will be available in the future.

Acknowledgements

Calculations have been carried out on the IBM SP5/512 machine at CINECA-Bologna and on the computational facilities at INAF-IASF Bologna. We warmly thank M. Genghini and C. Gheller for the technical support related to the machines used. We acknowledge the use of the code RECFAST. The ASI contract I/016/07/0 "COFIS" is acknowledged.

References

Berzins, M. 1990, in *Scientific Software Systems*, eds. J. C. Mason & M. G. Cox, 59, Chapman & Hall, p. 59
Berzins, M., Dew, P. M., & Furzeland, R. M. 1989, *Appl. Numer. Math.*, 5, 375
Burigana, C., Danese, L., & de Zotti, G. 1991, *A&A*, 246, 49
Burigana, C., Danese, L., & de Zotti, G. 1991, *ApJ*, 379, 1
Burigana, C., de Zotti, G., & Danese, L. 1995, *A&A*, 303, 323
Burigana, C. & Salvaterra, R. 2003, *MNRAS*, 342, 543
Burigana, C., et al. 2008, *MNRAS*, 385, 404
Chang, J. S. & Cooper, G. 1970, *J. Comput. Phys.*, 6, 1
Danese, L., Burigana, C. 1993, in *Present and Future of the Cosmic Microwave Background*, Lecture in Physics, Vol. 429, eds. J. L. Sanz et al., Springer, Heidelberg (FRG), p. 28
Dew, P. M., Walsh, J. 1981, *ACM Trans. Math. Software*, 7, 295
Fixsen, D. J., et al. 1996, *ApJ*, 473, 576
Fixsen, D. J., Mather, J. C. 2002, *ApJ*, 581, 817
Gould, R. J. 1984, *ApJ*, 285, 275
Karzas, W. J., Latter, R. 1961, *ApJS*, 6, 167
Kompaneets, A. S. 1956, *Zh. Eksp. Teor. Fiz.*, 31, 876 [Sov. Phys. JEPT, 4, 730, (1957)]
Mather, J. C. 2009, in *Questions of Modern Cosmology – Galileo's Legacy*, eds. M. D'Onofrio & C. Burigana, Springer, Sect. 5.3.1, p. 435
The Numerical Algorithms Group Ltd 2009, Oxford, UK
Peyraud, N. 1968, *Physics Letters A*, 27, 410
Procopio, P. & Burigana, C. 2009, *A&A*, 507, 1243
Rybicki G. B. & Lightman A. P. 1979, *Radiative processes in astrophysics*, Wiley, New York
Seager, S., Sasselov, D. D., Scott, D. 1999, *ApJ*, 523, L1
Skeel, R. D. & Berzins, M. 1990, *SIAM J. Sci. Statist. Comput.*, 11(1), 1
Zeldovich, Y. B. & Levich, E. V. 1970, *Zh. Eksp. Teor. Fiz.*, 11, 57
Zizzo, A. & Burigana, C. 2005, *New Astronomy*, 11, 1

Advances in Plasma Astrophysics
Proceedings IAU Symposium No. 274, 2010   © International Astronomical Union 2011
A. Bonanno, E. de Gouveia Dal Pino & A.G. Kosovichev, eds.   doi:10.1017/S1743921311007289

Statistical tools of interstellar turbulence: connecting observations with theory

B. Burkhart & A. Lazarian

Astronomy Department, University of Wisconsin, Madison, 475 N. Charter St., WI 53711, USA
email: `burkhart@astro.wisc.edul`

Abstract. MHD Turbulence is a critical component of the current paradigms of star formation, particle transport, magnetic reconnection and evolution of the ISM, to name just a few. Progress on this difficult subject is made via numerical simulations and observational studies, however in order to connect these two, statistical methods are required. This calls for new statistical tools to be developed in order to study turbulence in the interstellar medium. Here we briefly review some of the recently developed statistics that focus on characterizing gas compressibility and magnetization and their uses to interstellar studies.

Keywords. turbulence, methods:statistical numerical, ISM: general

---

1. Introduction

The paradigm of the interstellar medium has undergone major shifts in the past two decades thanks to the combined efforts of high resolution surveys and the exponential increase in computation power allowing for more realistic numerical simulations. The ISM is now known to be highly turbulent and magnetized, which affects ISM structure, formation, and evolution. Magnetohydrodynamic turbulence is essential to many astrophysical phenomena such as star formation, cosmic ray dispersion, magnetic reconnection, and many transport processes. However the study of turbulence is complicated by the fact that no complete theory for turbulence exists.

To this date, turbulence has always been understood in a statistical manner - showing the 'order from chaos.' The classical picture of Kolmogrov 1941 depicts turbulent flows as composed of eddies which transfer energy across a range of scales, typically from the larger injection scale down to what is know as the dissipative range. The eddies are unstable and break into smaller and smaller eddies, transferring their kinetic energy in a manner that conserves it over time. This picture of the 'energy cascade' is highly dependent on the characteristic dimensionless Reynolds number, which describes the ratio of the inertial to viscous forces. By simple dimensional analysis it can be shown that the energy spectrum scales as $k^{-5/3}$. More current research shows that turbulence is intermittent and not statistically self-similar, as is assumed in the Kolmogrov picture. To add to the complexity, the ISM is in a plasma state and gas dynamics are governed by the MHD equations, which changes the the scaling relationships via interactions of MHD waves and Alfvénic shearing.

In addition to the Reynolds number, other important parameters of ISM MHD turbulence include the sonic Mach number, which describes the compressibility of the gas as $\mathcal{M}_s \equiv \mathbf{v}/C_s$, and the Alfvénic Mach number $\mathcal{M}_A \equiv \mathbf{v}/v_A$, where $v_A = |\mathbf{B}|/\sqrt{\rho}$ is the Alfvénic velocity, $\mathbf{B}$ is magnetic field and ρ is density. These parameters are not always easy to characterize observationally, with the Alfvénic Mach number being particularly difficult due to cumbersome observational measurements of vector magnetic field.

Several techniques have been around for decades in order to study ISM turbulence and its properties. Many of these hinge on either density fluctuations, via scintillation in ionized media, Radio position-position-velocity data (PPV) and column density maps

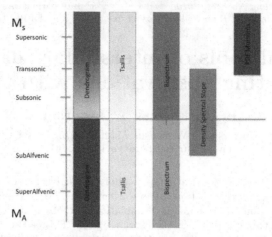

Figure 1. Cartoon showing different statistics studied with their dependencies on the sonic and Alfvénic Mach numbers. The different colors indicate different staitistics. The intensity of the colors indicate the confidence the statistics can provide the parameter on the y-axis in our simulations.

for neutral media. The advantage of spectroscopic data is that it contains information about the turbulent velocity field as well as the density fluctuations, however separation of the density and velocity fields is non trivial (Lazarian 2009). When invoking the phenomena of turbulence many researchers base their analysis around the slope of the log-log spatial power spectrum. While the power spectrum gives information about the energy per wavenumber (or frequency), it only contains the Fourier amplitudes and completely ignores the phases. This alone is motivation for the development of complimentary techniques.

In general, the best strategy for studying a difficult subject like interstellar turbulence is to use a synergetic approach, combining theoretical knowledge, numerical simulations, and observational data via statistical studies. In this way one can obtain the most complete and reliable picture of the physics of turbulence. Here we seek to extend the statistical comparison between numerical and observational turbulence by reviewing statistical tools that can greatly compliment the information provided by the power spectrum. In particular we focus the review on tools that can provide information on $\mathcal{M}_s$ and $\mathcal{M}_A$. This review will highlight several different tools (see Figure 1) studied in the works of Kowal *et al.*, (2007), Burkhart *et al.*, (2009), (2010), (2010b), Esquivel *et al.*, (2010), Toffelmire *et al.*, (2010), Chepurnov & Lazarian (2009), Rosolowsky *et al.*, (2008), and Goodman *et al.*, (2009), which represent a mixture of numerical and observational studies. We focus on statistics that have application for observable data (PPV or column density).

2. Supersonic versus Subsonic Turbulence

The sonic Mach number describes the ratio of the flow velocity to the sound speed, and thus is a measure of the compressibility of the medium. Turbulence that is supersonic displays very different characteristics then subsonic turbulence in terms of the spectral slope and density/ velocity fluctuations. Because the physical environment of compressible turbulence is very different from incompressible, this parameter is extremely important for many different fields of astrophysics including, but not limited to, star formation and cosmic ray acceleration.

2.1. *Higher Order Moments of Column Density*

Moments of the density distribution can be used to roughly determine the gas compressibility through shock density enhancements. As the ISM media transitions from subsonic to supersonic and becomes increasingly supersonic, shocks create enhanced mean value and variance of the density PDF. In addition, as the shocks become stronger, the PDF is skewed and becomes more kurtotic then Gaussian. However, Kowal *et al.*, (2007) showed that this method is not very effective for subsonic cases, as these distributions are roughly Gaussian. For areas where Mach numbers approach and exceed unity, higher order moments of column density PDFs can be used as a measure of compressibility.

2.2. *Spectrum*

The power spectrum has been used for studies of both observational and numerical turbulence for decades. In addition to providing information on the energy cascade, the spectrum can also be used to obtain compressibility in column density maps. As turbulence transitions from subsonic to supersonic, density enhancements due to shocks create small scale structures which shallow out the spectral slope. Additionally, the presence of a strong magnetic field can create Alfvénic shearing, which can steepen the slope via destruction of small scale structures. Subsonic low magnetic field spectral slope values are close to what is predicted for hydrodynamic Kolmogorov turbulence and the addition of a strong magnetic field and shocks cause deviations from the predicted -11/3. † Burkhart *et al.*, (2010) compared the spectral slope found for the SMC galaxy with simulations of turbulence and found that the SMC's slope of -3.3 matches well with transsonic type turbulence. This Mach number range has been independently confirmed by observational methods to obtain the sonic number which utilize the ratio of the spin the kinetic temperature and also the higher order moments.

2.3. *Bispectrum*

While the power spectrum has been used extensively in ISM studies, higher order spectrum have been more rare. The bispectrum, or Fourier transform of the 3rd order autocorrelation function, has been applied to isothermal ISM turbulence simulations and the SMC only recently (Burkhart *et al.*, 2009, 2010) although it is extensively used in other fields including cosmology and biology.

The bispectrum preserves both the amplitude and phase and provides information on the interaction of wave modes. Completely randomized modes will show a bispectrum of zero, while mode coupling will show non-zero bispectrum. Shocks and high magnetic field have been shown to increase mode coupling in the bispectrum. Due to their ability to shallow out the density energy spectrum, shocks greatly enhance the small scale wave-wave coupling. With simulations with the same sonic number, mode correlation is shown to increase with an increase in magnetic field, however this is less clear for column density as it is for 3D density.

3. Magnetization of Turbulence

The Alfvén number is the dimensionless ratio of the flow velocity to the Alfvén speed. As the Alfvén speed depends on the magnetic field, this ratio can provide information on the strength of the magnetic field relative to the velocity and density. The Alfvénic number is critical in several fields including interplanetary studies and star formation. The solar wind is known to be a super-Alfvénic flow while the Alfvénic number in star forming regions is still hotly debated.

3.1. *Tsallis PDFs of PPV and Column Density*

PDFs of increments (of density, magnetic field, velocity etc.) are a classic way to study turbulence since the phenomena is scale dependent. The Tsallis function was formulated in Tsallis (1988) as a means to extend traditional Boltzmann-Gibbs mechanics to fractal and multifractal systems.

$$R_q = a \left[1 + (q-1)\frac{\Delta f(x,r)^2}{w^2} \right]^{-1/(q-1)} \tag{3.1}$$

† For incompressible turbulence, the Kolmogorov power spectrum in three dimensions (3D) is $k^{-11/3}$, in 2D it is $k^{-8/3}$, and 1D $k^{-5/3}$ for the same energy spectrum E(k).

The Tsallis distribution (Equation 3.1) can be fit to PDFs of increments, that is, $f(x,r) = G(x + r) - G(x)$, where G(x) is a particular field (for example, turbulent density, velocity or magnetic field). The Tsallis fit parameters (q, a, and w in Equation 3.1) describe the width, amplitude, and tails of the PDF. These parameters have been shown to have dependencies on both sonic and Alfvénic Mach number. Tsallis parameters are able to distinguish between sub, tras and supersonic turbulence as well as gauge wether the turbulence is sub-Alfvénic or super-Alfvénic. The parameter that describes the width of the PDF distribution is particularly sensitive.

3.2. *Dendrograms of Position-Position-Velocity (PPV) data*

A Dendrogram (from the Greek dendron tree,- gramma drawing) is a hierarchical tree diagram that has been used extensively in other fields, particular galaxy evolution and biology. It is a graphical representation of a branching diagram, and for our particular purposes with PPV data, quantifies how and where local maxima of emission merge with each other. The dendrogram was first used on ISM data in Rosolowsky *et al.*, 2008 and Goodman *et al.*, 2009 in order to characterize self-gravitating structures in star forming molecular clouds.

Burkhart *et al.*, 2010 used the dendrogram on synthetic PPV cubes and found it to be rather sensitive to magnetic density/velocity enhancements. They looked at the moments of the distribution of local maxima in emission found in the tree diagram. These moments showed clear signs of being dependent on Mach numbers, with the particular strength being the Alfvén number. When high frequency filtering is applied in order to mask small scale enhancements due to shocks, the magnetic enhancements fully dominate the moments of the tree diagram distribution. The dendrogram is also able to distinguish between simulations that show varying degrees of gravitational strength. It is also very encouraging that this statistic is working in PPV space, while other statistics studied utilize the column density maps. This further motivates the synergetic approach of using these statistics.

4. Conclusions

The last decade has seen major increases in the knowledge of the ISM and of its turbulent nature thanks to high resolution observations and advanced numerical simulations. This calls for new advances in statistical tools in order to best utilize the wealth of observational data in light of numerical and theoretical predictions. Recently several authors have explored new tools for studying turbulence beyond the power spectrum. While these proceedings do not cover all the useful tools in the literature, we attempt to provide some review on tools that describe the gas compressibility and the Alfvénic Mach number by utilizing density fluctuations created by shocks and magnetic density enhancements.

References

Burkhart, B., Falceta-Goncalves, D., Kowal, G., & Lazarian, A., 2009, *ApJ*, 693, 250
Burkhart, B., Stanimirovic, S., Lazarian, A., & Kowal, G., 2010, *ApJ*, 708, 1204
Burkhart, B., Goodman, A., Lazarian, A., & Rosolowsky, E., 2010, in prep.
Chepurnov, A., Lazarian, A., Gordon, J., & Stanimirovic., S., 2008, *ApJ*, 688, 1021
Esquivel, A. & Lazarian, A., 2010, *ApJ*, 710, 125
Goodman, A. A., Rosolowsky, E. W., Borkin, M. A., Foster, J. B., Halle, M., Kauffmann, J. &
      Pineda, J. E., 2009, *Nature letters*, 457, 63
Lazarian, A. 2009, *Space Science Reviews*, 143, 357
Kowal, G., Lazarian, A., & Beresnyak, A., 2007, *ApJ*, 658, 423
Toffelmire B., Burkhart, B., & Lazarian, A., 2010, submitted.
Rosolowsky, E. W., Pineda, J. E., Kauffmann, J., & Goodman, A. A. 2008, *ApJ*, 679, 1338

Advances in Plasma Astrophysics
Proceedings IAU Symposium No. 274, 2010
A. Bonanno, E. de Gouveia Dal Pino & A.G. Kosovichev, eds.
© International Astronomical Union 2011
doi:10.1017/S1743921311007290

An XMM-Newton view of a small sample of Seyfert 1 Galaxies

M. V. Cardaci[1,2] G. F. Hagele[1,2], M. Santos-Lleó[3], Y. Krongold[4], A. I. Díaz[1] and P. Rodriguez-Pascual[3]

[1]Universidad Autónoma de Madrid, Cantoblanco, 28049-Madrid, Spain

[2]Facultad de Cs. Astronómicas y Geofísicas, Universidad Nacional de La Plata, Paseo del Bosque s/n, 1900 La Plata, Argentina

[3]XMM-Newton Science Operation Centre, ESAC, ESA, PO Box 50727, 28080 Madrid, Spain

[4]Instituto de Astronomía, Universidad Nacional Autónoma de México, Apartado Postal 70-264, 04510 México DF, México

email: monica.cardaci@uam.es

Abstract. We present a detailed analysis of all the X-ray data taken by the XMM-Newton satellite of a small sample of five Seyfert 1 galaxies: ESO 359-G19, HE 1143-1810, CTS A08.12, Mkn 110, and UGC 11763. Our aim is to characterize the different components of the material that print the absorption and emission features in the X-ray spectra of these objects. The continuum emission was studied through the EPIC spectra taking advantage of the spectral range of these cameras. The high resolution RGS spectra were analyzed in order to characterize the absorbing features and the emission line features that arise in the spectra of these sources.

Keywords. galaxies: active, galaxies: Seyfert, galaxies: individual (UGC 11763, ESO 359-G19, HE 1143-1810, CTS A08.12, Mkn 110), X-rays: galaxies

1. Introduction

The *XMM-Newton* satellite is mainly devoted to perform observations in the X-ray band. On-board this satellite are three EPIC cameras that take images and medium resolution spectra, two high resolution spectrographs (RGS), and one optical monitor that can take images or spectra in some optical-UV bands.

A Seyfert 1 is a radio quiet active galactic nucleus (AGN). Their hard X-ray radiation can be characterised with a power law plus some lines produced by iron, for example, the Fe Kα line at 6.4 keV. The hard power law describes also the continuum in the soft band only in about the 50% of the objects. For the rest it is necessary to add another component to account for the extra flux observed. On this soft band one can find as well the transitions involving He-like and H-like atomos of, among other, Oxygen, Neon, and also absorptions of heavier elements that are produced in clouds of partially ionized gas in movement.

The aim of this work is to be able to characterize the X-ray continuum emission of the Seyfert 1 AGNs and to identify possible emission and absorption features in the X-ray spectra and infer the physical conditions of the material in which they are produced. To do this, we select a small sample of Seyfert 1 galaxies and we perform detailed analysis of all data taken during the XMM-Newton observations.

2. The sample

We selected a small sample of five objects (Table 1) with a reliable classification as Seyfert 1, that are located at galactic coordinates where the Hydrogen column densities

Table 1. Journal of observations.

	nH (10^{20} cm^{-2})	Obs. date	z	T_{exp} (s)
UGC 11763	4.67	16 May 2003	0.063	39000
ESO 359-G19	1.02	09 March 2004	0.056	24000
HE 1143-1810	3.40	08 June 2004	0.033	31000
CTS A08.12	4.07	30 Oct. 2004	0.029	46000
Mkn 110	1.42	15 Nov. 2004	0.035	47000

are low (about 10^{20} cm^{-2}, which is important for having mainly unabsorbed soft X-rays information), and that are also bright X-rays objects (as seen by other missions). This kind of objects are variable in scale times of days. We pick two objects in a moderate to high historical flux state (HE 1143-1810 and Mkn 110) and the other three in a quite low historical flux state.

3. Results and Conclusions

From the detailed analysis of the X-ray data taken by the *XMM-Newton* satellite for these five objects we find that the parameters that define all the continuum components take values that are well within the ranges found in other Seyfert 1 and QSO (Piconcelli *et al.* 2005). The indexes of the hard power laws range beteewen 1.2 and 1.8; the lowest values corresponding to the two brightest objects. The soft excess is modelled with a black body component for the lower flux objects with temperatures well within the ranges found in other Seyfert 1 and QSO (kT~80 - 130 keV).

The Fe-Kα line was detected in all the five objects, and it is in general weak. A few emission lines were also detected in the soft X-ray band, mainly Oxygen lines such as the OVII Heα triplet, detected on four of the objects (even though not all the components are always found). Only in UGC 11763 Ne IX and Fe XVIII lines have been found. In fact UGC 11763 is the only one that shows significant spectral structure, with features both in absorption and emission. Two more objects (CTS A08.12 and Mkn 110) show hints of weak absorption, although not significant.

The absorbing material is mainly detected thanks to the presence of a broad absorption feature around 15-17Å (see Figure 1), attributed to an unresolved transition array (UTA) resulting from a blend of lines from iron L. The photoinization code used to model the absorption components was developed by Krongold *et al.*, (2003) and the main parameters are: the ionization parameter (U), the column density of the absorbing media and the velocity of the material.

Analysing the absorption features in the UGC11763 spectra we find two partially ionized absorbing components, with distinct ionization states (log U $\sim$ 1.65 and 2.6), and with the same velocity respect to the nucleus, (Cardaci *et al.*, 2009).

Placing the absorbing components on the thermal equilibrium curve (see Figure 1), we find that both components lay on stable parts of the curve which is consistent with having the same gas pressure. Adding also the fact that they are kinematically indistinguishable, we conclude that they could constitute two phases of the same medium.

This is a very interesting result because the temperature of the low ionization component (T~1.8×10^5 K) is an order of magnitude higher than those found in other Seyfert galaxies meaning that the ions producing the UTA feature are different: in our case Fe XIII to Fe XV and in the other galaxies Fe VII to Fe XII. We have to note that only gas at such high temperature could coexist in pressure equilibrium with the High Ionization Component.

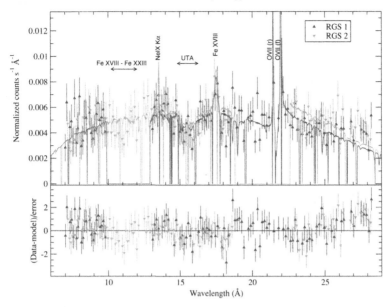

Figure 1. High resolution X-ray spectra -in the rest frame- of UGC 11763 as obtained with RGS1, (blue up triangles), and RGS2 (cyan down triangles), and binned to 15 channels per bin. Solid lines are the convolution of the best fit model with instrument responses.

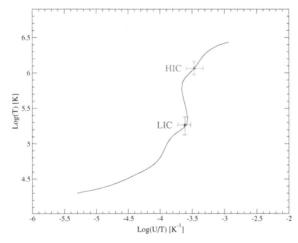

Figure 2. Thermal stability curve of UGC 11763 obtained using the SED based on the simultaneous multiwavelength data obtained in this observation and assuming a column density of $\log N_H = 21.5 \, \mathrm{cm}^{-1}$. We have indicated the position of the two components of photoionized absorbing material.

4. Summary

Regarding the absorbing material around the central engine of the AGNs studied, we find that only one out of five objects shows clear signs of partially ionized (warm) material.

We could characterize the physical conditions of material around UGC 11763. For which we find two absorbing components that are consistent with their being in pressure equilibrium, i.e., two phases of the same media (UTA of higher ionization than those found in other AGNs).

References

Cardaci, M. V., Santos-Lleó, M., Krongold, Y., Hägele, G. F., Díaz, A. I., & Rodríguez-Pascual, P. 2009, *A&A*, 432, 15

Krongold, Y., Nicastro, F., Brickhouse, N. S., Elvis, M., Liedahl, D. A. & Mathur, S. 2003, *ApJ*, 597, 832

Piconcelli, E., Jimenez-Bailón, E., Guainazzi, M., Schartel, N., Rodríguez-Pascual, P. M. & Santos-Lleó, M. 2005, *A&A*, 432, 15

Advances in Plasma Astrophysics
Proceedings IAU Symposium No. 274, 2010
A. Bonanno, E. de Gouveia Dal Pino & A.G. Kosovichev, eds.

© International Astronomical Union 2011
doi:10.1017/S1743921311007307

How can vorticity be produced in irrotationally forced flows?

Fabio Del Sordo and Axel Brandenburg

NORDITA, Roslagstullsbacken 23, SE-10691 Stockholm, Sweden; and
Department of Astronomy, Stockholm University, SE 10691 Stockholm, Sweden

Abstract. A spherical hydrodynamical expansion flow can be described as the gradient of a potential. In that case no vorticity should be produced, but several additional mechanisms can drive its production. Here we analyze the effects of baroclinicity, rotation and shear in the case of a viscous fluid. Those flows resemble what happens in the interstellar medium. In fact in this astrophysical environment supernovae explosion are the dominant flows and, in a first approximation, they can be seen as spherical. One of the main difference is that in our numerical study we examine only weakly supersonic flows, while supernovae explosions are strongly supersonic.

Keywords. Galaxies: magnetic fields – ISM: bubbles

Turbulence in the interstellar medium (ISM) is mainly driven by supernovae explosions, which are among the most dramatic events in terms of release of energy. Those explosions are also very important because they can affect scales up to ~ 100 pc. Moreover they are able to inject in the ISM enough energy to sustain turbulent flows with velocities of ~ 10 km/s. It is well known that turbulence is one of the key ingredients to be taken in account when discussing many astrophysical process – especially in the production of magnetic fields. This is indeed one of our ultimate goals, even though here we do not take any magnetic field into account. As a first approximation, a supernova explosion can be regarded as a purely spherical expansion wave. Thus, we choose a setup consisting of purely potential forcing: we simulate spherical expansions, as already done by Mee & Brandenburg (2006). For our numerical experiments we use the PENCIL CODE, http://pencil-code.googlecode.com/. We have recently extended this work to include rotation, shear, and baroclinicity; see Del Sordo & Brandenburg (2010). Here we report on some highlights of their work.

We analyze flows that are only weakly supersonic and use a constant and uniform viscosity in an unstratified medium. In our model we solve the Navier-Stokes equations in the viscous case. We consider uniform viscosity in an unstratified medium. We force our system to be only weakly supersonic and we use a potential forcing $\nabla\phi$ where ϕ is given by randomly placed Gaussian of radius R around the position $\boldsymbol{x}_{\rm f}(t)$. We use two different forms for the time dependence of the forcing position $\boldsymbol{x}_{\rm f}$. In the first case we consider a δ-correlated forcing, that is, every timestep has a $\boldsymbol{x}_{\rm f}$ completely independent from the previous. Then we also study the situation in which the forcing remains constant during a time interval $\delta t_{\rm force}$.

Next, we add to the system one of three effects that we want to analyze, taking into account each of them separately. We start by considering the action of rotation under isothermal condition. Under the influence of rotation the system is subject to the action of the Coriolis force. That is, we add the term $2\boldsymbol{\Omega} \times \boldsymbol{u}$ in the evolution equation of velocity. In our simulations we investigate flows with Reynolds numbers (based on the wavenumber of the energy-carrying eddies) of up to 150. The aim of this investigation is

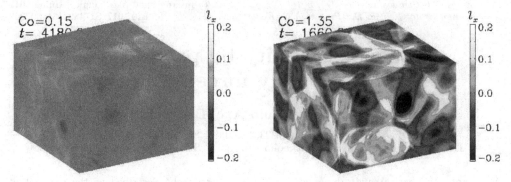

Figure 1. Vertical component of vorticity on the periphery of the periodic domain for two values of the Coriolis number. Note that significant amounts of vorticity are only being produced when Co is of the order of unity.

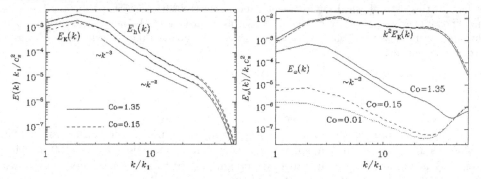

Figure 2. Left: spectra of kinetic energy and enthalpy for two values of the Coriolis number for Re = 25 and St_{force} = 0.4. The two straight lines give the slopes -2 and -3, respectively. Right: Enstrophy spectra, $E_\omega(k)$, compared with $k^2 E_K(k)$, for Re = 25, St_{force} = 0.4, and three values of the Coriolis number.

to quantify the production of vorticity. We find that vorticity is indeed produced with both kinds of forcing we have used for driving the spherical expansion. Nevertheless the case of δ-correlated forcing seems to be more prone to spurious production of vorticity that we believe is due to numerical artifact.

We find that significant vorticity is only being produced when the Coriolis number, Co $= 2\Omega/u_{rms}k_f$ is about unity; see Fig. 1 for Co = 0.15 and 1.35. For both cases we show in the left-hand panel the spectra of kinetic energy and enthalpy, $E_K(k)$ and $E_h(k)$, respectively. There is no clear inertial range, but in all cases the energy spectra show a clear viscous dissipation range. There can easily be spurious vorticity generation, possibly still due to marginally sufficient resolution. The possibility of a spurious vorticity is indeed verified by the right-hand panel of Fig. 2, where we compare enstrophy spectra at different Coriolis numbers. Note that for large values of Co, the enstrophy spectrum decays like k^{-3}. However, for smaller values of Co the level of enstrophy at the mesh scale remains approximately unchanged and is thus responsible for the spurious vorticity found above for small values of Co and not too small values of Re. For larger values of Co, the production of vorticity is an obvious effect of rotation in an otherwise potential velocity field, and it is most pronounced at large length scales, as can also be seen in the right-hand panel of Fig. 2.

In the presence of shear, we find, in analogy with the case of rotation, production of vorticity proportional to the magnitude of the shear. However our results indicate that

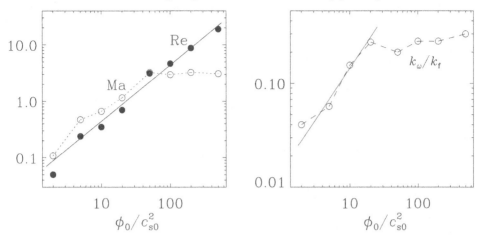

Figure 3. Dependence of Ma and Re, as well as of the normalized vorticity, k_ω/k_f, on ϕ_0 for $\nu/c_s R = 1$.

under the typical physical conditions in the interstellar medium in our Galaxy, neither rotation nor shear would be strong enough to produce significant amounts of vorticity (Del Sordo & Brandenburg (2010)).

Finally, we relax the isothermal condition to let the system evolve under the action of the baroclinic term. In this situation we have non-parallel gradients of pressure and density. The baroclinic term is proportional to the cross product of the two gradients, resulting from the curl of the term $\rho^{-1}\nabla p$. In Fig. 3 we show the dependence of various quantities on the forcing amplitude ϕ_0 normalized by the reference sound speed c_{s0}. The Mach number saturates at about Ma $= 3$, and the rms value of the entropy gradient increases up until this point. The amount of vorticity production in terms of k_ω/k_f is about 0.3 for $\phi_0/c_{s0}^2 \gtrsim 20$. For smaller values, on the other hand, there is an approximately linear increase with $k_\omega/k_f \approx 0.014\phi_0/c_{s0}^2$.

Given that in our Galaxy the Mach number of the turbulence is about unity (Beck *et al.* (1996)) it is clear that the baroclinic term is much more efficient in driving the production of vorticity. The fact that the highest amount of vorticity is observed when shock fronts encounter each other suggests that supersonic conditions need to be investigated more deeply.

Regarding dynamo action, as pointed out by Brandenburg & Del Sordo (2009), the presence of vorticity does not seem to affect the diffusion of magnetic fields differently than a complete irrotational turbulence. Nevertheless, vorticity plays an important role in dynamo processes, so it is important to address the problem of the generation of vorticity and the possible role of other effects. In future work we will address the connection between vorticity generation and the dynamo effect for magnetic fields.

References

Beck, R., Brandenburg, A., Moss, D., Shukurov, A., & Sokoloff, D. *ARA&A*, 1996, 34, 155
Brandenburg, A. & Del Sordo, F. (in press) 2009 Turbulent diffusion and galactic magnetism
    Highlights of Astronomy, Vol. **15** E. de Gouveia Dal Pino CUP, arXiv:0910.0072
Del Sordo, F. & Brandenburg, A. 2010, *A&A*, submitted, arXiv:1008.5281
Mee, A. J. & Brandenburg, A. *MNRAS*, 2006, 370, 415

Advances in Plasma Astrophysics
Proceedings IAU Symposium No. 274, 2010
A. Bonanno, E. de Gouveia Dal Pino & A. G. Kosovichev, eds.

© International Astronomical Union 2011
doi:10.1017/S1743921311007319

The fate of magnetic fields in colliding galaxies

Hanna Kotarba[1], Harald Lesch[1], Klaus Dolag[2] and Thorsten Naab[2]

[1] University Observatory Munich, Ludwig Maximilians University,
Scheinerstrasse 1, D-81679 Munich, Germany
email: kotarba@usm.lmu.de

[2] Max Planck Institute for Astrophysics,
Karl-Schwarzschild-Str. 1, D-85741 Garching, Germany

Abstract. The evolution and amplification of large-scale magnetic fields in galaxies is investigated by means of high resolution simulations of interacting galaxies. The goal of our project is to consider in detail the role of gravitational interaction of galaxies for the fate of magnetic fields. Since the tidal interaction up to galaxy merging is a basic ingredient of cold-dark matter (CDM) structure formation models we think that our simulations will give important clues for the interplay of galactic dynamics and magnetic fields.

Keywords. galaxies: ISM, magnetic fields, interactions, methods: numerical, N-body simulations

1. Introduction

Radio observations reveal beyond any doubt that galaxies are permeated by ordered large-scale and irregular small scale magnetic fields, with spatial scales ranging from sizes of stars up to the spiral arm lengths, disk thickness and halo radii (e.g. Beck & Hoernes, 1996, Chyży *et al.*, 2007, Beck, 2009, Vollmer *et al.*, 2010). Typical field strengths are of order of several μG, ranging from a few μG in dwarfs (e.g. Chyży *et al.*, 2003) up to 30 μG in the star-forming regions of grand-design spiral galaxies (Fletcher *et al.*, 2004). Magnetic fields have also been observed at high red-shifts like e.g. in damped Ly-α systems observed at red-shifts of $z \approx 0.4 - 2$. These systems, which have often been interpreted as large progenitors of present-day spirals (see Wolfe *et al.*, 2005 for a review), seem to host similar μG magnetic fields (e.g. Bernet *et al.*, 2008), or even magnetic fields of several tens of μG (Wolfe *et al.*, 2008). The conversion of kinetic and turbulent energy into magnetic energy is a well investigated process in plasma astrophysics. The general term for this kind of energy transfer is "dynamo". It is commonly believed that dynamo processes like the turbulent dynamo and the $\alpha\Omega$-dynamo play a significant role in the evolution of magnetic fields (e.g. Kulsrud, 1999, Arshakian *et al.*, 2009, and references therein). Those dynamo processes, however, rely on the presence of a "seed field" already existing at the onset of dynamo action. Those seed fields may be primordial, or a consequence of battery mechanisms in proto-galaxies or in the first stars. Primordial fields or battery processes could account for seed fields of $10^{-21} - 10^{-18}$ G (see e.g. Rees, 1987 and references therein), which would have to be amplified by more than ten orders of magnitude to reach the observed present-day values in galaxies. Therefore, other scenarios have been proposed. Within the most promising scenario, seed magnetic fields are built up during the lifetime of stars and expelled via relativistic winds during supernovae (SN) explosions (Rees, 1987). Those Crab-type SN remnants host magnetic fields of $\approx 10^{-4}$ G, pervading a volume of several cubic parsecs. Given that we would expect a huge number of stars forming in a protogalaxy, the mean field strength that could permeate the galaxy would be by orders of magnitude higher than that generated by the processes mentioned before. The galactic dynamo process could then reorder and further amplify this stellar seed field, while new born stars continue to seed magnetic fields into the galaxy. The seeding, reordering and amplification of those stellar seed fields has been recently shown by Hanasz (2009).

Since the evolution of large-scale magnetic fields is a part of galaxy evolution in general, the observations of remarkably strong magnetic fields even in very young galaxies raise the question about the origin and the evolution of the magnetic fields in our universe, especially in the context of hierarchical structure formation models. In these scenarios all present-day galaxies have undergone several major and minor mergers at early epochs and thereafter continuous accretion of gas and smaller galactic subunits (White & Rees, 1978, White & Frenk 1991, see Benson (2010) for a recent review). In other words, gravitational interactions are a key ingredient for the first phases of galaxy evolution. We can crudely estimate the amount of free energy during an interaction of two galactic subunits to be proportional to their relative velocity squared, i.e. $E_{\text{free}} \sim v_{\text{rel}}^2$. Obviously, some of this energy released during the interaction is converted into thermal energy of hot gas. High energy particles also carry away some of the energy. However, it is reasonable to assume that at least some of this energy is converted into magnetic field energy during the compression of gas and the formation of tidal structures. As the amount of E_{free} can be very large during a major merger, the amount of energy converted into magnetic energy can be significant. Moreover, burst of star formation driven by the interaction (see e.g. Naab *et al.*, 2006, Hopkins *et al.*, 2008 and references therein) could continuously seed new magnetic field, thus intensifying the amplification of magnetic fields even more.

Given the importance of the formation process for the present-day appearance of galaxies, there is a growing interest in the evolution of magnetic fields in the early universe (e.g. de Souza & Opher 2010) and during galaxy formation (e.g. Dubois & Teyssier, 2010, Wang & Abel, 2009, Schleicher *et al.*, 2010)

The dramatic impact of mergers on the gas flows will directly affect the dynamics of the magnetic fields of the systems. Since gas and magnetic field are tightly coupled, the magnetic field traces the gas motion and will be amplified by shocks and radial gas inflow. It is the aim of our work to shed light on the complicated nature of the evolution of magnetic fields in interacting systems. Understanding those systems is the first step towards a more complete understanding of the evolution of magnetic fields in the early universe.

2. Simulations

The simulations presented here were performed with the N-body/SPH-code GADGET (Springel, 2005). The dynamics of the Lagrangian fluid elements are followed using a SPH formulation including the evolution of magnetic fields which was implemented and tested by Dolag & Stasyszyn 2009. See Kotarba *et al.* (2009) and Kotarba *et al.* (2010) for more details.

The Antennae galaxies.

A perfect example of the strong coupling of gas and magnetic fields is the interacting system NGC4038/4039 (the "Antennae galaxies"). We have simulated this system including magnetic fields for the first time (Kotarba *et al.*, 2010a). The simulations demonstrated that even an initial magnetic field as small as 10^{-9} G is efficiently amplified during the interaction of two equal-mass spiral galaxies. The strength of the magnetic

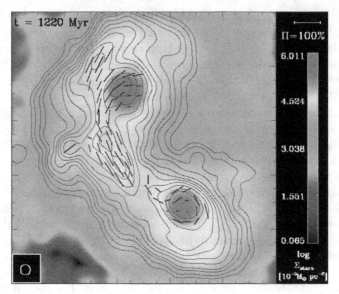

Figure 1. Artificial radio map of the simulated Antennae galaxies.

field thereby saturates at a value of 10 μG, independent of the initial magnetic field strength. This value compares well with the observations.

Fig. 1 shows an artificial radio map calculated at time of best match between the simulated gas and stellar distributions and observations. The radio map is in convincing morphological agreement with synchrotron observations of the Antennae system. Thus, the underlying numerics of the applied N-body/SPH code GADGET showed to be capable of following the evolution of magnetic fields in a highly nonlinear environment.

Saturation phenomena like that seen in our simulations usually occur when some kind of equipartition is reached. In case of the magnetic field, one usually assumes equipartition between the magnetic field energy density, the energy density of cosmic rays (CRs) and the energy density of the ionized gas. As we did not include CRs in our simulations, we have considered the energy densities of the gas. We have found that the estimated turbulent energy density of the gas is of the same order of magnitude as the magnetic field energy density. In view of the turbulent dynamo models which assume turbulence to be the main source of magnetic field amplification, this result seems reasonable. However, as we do not resolve the turbulent spectra of the gas in our simulations, but, instead, estimate the turbulence to be proportional to the rms velocity squared inside the SPH smoothing kernels, more studies are needed to assess the reason for the saturation of the magnetic field.

Ménage à trois.

We have performed simulations of three colliding galaxies to proof the saturation and equipartition phenomena seen in Kotarba *et al.*, 2010a on the basis of a more general setup (Kotarba *et al.*, 2010b). In these simulations, the magnetic field saturates at a value of about 1 G - which is again consistent with observations -, independent of the initial value. Again, we find equipartition between the magnetic and the turbulent energy density.

Furthermore, the inclusion of the IGM in the simulations of colliding galaxies allows also for studying the magnetization of the IGM itself. We find that the shocks driven into the IGM by the interaction are the stronger, the higher the initial magnetic field within the galaxies and the IGM, suggesting that the shocks are gaining higher Mach numbers due to magnetic pressure support. Moreover, we find that there exists also a saturation magnetic field within the IGM, which has a mean value of 10^{-8} G. Again, this value agrees with observations.

Movies of the evolution of the magnetic field, the gas density, the temperature and the rms velocity als well as artificial radio maps for the "ménage à trois" can be found at http://www.usm.uni-muenchen.de/people/kotarba/public.html.

3. Conclusion

More detailed studies of galaxy interactions are needed to gain stronger insight into the nature of magnetic field saturation in interacting systems. However, on the basis of the performed simulations, we are confident that we are close to solve this issue.

Summing up our simulations clearly show that galactic interactions are efficient drivers of magnetic field amplification. All of the achieved saturation values agree with observations, thus indicating that these values develop naturally in real galaxies. On the basis of this work, we conclude that any galactic encounter in the history of the universe should have contributed to the magnetization of the universe. Within the CDM hierarchical structure formation model all galaxies build up through several major and minor mergers and the accretion of gas and smaller galactic subunits. Thus, within these models, the magnetization of the universe should be a natural part of galaxy formation and evolution.

References

Arshakian, T. G., Beck, R., Krause, M., & Sokoloff, D. 2009, *A&A*, 494, 21-32

Beck, R. & Hoernes, P. 1996, *NATURE*, 379, 47-49

Beck, R. 2009, In: Magnetic Fields in the Universe II: From Laboratory and Stars to the Primordial Universe, ed. A. Esquivel, J. Franco, G. Garcï-Segura, E. M. de Gouveia Dal Pino, A. Lazarian, S. Lizano, & A. Raga, *RevMexAA*, 36, 1-8

Benson, A. J. 2010, *Physics Reports*, 495, 33-86

Bernet, M. L., Miniati, F., Lilly, S. J., Kronberg, P. P., & Dessauges-Zavadsky, M. 2008, *NATURE*, 454, 302-304

Chyży, K. T., Knapik, J., Bomans, D. J., Klein, U., Beck, R., Soida, M., & Urbanik, M. 2003, *A&A*, 405, 513-524

Chyży, K. T., Bomans, D. J., Krause, M., Beck, R., Soida, M., & Urbanik, M. 2007, *A&A*, 462, 933-941

Dolag, K. & Stasyszyn, F. 2009, *MNRAS*, 398, 1678-1697

Dubois, Y. & Teyssier, R. 2010, *A&A*, 523, A72+

Fletcher, A., Beck, R., Berkhuijsen, E. M., Horellou, C., & Shukurov, A. 2004, In: How Does the Galaxy Work? A Galactic Tertulia with Don Cox and Ron Reynolds., ed. E. J. Alfaro, E. P+xiz, J. Franco, *ASSL* (Dordrecht: Kluwer), 315, 299-+

Hanasz, M., Wóltański, D., & Kowalik, K. 2009, *ApJL*, 706, 155-159

Hopkins, P. F., Hernquist, L., Cox, T. J., Dutta, S. N. & Rothberg, B. 2008, *ApJ*, 679, 156-181

Kotarba, H., Lesch, H., Dolag, K., Naab, T., Johansson, P. H., & Stasyszyn, F. A. 2009, *MNRAS*, 397, 733-747

Kotarba, H., Karl, S. J., Naab, T., Johansson, P. H., Dolag, K., Lesch, H., & Stasyszyn, F. A. 2010a, *ApJ*, 716, 1438-1452

Kotarba, H., Lesch, H., Dolag, K., Naab, T., Johansson, P. H., Donnert, J., & Stasyszyn, F. A. 2010b, *ArXiv e-prints*, arXiv1011.5735

Kulsrud, R. M. 1999, *ARAA*, 37, 37-64

Naab, T., Jesseit, R., & Burkert, A. 2006, *MNRAS*, 372, 839-852

Rees, M. J. 1987, *QJRAS*, 28, 197-206

Schleicher, D. R. G., Banerjee, R., Sur, S., Arshakian, T. G., Klessen, R. S., Beck, R., & Spaans, M. 2010, *A&A*, 522, A115+

de Souza, R. S. & Opher, R. 2010, *PhRvD*, 81, 6, 7301

Springel, V. 2005, *MNRAS*, 364, 1105-1134

Vollmer, B., Soida, M., Chung, A., Beck, R., Urbanik, M., Chyży, K. T., Otmianowska-Mazur, K., & van Gorkom, J. H. 2010, *A&A*, 512, 36+

Wang, P. & Abel, T. 2009, *ApJ*, 696, 96-109

White, S. D. M. & Frenk, C. S. 1991, *ApJ*, 379, 52-79

White, S. D. M. & Rees, M. J. 1978, *MNRAS*, 183, 341-358

Wolfe, A. M., Gawiser, E., & Prochaska, J. X. 2005, *ARAA*, 43, 861-918

Wolfe, A. M., Jorgenson, R. A., Robishaw, T., Heiles, C., & Prochaska, J. X. 2008, *NATURE*, 455, 638-640

Advances in Plasma Astrophysics
Proceedings IAU Symposium No. 274, 2010
A. Bonanno, E. de Gouveia Dal Pino & A.G. Kosovichev, eds.
© International Astronomical Union 2011
doi:10.1017/S1743921311007320

3D numerical simulations of magnetic field evolution in barred galaxies and in spiral galaxies under influence of tidal forces

Katarzyna Otmianowska-Mazur[1], Katarzyna Kulpa-Dybeł[1], Barbara Kulesza-Żydzik[1], Hubert Siejkowski[1] and Grzegorz Kowal[1,2]

[1] Astronomical Observatory, Jagiellonian University,
ul Orla 171, 30-244, Kraków, Poland
email: otmian@oa.uj.edu.pl

[2] Instituto de Astronomia, Geofisica e Ciencias Atmosfericas, Universidade de Sao Paulo, Rua do Matao 1226, EP 05508-900, Sao Paulo, Brazil
email: kowal@astro.iag.usp.br

Abstract. We present the results of the three-dimensional, fully non-linear MHD simulations of the large-scale magnetic field evolution in a barred galaxy with the back reaction of magnetic field to gas. We also include the process of the cosmic-ray driven dynamo. In addition, we check what physical processes are responsible for the magnetic field evolution in the tidally influenced spiral galaxies. We solve the MHD equations for the gas and magnetic field in a spiral galaxy with gravitationally prescribed bulge, disk and halo which travels along common orbit with the second body. In order to compare our modeling results with the observations we also construct the maps of high-frequency (Faraday rotation-free) polarized radio emission from the simulated magnetic fields. The model accounts for the effects of projection and limited resolution.

We found that the obtained magnetic field configurations are highly similar to the observed maps of the polarized intensity of barred galaxies, because the modeled vectors form coherent structures along the bar and spiral arms. We also found a physical explanation of the problem of inconsistency between the velocity and magnetic fields character present in this type of galaxies. Due to the dynamical influence of the bar, the gas forms spiral waves which go radially outward. Each spiral arm forms the magnetic arm which stays much longer in the disk than the gaseous spiral structure. The modeled total energy of magnetic field and magnetic flux grows exponentially due to the action of the cosmic-ray driven dynamo. We also obtained the polarization maps of tidally influenced spiral galaxies which are similar to observations.

Keywords. galaxies: evolution, magnetic fields, ISM: cosmic rays, methods: numerical, MHD

1. Introduction

The radio synchrotron observations of disks of nearby galaxies (spiral and barred) indicate that magnetic fields are present in a big number of them (Beck 2009). Moreover, they show that the total strength of the magnetic field is in the range of 10-30μG in the spiral arms. In the interstellar medium (ISM) such strength is dynamically important, because the magnetic field has similar mean energy density to the gas and to cosmic rays. In the spiral arms we mainly observe a random component of the magnetic field, while the maxima of polarized emission distributed along gaseous arms are visible in interarm regions (Beck 2009).

The sample of 20 radio emission maps of barred galaxies (Beck *et al.* 2002) show that the central region of the galaxy NGC 1365 has the strongest polarized emission. In the outer disk of this galaxy the polarized magnetic arms are well visible between the bar and spiral arms similarly to spiral galaxies. Furthermore, the synchrotron observations of barred galaxies indicate the presence of regions located upstream the dust lanes, in which the polarization vectors change quickly their pitch angles what results in creation of depolarization valleys within the telescope beam. The same observations of vanishing polarized intensity are observed near shear shock areas. In several barred galaxies the topology of magnetic field is very similar to NGC 1365 (e.g. NGC 1672, NGC 7552 and NGC 1097- Beck et al 2002). In spiral galaxies the synchrotron observations also reveal the presence of strong polarized halos extending few kiloparsecs from the disks and having the X-shape topology.

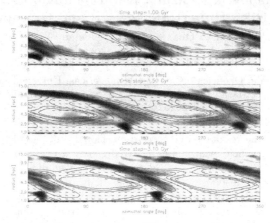

Figure 1. Distribution of polarized intensity and B-vectors superimposed on the gas density in the frame of azimuthal angle in the disk and the logarithm of distance from the center for three time steps: 1.0 Gyr, 1.5 Gyr, 3.1 Gyr. The map has been convolved to the resolution 40″.

The development of the magnetic field in barred galaxies was studied numerically by many authors (e.g. Otmianowska-Mazur *et al.* 2002, Moss *et al.* 2007, Kulesza-Żydzik *et al.* 2009 - Paper I, Kulesza-Żydzik *et al.* 2010 - Paper II), however, the models including the classical dynamo theory in this calculations did not show the promising results (Moss *et al.* 2007). In opposition to them in our two papers (Paper I & II) we presented numerical models which are capable to explain the drift of magnetic field arms into the interarm region.

In our simulations, made with the help of Godunov code (Kowal *et al.* 2009, see also Paper I & II), we apply the nonaxisymmetric gravitational potential of a bar rotating as a solid body and a galactic disk rotating differentially. The drift of magnetic arms can be explained as follows. In the beginning, due to a faster rotation of the bar than the disk, gas in a differentially rotating disk concentrates in the spiral arms. Due to the shear and compression these gaseous arms induce similar magnetic arms distributed along them. The further rotation of the bar, being in agreement with its pattern speed, produces the new package of spiral arms together with the new magnetic ones. The "old" magnetic arms stay between the spiral arms together with the gas in the disk. Such mechanism could happen even few times during the evolution time of a barred galaxy (see Paper I & II). In the present 3D MHD numerical simulations of a barred galaxy we additionally include the cosmic-ray driven dynamo process (see Hanasz *et al.* 2009). We would like to learn how such realistic mechanism as supernova actions in the disk can change the magnetic field evolution in our modeled barred galaxy. Also, in order to check how the process of shifting of the magnetic arms works in spiral galaxies under the influence of tidal forces, we constructed a model of spiral galaxy orbiting around the center of mass where we do not apply cosmic ray driven dynamo process.

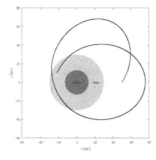

Figure 2. The position of the big galaxy together with the orbit of the small galaxy.

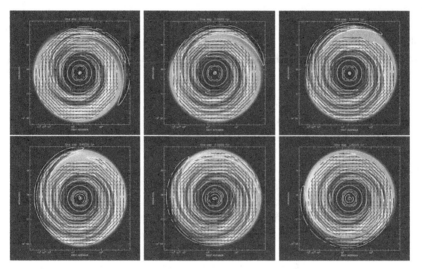

Figure 3. The polarization maps of the big galaxy superimposed onto the density color maps at six time steps 3.1 Gyr 3.2 Gyr, 3.3 Gyr, 3.4 Gyr, 3.5 Gyr, 3.6 Gyr.

2. Results of magnetic field evolution in barred galaxies with CR dynamo

First of all, we describe the problem of shifting of magnetic arms into the inter-arm region. The barred galaxy in our model is simulated in the Cartesian grid with the size of 30 kpc x 30 kpc x 7.5 kpc in X, Y and Z direction, respectively (Kulpa-Dybeł *et al.* this Volume - see also the exponential growth of the magnetic flux). The mass of the stars in the whole galaxy is assumed to be $1.75 \times 10^{11} M_\odot$. In addition, we apply the mass of the bar which equals $1.5 \times 10^{10} M_\odot$ with the size of 6 kpc x 3 kpc x 2.5 kpc. The bar rotates with the speed of 30 km/s. Using the simulated magnetic field and cosmic rays distribution we computed the polarization maps which can be compared directly with the polarization maps of the real galaxies.

In Fig. 1 we present our results in a graph showing the distribution of polarized intensity and B-vectors orientation superimposed on the gas density in the frame of azimuthal angle in the disk and the logarithm of distance from the center for three time stages: 1.0 Gyr, 1.5 Gyr, 3.1 Gyr. The map at the first time step (1.0 Gyr) reveals that the gaseous spiral arms just induced the magnetic arms which are exactly distributed along them. The next stage at 1.5 Gyr shows a shift of the magnetic arm from the left side of the gaseous arms to the interarm region. The last graph presents that magnetic field is mainly in the interam regions, however a part of the polarized intensity is also visible along the spiral arms in the region close to the center. Such behavior of magnetic arms shows that the process of shifting them to the interarm region is present.

3. Results of the magnetic field evolution in the spiral galaxies under the tidal force without the CR dynamo

In order to check how the process of formation of the spiral and magnetic arms looks in spiral galaxies under the influence of tidal motions we prepare a model of the disk galaxy gravitationally disturbed by a second small body.

Fig. 2 reveals the orbit of the small body which starts its movement at the right side of the picture and ends in the halo of the big galaxy presented by disk (8.0×10^{10} $M_\odot$) and halo (7.3×10^{11} $M_\odot$) as color plots. Initially, the disk rotates differentially with the circular velocity of 200 km/s. The size of our big modeled galaxy is 30 kpc$\times$ 30 kpc $\times$ 7.5 kpc with the resolution of $256 \times 256 \times 64$ in X, Y and Z, respectively. In the beginning we assume the hydrostatic equilibrium with $p_{mag}/p_{gas} = 10^{-6}$ with the axisymmetrical magnetic field component.

Fig. 3 shows six maps of the computed polarization vectors superimposed onto the density maps at six time stages in Gyrs: 3.1, 3.2, 3.3, 3.4, 3.5, 3.6. We see that tidal forces induce the spiral arm very quickly, and the magnetic structures are also formed due to the compression and shear connected with the gas motion. In the first presented snapshot we observe the magnetic arms distributed along the spiral ones, polarized vectors in the interarm region and the region without magnetic field along the second spiral arm. In the second picture (3.2 Gyr) the big region with no magnetic field is present. In the third map (3.3 Gyr) we can see that one magnetic arm could be a bit faster than the gaseous one going into the inter-arm region down and for the next two time steps (3.4 Gyr, 3.5 Gyr) this process continues. In the end at the time 3.6 Gyr we can see only the new magnetic arm distributed along the spiral arm. We can observe here also the magnetic field in the whole disk. In order to demonstrate that the process of shifting of magnetic arms into inter-arm regions exists, although not so strong as in the barred galaxies, we plan to make an analysis of the behavior of the spiral waves together with the magnetic arms.

4. Conclusions

In the calculations of magnetic field evolution in barred galaxies, our model reproduces the large-scale structures of magnetic field observed in such galaxies. In the same time our simulations solve the problem of the shift of the magnetic arms to the interarm region. Our second experiment shows that under the influence of a tidal force, the galaxies form magnetic and spiral arms, reproducing the large-scale structures of magnetic field observed in spiral galaxies.

References

Beck, R. *et al.* 2002, *A&A*, 391, 83
Beck, R. 2009, *Astrophys. Space Sci. Trans.*, 5, 43
Hanasz, M., Wóltański, D., & Kowalik, K. 2009, *ApJ*, 706, 155
Kowal, G., Lazarian, A., Vishniac, E. T., & Otmianowska-Mazur, K. 2009, *ApJ*, 700, 63
Kulesza-Żydzik, B., Kulpa-Dybeł, K., Otmianowska-Mazur, K., Kowal, G., & Soida, M. 2009, *A&A*, 498L, 21
Kulesza-Żydzik, B., Kulpa-Dybeł, K., Otmianowska-Mazur, K., Soida, M., & Urbanik M. 2010, *A&A* 522, 61
Moss, D., Snodin, A., Englmaier, P., Shukurov, A., Beck, R., & Sokoloff, D. 2007, *A&A* 465, 15

Advances in Plasma Astrophysics
Proceedings IAU Symposium No. 274, 2010
A. Bonanno, E. de Gouveia Dal Pino & A.G. Kosovichev, eds.

© International Astronomical Union 2011
doi:10.1017/S1743921311007332

Ferromagnetic properties of charged vector bosons condensate in the early universe

Gabriella Piccinelli

Centro Tecnológico, FES Aragón, Unversidad Nacional Autónoma de México
Av. Rancho Seco S/N, Bosques de Aragón, Nezahualcóyotl
Estado de México 57130, Mexico
email: itzamna@unam.mx

Abstract. Bose-Einstein condensation in the early universe is considered. The magnetic properties of a condensate of charged vector bosons are studied, showing that a ferromagnetic state is formed. As a consequence, the primeval plasma may be spontaneously magnetized inside macroscopically large domains and primordial magnetic fields can be generated.

Keywords. Early universe, magnetic fields, elementary particles.

---

1. Introduction

Cosmological Magnetic fields. Magnetic fields seem to be pervading the entire universe: they have been observed in galaxies, clusters and high redshift objects (for observational reviews see, e.g., Kronberg 1994 and Carilli & Taylor 2002). Although at present there is no conclusive evidence about their origin, their existence in the early universe cannot certainly be ruled out. They use to be appreciated in cosmology since they help to solve some problems, in particular in the baryogenesis process (e.g., Sánchez, Ayala & Piccinelli 2007 and references therein and Piccinelli & Ayala 2004, for a review).

The determination of the physical processes able to generate magnetic fields represents a long standing cosmological problem. The general approach is to identify mechanisms for the generation of seed fields which can later be amplified into fields on larger scales, but nowadays no completely successful mechanism has been proposed.

Bose-Einstein condensates (BEC). Another interesting phenomenon that may be present in the early Universe is Bose-Einstein condensation: a quantum phenomenon of accumulation of identical bosons in the same state, which is their lowest energy (zero momentum) state. Under these conditions they behave as a single macroscopic entity described by a coherent wave function rather than a collection of separate independent particles. Even though Bose-Einstein condensation had been foreseen long time ago (1925), it took seventy years to make the first experimental observation, which was performed in a dilute gas of rubidium (Anderson *et al.* 1995). Difficulties in performing this observation were created by the extremal conditions necessary for the condensation. Indeed, the Bose-Einstein condensation takes place when the inter-particle separation is smaller than their de Broglie wavelength, $\lambda_{dB} \sim 2\pi/\sqrt{2mT}$, so the system must be cooled down to a very low temperature at ordinary densities. In the early universe, the existence of condensates depends on an interplay between density and temperature (see Dolgov, Lepidi & Piccinelli 2010, for a discussion on the conditions for which condensation could take place).

In the recent years the study of BEC became an active area of research in different fields of physics from plasma and statistical physics (for a review see Pethick & Smith

2002 and references therein) to astrophysics and cosmology. For instance, neutral scalar BEC have been proposed as dark matter (Ji & Sin 1994, Matos & Guzmán 1999).

The presence of charged BEC has interesting consequences in gauge field theories. For instance, in Dolgov, Lepidi & Piccinelli (2009), electrodynamics of charged fermions and condensed scalar bosons was considered. The screening of impurities in such plasma was found to be essentially different from the case when the condensate is absent. A similar problem was considered also in the framework of an effective field theory (Gabadadze & Rosen (2008), (2009), (2010)), analyzing the thermodynamical properties of the system and focusing on the astrophysics of helium white dwarfs. The possible condensation of helium nuclei, previous to crystallization, would affect the cooling process and leave observational signatures.

The condensation of gauge bosons of weak interactions was considered in the pioneering papers by Linde (1976), (1979), where it was argued that, at sufficiently high leptonic chemical potential, a classical W_j boson field could be created and, under certain conditions, there could be no symmetry restoration in the early universe.

As we have seen in the previous applications, the condensate can be made of either scalar or vector bosons. In both cases, bosons are in the lowest energy state, but in the vector case they have an additional degree of freedom: spin. The state of rest and identicalness inherent to the condensed particles suggests that they could present an ordered configuration of their spin magnetic moments.

In this work we consider the BEC of charged W-bosons, which may be formed if the cosmological lepton asymmetry happened to be sufficiently high, i.e. if the chemical potential of neutrinos was larger than the W boson mass at this temperature, and study its magnetic properties. We concentrate on the condensation below electroweak phase transition, although the phenomenon could also be considered above it.

2. The model

We consider a simple example of electrically neutral plasma made of fermions (electrons and neutrinos), gauge and scalar bosons, with zero baryonic number density but with a high leptonic one. This implies a (large) chemical potential for leptons and W bosons and, consequently, the formation of a charged vector BEC. For simplicity we work with only one family of leptons, but this does not influence the essential features of the result. Quarks may be essential for the condition of vanishing of all gauge charge densities in plasma and for the related cancellation of the axial anomaly but we work in the lowest order of the perturbation theory, where the anomaly is absent.

A caveat for a large lepton asymmetry arises from the big bang nucleosynthesis (BBN) with strongly mixed neutrinos. It is shown in Dolgov et al. (2002) that leptonic chemical potentials of all neutrino flavors are restricted by $|\mu_\nu/T| < 0.07$ at the BBN epoch. However, it should be noted that the entropy release from the electroweak epoch down to the BBN epoch diminishes the lepton asymmetry, by the ratio of the particle species present in the cosmological plasma at these two epochs, which is approximately 10. This helps to alleviate the bound and to allow favorable conditions for condensation in the early universe, still compatible with BBN (see Dolgov, Lepidi & Piccinelli 2010, for a detailed discussion on the different possibilities).

The essential reactions are the direct and inverse decays of W: $W^+ \leftrightarrow e^+ + \nu$. The equilibrium with respect to these processes imposes the equality between the chemical potentials: $\mu_W = \mu_\nu - \mu_e$, and the condition of electroneutrality reads: $n_{W^+} - n_{W^-} - n_{e^-} + n_{e^+} = 0$.

Considering the Lagrangian of the minimal electroweak model, we derive the equations of motion that allow to take into account the spin-spin interactions of the vector bosons.

3. Spin-spin interactions of the condensed bosons

The spins of the individual vector bosons can be either aligned or anti-aligned, forming, respectively, ferromagnetic and anti-ferromagnetic states, see, e.g., Pethick & Smith (2002). The realization of one or the other state is determined by the spin-spin interaction between the bosons. In the lowest angular momentum state, $l = 0$, a pair of bosons may have either spin 0 or 2. Depending on the sign of the spin-spin coupling, one of those states would be energetically more favorable and would be realized at the condensation.

In solid state physics, the dominant contribution to the spin-spin interactions comes from quantum exchange effects; for bosons these forces are not essential and the spin-spin interaction is determined by the electromagnetic interaction between their magnetic moments and by their self-interactions (Dolgov, Lepidi & Piccinelli 2010).

Electromagnetic interactions can be found from the analogue of the Breit equation for electrons, which leads to the spin-spin potential:

$$U_{em}^{spin}(r) = \frac{e^2}{4\pi m_W^2}\left[\frac{(\mathbf{S}_1 \cdot \mathbf{S}_2)}{r^3} - 3\frac{(\mathbf{S}_1 \cdot \mathbf{r})(\mathbf{S}_2 \cdot \mathbf{r})}{r^5} - \frac{8\pi}{3}(\mathbf{S}_1 \cdot \mathbf{S}_2)\delta^{(3)}(\mathbf{r})\right],$$

where $\mathbf{S} = -i\,\mathbf{W}^\dagger \times \mathbf{W}$ is the spin operator of vector particles and m_W its mass.

To calculate the contribution of this potential into the energy of two W-bosons we have to average it over their wave function. In particular, in the condensate case, it is a S-wave function that is angle independent. Hence the contributions of the first two terms in the previous equation mutually cancel out and only the third one remains, which has negative coefficient. Thus the energy shift induced by the spin-spin interaction is:

$$\delta E = \int \frac{d^3r}{V}U_{em}^{spin}(r) = -\frac{2\,e^2}{3\,V m_W^2}(\mathbf{S}_1 \cdot \mathbf{S}_2),$$

where V is the normalization volume.

Since $S_{tot}^2 = (S_1 + S_2)^2 = 4 + 2S_1S_2$, the average value of S_1S_2 is $S_1S_2 = S_{tot}^2/2 - 2$. For $S_{tot} = 2$ this term is $S_1S_2 = 1 > 0$, while for $S_{tot} = 0$ it is $S_1S_2 = -2 < 0$. Thus, for this interaction, the state with maximum total spin is more favorable energetically and W-bosons should condense with aligned spins.

Quartic self-coupling of W. Due to its non-Abelian character, the weak sector presents a quartic self-coupling in the Lagrangian:

$$L_{4W} = -\frac{e^2}{2\sin^2\theta_W}\left[(W_\mu^\dagger W^\mu)^2 - W_\mu^\dagger W^{\mu\dagger}W_\nu W^\nu\right] = \frac{e^2}{2\sin^2\theta_W}\left(\mathbf{W}^\dagger \times \mathbf{W}\right)^2,$$

where θ_W is the Weinberg angle. It is assumed here that $\partial_\mu W^\mu = 0$ and thus only the spatial 3-vector $\mathbf{W}$ is non-vanishing, while $W_t = 0$.

The interaction potential is given by the Fourier transform of this term which, with proper (nonrelativistic) normalization, leads to:

$$U_{4W}^{(spin)} = \frac{e^2}{8m_W^2\sin^2\theta_W}(\mathbf{S}_1 \cdot \mathbf{S}_2)\delta^{(3)}(\mathbf{r}).$$

Thus, the contribution of the quartic self-coupling of W to the spin-spin interaction has opposite sign to the electromagnetic term and tends to favour an antiferromagnetic state. Nonetheless, in the standard model, the electromagnetic term is dominant.

4. Discussion and Conclusions

Generation of primordial magnetic fields. We have studied the Bose-Einstein condensation of charged weak bosons, driving special attention to the magnetic properties of the condensate. The spin-spin interaction has two contributions: the electromagnetic one - leading to a ferromagnetic state - and a contact quartic self-interaction that produces an antiferromagnetic state. We have found that the former dominates, generating a macroscopically large ferromagnetic configuration. We then expect that the primeval plasma, where such bosons condensed, can be spontaneously magnetized. The typical size of the magnetic domains is determined by the cosmological horizon at the moment of the condensate evaporation. The latter takes place when the neutrino chemical potential, which scales as temperature in the course of cosmological cooling down, becomes smaller than the W mass at this temperature. Large scale magnetic fields created by the ferromagnetism of W-bosons might survive after the decay of the condensate due to the conservation of the magnetic flux in plasma with high electric conductivity. Such magnetic fields at macroscopically large scales could be the seeds of the observed larger scale galactic or intergalactic magnetic fields.

Some considerations: The long range interactions between magnetic moments can, in principle, be screened by plasma physics, while the local quartic interaction cannot. This could change the relative strength of these two effects. However, in the broken phase, the problem is reduced to that of pure QED, where it is known that magnetic forces are not screened. On the other hand, the situation is not clear in non-Abelian theory and it could happen that, in the symmetric phase, the screening might inhibit magnetic spin-spin interaction. Interactions with relativistic electrons and positrons are neglected. In principle, they could distort the spin-spin interactions of W by their spin or orbital motion and thus destroy the attraction of parallel spins of W. However, it looks hardly possible because electrons are predominantly ultra-relativistic and they cannot be attached to any single W boson to counterweight its spin. Finally, the scattering of electrons (and quarks) on W-bosons may lead to the spin flip of the latter, but in thermal equilibrium this process does not change the average value of the spin of the condensate.

References

Kronberg, P. P. 1995, *Rep. Prog. Phys.*, 57, 325; Carilli, C. L. & Taylor, G. B., 2002, *Ann. Rev. Astron. Astrophysics*, 40, 319

Sánchez, A. Ayala, A., & Piccinelli G. , 2007, *Phys. Rev.*, D75, 043004; Piccinelli, G. & Ayala, A., 2004, in: N. Breton, J. L. Cervantes-Cota, M. Salgado, (eds.), *The Early Universe and Observational Cosmology*, (Lect.Notes Phys. 646: Springer Verlag), p. 293

Anderson, M., Ensher, J., Matthews, M., Wieman, C., & Cornell, E., 1995, *Science*, 269, 198

Dolgov, A. D., Lepidi, A., & Piccinelli, G, 2010, *JCAP*, 08, 031

Pethick, C. J. & Smith, H., 2002, *Bose-Einstein condensation in dilute gases*, (Cambridge University Press, Cambreidege U.K.)

Ji, S. U. & Sin, S-J., 1994, *Phys. Rev.* D50, 3655; Matos, T. & Guzmán, F. S., 1999, *F. Astron. Nachr.*, 320, 97

Dolgov, A. D., Lepidi, A., & Piccinelli, G, 2009, *JCAP*, 02, , 016; *Phys. Rev.*, D80, 125009

Gabadadze, G. & Rosen, R. A., 2008, *JCAP*, 10, 030; 2009, *JCAP*, 02, 016; 2010, *JCAP*, 04, 028

Linde, A. D., 1976, *Phys. Rev.*, D14, 3345; 1979, *Physics Letters*, 86B, 39

Dolgov, A. D., Hansen, S. H., Pastor,S., Petcov, S. T., Raffelt, G. G., & Semikoz, D. V., 2002, *Nucl. Phys.*, B632, 363

Advances in Plasma Astrophysics
Proceedings IAU Symposium No. 274, 2010
A. Bonanno, E. de Gouveia Dal Pino & A.G. Kosovichev, eds.

© International Astronomical Union 2011
doi:10.1017/S1743921311007344

3D model of magnetic fields evolution in dwarf irregular galaxies

Hubert Siejkowski[1], Marian Soida[1], Katarzyna Otmianowska-Mazur[1], Michał Hanasz[2] and Dominik J. Bomans[3]

[1] Astronomical Observatory, Jagiellonian University, ul. Orla 171, 30-244 Kraków, Poland
email: h.siejkowski@oa.uj.edu.pl

[2] Torun Centre for Astronomy, Nicolaus Copernicus University, 87-148 Toruń/Piwnice, Poland

[3] Astronomical Institute of Ruhr-University Bochum, Univeristatsstr. 150/NA7, D-44780 Bochum, Germany

Abstract. Radio observations show that magnetic fields are present in dwarf irregular galaxies (dIrr) and its strength is comparable to that found in spiral galaxies. Slow rotation, weak shear and shallow gravitational potential are the main features of a typical dIrr galaxy. These conditions of the interstellar medium in a dIrr galaxy seem to unfavourable for amplification of the magnetic field through the dynamo process. Cosmic-ray driven dynamo is one of the galactic dynamo model, which has been successfully tested in case of the spiral galaxies. We investigate this dynamo model in the ISM of a dIrr galaxy. We study its efficiency under the influence of slow rotation, weak shear and shallow gravitational potential. Additionally, the exploding supernovae are parametrised by the frequency of star formation and its modulation, to reproduce bursts and quiescent phases. We found that even slow galactic rotation with a low shearing rate amplifies the magnetic field, and that rapid rotation with a low value of the shear enhances the efficiency of the dynamo. Our simulations have shown that a high amount of magnetic energy leaves the simulation box becoming an efficient source of intergalactic magnetic fields.

1. Introduction

Dwarf irregular galaxies have relatively simple structure. They are smaller, less massive and have lower luminosity than spirals and ellipticals. Their rotation speed is very low and the rotation curve could be very complex (e.g. NGC 4449). The structure of a galaxy is often disturbed by a strong burst of star formation. Weak gravitational potential and slow rotation cause that supernovae explosions can substantially influence the gas distribution and global velocity pattern. Energy injected by a starbursting events is enough to drive a gas outflow from a dwarf galaxy. Together with gas also metals and magnetic fields are transported.

Radio observations (Chyży *et al.* 2000, 2003; Kepley *et al.* 2010; Klein *et al.* 1991, 1992) show that dIrr galaxies can have relatively strong magnetic fields. The typical total magnetic field strength is 5–15 μG with a uniform component about 5 μG. The observed magnetic fields suggest that a dynamo process should operate in these galaxies. We investigate the cosmic-ray driven dynamo model in the environment of a typical dwarf irregular galaxy.

2. Model description and initial setup

The CR-driven dynamo model consists of the following elements based on Hanasz *et al.* (2006) and references therein. We assume:

• the cosmic ray component described by the diffusion-advection transport equation and we adopt anisotropic diffusion;

• localized sources of CR, i.e. random explosions of supernovae in the disk volume. The cosmic ray input of individual SN remnant is 10% of the canonical kinetic energy output (10^{51} erg);

• resistivity of the ISM to enable the dissipation of the small-scale magnetic fields (see Hanasz & Lesch 2003). In the model, we apply the uniform resistivity and neglect the Ohmic heating of gas by the resistive dissipation of magnetic fields;

• shearing boundary conditions, tidal and Coriolis forces;

• realistic vertical disk gravity following the model by Ferrière (1998) with rescaled disk and halo masses by one order of magnitude.

The 3D cartesian domain size is 0.5 kpc × 1 kpc × 8 kpc in x, y, z coordinates corresponding to the radial, azimuthal, and vertical directions, respectively, with a grid size $(20 \text{ pc})^3$. The boundary conditions are sheared-periodic in x, periodic in y, and outflow in z direction. The positions of SNe are chosen randomly with a uniform distribution in the xy plane and a Gaussian distribution in the vertical direction. In addition, the SNe activity is modulated during the simulation. The applied value of the perpendicular CR diffusion coefficient is $K_\perp = 10^3 \text{ pc}^2 \text{ Myr}^{-1}$ and the parallel one is $K_\parallel = 10^4 \text{ pc}^2 \text{ Myr}^{-1}$ (see Hanasz *et al.* 2009). The initial state of the system represents the magnetohydrostatic equilibrium with the horizontal, purely azimuthal magnetic field with $p_{mag}/p_{gas} = 10^{-4}$.

3. Results

3.1. *Rotation and shear*

We studied dependence of the magnetic field amplification on the parameters describing the rotation curve, namely, the shearing rate q and the angular velocity Ω. The evolution in the total magnetic field energy E_B and total azimuthal flux B_ϕ for different values of Ω is shown in Fig. 1, left and right panel, respectively. Models with higher angular velocities, starting from 0.03 Myr^{-1}, initially exhibit exponential growth of the magnetic field energy E_B till 1 200 Myr followed by a saturation. The saturation values of the magnetic energy for these three models are similar and E_B exceeds the value 10^4 in the normalized units. The magnetic energy in the models R.01Q1† (slow rotation, moderate shear) and R.02Q1 (slow rotation, moderate shear) grows exponentially during the whole simulation and does not reach the saturation level. The total azimuthal magnetic flux evolution (Fig. 1, right) shows that a higher angular velocity leads to a higher amplification. There is no amplification in model R.01Q1.

3.2. *Star formation*

We checked how the frequency and modulation of SNe influence the amplification of magnetic fields. The evolution in total magnetic field energy and total azimuthal flux for different supernova explosion frequencies are shown in Fig. 2. The total magnetic energy evolution for all models is similar, but the differences are apparent in the evolution of azimuthal flux. The most efficient amplification of B_ϕ appears for SF10R.03Q.5 (medium SFR, moderate rotation, low shear) and SF10R.03Q1 (moderate shear), and for other models the process is less efficient. In addition, for models SF30R.03Q.5 (high SFR, low shear) and SF30R.03Q1 (high SFR, moderate shear), we observe a turnover in magnetic field direction. The results suggest that the dynamo requires higher frequencies of

---

† Letter R stands for angular velocity (rotation) given in Myr^{-1}, Q shearing rate ($q = -d \ln \Omega / d \ln R$) and SF for star formation given in $\text{kpc}^{-2} \text{ Myr}^{-1}$.

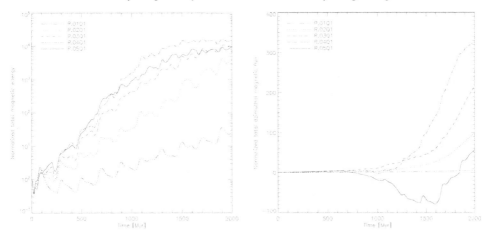

Figure 1. Evolution of the total magnetic energy E_B (left panel) and the total azimuthal flux B_ϕ (right) for models with different rotation. Both quantities are normalized to the initial value.

supernova explosions to create more regular fields, although, if the explosions occur too frequently because of a strong wind transporting magnetic field out of the disk.

3.3. *Outflow of magnetic field*

To measure the total production rate of the magnetic field energy during the simulation time, we calculated the outflowing E_B^{out} through the xy top and bottom domain boundaries. To estimate the magnetic energy loss, we computed the vertical component of the Poynting vector. Its value is computed in every cell belonging to the top and bottom boundary planes and then integrated over the entire area and time. For models with a low dynamo efficiency most of the initial magnetic field energy is transported out of the simulation box. In some cases (i.e., all models except R.01Q0 and R.05Q0 with zero shear), we find that the energy loss E_B^{out} is comparable to the energy remaining inside the domain $\bar{E}_B^{end}$. In these models, the ratio $E_B^{out}/\bar{E}_B^{end}$ varies from 0.03 to 0.96 and is highly dependent on the supernova explosion frequency. The results show that the outflowing magnetic energy is substantial (see Siejkowski *et al.* 2010) suggesting, that irregular galaxies can be efficient sources of intergalactic magnetic fields.

4. Conclusions

We have described the evolution of the magnetic fields in irregular galaxies in terms of a cosmic-ray driven dynamo (Siejkowski *et al.* 2010). The amplification of magnetic fields have been studied under different conditions characterized by the rotation curve (the angular velocity and the shear) and the supernovae activity (its frequency and modulation) typical for irregular galaxies. We have found that:

• in the presence of slow rotation and weak shear in irregular galaxies, the amplification of the total magnetic field energy is still possible;

• shear is necessary for efficient action of CR-driven dynamo, but the amplification itself depends weakly on the shearing rate;

• higher angular velocity enables a higher efficiency in the CR-driven dynamo process;

• the efficiency of the dynamo process increases with SNe activity, but excessive SNe activity reduces the amplification;

• for high SNe activity and rapid rotation, the azimuthal flux reverses its direction;

• the outflow of magnetic field from the disk is high, suggesting that dIrr galaxies may

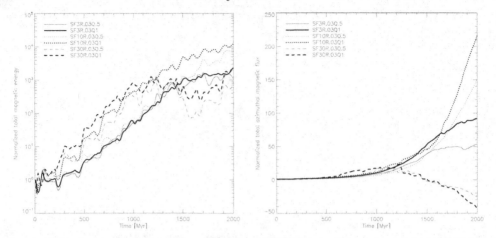

Figure 2. Evolution of the total magnetic energy E_B (left panel) and the total azimuthal flux B_ϕ (right) for models with different supernova explosion frequency and shearing rate. Both quantities are normalized to the initial value.

magnetize the intergalactic medium as predicted by Kronberg *et al.* (1999) and Bertone *et al.* (2006).

The performed simulations indicate that the CR-driven dynamo can explain the observed magnetic fields in dwarf irregular galaxies. In future work we plan to determine the influence of other ISM parameters and perform global simulations of these galaxies.

Acknowledgements

This work was supported by Polish Ministry of Science and Higher Education through grants: 92/N-ASTROSIM/2008/0 and 3033/B/H03/2008/35, and by the DFG Research Group FOR1254. Presented computations have been performed on the GALERA super-computer in TASK ACC in Gdańsk.

References

Bertone, S., Vogt, C., & Enßlin, T. 2006, *MNRAS*, 370, 319
Chyży, K. T., Beck, R., Kohle, S., Klein, U., & Urbanik, M. 2000, *A&A*, 355, 128
Chyży, K. T., Knapik, J., Bomans, D. J., Klein, U., Beck, R., Soida, M., & Urbanik, M. 2003, *A&A*, 405, 513
Ferrière, K. 1998, *ApJ* 497, 759
Hanasz, M., Kowal, G., Otmianowska-Mazur, K., & Lesch, H. 2006, *AN* 327, 469
Hanasz, M. & Lesch, H. 2003, *A&A*, 404, 389
Hanasz, M., Otmianowska-Mazur, K., Kowal, G., & Lesch, H. 2009, *A&A*, 498, 335
Kepley, A. A., Mühle, S., Everett, J., Zweibel, E. G., Wilcots, E. M., & Klein, U. 2010, *ApJ*, 712, 536
Klein, U., Giovanardi, C., Altschuler, D. R., & Wunderlich, E. 1992, *A&A*, 255, 49
Klein, U., Weiland, H., & Brinks, E. 1991, *A&A*, 246, 323
Kronberg, P. P., Lesch, H., & Hopp, U. 1999, *ApJ*, 551, 56
Siejkowski, H., Soida, M., Otmianowska-Mazur, K., Hanasz, M., & Bomans, D. J. 2010, *A&A*, 510, A97

Advances in Plasma Astrophysics
Proceedings IAU Symposium No. 274, 2010
A. Bonanno, E. de Gouveia Dal Pino & A.G. Kosovichev, eds.

© International Astronomical Union 2011
doi:10.1017/S1743921311007356

General relativistic magnetic perturbations and dynamo effects in extragalactic radiosources

L. C. Garcia de Andrade[1]

[1]Departamento de Física Teórica-IF
Universidade do Estado do Rio de Janeiro
Rua São Francisco Xavier, 524
Cep 20550-003, Maracanã, Rio de Janeiro, RJ, Brasil

Abstract. By making use of the MHD self induction equation in general relativity (GR), recently derived by Clarkson and Marklund (2005), it is shown that when Friedmann universe possesses a spatial section whose Riemannian curvature is negative, the magnetic energy bounds computed by Nuñez (2002) also bounds the growth rate of the magnetic field given by the strain matrix of dynamo flow. Since in GR-MHD dynamo equation, the Ricci tensor couples with the universe magnetic field, only through diffusion, and most ages are highly conductive the interest is more theoretical here, and only very specific plasma astrophysical problems can be address such as in laboratory plasmas. Magnetic fields and the negative curvature of some isotropic cosmologies, contribute to enhence the amplification of the magnetic field. Ricci curvature energy is shown to add to strain matrix of the flow, to enhance dynamo action in the universe. Magnetic fluctuations of the Clarkson-Marklund equations for a constant magnetic field seed in highly conductive flat universes, leads to a magnetic contrast of $\frac{\delta B}{B} \approx 2$, which is well within observational limits from extragalactic radiosources of $\frac{\delta B}{B} \approx 1.7$. In the magnetic helicity fluctuations the magnetic contrast shows that the dynamo effects can be driven by these fluctuations.

---

1. Introduction

Since the early work by Gamow and Teller (1939), where they discovered an interesting relation between the Riemannian spatial hypersurfaces of negative curvature and the expansion of the open universe, this led naturally one to argue if that there is a similar relationship between there is a similar relation between the curvature and the expansion. In this paper, one shows that even when curvature vanishes, there is a constraint between the expansion and magnetic helicity. When the curvature of the spatial section is negative, a relationship is obtained between expansion, and Ricci curvature. Actually is easy to shown from dynamo equation that the expansion possesses a lower bound by the Ricci curvature. More recently, dynamo effects in cosmology have been adressed recently by Kleides et al.(2007), by using a anisotropic cosmological model in ideal highly conductive plasmas. In the realm of isotropic cosmology Brandenburg et al.(1998) have considered that decay of magnetic field can be slowed when turbulence is switch on. In this paper the isotropic case is further investigated in the case of hydromagnetic dynamo by making use of a recently derived , GR-MHD dynamo equation of Marklund and Clarkson (2005). This case is examined by using the approach developed by Nuñez (1995) in Sobolev spaces. Another example is that of a magnetic field helicity in cosmlogy. the paper is organized as follows: In section II the magnetic helicity model is addressed while in section III, the magnetic energy inequalities and the contribution of the Ricci term integral are given.

Section IV addresses the issue of magnetic perturbations on extra radiogalactic sources. Conclusions are presented in section V.

2. Maqnetic helicity in non-ideal plasma cosmology

Three dimensional dynamos in Riemannian space can be given by the self-induction equation (Arnold *et al.*, 1981).

$$\frac{\partial B}{\partial t} = -\{v, B\} + div(v)B + \eta \Delta B \tag{2.1}$$

where η is the plasma resistivity and $\Delta := \nabla^2$ is the Laplacian operator, where $\{v, B\} = -curl(v \times B)$. In Chicone *et al.*(1999) the Riemannian manifold was confined to incompressible dynamo flows, where the second term on the RHS of expression (3.2) vanishes, and along with the divergence-free condition of the magnetic field

$$div B = 0 \tag{2.2}$$

they form a solenoidal vector field in Riemannian manifold. On the other MHD-GR dynamo equation in pseudo-Riemannian spacetime given by its 3-spatial section

$$\dot{\mathbf{B}} - \nabla \times [\mathbf{v} \times \mathbf{B}] - \eta \Delta \mathbf{B} = -\eta \left[\frac{2}{3}\theta \mathbf{B} - Ric.\mathbf{B} \right] - \left(1 + \frac{2}{3}\eta\theta \right) \frac{2}{3}\theta \mathbf{B} - \eta \nabla \times (\mathbf{a} \times \mathbf{B}) + \eta \Theta \tag{2.3}$$

where Ric represents Ricci tensor and θ is the expansion of the Friedmann universe. In this case shear and global vorticity vanish. As is easily noted from this equation, the Ricci curvature has no effect in dynamo action in most highly conductive cosmological models, since then the diffusion constant cefficient η vanishes. The symbol Θ is given by

$$\Theta = -\frac{2}{3}\dot{\theta}\mathbf{B} - \frac{2}{3}\theta\dot{\mathbf{B}} - \mathbf{a} \times [\nabla \times \mathbf{B} - \mu_0\mathbf{j} + \mathbf{a} \times \mathbf{B}] \tag{2.4}$$

By Fourier analyzing the above equation by the expression

$$\mathbf{B} = \mathbf{B}_0 \exp[\gamma t - i k \mathbf{x}] \tag{2.5}$$

yields

$$div \mathbf{B} = i \mathbf{k} \cdot \mathbf{B} = 0 \tag{2.6}$$

for the solenoidal equation and

$$\mathbf{k} \times (\mathbf{v} \times \mathbf{B}) = 0 \tag{2.7}$$

$$\mathbf{a} \times (\mathbf{a} \times \mathbf{B}) \approx 0 \tag{2.8}$$

where in the last equation one had assumed that the acceleration of the universe $\mathbf{a}$ is low. Magnetic helicity is given by the constraint

$$\nabla \times \mathbf{B} = \lambda \mathbf{B} \tag{2.9}$$

and it is used in the above computations yields

$$\dot{\mathbf{B}} - \nabla \times [\mathbf{v} \times \mathbf{B}] + \eta \lambda^2 \mathbf{B} = -\eta \left[\frac{2}{3}\theta \mathbf{B} - Ric.\mathbf{B} \right] - \left(1 + \frac{2}{3}\eta\theta \right) \frac{2}{3}\theta \mathbf{B} - \eta \nabla \times (\mathbf{a} \times \mathbf{B}) + \eta \Theta \tag{2.10}$$

In the flat Friedmann universe, where the spatial section of the spacetime is zero, r $Ric = 0$, this equation reduces to the one

$$\frac{1}{2}\dot{\mathbf{B}}^2 = \eta[\lambda^2 + \frac{4}{9}\theta^2 - \frac{2}{3}\dot{\theta}]\mathbf{B}^2 - \frac{2}{3}\theta\mathbf{B}^2 \tag{2.11}$$

where the convective term $\nabla \times [\mathbf{v} \times \mathbf{B}]$ vanishes since by Fourier transformation

$$\mathbf{k} \times [\mathbf{v} \times \mathbf{B}] = i[(\mathbf{k}.\mathbf{B})\mathbf{v} - (\mathbf{k}.\mathbf{v})\mathbf{B}] \tag{2.12}$$

since both velocity and magnetic fields are solenoidal. Then, considering that the magnetic field decays as $B \approx s^{-2}$, one obtains the self-induction equation as

$$\dot{\theta}_1 - \frac{9}{4}\theta_1 = 0 \tag{2.13}$$

where the old expansion is related to the new as

$$\theta = \theta_1 - \frac{2}{3}\lambda^2 \tag{2.14}$$

Therefore solution of equation (2.13) is

$$\theta = \theta_0 \sinh\left[\frac{3}{2}t\right] + \frac{2}{3}\lambda^2 \tag{2.15}$$

which shows that magnetic helicity enhances the expansion of the universe, when magnetic field decay. Now let us briefly examine and put some bounds on the growth of the magnetic field, when the Riemannian spatial curvature of the spatial section is negative as in 3D case of fast dynamos in Riemannian compact manifolds by Chicone and Latushkin (1999). In this case the FRW metric is open and the self-induction equation is

$$\frac{\dot{B}}{B} = \eta[-\lambda^2 + \frac{2}{3}\dot{\theta} - Ric - \frac{4}{9}\theta^2] - \frac{2}{3}\theta \geqslant 0 \tag{2.16}$$

the equal sign corresponds to marginal dynamos. Since $Ric \leqslant 0$, in the case of Riemannian negative curvature, the Ric term contributes to enhance the dynamo effect as in Kleidis et al result, but basic difference is that here the plasma is non-ideal and the spacetime isotropic since one imagines, that here dynamo is kinematic and not hydromagnetic as in the next section. In the case of negative Riemann spatial or Gaussian Ricci curvature, and the constraint of slow expansion, $\theta << 1$ and $\dot{\theta} << 1$ yields

$$Ric < \frac{3}{2}\eta\theta \tag{2.17}$$

which yields an explicitly constraint between expansion the diffusion and the Ricci scalar.

3. Bounds on hydromagnetic dynamo growth rate cosmology

In this section one shall consider the bounds on the magnetic energy of the hydromagnetic dynamos in spatially curved cosmlogical isotropic models. One shall demonstrate from the above GR-MHD relativistic plasma dynamo equations for finite magnetic Reynolds numbers $R_m = \eta^{-1}$, the Ricci scalar volume integral is fundamental for these bounds. Let us consider the magnetic energy time evolution as

$$\frac{1}{2}\frac{\partial \int B^2 dV}{\partial t} = \int (\mathbf{B}.\nabla)\mathbf{v}.\mathbf{B}dV - \eta\int |\nabla B|^2 dV - \eta\int Ric.\mathbf{B}.\mathbf{B}dV \tag{3.1}$$

Since under the plasma resistivity, the only new term compared to non-GR plasmas is the one with Ricci tensor, then to investigate the magnetic energy bounds, it is enough to investigate this bound contribution as

$$-\eta\int Ric.\mathbf{B}.\mathbf{B}dV = -\eta\int R_{ij}B^i B^j dV = -6\eta\int \frac{K}{a^2}\delta_{ij}B^i B^j dV = +6\eta\int \frac{1}{a^2}B^2 dV \geqslant 0 \tag{3.2}$$

where K is the Gaussian negative curvature of the negatively curved Riemannian spatial section like in Gamow-Teller universe. Thus since this contribution is positive, when the universe shrinks or collapses the curvature term will greatly enhance dynamo action and the bounds for the growth rate. Since, as shown by Nuñez (2002) in his study of the dissipation of kinetic energy in the hydromagnetic dynamo, the following inequality applies

$$\left| \int \mathbf{B}.\nabla\mathbf{v}.\mathbf{B}dV \right| \leqslant \frac{1}{2}||(\nabla\mathbf{v})^t + \nabla\mathbf{v}||_\infty) \int B^2 dV \tag{3.3}$$

Inserting the new Ricci curvature energy into this expression yields

$$\gamma \leqslant \left(\frac{1}{2}||(\nabla\mathbf{v})^t + \nabla\mathbf{v}||_\infty + < \frac{6\eta}{a^2} > \right) \tag{3.4}$$

where to obtain this term, one had to assume that the gradients of the magnetic field would be highly suppressed by the overall magnetic energy at all. The brackets on the RHS of the inequality means a mean of averaged value over the volume of a small portion of the universe. Thus in general the Ricci curvature adds to strain matrix to enhance dynamo action in the presence of diffusion. This seems to agree with Kleides et al result in the case of diffusive isotropic cosmolgical models.

4. Magnetic fluctuations and dynamo effects in GR extragalactic radiosources

As pointed out by Ruzmaikin *et al.*(1981), there is a gap between the observational value of the contrast obtained by the joint catalogue of polarization properties of extra galactic radiosources $\frac{\delta B}{B} \approx 1.7$ and the one estimated by Ruzmaikin *et al.*(1981) as $\frac{\delta B}{B} \approx 10^3$. In this section an estimate more realistic and well within observational values is given by the contrast $\frac{\delta B}{B} \approx 2$. These computations are performed in the magnetic helicity-free case. To start let us use the magnetic field perturbation given by

$$B = B_0 + \delta B \tag{4.1}$$

where δB is the magnetic fluctuation and B_0 is here the constant magnetic field that may seed the galactic dynamo. Substitution of this fluctuation into the above expression for the magnetic energy density, in the case of a highly conducting ($\eta = 0$) universe yields

$$\frac{\delta B}{B} = 1 + exp\left[-\frac{2}{3} \int \delta\theta dt \right] \tag{4.2}$$

where one has addopted a constant initial expansion of the universe, to simplify computations. For a very small universe fluctuation in expansion and for a finite time interval, this result reduces to

$$\frac{\delta B}{B} = 2 - \frac{2}{3} \int \delta\theta dt] \approx 2 \tag{4.3}$$

as one wishes to prove. Let us now consider the case where the magnetic helicity fluctuations are present. In this case computations leads to

$$\frac{\delta B}{B} = \frac{1}{\sqrt{2 + \frac{1}{3}\delta\theta\eta}}\delta\lambda \tag{4.4}$$

which reduces to

$$\frac{\delta B}{B} = \frac{1}{\sqrt{2\eta}}\delta\lambda \tag{4.5}$$

This last expression can be used to place a limit on magnetic helicity from the magnetic contrast and magnetic Reynolds numbers.

5. Conclusions

Gravitational magnetic fluctuations in terms of universe expansin fluctuations can be obtained from GR-MHD dynamo equation. Physical applications lead us to find contrasts to the magnetic fields or magnetic field fluctuations that are well within experimental values for the extra galactic radio sources. Bounds for hydromagnetic energy based on these gr dynamo equation are also found.

6. Acknowledgements

Several discussions with A Brandenburg, Yu Latushkin, D Sokoloff and R Beck are highly appreciated. I also thank financial supports from UERJ and CNPq.

References

Gamow, G. & Teller, E., 1939, *Phys. Rev.*; Widrow, L. *Rev. Mod. Phys.*, 2002, 74, 75.
Kleides, K., Kuiroikidis, A., Papadopoulos D., & Vlahos, F., *arxiv 07124239*.
Brandenburg, A., 1998, *Phys. Rev.* D
Marklund, M. & Clarkson, C. 2005, *MNRAS*, 358, 892.
Nuñez, M., *J Math Phys.* 2003, 38,1538 (1997). Nuñez, M., J Phys A,8903.
Arnold, V., Zeldovich, Y. B., Ruzmaikin, A., & Sokoloff, D. D., 1981, *JETP*, 1981, 81 & 1982, Doklady Akad. Nauka, SSSR, 266, 1357.
Chicone C. & Latushkin, Y., 1999, American Mathematical Society, AMS.
Nuñez, M. 2002, Geophys and Astr. Fluid Dynamics, 96, 345.

Advances in Plasma Astrophysics
Proceedings IAU Symposium No. 274, 2010
A. Bonanno, E. de Gouveia Dal Pino & A. G. Kosovichev, eds.

© International Astronomical Union 2011
doi:10.1017/S1743921311007368

Cosmic ray driven dynamo in barred and ringed galaxies

K. Kulpa-Dybeł[1], K. Otmianowska-Mazur[1], B. Kulesza-Żydzik[1], G. Kowal[1,2], D. Wóltański[3], M. Hanasz[3] and K. Kowalik[3]

[1] Astronomical Observatory, Jagiellonian University, ul Orla 171, 30-244 Kraków, Poland
[2] Núcleo de Astrofsica Teorica, Universidade Cruzeiro do Sul-Rua Galvão Bueno 868, CEP 01506-000 São Paulo, Brazil
[3] Centre for Astronomy, Nicholas Copernicus University, PL-87148 Piwnice/Toruń Poland

Abstract. We study the global evolution of the magnetic field and interstellar medium (ISM) of the barred and ringed galaxies in the presence of non-axisymmetric components of the potential, i.e. the bar and/or the oval perturbations. The magnetohydrodynamical dynamo is driven by cosmic rays (CR), which are continuously supplied to the disk by supernova (SN) remnants. Additionally, weak, dipolar and randomly oriented magnetic field is injected to the galactic disk during SN explosions. To compare our results directly with the observed properties of galaxies we construct realistic maps of high-frequency polarized radio emission. The main result is that CR driven dynamo can amplify weak magnetic fields up to few μG within few Gyr in barred and ringed galaxies. What is more, the modelled magnetic field configuration resembles maps of the polarized intensity observed in barred and ringed galaxies.

Keywords. galaxies: evolution, magnetic fields, ISM: cosmic rays, methods: numerical

1. Introduction

To explain the observational properties of the magnetic field in barred and ringed galaxies the dynamo action is necessary. It is thought that the CR driven dynamo can be responsible for the following effects: amplification of galactic magnetic fields up to several μG within a lifetime of a few Gyr; large magnetic pitch angles of about $-35°$; symmetry (even, odd); maintenance of the created magnetic fields in a steady state; magnetic field which does not follow the gas distribution, i.e. magnetic fields in NGC 4736 crossing the inner gaseous ring without any change of their direction (Chyży & Buta 2008) or magnetic arms in NGC 1365 which are located between gaseous spiral (Beck et al. 2002).

2. Cosmic ray driven dynamo

CR driven dynamo is based on the following effects (details see Hanasz et al 2009 and references therein, and Hanasz et al, this volume): The CR component described by We numerically investigated the CR driven dynamo model in a computational domain which covers 30 kpc $\times$ 30 kpc $\times$ 7.5 kpc of space with 300 $\times$ 300 $\times$ 75 cells of 3D Cartesian grid, what gives 100 pc of spatial resolution in each direction. the diffusion-advection transport is appended to the set of resistive MHD equations. The CR energy is continuously supplied to the disk by SN remnants. In all models we assume that 10% of 10^{51} erg SN kinetic energy output is converted into the CR energy, while the thermal energy from SN explosions is neglected. Additionally, no initial magnetic field is present but the weak and randomly oriented magnetic field is introduced to the disk in 10% of SN explosions. Following Giacalone & Jokipii (1999) we assume that the CR gas diffuses anisotropically along magnetic field lines. In order to allow the topological evolution of

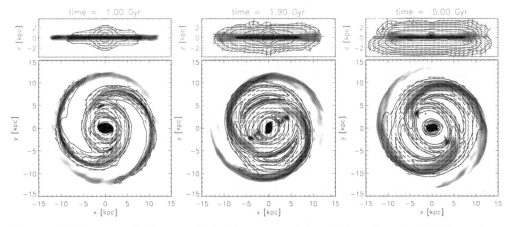

Figure 1. Face-on and edge-on polarization maps at $\lambda = 6.2$ cm for selected times steps. Polarized intensity (contours) and polarization angles (dashes) are superimposed onto column density plots (grey-scale). All maps have been smoothed down to the resolution $40''$. The black color represents the regions with the highest density.

magnetic fields we apply a finite uniform magnetic resistivity of the ISM. All numerical simulation have been performed with the aid of the Godunov code (Kowal *et al.* 2009).

3. Barred galaxy

Figure 1 shows the distributions of polarization angles (vectors) and polarized intensity (contours) superimposed onto the column density maps. On the face-on maps the magnetic field initially follows the gas distribution, as can easily be seen for time $t = 1.0$ Gyr, where the magnetic field strength maxima are aligned with the gaseous ones. However, at later time-steps, the magnetic arms begin to detach themselves from the gaseous spirals and drift into the inter-arm regions. In the edge-on maps the extended structures of the polarization vectors are present. This configuration of the magnetic field vectors bears some resemblance to the extended magnetic halo structures of the edge-on galaxies (so called X-shaped structures).

The face-on and edge-on distribution of the toroidal magnetic field is displayed in Figure 2 (left and middle panels). At $t = 0.5$ Gyr the toroidal magnetic field is mainly disordered as it is introduced to the disk through randomly oriented SN explosions. At later time ($t = 5.1$ Gyr) most of the toroidal magnetic field becomes well ordered. Moreover, we can distinguish regions with negative and positive toroidal magnetic field which form an odd (dipole-type) configuration of the magnetic field with respect to the galactic plane. In Figure 2 (right panel) the exponential growth of the total magnetic energy and the total azimuthal flux due to the CR driven dynamo action in the barred galaxy is shown. At time $t = 5$ Gyr the CR dynamo action saturates and the magnetic field reaches equipartition.

4. Ringed galaxy

CR driven dynamo also works in the case of the ringed galaxy NGC 4736. The exponential growth of the magnetic energy is even faster than in the case of the barred galaxy (Figure 3, right panel). The obtained distribution of the gas density (Figure 3, left panel) as well as the distribution of polarization angles (vectors) and polarized intensity (contours) bear some resemblance to the observation of NGC 4736. To get better

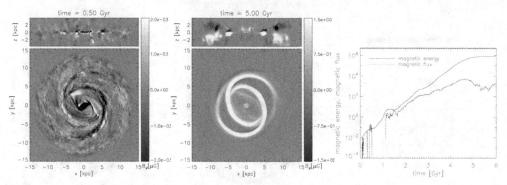

Figure 2. *Left and middle panel:* Distribution of the toroidal magnetic field for selected times steps. The white color represents the regions with the positive toroidal magnetic field, while black with negative. *Right panel:* The time dependence of the total magnetic field energy B^2 (solid line) and the mean B_ϕ flux (dashed) calculated for the barred galaxy model.

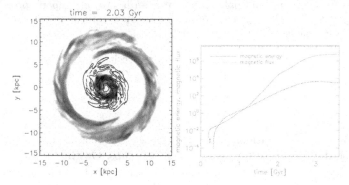

Figure 3. *Left panel:* The face-on polarization map at $\lambda = 6.2$ cm at time $t = 2.03$ Gyr superimposed onto gaseous map. The map has been smoothed to the resolution $40''$. The black color represents the regions with the highest density, white with the smallest. *Right panel:* The time dependence of the total magnetic field energy B^2 (solid line) and the mean B_ϕ flux (dashed) calculated for the ringed galaxy model.

results a more sophisticated numerical model of the NGC 4736 is planned. Namely, our model of the ringed galaxy consists of four components: the large and massive halo, the central bulge, the outer disc and the oval distortion. However, following observations of the NGC 4736 this galaxy possesses one more component: the very small bar. Thus, to get better results we have to include the additional small bar in our simulations.

Acknowledgements

This work was supported by Polish Ministry of Science and Higher Education through grants: 92/N-ASTROSIM/2008/0 and 3033/B/H03/2008/35. The computations presented here have been performed on the GALERA supercomputer in TASK Academic Computer Centre in Gdańsk.

References

Beck, R. *et al.* 2002, *A&A*, 391, 83
Chyży, K. T. & Buta, R. J., 2008 2008, *ApJ*, 677, L17
Giacalone, J. & Jokipii, R. J. 1999, *ApJ*, 520, 204
Hanasz, M., Wóltański, D., & Kowalik, K. 2009, *ApJ*, 706, 155
Kowal, G., Lazarian, A., Vishniac, E. T., & Otmianowska-Mazur, K. 2009, *ApJ*, 700, 63

Plasma astrophysics in numerical simulations

A. Nordlund

Advances in Plasma Astrophysics
Proceedings IAU Symposium No. 274, 2010
A. Bonanno, E. de Gouveia Dal Pino & A. G. Kosovichev, eds.

© International Astronomical Union 2011
doi:10.1017/S174392131100737X

Simulations of astrophysical dynamos

Axel Brandenburg

NORDITA, AlbaNova University Center, Roslagstullsbacken 23, SE-10691 Stockholm, Sweden;
Department of Astronomy, Stockholm University, SE 10691 Stockholm, Sweden

Abstract. Numerical aspects of dynamos in periodic domains are discussed. Modifications of the solutions by numerically motivated alterations of the equations are being reviewed using the examples of magnetic hyperdiffusion and artificial diffusion when advancing the magnetic field in its Euler potential representation. The importance of using integral kernel formulations in mean-field dynamo theory is emphasized in cases where the dynamo growth rate becomes comparable with the inverse turnover time. Finally, the significance of microscopic magnetic Prandtl number in controlling the conversion from kinetic to magnetic energy is highlighted.

Keywords. Sun: magnetic fields

1. Introduction

There are two important aspects connected with astrophysical dynamos compared with dynamos on a bicycle. Firstly, they are self-excited and do not require any permanent magnets. Secondly, they are *homogeneous* in the sense that the medium is conducting everywhere in the dynamo proper and there are no wires or insulators inside. Self-excited dynamos were invented by the Danish inventor Søren Hjorth, who received the patent for this discovery in 1854, some 12 years before Samuel Alfred Varley, Ernst Werner von Siemens and Charles Wheatstone announced such an invention independently of each other. Von Siemens is known for having recognized its industrial importance in producing the most powerful generators at the time, for which he, in turn, received a patent in 1877.

The idea that homogeneous dynamos might work in the Sun, was first proposed by Larmor (1919) in a one-page paper. However, some 14 years later, Cowling (1933) showed that axisymmetric dynamos cannot work in a body like the Sun. At the time it was not clear whether this failure was genuine, or whether it was critically connected with Cowling's assumption of axisymmetry. The suspicion that the third dimension might be critical was not particularly emphasized when Larmor (1934) tried to defend his early suggestion with the words "the self-exciting dynamo analogy is still, so far as I know, the only foundation on which a gaseous body such as the Sun could possess a magnetic field: so that if it is demolished there could be no explanation of the Sun's magnetic field even remotely in sight."

The essential idea about the operation of the solar dynamo came from Parker (1955), who developed the notion that cyclonic events would tilt a toroidal field systematically in the poloidal direction, closing thereby a critical step in the dynamo cycle. While this concept is still valid today, it still required the existence proof by Herzenberg (1958) that began to convince critics that Cowling's antidynamo theorem does not extend to the general case of three dimensions.

Nevertheless, subsequent progress in modeling the solar dynamo appears to have been suspended until the foundations of a mean-field treatment of the induction equation were developed by Steenbeck *et al.*, (1966). In the following years, a large number of models were computed covering mostly aspects of the solar dynamo (Steenbeck & Krause, 1969a; Parker, 1970a; Parker, 1970b; Parker, 1970c; Parker, 1971b; Parker, 1971d; Parker,

1971f), but in some cases also terrestrial dynamos (Steenbeck & Krause, 1969b; Parker, 1971c) and the galactic dynamo (Parker, 1971a; Parker, 1971e; Vainshtein & Ruzmaikin, 1971; Vainshtein & Ruzmaikin, 1972). These developments provided a major boost to dynamo theory given that until then work on the galactic dynamo, for example, focussed on aspects concerning the small-scale magnetic field (Parker, 1969), but not the global large-scale fields on the scale of the entire galaxy. In fact, also regarding small-scale dynamos, there were important developments made by Kazantsev (1968), but they remained mostly unnoticed in the West, even when the first direct simulations by Meneguzzi *et al.* (1981) demonstrated the operation of such a dynamo in some detail. In fact, in some of these dynamos, the driving of the flow involved helicity, but its role in helping the dynamo remained unconvincing, because no large-scale field was produced. We now understand that this was mainly because there was not enough scale separation between the scale of the domain and the forcing scale, and that one needs at least a ratio of 3 (Haugen *et al.*, 2004).

Simulations in spherical geometry were much more readily able to demonstrate the production of large-scale magnetic fields (Gilman, 1983; Glatzmaier, 1985), but even today these simulations produce magnetic fields that propagate toward the poles (Käpylä *et al.*, 2010) and not toward the equator, as in the Sun. We can only speculate about possible shortcomings of efforts such as these that must ultimately be able to reproduce the solar cycle.

Several important developments happened in the 1980s. Firstly, it became broadly accepted that the magnetic field inside the Sun might be in a fibril state (Parker, 1982), i.e. the filling factor is small and most of the field is concentrated into thin flux tubes, as manifested by the magnetic field appearance in the form of sunspots at the surface. However, such tubes would be magnetically buoyant, and are expected to rise to the surface on a time scale of some 50 days (Moreno-Insertis, 1983; Moreno-Insertis, 1986). This time is short compared with the cycle time and might lead to excessive magnetic flux losses, which then led to the proposal that the magnetic field would instead be generated in the overshoot layer beneath the convection zone. This idea is still the basic picture today, although simulations of convection generally produce magnetic fields that are distributed over the entire convection zone.

Yet another important development in the 1980s was the proposal that the α effect might actually be the sum of a kinetic and a magnetic part and that the magnetic part can be estimated by solving an evolution equation for the magnetic helicity density. The importance of this development was obscured by the excitement that the two evolution equations for poloidal and toroidal field, supplemented by a third equation for the magnetic helicity density, could produce chaos (Ruzmaikin, 1981). The connection to what was to come some 10–20 years later was not yet understood at that point. Simulations of helical MHD turbulence in a periodic domain demonstrated that in a periodic domain the α effect might be quenched in an Re_M-dependent fashion like

$$\alpha = \frac{\alpha_0}{1 + \mathrm{Re}_M \overline{\boldsymbol{B}}^2 / B_{\mathrm{eq}}^2}. \tag{1.1}$$

If this were also true of astrophysical dynamos, α would be negligibly small and would not be relevant for explaining the magnetic field in these bodies. Such quenching is therefore nowadays referred to as catastrophic quenching. However, there is now mounting evidence that this type of α quenching is a special case of a more general formula (Brandenburg, 2008)

$$\alpha = \frac{\alpha_0 + \mathrm{Re}_M \left[\eta_t \mu_0 \overline{\boldsymbol{J}} \cdot \overline{\boldsymbol{B}} / B_{\mathrm{eq}}^2 - (\boldsymbol{\nabla} \cdot \overline{\boldsymbol{F}}_{\mathrm{C}}) / 2 k_{\mathrm{f}}^2 B_{\mathrm{eq}}^2 - (\partial \alpha / \partial t) / 2 \eta_t k_{\mathrm{f}}^2 \right]}{1 + \mathrm{Re}_M \overline{\boldsymbol{B}}^2 / B_{\mathrm{eq}}^2}, \tag{1.2}$$

which comes from magnetic helicity conservation. Note that in this equation there are 3 new terms that all scale with Re_M and are therefore important. Even in a closed or periodic domain, the first and third terms in squared brackets contribute, the most promising way out of catastrophic quenching is through magnetic helicity fluxes (Blackman & Field, 2000; Kleeorin et al., 2000). These developments are still ongoing and we refer here to some recent papers by (Shukurov et al., 2006; Brandenburg et al., 2009; Candelaresi et al., 2010).

2. Simulating dynamos

2.1. Roberts flow

Much of the theoretical understanding of dynamos is being helped by numerical simulations. In fact, nowadays one of the simplest dynamos to simulate is the Roberts flow dynamo. A kinematic dynamo can be simulated by adopting a velocity field of the form

$$U = \nabla \times \psi z + k_{\mathrm{f}} \psi z, \qquad (2.1)$$

where $\psi = (U_0/k_0) \cos k_0 x \cos k_0 y$. In order to give some idea about the ease at which reasonably accurate solutions can be obtained we give in Table 1 the numerically obtained critical values of the magnetic Reynolds number, $\mathrm{Re}_M = u_{\mathrm{rms}}/\eta k_{\mathrm{f}}$, for a given resolution.

Table 1. Critical values of Re_M for the Roberts flow dynamo at low resolution (from 8^3 to 32^3 mesh points) and different spatial order of the numerical scheme.

Resolution	$\mathrm{Re}_M c$		
	2nd order	6th order	8th order
8^3	5.23	5.15	5.16
16^3	5.517	5.514	5.518
32^3	5.522	5.522	5.521

Even turbulent dynamos are nowadays easy to simulate and meaningful results have been obtained already at relatively low resolution, provided the flow is helical (Brandenburg, 2001). However, there are also examples where numerical aspects can have a major effect on the outcome of such simulations. In the following we discuss two examples: magnetic hyperdiffusion and the use of Euler potential with artificial magnetic diffusion.

2.2. Magnetic hyperdiffusion in helicity-driven dynamos

Dynamos work by maintaining the magnetic field against Ohmic decay via magnetic induction. It is then not surprising that the result can be sensitive to the numerical treatment of magnetic diffusion. One example is the consideration of magnetic hyperdiffusion of the form

$$\eta J \rightarrow (-1)^{n-1} \nu_n \nabla l a^{2n-2} J \qquad (2.2)$$

instead of the regular ηJ term. In helical dynamos in closed or periodic domains the saturation time is given by $\eta k_1^2 \Delta t = 1$, but with hyperdiffusion this condition becomes $\eta_n k_1^{2n} \Delta t = 1$; see Figure 1. This is exactly what one expects from a scheme like hyperdiffusion that enhances the effective magnetic Reynolds number. However, another important modification is that the saturation amplitude increases from Brandenburg & Sarson, 2002

$$\frac{\overline{B}^2}{B_{\mathrm{eq}}^2} \approx \frac{k_{\mathrm{f}}}{k_1} \quad \text{to} \quad \frac{\overline{B}^2}{B_{\mathrm{eq}}^2} \approx \left(\frac{k_{\mathrm{f}}}{k_1}\right)^{2n-1}. \qquad (2.3)$$

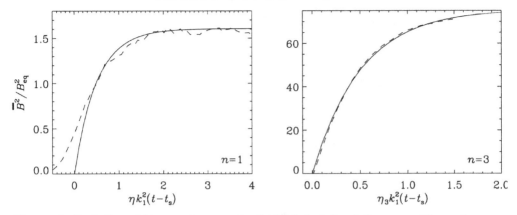

Figure 1. Evolution of large-scale magnetic field (dashed lines) for runs with regular magnetic diffusion ($n = 1$, left, $u_{\rm rms}/\eta k_1 = 45$) and magnetic hyperdiffusion ($n = 3$, right, $u_{\rm rms}/\eta_3 k_1 = 3200$), at 16^3 resolution and $k_f/k_1 = 3$, compared with the prediction (solid lines) of Brandenburg & Sarson, 2002, where t_s is the saturation time of the small-scale field.

This is a caveat that is important to keep in mind when employing magnetic hyperdiffusion for astrophysical simulations. This has a major effect on the saturation of large-scale magnetic fields that is at first glance surprising. However, once one realizes that the saturation in a periodic domain is governed by the magnetic helicity equation applied to the steady state, i.e. by

$$\frac{\mathrm{d}}{\mathrm{d}t} < \boldsymbol{A} \cdot \boldsymbol{B} > = -2\eta < \boldsymbol{J} \cdot \boldsymbol{B} >, \tag{2.4}$$

where we split the right-hand side into contributions from large- and small-scale fields, i.e. $\boldsymbol{B} = \overline{\boldsymbol{B}} + \boldsymbol{b}$ and $\boldsymbol{J} = \overline{\boldsymbol{J}} + \boldsymbol{j}$, so that $< \boldsymbol{J} \cdot \boldsymbol{B} > = < \overline{\boldsymbol{J}} \cdot \overline{\boldsymbol{B}} > + < \boldsymbol{j} \cdot \boldsymbol{b} >$, we have for fully helical large- and small-scale fields in the steady state,

$$0 = -2\eta < \overline{\boldsymbol{J}} \cdot \overline{\boldsymbol{B}} > -2\eta < \boldsymbol{j} \cdot \boldsymbol{b} > \approx \pm 2\eta \left(k_1 < \overline{\boldsymbol{B}}^2 > -k_f < \boldsymbol{b}^2 > \right), \tag{2.5}$$

with $< \boldsymbol{b}^2 > \approx B_{\rm eq}^2$, it becomes clear that the use of magnetic hyperdiffusion picks up the k_1 and k_f factors at correspondingly higher powers, leading thus to Equation 2.3. The upper and lower signs of the term on the right-hand side of Equation 2.5 apply to small-scale forcings with positive and negative helicity, respectively.

2.3. MHD with Euler potentials

Until recently, the use of Euler potentials (EP) has been a popular choice for solving the MHD equations numerically using Lagrangian methods (Price & Bate, 2007; Rosswog & Price, 2007).

The representation of $\boldsymbol{B}$ in terms of EP, α and β, as

$$\boldsymbol{B} = \boldsymbol{\nabla}\alpha \times \boldsymbol{\nabla}\beta \tag{2.6}$$

is a nonlinear one, which solves the induction equation in the case $\eta = 0$, $\partial \boldsymbol{B}/\partial t = \boldsymbol{\nabla} \times (\boldsymbol{U} \times \boldsymbol{B})$, provided $\mathrm{D}\alpha/\mathrm{D}t = \mathrm{D}\beta/\mathrm{D}t = 0$. The problem is that the magnetic field tends to develop sharp structures that will not be properly resolved by the numerical scheme. The hope has been that the overall properties of the magnetic field at larger scales would then still be approximately correct. However, this turns out not to be the case. And, more importantly, as one increases the resolution, the solution converges, but it is simply the wrong solution. This has been demonstrated in detail in a separate paper

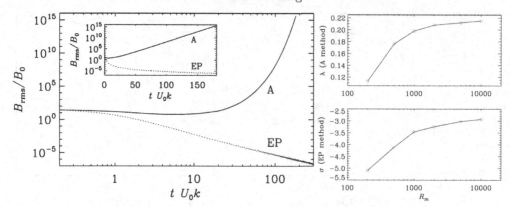

Figure 2. Left: Comparison of the evolution of $B_{\rm rms}$ for the modified Galloway–Proctor flow with point-wise zero helicity for methods A and EP using 256^3 meshpoints and $\mathrm{Re}_M = 10^4$. Note the power law scaling for the EP method and the exponential scaling for the A method. Right: Re_M dependence of the exponents λ and σ characterizing the evolution of $B_{\rm rms} \sim e^{\lambda t}$ for the A method and $B_{\rm rms} \sim t^\sigma$ for the EP method for the modified Galloway–Proctor flow with point-wise zero helicity for methods A and EP using 256^3 meshpoints. Adapted from Brandenburg (2010).

(Brandenburg, 2010) where, among other cases, solutions of the Galloway–Proctor fast dynamo flow were considered. In Figure 2 we show that the solution of the induction equation using EP leads to algebraically decaying solutions of the form

$$B_{\rm rms} \sim t^\sigma \quad \text{(EP method)}, \tag{2.7}$$

where, with increasing resolution, σ converges toward a value around -3, while with the usual vector potential method (A method) one solves $\partial \boldsymbol{A}/\partial t = \boldsymbol{U} \times \boldsymbol{B} - \eta \boldsymbol{J}$ with $\boldsymbol{B} = \boldsymbol{\nabla} \times \boldsymbol{A}$ and $\boldsymbol{J} = \boldsymbol{\nabla} \times \boldsymbol{B}$, and finds instead

$$B_{\rm rms} \sim e^{\lambda t} \quad \text{(A method)}, \tag{2.8}$$

where λ also converges, and its value is about 0.22 for $\mathrm{Re}_M = 10^4$; see Figure 2.

It is quite remarkable that by changing the properties of the solution at small scales only slightly, one can produce rather dramatic effects. This includes cases of magnetic hyperdiffusivity, where the large-scale field amplitude can be quite different, albeit in agreement with the theory applied to the hyperdiffusive case (Brandenburg & Sarson, 2002). Another example is that of artificial diffusion in solutions for the EP, where the obtained results bear no resemblance with those obtained using the A method.

2.4. Quantitative comparison between simulations and mean-field theory

Mean-field theory has the potential of being a quantitatively accurate and hence predictive theory. In order to establish this in particular cases, it is important to consider as many contact points between theory and simulations as possible. One thing we can do is to determine α effect, turbulent diffusion, and other effects from simulations, and to compare the thus calibrated mean-field model with simulations. This provides an important resource for ideas of what one might have been missing in various contexts. Here we just mention the case of the Roberts flow, for which α_{ij} and η_{ij} have been determined using the test-field methods (Schrinner et al., 2007). One might then expect that the growth rate obtained from the underlying dissipation rate,

$$\lambda = \alpha k - (\eta_{\rm t} + \eta)k^2 \tag{2.9}$$

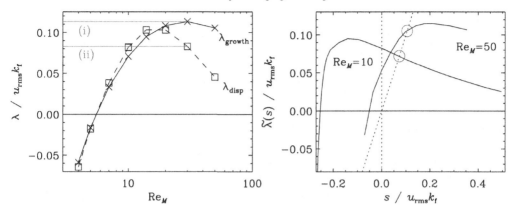

Figure 3. Left: Re_M dependence of the growth rate for the Roberts flow as obtained from a direct calculation ($\lambda_{\mathrm{growth}}$) compared with the result of the dispersion relation, $\lambda_{\mathrm{disp}} = \alpha k_z - (\eta + \eta_{\mathrm{t}})k_z^2$, using a cubic domain of size L^3, where $k_1 = 2\pi/L$ and $k_{\mathrm{f}} = \sqrt{2}k_1$. For this range of Re_M, the most unstable mode is the largest one that fits in the box ($k_z = k_1$). The two horizontal lines in gray mark the values of $\lambda_{\mathrm{growth}}$ and λ_{disp} at $\mathrm{Re}_M = 30$, denoted by (i) and (ii), respectively. Right: Laplace-transformed effective growth rate, $\tilde{\lambda}(s) = \tilde{\alpha}(s)k - [\eta + \tilde{\eta}_{\mathrm{t}}(s)]k^2$, for the Roberts flow with $\mathrm{Re}_M = 10$ and 50. Note the different signs of the slope at the intersection with the diagonal (denoted by circles). Adapted from Hubbard & Brandenburg (2009).

should agree with the value obtained from the direct calculation. This is however only the case for $\lambda = 0$, but not for $\lambda \neq 0$. However, it would be a mistake to assume that there is something wrong with the test-field methods. Instead, what is wrong here is just the assumption of homogeneity and stationarity. Obviously, when $\lambda \neq 0$, the solution is exponentially growing or decaying like $e^{\lambda t}$, so it is clearly not steady!

When the assumption of steadiness is no longer satisfied, one has to return to the underlying integral relation between α and the mean field, i.e. $\alpha(z,t)\overline{B}(z,t)$ has to be replaced by a convolution, so

$$\alpha(z,t)\overline{B}(z,t) \to \int \alpha(z-z', t-t')\overline{B}(z',t')\,\mathrm{d}z'\,\mathrm{d}t', \qquad (2.10)$$

and similarly for the magnetic diffusivity. One way of dealing with this complication is to note that the convolution in real space corresponds to a multiplication in Fourier space. In other words, we can write

$$\tilde{\mathcal{E}}(k,\omega) = \tilde{\alpha}(k,\omega)\tilde{B}(k,\omega), \qquad (2.11)$$

where $\tilde{\mathcal{E}}(k,\omega) = \int \overline{\mathcal{E}}(z,t)\,e^{-i(kz-\omega t)}\,\mathrm{d}z\,\mathrm{d}t$ is the Fourier-transformation of $\overline{\mathcal{E}}(z,t)$ (and likewise for the other fields). Of course, the value of ω that is of interest is $\omega = \mathrm{i}\lambda$, where λ is the then self-consistently obtained growth rate. This has been described in detail by Hubbard & Brandenburg (2009), who motivated their study using the example of the Roberts flow, where the discrepancy between the numerical solution and that obtained for $\lambda = 0$ is shown. A reasonable fit to their data is $\alpha(k,\omega) = 1/(1 - \mathrm{i}\omega\tau)$ for $k \to 0$.

3. Low magnetic Prandtl number and application to accretion discs

In many astrophysical bodies the magnetic Prandtl number, $\mathrm{Pr}_M = \nu/\eta$, is either large or small, but not around unity. Again, from a numerical point of view, it is surprising that the ratio ν/η is important even though

$$\nu \to 0, \quad \eta \to 0. \qquad (3.1)$$

In many numerical treatments it is implicitly assumed that the exact values of ν and η do not explicitly matter, because it should not matter how long each of the two turbulent cascades is. This should indeed be true provided the dynamics we are interested in takes place entirely on the large scales. But for helical large-scale dynamos, this is evidently not the case, because in the kinematic regime, the magnetic energy spectrum follows the Kazantsev $k^{3/2}$ spectrum and reaches a peak at the resistive scale, corresponding to the wavenumber $k_\eta = (J_{\rm rms}/\eta)$. However, this applies first of all only to the kinematic regime and, seemingly, it is relevant only for the small-scale dynamo regime. This has been discussed recently in connection with the contrasting case of large-scale dynamos, where the excitation conditions are not affected by the value of Pr_M (Brandenburg, 2009; Brandenburg, 2011). Nevertheless, even in this case the dissipation rates of kinetic and magnetic energies, ϵ_K and ϵ_M, respectively, do depend on the value of Pr_M. Although it seems fairly clear that the ratio ϵ_K/ϵ_M increases with Pr_M in power law fashion proportional to Pr_M^n, the exponent n is not well constrained. Earlier work covering the range $10^{-3} \leqslant \mathrm{Pr}_M \leqslant 1$ at 512^3 resolution suggested $n = 1/2$, although additional data covering also the range $1 < \mathrm{Pr}_M \leqslant 10^3$ given an exponent closer to $n = 0.6$ or even $n = 2/3$. There is at present no theory for the value of the exponent n.

Applying these findings to accretion discs, we should first recall that the work of Lesur & Longaretti (2007) suggested that the onset of the magneto-rotational instability shows a strong dependence on Pr_M. However, one should expect that when large-scale dynamo action is possible, this condition may change and the onset would then be independent of Pr_M. This is indeed what Käpylä & Korpi (2010) find. The question is of course what is the relevant large-scale dynamo mechanism in this case. One proposal is the incoherent α–shear dynamo, which works through constructive amplification of the mean field in the direction of mean shear. Yet another possible mechanism is the shear–current dynamo (Rogachevskii & Kleeorin, 2003), although this result has not yet been confirmed.

4. Conclusions

Simulating simple dynamos on the computer is nowadays quite simple. Nevertheless, we have seen here examples that illustrate that things can also go quite "wrong". In the case of magnetic hyperdiffusion, it is clear what happens (Brandenburg & Sarson, 2002), so that magnetic hyperdiffusion can also be used to ones advantage, as was demonstrated in Brandenburg *et al.* (2002). However, in the case of Euler potentials it is not clear what happens and whether this method can be used to simulate even the ideal MHD equations, given that each numerical scheme will introduce some type of diffusion. In this short review, we have also attempted to clarify why numerical calculations of α effect and turbulent diffusion using the standard test-field method (Schrinner *et al.*, 2007) would yield values that can only reproduce a correct growth rate in the case of vanishing growth. In all other case, a representation in terms of integral kernels has to be used. Finally, we have discussed some effects of using magnetic Prandtl numbers that are different from unity. It turns out that in the steady state, the rate of transfer from kinetic to magnetic energy depends on the value of Pr_M. This is somewhat unexpected, because the onset condition for dynamo action does not depend on Pr_M Brandenburg, 2009, and yet the actual efficiency of the dynamo, as characterized by the work done against the Lorentz force, $- < U \cdot (J \times B) >$, does depend on Pr_M and is proportional to Pr_M^{-n} (with n between 1/2 and 2/3) for large values of Pr_M.

Understanding the limits of numerical simulations is just as important as appreciating its powers. As the example with the problem with mean-field and simulated growth rates shows, understanding the initial mismatch can be the key to a more advanced and more accurate theory that will ultimately be needed when describing some of the yet unexplained properties of astrophysical dynamos.

References

Blackman, E. G. & Field, G. B., MNRAS 2000, *318*, 724

Brandenburg, A., *ApJ*, 2001, *550*, 824

Brandenburg, A. *Astron. Nachr.*, 2008, *329*, 725

Brandenburg, A. *ApJ*, 2009, *697*, 1206

Brandenburg, A. *MNRAS*, 2010, *401*, 347

Brandenburg, A. *Astron. Nachr.*, 2011, *332*, 51 (arXiv:1008.4226)

Brandenburg, A. & Sarson, G. R. *Phys. Rev. Lett.*, 2002, *88*, 055003

Brandenburg, A., Candelaresi, S., & Chatterjee, P. MNRAS2009, *398*, 1414

Brandenburg, A., Dobler, W., & Subramanian, K. *Astron. Nachr.,*, 2002, *323*, 99

Candelaresi, S., Hubbard, A., Brandenburg, A., & Mitra, D. 2010 (arXiv:1010.6177)

Cowling, T. G. *MNRAS*, 1933, *94*, 39

Gilman, P. A. *ApJS*, 1983, *53*, 243

Glatzmaier, G. A. *ApJ*, 1985, *291*, 300

Herzenberg, A.*Proc. Roy. Soc. Lond.*, 1958, *250A*, 543

Haugen, N. E. L., Brandenburg, A., & Dobler, W. *Phys. Rev. E*, 2004, *70*, 016308

Hubbard, A. & Brandenburg, A. *ApJ*, 2009, *706*, 712

Käpylä, P. J. & Korpi, M. J. 2010 (arXiv:1004.2417)

Käpylä, P. J., Korpi, M. J., Brandenburg, A., *et al. Astron. Nachr.*, 2010, *331*, 73

Kazantsev, A. P. *Sov. Phys. JETP*, 1968, *26*, 1031

Kleeorin, N., Moss, D., Rogachevskii, I., & Sokoloff, D. *A&A*, 2000, *361*, L5

Larmor, J. 1919*Rep. Brit. Assoc. Adv. Sci.* 159

Larmor, J. *MNRAS*, 1934, *94*, 469

Lesur, G. & Longaretti, P.-Y. *A&A*, 2007, *378*, 1471

Meneguzzi, M., Frisch, U., & Pouquet, A. *Phys. Rev. Lett.*, 1981, *47*, 1060

Moreno-Insertis, F. *A&A*, 1983,*122*, 241

Moreno-Insertis, F. *A&A*, 1986, *166*, 291

Parker, E. N. *ApJ*, 1955, *122*, 293

Parker, E. N. *ApJ*, 1982, *256*, 302

Parker, E. N. *ApJ*, 1969, *157*, 1129

Parker, E. N. *ARA&A*, 1970a, *8*, 1

Parker, E. N. *ApJ*, 1970b, *160*, 383

Parker, E. N. *ApJ*, 1970c, *162*, 665

Parker, E. N. *ApJ*, 1971a, *163*, 255

Parker, E. N. *ApJ*, 1971b, *163*, 279

Parker, E. N. *ApJ*, 1971c, *164* 491

Parker, E. N. *ApJ*, 1971d *165*, 139

Parker, E. N. *ApJ*, 1971e *166*, 295

Parker, E. N. *ApJ*, 1971f *168*, 239

Price, D. J. & Bate, M. R. *MNRAS*, 2007, *377*, 77

Rogachevskii, I. & Kleeorin, N. *Phys. Rev. E*, 2003, *68*, 036301

Rosswog, S. & Price, D. *MNRAS*, 2007, *379*, 915

Ruzmaikin, A. A. 1981 *Comments Astrophys.*, *9* 85

Schrinner, M. Rädler, K.-H., Schmitt, D., *et al. Geophys. Astrophys. Fluid Dyn.*, 2007, *101*, 81

Shukurov, A., Sokoloff, D., Subramanian, K., & Brandenburg, A. *A&A*, 2006, *448*, L33

Steenbeck, M. & Krause, F. *Astron. Nachr.*, 1969a, *291*, 49

Steenbeck, M. & Krause, F. *Astron. Nachr.*, 1969b, *291*, 271

Steenbeck, M., Krause, F., & Rädler, K.-H. 1966, *Z. Naturforsch.*, *21a*, 369

Vainshtein, S. I. & Ruzmaikin, A. A. 1971, *Astron. Zh.*, *48*, 902

Vainshtein, S. I. & Ruzmaikin, A. A. 1972, *Astron. Zh.*, *49*, 449

Advances in Plasma Astrophysics
Proceedings IAU Symposium No. 274, 2010
A. Bonanno, E. de Gouveia Dal Pino & A. G. Kosovichev, eds.
© International Astronomical Union 2011
doi:10.1017/S1743921311007381

Relativistic jets and current driven instabilities

A. Ferrari, A. Mignone and M. Campigotto

Dipartimento di Fisica Generale, Università di Torino,
Via Pietro Giuria 1, 10125, Torino, Italy
email: ferrari@ph.unito.it

Abstract. We review the present results on the study of the propagation of relativistic collimated outflows characteristics of active galaxies and active stars. Magnetic fields, namely their azimuthal components, gives rise to current driven instabilities whose nonlinear development can actually be connected to the complex morphologies observed in astrophysical jets.

1. Introduction

The first direct evidence of relativistic effects in astrophysics was the discovery of extended radio sources with supersonic jets showing also apparent superluminal motions Cohen *et al.* (1971). Active galaxies in general, in particular FR II radio galaxies, quasars and blazars, are characterized by the dynamics of powerful outflows with speed close to the velocity of light; in addition the matter of these outflows is relativistic by itself with large Lorentz factors in order to produce non thermal radiation by synchrotron and inverse Compton processes from the radio to the gamma-ray band. The same physics, although on a much smaller scale, underlies the morphologies and radiation characteristics of galactic relativistic stars and X-ray binaries. Recently relativistic jets have been proposed as the basic ingredient to interpret gamma-ray bursts.

The study of the physics of relativistic jets follows three major lines of investigation: (i) acceleration and collimation, (ii) propagation and stability in the interaction with the intergalactic plasma, (iii) termination and momentum/energy release into the external ambient. The current understanding of line (i) is through the so-called magneto-centrifugal mechanism in which a magnetized differentially rotating accretion disk generates Lorentz forces that accelerate a charged particle beam to relativistic velocity along the rotation axis of the system and at the same time keep the beam collimated by the azimuthal magnetic field component. This azimuthal component is convected by the beam while propagating to large scales and is crucial in defining its stability. Several papers have analyzed the relativistic jet stability (for a review see Massaglia *et al.*, 2008). Beam instabilities can be divided in two classes: instabilities due to interaction of the relativistic beam with the external medium and intrinsic instabilities related to the magneto-hydrodynamic structure of the beam. External instabilities are pressure-driven and produce mixing with the ambient with transfer of linear momentum and slowing-down of the flow. Intrinsic instabilities are typically current-driven more prone to affect the morphology of the beam and its radiation emission characteristics.

In this paper we shall concentrate on the case of intrinsic current-driven instabilities, referring to previous papers for external instabilities that are likely to be more important in the termination phases of the jet, (Rossi *et al.*, 2008). Current driven instabilities have been studied in the non-relativistic linear limit by Appl & Camenzind (1993), while relativistic linear studies have been presented by Istomin & Pariev (1994), (1996). Linear studies are strongly affected by the assumed boundary conditions and cannot follow the complex nonlinear effects that may damp unstable modes. In the following we discuss a series of fully three-dimensional numerical simulations of the propagation of relativistic plasma jets with both axial and azimuthal magnetic field components. Interesting effects of relevance for astrophysical applications are obtained and analyzed.

2. Numerical Approach

We study the dynamical evolution of current-carrying jets by solving the three-dimensional time-dependent special relativistic MHD equations:

$$
\begin{aligned}
\frac{\partial}{\partial t}(\rho\gamma) + \nabla \cdot (\rho\gamma\boldsymbol{v}) &= 0\,, \\
\frac{\partial \boldsymbol{m}}{\partial t} + \nabla \cdot \left[w\gamma^2\boldsymbol{vv} - \boldsymbol{BB} - \boldsymbol{EE} + \mathsf{I}\,(p + p_E) \right] &= 0\,, \\
\frac{\partial \boldsymbol{B}}{\partial t} + \nabla \times \boldsymbol{E} &= 0\,, \\
\frac{\partial}{\partial t}\left(\gamma^2 w - p + \frac{p_E}{2}\right) + \nabla \cdot \boldsymbol{m} &= 0\,,
\end{aligned}
\tag{2.1}
$$

where ρ is the rest-mass density, γ the Lorentz factor of the flow, $\boldsymbol{m} = w\gamma^2\boldsymbol{v} + \boldsymbol{E} \times \boldsymbol{B}$ is the momentum density, $\boldsymbol{B}$ is the magnetic field, $\boldsymbol{E} = -\boldsymbol{v} \times \boldsymbol{B}$ is the electric field (in the infinite conductivity limit) and w is the gas enthalpy. The total pressure $p + p_E$ accounts for thermal and electromagnetic contributions with $p_E = \boldsymbol{B}^2/2 + \boldsymbol{E}^2/2$. The gas enthalpy w is related to ρ and p via the TM equation of state (Mignone & McKinney, 2007).

We analyze two different configurations consisting, namely, of a full jet (§2.1) and an infinitely long periodic jet (§2.2). The system of conservation laws (2.1) is solved using the high-resolution, Godunov-type, PLUTO code (Mignone *et al.*, 2007) for astrophysical gasdynamics.

2.1. *Full Jet Simulations*

In the global simulations, we inject a relativistic beam at the lower $y-$ boundary into a stationary medium with uniform density and pressure. The inflow values are prescribed in terms of the jet-to-ambient density contrast η, Lorentz factor γ, sonic and Alfvénic Mach numbers. We consider two different magnetic field topologies. In the first one, the magnetic field is purely toroidal and it is carried along with the beam. Radial profiles are chosen to guarantee force balance between Lorentz force and pressure gradient inside the jet by solving the radial momentum equilibrium equation (2.2), see also Mignone *et al.* (2010). In the second case, the field is constant and purely poloidal in the direction of jet propagation and it threads the ambient medium as well.

The computational domain is defined by $x, z \in [-L/2, L/2]$, $y \in [0, L_y]$ where $L = 56$ and $L_y = 80$. A uniform grid resolution is employed in y and for $|x|, |z| < 10$ while geometrically stretched cells are used otherwise. The overall resolution is $640 \times 1600 \times 640$, corresponding to the largest resolution numerical simulation of three dimensional relativistic jet carried so far.

2.2. *Infinite Periodic Jet*

The initial condition consists of an equilibrium configuration describing an infinitely long plasma column moving at relativistic speed in the vertical z direction. For simplicity, we study the equilibrium by imposing cylindrical symmetry around the axis of propagation (i.e., $\partial/\partial\phi = \partial/\partial z = 0$) and by setting the radial components of velocity and magnetic field to zero, $v_r = B_r = 0$. With these assumptions the Lorentz and electric forces have only one non-vanishing component along the radial direction and a stationary configuration requires $\nabla \times E = 0$ as well as the solution of the r-component of the momentum equation:

$$\frac{\partial p}{\partial r} - \frac{w\gamma^2 v_\phi^2}{r} = \left[E\nabla \cdot E + (\nabla \times B) \times B \right] \cdot \hat{e}_r , \qquad (2.2)$$

where $\hat{e}_r$ is the unit vector in the radial direction. The previous equation states a mutual balance between pressure, centrifugal, electrical and Lorentz forces. In the following we will assume that pressure forces are negligible ($p = \text{cost}$) and that the longitudinal component of magnetic field is constant, $\partial_r B_z = 0$. Furthermore, we prescribe the radial profiles of v_z, v_ϕ, B_z so that direct integration of Eq. (2.2) leads to the following quadratic equation in B_ϕ:

$$B_\phi^2 = (v_\phi - KB_\phi)^2 B_z^2 + Q , \qquad \text{with} \qquad Q = \frac{1}{r^2} \int_0^r 2w\gamma^2 v_\phi^2 r\, dr , \qquad (2.3)$$

where $K = v_z/B_z$. The explicit form of $v(r)$ and B_z must be properly chosen in order to match the parameter definition,

$$q = \left.\frac{B_\phi}{B_z}\right|_{r=\frac{1}{2}} , \qquad \gamma = \left.\frac{1}{\sqrt{1 - v_z^2 - v_\phi^2}}\right|_{r=0} , \qquad M_s^2 = \frac{w\gamma^2 v^2}{p} , \qquad M_A^2 = \frac{w\gamma^2 v^2}{\langle B^2 \rangle} , \qquad (2.4)$$

that is, the topology of magnetic field q, the Lorentz gamma factor, the kinetic to thermal energy ratio and the kinetic to (average) magnetic energy ratio. We choose the profile of the azimuthal velocity to be

$$v_\phi(r) = v_{\phi,M} \frac{r}{b} \left[1 - e^{(r-1)^3/a^3} \right] , \qquad v_z(r) = \frac{v_z(0)}{\cosh(r^4)} , \qquad (2.5)$$

where $a = 0.4$ an $b = 0.43$ are chosen in such a way that the azimuthal velocity has a maximum in $r \approx \frac{1}{2}$ where $v_\phi = v_{\phi,M}$. If B_z is known, the explicit value of $v_{\phi,M}$ could be found by solving Eq. (2.3) in $r = \frac{1}{2}$. In practice, however, B_z is not known a priori but depends on the velocity through the definition of M_A (Eq 2.4) and one has to iterate on the values of B_z until the prescribed kinetic to magnetic energy ratio is satisfied.

The computational domain consists of the three-dimensional box $x, y \in [-10, 10]$, $z \in [0, 10]$ in units of the column radius. The grid resolution is uniform in z and for $|x|, |y| < 5$, while it becomes geometrically stretched outside of this region. Employing 20 zones on the column radius and 44 zones to reach the outer boundaries in each directions, we end up with an overall resolution of $288 \times 288 \times 200$ computational cells. Periodic boundary conditions are used in the vertical direction while open boundaries hold at the remaining sides.

3. Results

3.1. *Full Jet Simulations*

As shown in (Mignone *et al.*, 2010), the propagation of three-dimensional jet carrying a strong toroidal field bears little resemblance with the corresponding 2-D axisymmetric simulations showing prominent nose cone structures, (Komissarov, 1999; Leismann *et al.*, 2005). Here, instead, the fully developed three dimensional structures becomes unstable

	A1	B1	B2	B3	B4	B5	C1	C2
γ	2	5	5	5	5	5	10	10
M_A	1	3	3	3	1	1	3	3
q	4	1	4	10	1	4	4	10

Table 1. Simulation cases for the periodic jet runs. Here γ, M_A and q have the meaning given in Eq. (2.4). In all cases, the sonic Mach numbers M_s and the density contrast are held fixed to 30 and 1, respectively.

in proximity of the jet head and the trajectory progressively bends away from the main longitudinal axis. Overall, the backflowing material forms a very asymmetric cocoon as a result of the changes in the direction of the jet head. This is clear by inspecting

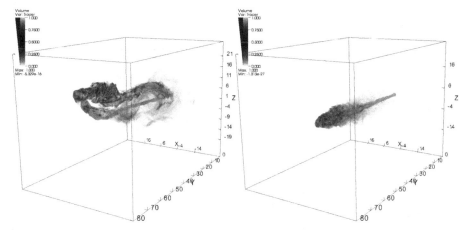

Figure 1. Three-dimensional rendering of the passive scalar in the full jet simulation at $t = 770$ for the toroidal jet simulation (left panel) and poloidal (constant vertical field) simulation (right panel).

the left panel in Fig 1 showing the spatial distribution of the passive scalar at $t = 770$ (in simulation units). The wandering of the jet head, induced by kink-instability effects, may create multiple sites where the jet impacts on the external medium forming strong shocks. This behaviour may originate the multiple hotspots that are observed in several radiogalaxies.

The toroidal field seems to protect the jet core from any interaction with the surrounding, thus preventing momentum transfer to the external medium. This is due to the presence of low-order modes ($m < 2$) only, producing oval deformations but not inducing any mixing. The tension of the toroidal field, in fact, acts as a strong stabilizing factor for high-order modes and allows to maintain a highly relativistic spine all along its length, (Mignone *et al.*, 2010).

In the case of a purely constant longitudinal field, on the other hand, the results are more similar to the hydro cases discussed by Rossi *et al.*, (2008). Indeed, the presence of a magnetic field does not seem to introduce significant differences with respect to the unmagnetized case and the jet propagates along the main direction axis without suffering from any deflection, see the right panel in Fig 1.

3.2. *Periodic Jet Simulations*

We have carried out numerical simulations using the cases illustrated in Table 3.2. As a general trend, our results indicate that the presence of a relatively larger toroidal magnetic field component ($q \gtrsim 1$) leads to the onset of instability which, in most cases, is triggered by the $m = 1$ (kink) unstable mode.

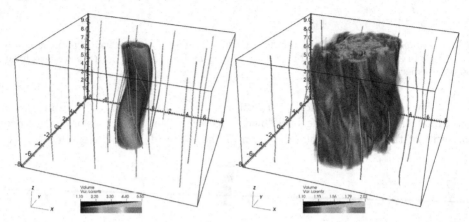

Figure 2. Three-dimensional rendering of the Lorentz factor in the periodic jet simulation for case B2 at $t = 75$ (left panel) and $t = 225$ (right panel). Magnetic field lines are overplotted.

This behavior can be observed in the left panel of Fig 2 showing prominent wiggling and column deformation at $t = 75$ for case B2. The nonlinear development (right panel at $t = 200$) shows that the jet structure is re-organized on a larger layer with considerable reduction of the Lorentz factor ($\gamma \sim 2$ at the end of simulation) and a more chaotic distribution. A similar behavior is also observed for larger values of q (case B3) and at larger Lorentz factors (cases C1 and C2). On the other hand, when toroidal and poloidal components are comparable ($B_z \approx B_\phi$ at $r = \frac{1}{2}$ in cases B1 and B4) the column does not show any evidence for instability. See, for instance, the left panel in Fig 3. For larger magnetic field strengths (case B5), we observe a stable behavior even

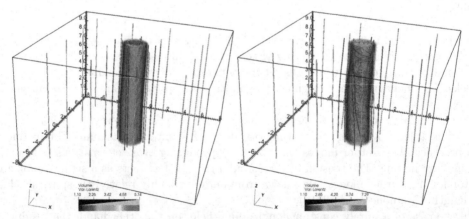

Figure 3. Same as Fig 2 showing the stability of the column for case B1 (left panel, $t = 200$) and B5 (right panel, $t = 320$).

in long term integration, as shown in the right panel of Fig 3. In the regime of strong magnetizations, in fact, the system can be well described by the force-free approximation. In this limit, our simulation results are in agreement with the findings of Istomin & Pariev (1994), (1996) who have shown that a jet with a longitudinal electric current remains stable with respect to helical as well as axially symmetric (pinch) modes. We believe that the induced stability owes to the stabilizing action of the electric field and, therefore, it has to be considered as a purely relativistic effect. To support our conclusion, we notice that corresponding weakly relativistic cases (e.g. case A1 with $\gamma = 2$) are still unstable, as expected on a classical basis.

4. Summary

Kink modes in current-driven instabilities appear to be very important in the morphological evolution of relativistic jets. In the full-jet simulations the irregular oscillations in the propagation direction and the spreading of the jet's head are reminiscent of the jets in extended radio galaxies. Large Poynting fluxes associated with azimuthal magnetic field components produce strong kinks and complex head wigglings. On the other hand, large kinetic fluxes avoid kinks and the jet's head proceeds uprightly. The spine of the jet is always highly relativistic on the average, while shocks create intermittent structures. In a separate set of simulations (Rossi *et al.*, 2010) we have verified that when the jet encounters a low density region outside the associated galaxy its structure becomes again straight and organized and kinks disappear.

In order to explore the instability mechanism as a function of the physical parameters (in particular of the Kinetic to Poynting flux ratio M_A), we have performed simulations of an infinite periodic jet. The results indicate that the instability is strong in the non-relativistic or relativistic cases for $M_A \gtrsim 1$, provided the magnetic azimuthal component is stronger than the longitudinal ($q > 1$). Instability becomes weaker for larger field strengths ($M_A \sim 1$) and tends to disappear when the magnetic field is strong enough (M_A small) to move the system towards the force-free field limit (Istomin & Pariev, 1994).

Acknowledgement

The three-dimensional numerical simulations of the infinite periodic jet have been carried out using the `matrix` cluster available at the Caspur computing facility.

References

Appl, S. & Camenzind, M. 1992, *A&A*, 256, 354

Cohen, M., Canno, W., Purcell, G., Shaffer, D., Broderick, J., Kellermann, K., & Jauncey, D. 1971, *ApJ*, 170, 207

Istomin, Y. N. & Pariev, V. I. 1994, *MNRAS*, 267, 629

Istomin, Y. N. & Pariev, V. I. 1996, *MNRAS*, 281, 1

Komissarov S., 1999, *MNRAS*, 308, 1069

Leismann T., Anton L., Aloy M. A., Muller W., Mart M., Miralles J. A., & Ibanez J. M., 2005, *A&AS*, 436, 503

Massaglia, S., Bodo, G., Mignone, A., & Rossi, P. 2008, *Jets from Young Stars III - Numerical MHD and Instabilities*, Lecture Notes in Physics 754, Springer

Mignone, A., Bodo, G., Massaglia, S., Matsakos, T., Tesileanu, O., Zanni, C., & Ferrari, A. 2007, *ApJS*, 170, 228

Mignone, A. & McKinney, J. C. 2007, *MNRAS*, 378, 1118

Mignone, A., Rossi, P., Bodo, G., Ferrari, A., & Massaglia, S. 2010, *MNRAS*, 402, 7

Rossi, P., Mignone, A., Bodo, G., Massaglia, S., & Ferrari, A. 2008, *A&AS*, 488, 795

Rossi, P., Mignone, A., Bodo, G., Massaglia, S., & Ferrari, A. 2010, in preparation

Advances in Plasma Astrophysics
Proceedings IAU Symposium No. 274, 2010
A. Bonanno, E. de Gouveia Dal Pino, A. Kosovichev, eds.
© International Astronomical Union 2011
doi:10.1017/S1743921311007393

Global MHD simulations of disk-magnetosphere interactions: accretion and outflows

M. M. Romanova[1], R. V. E. Lovelace[2], G. V. Ustyugova[3] and A. V. Koldoba[3]

[1]Department of Astronomy, Cornell University,
Ithaca, NY 14853-6801
email: `romanova@astro.cornell.edu`

[2]Department of Astronomy and Department of Applied and Engineering Physics,
Cornell University, Ithaca, NY 14853-6801
email: `RVL1@cornell.edu`

[3]Keldysh Institute of Applied Mathematics
Russian Academy of Sciences, Moscow, Russia
email: `ustyugg@rambler.ru`; `koldoba@rambler.ru`

Abstract. We outline recent progress in understanding the accretion of plasma to rotating magnetized stars obtained from global axisymmetric (2D) and 3D magnetohydrodynamic (MHD) simulations in three main areas: **(1.) Formation of jets from disk accretion onto rotating magnetized stars:** From simulations where the viscosity and magnetic diffusivity within the disk are described by alpha models, we find long-lasting conical outflows/jets from the disk/magnetosphere boundary in both the case where the star is *slowly rotating* and where it is *rapidly rotating* (the "propeller regime"). Most of the mass flux in the outflows is in a hollow cone but inside this cone there is a low-density high-velocity magnetically dominated flow along the open polar field lines of the star. The outflows occur under conditions where the poloidal magnetic flux of the star is bunched up by the accretion disk near the disk/magnetosphere boundary. Recent simulations show that the conical outflows become well-collimated for axial distances of $\lesssim 20$ times the inner disk radius. Exploratory 3D simulations show that conical winds are axisymmetric about the rotational axis (of the star and the disk), even when the dipole field of the star is significantly misaligned. **(2.) Formation of intrinsically one-sided jets from disk accretion to rotating magnetized stars:** There is strong observational evidence for an asymmetry between the approaching and receding jets from a number of young stars. We discuss the first MHD simulations of the formation asymmetric or one-sided jets arising from disk accretion to a rotating star with an asymmetric (dipole plus quadrupole) magnetic field. **(3.) Global axisymmetric and 3D simulations of the magnetorotational instability (MRI) in disk accretion onto magnetized stars:** In the axisymmetric simulations we observe cases where there is episodic or quasi-periodic burst of accretion similar to that observed in one X ray source. In 3D MHD simulations of accretion onto stars with tilted dipole fields using our Godunov-type code based on the "cubed sphere" grid we find that the density distribution is much less smooth than in the case of the laminar accretion flow described by α−viscosity. Instead, large turbulent cells dominate the flows and are strongly elongated in the azimuthal direction.

Keywords. accretion, accretion disks; MHD; stars: magnetic fields

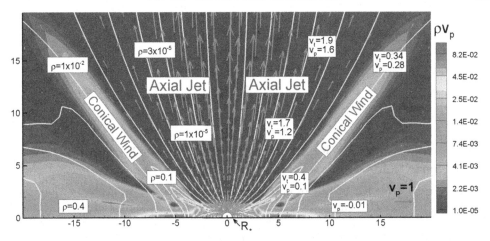

Figure 1. Matter flux ρv_p (background), sample field lines, and poloidal velocity vectors in a propeller-driven outflow at time $T = 1400$ (rotation periods at the inner disk radius). Sample numerical values are given for the poloidal v_p and total v_t velocity, and for the density ρ for different parts of the simulation region. Unit density corresponds to $\rho_0 = 4.1 \times 10^{-13}\,\mathrm{g\,cm}^{-3}$

1. Outflows/jets from the disk-magnetosphere boundary

Outflows and jets are observed from many disk accreting objects ranging from young stars to systems with white dwarfs, neutron stars, and black holes (e.g., Livio 1997). In addition to steady outflows there are episodic outbursts associated with periods of enhanced accretion (e.g., Cabrit *et al.* 1990). A large body of observations exists for outflows from young stars at different stages of their evolution, ranging from protostars, where powerful collimated outflows are observed, to classical T Tauri stars (CTTSs), where the outflows are weaker and often less collimated (see review by Ray *et al.* 2007). Different theoretical models have been proposed to explain the outflows from protostars and CTTSs (see review by Ferreira, Dougados, & Cabrit 2006). The models include those where the outflow originates from a radially distributed disk wind (Königl & Pudritz 2000; Casse & Keppens 2004) or from the innermost region of the accretion disk (Lovelace, Berk & Contopoulos 1991). Further, there is the X-wind model (Shu *et al.* 1994) where most of the outflow originates from the disk-magnetosphere boundary. The maximum velocities in the outflows are usually of the order of the Keplerian velocity of the inner region of the disk. This favors the models where the outflows originate from the inner disk region, or from the disk-magnetosphere boundary (if the star has a dynamically important magnetic field).

We have carried out systematic axisymmetric and limited 3D MHD simulations of outflows/jets from the disk-magnetosphere boundaries of rotating magnetized stars (Romanova *et al.* 2009). The disk is at a low temperature and is modeled by an alpha viscosity and a second alpha magnetic diffusivity and a high temperature low-density disk corona. We found outflows in two main cases: (1) where the star rotates slowly but the poloidal field lines of the star are bunched at the disk-magnetosphere boundary, and (2) where the star rotates rapidly in the propeller regime and the condition for bunching is also satisfied. In both cases, two-component outflows are observed as shown in Figure 1. One component originates near the disk/magnetosphere boundary and has a narrow-shell conical shape, and we term it a "conical wind". The other component is a magnetically dominated high-velocity low-density wind which flows along the open polar field lines of the star and it is referred to as the "jet." The energy and angular momentum fluxes in the two components are typically comparable. Exploratory 3D simulations show that conical winds are axisymmetric about the rotational axis (of the star and the disk), even when the dipole field of the star is significantly misaligned.

Figure 2 shows the much stronger magnetic collimation found in our recent simulations for the case of a lower density corona *and* a significantly larger simulation region (Lii *et al.* 2010).

2. One-sided outflows/jets from rotating stars with complex magnetic fields

There is clear evidence, mainly from Hubble Space Telescope (HST) observations, of the asymmetry between the approaching and receding jets from a number of young stars. The objects include the jets in HH 30 (Bacciotti *et al.* 1999), RW Aur (Woitas *et al.* 2002), TH 28 (Coffey *et al.* 2004), and LkHα 233 (Pererin & Graham 2007). Specifically, the radial speed of the approaching jet may differ by a factor of two from that of the receding jet. For example, for RW Aur the radial redshifted speed is ~ 100 km/s whereas the blueshifted radial speed is ~ 175 km/s. The mass and momentum fluxes are also significantly different for the approaching and receding jets in a number of cases. It is possible that the observed asymmetry of the jets could be due to say differences in the gas densities on the two sides of the source. However, there is substantial observational evidence that young stars often have *complex* magnetic fields consisting of dipole, quadrupole, and higher order poles possibly misaligned with respect to each other and the rotation axis (e.g., Donati *et al.* 2008).

We have carried out MHD simulations of the formation of conical winds for the axisymmetric dipole/quadrupole field combinations for cases where the outflows are *not* required to be symmetrical about the equatorial plane (Lovelace *et al.* 2010). The left-hand panel of Fig. 3 shows the vacuum field and the right-hand panel a sample result from our simulations.

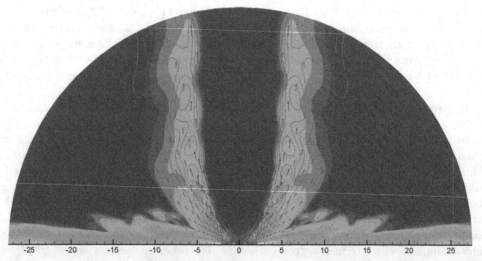

Figure 2. Axisymmetric simulations for a lower coronal density ($3 \times 10^{-4} \rho_{\text{disk}}$) *and* a larger simulation region show that the toroidal magnetic field gives strong collimation of the outflow (Lii *et al.* 2011).

3. Global MHD simulations of the MRI driven accretion onto magnetized stars

It is widely thought that the magnetorotational instability (MRI) in ionized weakly-magnetized accretion disks gives rise to turbulence which is responsible for the fast outward transport of angular momentum in the disk (Balbus & Hawley 1991, 1998). The MRI driven accretion has been observed in different simulations, local shearing box models, global axisymmetric models, and global 3D simulations (e.g., Balbus & Hawley 1998; Hawley *et al.* 2001; Beckwith, Hawley, & Krolik 2009). Fortuitously, the MRI turbulence found in simulations can be described approximately by an Shakura & Sunyaev (1973) α coefficient of $\sim 0.01 - 0.1$.

The global axisymmetric and 3D simulations of the MRI turbulence in disks has been done for the case of accretion disks around black holes and for cases where the initial magnetic field is inside the disk. However, the situation is different for the case of an accretion disk around a rotating star with a dynamically important magnetic field, or the case where the star has an effectively solid surface at which magnetic flux can accumulate and interact with an incoming MRI-driven flow. The details of such interactions are unknown and no simulations of such interaction have been done yet.

Long wavelength components ($\gg$ disk half-thickness) of the MRI instability may be directly contributing to the observed time variations in some sources. A number of anomalous millisecond pulsars (AMP) show episodic or quasi-periodic flares in the tails of their bursts. For example, the pulsar SAX J1808.4-3658 shows flaring activity with a quasi-period of about 1Hz (Patruno *et al.* 2009).

Recently, we have found interesting variability in axisymmetric simulations of the MRI-driven accretion onto magnetized star (Romanova *et al.* 2011a). The simulations show that the disk-magnetosphere interaction depends on the orientation of the poloidal magnetic field inside the disk relative to the direction of the B−field inside the magnetosphere. This field is much smaller than the azimuthal field, but it determines the character of

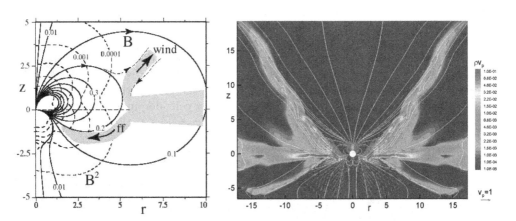

Figure 3. The **left-hand panel** shows the magnetic field lines $\Psi(r, z) =$ const and constant magnetic pressure lines for the case of an aligned dipole and quadrupole field where the flux function is $\Psi = \mu_d r^2 / R^3 + (3/4)\mu_q z r^2 / R^5$, where $R^2 = r^2 + z^2$ and μ_d is the dipole moment and μ_q is the quadrupole moment. Roughly, μ_q/μ_d is the distance at which dipole and quadrupole fields are equal. The funnel flow (ff) and the wind in this figure are suggested. The dashed lines are constant values of $\mathbf{B}^2$. The **right-hand panel** gives snapshot of the outflow in the case where $\tilde{\mu}_d = 10$, $\tilde{\mu}_q = 20$ at $t = 50$ as discussed by Lovelace *et al.* (2010). The color background shows the matter flux-density and the lines are the poloidal field lines. The vectors show the poloidal velocity.

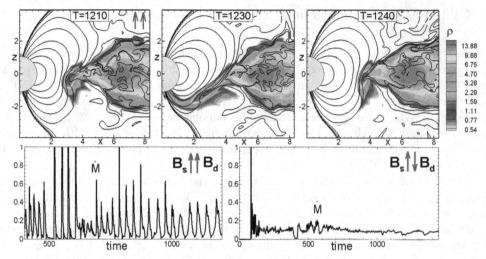

Figure 4. The top panels show three moments of time for a flaring episode of accretion for a case where the seed magnetic field of the disk is parallel to that of the star at the disk-magnetosphere boundary. The bottom panels show the accretion rate onto the star for cases of parallel (left panel) and antiparallel (right panel) fields. From Romanova *et al.* (2011a).

the disk-magnetosphere interaction and variability of the accretion rate onto the star. Namely, in the case where the poloidal field in the disk is parallel to that of the star, the matter accumulates for a significant time, then it goes to a funnel flow and accretes to the star. Fig. 4 shows episodes of plasma accumulation (top left panel), accretion (top middle panel) and new accumulation (top right panel). The corresponding accretion rate onto the star is shown in the bottom panel of Fig. 4. The bottom right panel of Fig. 4 shows the opposite case where the poloidal field in the disk is anti-parallel to that of the star. The smooth light-curve reflects the fact that the field of the star and the disk have frequent events of reconnection where matter of the disk moves onto the field lines of the star more easily and the accretion is quasi-steady.

In further recent work we performed 3D MHD simulations of accretion onto stars with tilted dipole fields (Romanova *et al.* 2011b) using our Godunov-type code based on the "cubed sphere" grid (Koldoba *et al.* 2002). We observed in these simulations that the density distribution is much less smooth than in the case of the laminar accretion flow described by α−viscosity. Instead, large turbulent cells dominate the flows and are

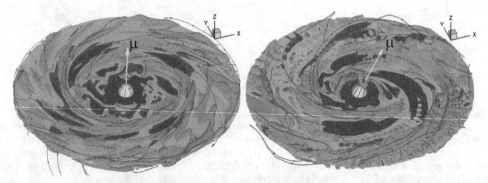

Figure 5. 3D MHD simulations of MRI-driven accretion onto a star with a dynamically-important magnetic field. Left panel: the dipole tilt angle is small, $\Theta = 5°$. Right panel: $\Theta = 30°$. From Romanova *et al.* (2011b).

strongly elongated in the azimuthal direction. Fig. 5 shows the full 3D view of the MRI-driven disks in cases of very low tilt of the dipole (left panel) and high tilt (right panel). One can see that in the case of a tilted dipole, matter accretes in funnel streams but the streams are less ordered compared with cases of laminar α−viscosity flow. Also, spiral paths of the flow are observed and we often see signs of one or two dominant spiral waves in the disk, and warping of the disk. Simulations were performed for different grid resolutions of our cubed sphere with the angular resolution varying from 61^2 up to 91^2 in each of 6 blocks. In addition, the grid was compressed in the vertical direction which made the initial vertical resolution of the disk of 110 cells across the disk. We observed that even at high resolution, large-scale structures are dominant.

We thank Dr. Alfio Bonanno for his lead in organizing this excellent symposium. This work was supported in part by NASA grants NNX08AH25G and NNX10AF63G and by NSF grant AST-0807129. MMR thanks NASA for use of the NASA High Performance Computing Facilities. AVK and GVU were supported in part by grant RFBR 09-02-00502a, Program 4 of RAS.

References

Bacciotti, F., Eisloffel, J., & Ray, T. P. 1999, *A&A*, 350, 917

Balbus, S. A. & Hawley, J. F. 1991, *ApJ*, 376, 214

Balbus, S. A. & Hawley, J. F. 1998, *Rev. Mod. Phs.*, Volume 70, 1-53

Beckwith, K., Hawley, J. F., & Krolik, J. H. 2009, *ApJ*, 707, 428

Cabrit, S., Edwards, S., Strom, S. E., & Strom, K. M. 1990, *ApJ*, 354, 687

Coffey, D., Bacciotti, F., Woitas, J., Ray, T. P., & Eislöffel, J. 2004, *ApJ*, 604, 758

Donati, J.-F. *et al.* 2008, *MNRAS*, 386, 1234

Ferreira, J, Dougados, C., & Cabrit, S. 2006, *A&A*, 453, 785

Hawley, J. F, Balbus, S. A., & Stone, J. M. 2001, *ApJ* Letters, 554, L49-L52

Koldoba, A. V., Romanova, M. M., Ustyugova, G. V., & Lovelace, R. V. E. 2002, *ApJ*, 576, L53-L56

Königl, A. & Pudritz, R. E. 2000, Protostars and Planets IV, Mannings, V., Boss, A. P., Russell, S. S. (eds.), University of Arizona Press, Tucson, p. 759

Lii, P., Romanova, M. M., & Lovelace, R. V. E. 2010, in preparation

Livio, M. 1997, Accretion Phenomena and Related Outflows; IAU Colloquium 163. ASP Conference Series; Vol. 121; ed. D. T. Wickramasinghe; G. V. Bicknell; and L. Ferrario, p. 845

Lovelace, R. V. E., Romanova, M. M., Ustyugova, G. V., & Koldoba, A. V. 2010, *MNRAS*, 408, 2083

Lovelace, R. V. E., Berk, H. L., & Contopoulos, J. 1991, *ApJ*, 379, 696

Patruno, A., Wijnands, R., van der Klis, M. 2009b, *ApJ*, 698, L60-L63

Perrin, M. D.,& Graham, J. R. 2007, *ApJ*, 670, 499

Ray, T., Dougados, C., Bacciotti, F., Eislffel, J., & Chrysostomou, A. 2007, Protostars and Planets V, B. Reipurth, D. Jewitt, and K. Keil (eds.), University of Arizona Press, Tucson, p. 231

Romanova, M. M., Ustyugova, G. V., Koldoba, A. V., & Lovelace, R. V. E. 2009, *ApJ*, 399, 1802

Romanova, M. M. *et al.* 2011a, in preparation

Romanova, M. M. *et al.* 2011b, in preparation

Shakura, N. I. & Sunyaev, R. A. 1973, *A&A*, 24, 337

Shu, F., Najita, J., Ostriker, E., Wilkin, F., Ruden, S., & Lizano, S. 1994, *ApJ*, 429, 781

Woitas, J., Ray, T. P., Bacciotti, F., Davis, C. J., & Eislöffel, J. 2002, *ApJ*, 580, 336

Advances in Plasma Astrophysics
Proceedings IAU Symposium No. 274, 2010
A.Bonanno, E. de Gouveia Dal Pino & A. G. Kosovichev, eds.
© International Astronomical Union 2011
doi:10.1017/S174392131100740X

Recent results from simulations of the magnetorotational instability

James M. Stone[1]

[1]Department of Astrophysical Sciences, Princeton University, Princeton NJ 08540, USA,
email: jmstone@princeton.edu

Abstract. The nonlinear saturation of the magnetorotational instability (MRI) is best studied through numerical MHD simulations. Recent results of simulations that adopt the local shearing box approximation, and fully global models that follow the entire disk, are described. Outstanding issues remain, such as a first-principles understanding of the dynamo processes that control saturation with no net magnetic flux. Important directions for future work include a better understanding of basic plasma processes, such as reconnection, dissipation, and particle acceleration, in the MHD turbulence driven by the MRI.

Keywords. accretion disks, instabilities, MHD, plasmas, turbulence, methods:numerical

1. Introduction

It has been nearly twenty years since Balbus & Hawley (1991) identified and first recognized the importance of the magnetorotational instability (MRI) for angular momentum transport in accretion flows. It has been over 15 years since the first fully three dimensional MHD simulation of the nonlinear regime of the MRI (Hawley, Gammie & Balbus 1995, hereafter HGB). In this time, there has been enormous progress in our understanding of the saturation mechanisms of the MRI, and at the same time new and puzzling questions have emerged that have yet to be understood. This paper provides a short review of some of those outstanding questions that have arisen in recent studies.

All of the work described here is based on numerical simulations using a variety of numerical algorithms as implemented in a variety of different codes. That such a wide range of codes have been used to study the MRI is important, because much greater confidence in a result is possible when it is confirmed by multiple workers using different methods. Many of the results I will present have been computed using a new Godunov method for MHD as implemented in the Athena code (Stone *et al.* 2008). Details of the numerical method required for studies of the MRI, for example the algorithms used for the shearing box source terms, and an orbital advection algorithm useful for global simulations, are described elsewhere (Stone & Gardiner 2010).

2. Local Shearing Box Simulations

By expanding the MHD equations in a local frame corotating with the disk at some fiducial radius, it is possible to study the MRI in a local Cartesian coordinate system in a patch of the disk which is small compared to the radius (HGB). By focusing all of the computational resources on a small patch of the disk, much greater resolution can be afforded. Typically, the patch has dimensions of $H \times 4H \times H$ in the radial, azimuthal, and vertical dimensions, where H is the thermal scale height. More recently, orbital advection algorithms have allowed studies in much wider radial domains, by eliminating

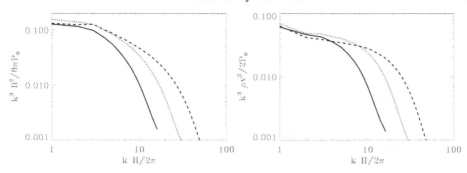

Figure 1. Power spectrum of the magnetic energy (left) and kinetic energy (right) averaged over orbits 50-100 in unstratified shearing box simulations of the MRI with net flux and no explicit dissipation. The solid, dotted, and dashed lines correspond to resolutions of $32/H$, $64/H$, and $128/H$ respectively.

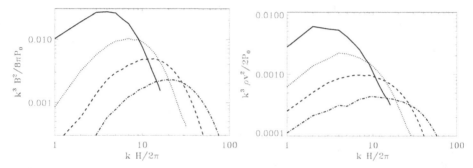

Figure 2. Same as Figure 1, but for the case of no net flux and resolutions up to $256/H$.

the background azimuthal flow from the stability constraint in the time step. Here we summarize results from two kinds of local, shearing box studies.

Unstratified Disks.

If the vertical extent of the domain is small, then the vertical component of gravity can be neglected, and there is no stratification of the density. Such models are appropriate for the midplane of the disk.

When the simulation domain contains *net vertical flux* of magnetic field, then HGB found the saturation amplitude increases with increasing flux. More recent studies have confirmed this result, and have also explored the dependence of the saturation amplitude on other parameters (Sano *et al.* 2003). Convergence of the stress is observed at very modest resolutions, about $32/H$. Increasing the radial extent of the domain decreases fluctuations in the stress caused by channel solutions but does not change the time-averaged value of the stress (Bodo *et al.* 2008). Figure 1 shows the Fourier power spectrum of the kinetic and magnetic energies at different resolutions in simulations with net vertical flux characterized by an initial $\beta = 1600$. Note that the power spectrum looks very similar to classical turbulence, with most of the power at small k, decreasing in a power law until the dissipation scale is reached.

One of the most important recent results was reported by Fromang & Papaloizou (2007) who found that the saturation level of the stress in unstratified disks with *no-net flux* and *no explicit dissipation* decreases as the resolution is increased. Figure 2 shows the Fourier power spectrum of the kinetic and magnetic energies at different resolutions simulations with no net flux. Note the results are vastly different from the net flux case (Figure 1). Now most of the power is at some intermediate wavenumber which changes with resolution, and the decrease in the total power is clearly evident as the resolution is increased.

In another important paper, Fromang *et al.* (2007) reported that the saturation amplitude with no net flux is constant with numerical resolution once explicit dissipation (viscosity and resistivity) is added. The interpretation is that fixing the dissipation fixes the magnetic Prandtl number $P_m = \nu/\eta$, where ν and η are the coefficients of viscosity and resistivity respectively, and that a converged solution is possible only at fixed P_m and Reynolds numbers. When only the grid provides dissipation, changing the resolution can change the effective Reynolds number, and there is no converged solution. In fact, Fromang *et al.* (2007) also found that sustained turbulence is not possible if the Reynolds number is too low.

It is interesting to note that a dynamo must be at work to explain the sustained turbulence observed in zero net flux simulations. The fact that the amplitude of the stress decreases with increasing resolution when there is no explicit dissipation, but that turbulence is still sustained, suggests that the dynamo is sensitive to the range of scales set by the box size and grid resolution. Recently Yousef *et al.* (2008) have reported that shear dynamos driven by turbulence are sensitive to the vertical box size. Is this true for the MRI? Work by Lesur & Ogilvie (2008) confirm that the saturation amplitude of the MRI with no net flux, and explicit dissipation, is increased when the vertical box size is increased. Recently we have found the same result *even with no explicit dissipation*. By increasing the box size in the vertical dimension to $2H$ we observe much larger amplitude stress in the saturated state. Even more remarkable is that this stress is independent of resolution even with no explicit dissipation, as shown in Figure 3. Clearly much work remains to understand the properties of the dynamo with no explicit dissipation.

Finally, there has been substantial effort to understand the effect of explicit dissipation on the MRI with net vertical flux. At low Reynolds numbers, Lesur & Longaretti (2007) found the stress was very strongly dependent on the Reynolds number. More recent studies of net toroidal fields by Simon & Hawley (2009) confirm this result, although they also suggest the Prandtl number dependence may be decreasing as the Reynolds number is increased. Such studies are challenging since very high resolutions are required to study the high Reynolds number regime, but there continues to be much effort in this direction.

Density stratified Disks.

Using the local shearing box, it is also possible to study the saturation of the MRI including the effect of vertical gravity, which produces stratification of the density. Early work (e.g. Miller & Stone 2000) showed that buoyancy produced a magnetized corona above the surface of the disk, and that magnetic energy rises from the midplane in a quasi-periodic fashion. New simulations are able to evolve stratified disks for much longer, with higher resolution, making these processes much clearer.

Figure 4 shows a space-time plot of the radial, azimuthal, and vertical components of the magnetic field (B_x, B_y, and B_z respectively), as well as the $x - y$ component of the Maxwell stress. These quantities are computed by averaging each variable over the $x-$ and $y-$directions, and then plotting the resulting vertical variation versus time. The calculation has no net flux, and no explicit dissipation.

Note the striking "butterfly" pattern evident in B_y. Clearly the dynamo at the midplane is creating toroidal field, which rises buoyantly out of the disk. Recently Gressel (2010) has modeled this dynamo using mean field theory and shown remarkable agreement between the spacetime plots of the model and simulations.

Also of interest are the time evolution of the stress as a function of resolution. Once again, the same level of stress is achieved, regardless of numerical resolution (Davis

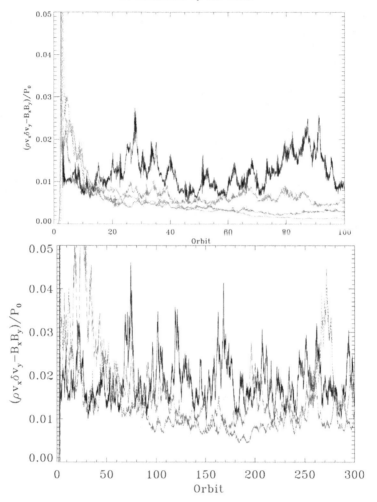

Figure 3. Time evolution of the stress at resolutions of $32/H$ (black), $64/H$ (red), $128/H$ (blue), and $256/H$ (green) for unstratified shearing box simulations of the MRI with no net flux and no explicit dissipation. The top panel is in a domain of $H \times 4H \times H$, while in the bottom the vertical box size is increased to $2H$. Note there is convergence in the latter.

et al. 2010; Shi & Krolik 2010). Stratified disks simulations have a vertical extent much larger than H, thus the fact the stress converges with resolution is in agreement with the result shown above for unstratified simulations with no net flux and no explicit dissipation in domains with large vertical extent. Davis *et al.* (2010) also show that sustained turbulence is achieved in stratified disks even at Reynolds numbers where the MRI is suppressed in unstratified disks. Taken together, these results demonstrate the importance of understanding the dynamo in the case of no net flux, and no explicit dissipation.

Finally, it is worth commenting that although the zero net flux case is of interest as an idealized model, it probably has no relevance for real disks. In the shearing box approximation, if there is no net flux initially, then there can never be any net flux at any time in the evolution. That an accretion disk could organize itself such that every $H \times 4H \times H$ domain in the disk had exactly no net flux for all time seems impossible.

Still, understanding the dynamics in the no net flux simulations could provide important insights to real disks.

3. Global Simulations

In addition to shearing box simulations, global simulations of the MRI in disks have been performed by many authors, for example Armitage (1998) and Hawley (2000) studied the saturation of the MRI in cylindrical disks with no stratification. Space limitations prevent a proper review of all we have learned about the MRI from global simulations here. Instead, one new direction being followed by several groups will be mentioned.

The dramatic increase in computational resources, combined with better algorithms (such as orbital advection to increase the timestep) means that very high resolution simulations of global disks can be performed. Here, high resolution means the same number of grid points per thermal scale height is the same as commonly used for local shearing box simulations. Figure 5 shows an image of the density from such a calculation, performed by K. Sorathia and C. Reynolds. Their study uses resolutions from $8/H$ up

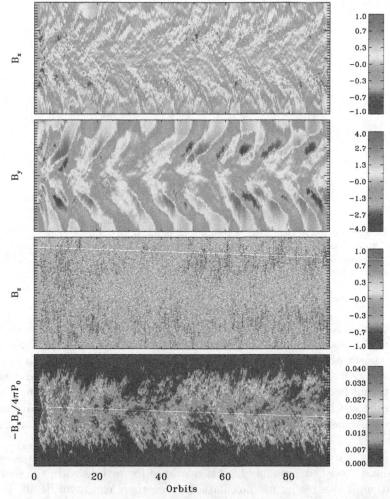

Figure 4. Space-time plots of various quantities in a density stratified shearing box simulation of the MRI.

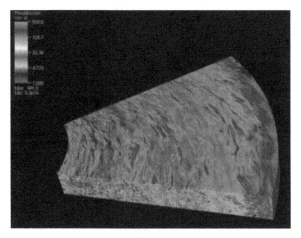

Figure 5 Image of the density after 30 orbits at the inner edge in a global cylindrical disk simulations of the MRI with a resolution of $16/H$.

to $32/H$ over the entire disk. By extracting small volumes equivalent to local shearing boxes, direct comparison can be made to previous work. Although this work is ongoing, an early result is that although the simulation begins with no net flux within each local volume, it quickly evolves to a state where there is a distribution of net fluxes, confirming the expectation above.

4. Summary and Future Work

At a meeting on plasma physics, it seems appropriate to focus on properties of the MRI that are connected to fundamental plasma processes. Recently, the properties of the MRI in weakly collisional plasmas (in the limit where the mean free path is large compared to the ion gyroradius) have begun to be explored. In the Braginskii regime, where kinetic effects are included through an anisotropic viscosity, it has been shown the growth rate of the MRI at long wavelengths for very weak fields is enhanced enormously (Quataert, Dorland, & Hammett 2002; Sharma, Hammett, & Quataert 2003; Islam & Balbus 2005). New simulations using the Braginskii equations of MHD have begun to explore this regime, which could be important for magnetic field amplification in protogalaxies.

A more sophisticated treatment of the kinetic plasma effects on the MRI was presented by Sharma *et al.* (2006). They found that at saturation, there was significant angular momentum transport by anisotropic pressure. In a two fluid study (with both electrons and ions) they found that anisotropic pressure was also responsible for significant heating of the electrons (Sharma *et al.* 2007), which has significant implications for weakly collisional accretion flows around compact objects. These studies have required an *ad hoc* treatment of micro-instabilities such as firehose and mirror modes, and much remains to be done to understand the effect of these on the MHD turbulence driven by the MRI in the kinetic regime. Ultimately, full PIC simulations of the MRI could settle important questions, like whether the combination of turbulence driven by the MRI and shear can accelerate particles, and whether energy dissipation occurs primarily in the ions or electrons, or equally in both.

In summary, although tremendous progress has been made towards understanding of angular momentum transport in accretion disks by the MRI using MHD simulations, much work remains to be done. For example, the properties of the dynamo driven by the MRI in the case of no net flux, as well as the butterfly diagram observed in stratified disks, remain to be fully understood. Fundamental astrophysics questions remain, such as what causes quasi-periodic oscillations in X-ray binaries. Finally, as discussed above, fundamental plasma problems remain.

References

Armitage, P. J. 1998, *ApJ*, 501, L189
Balbus, S. A. & Hawley, J. F. 1991, *ApJ*, 376, 214
Bodo, G., Mignone, A., Cattaneo, F., Rossi, P., & Ferrari, A. 2008, *A&A*, 487, 1
Davis, S. W., Stone, J. M., & Pessah, M. 2010, *ApJ* 713, 52
Fromang, S. & Papaloizou, J. 2007, *A&A*, 468, 1
Fromang, S., Papaloizou, J., Lesur, G., & Heinemann, T. 2007, *A&A*, 476, 1123
Gressel, O. 2010, *MNRAS*, 405, 41
Hawley, J. F., Gammie, C. F., & Balbus, S. A. 1995, *ApJ* 440, 742
Hawley, J. F. 2001, *ApJ* 554, 534
Islam, T. S. & Balbus, S. A. 2005, *ApJ* 633, 328
Lesur, G. & Longaretti, P.-Y. 2007, *A&A* 378, 1471
Lesur, G. & Ogilvie, G. I. 2008, *A&A* 488, 451
Miller, K. A. & Stone, J. M. 2000, *ApJ* 534, 398
Quataert, E., Dorland, W., & Hammett, G. W. 2002, *ApJ* 577, 524
Sano, T., Inutsuka, S., Turner, N., & Stone, J. M. 2004, *ApJ* 605, 321
Sharma, P., Hammett, G. W., & Quataert, E. 2003, *ApJ* 596, 1121
Sharma, P., Hammett, G. W., Quataert, E., & Stone, J. M. 2006, *ApJ* 637, 952
Sharma, P., Hammett, G. W., Quataert, E., & Stone, J. M. 2007, *ApJ* 667, 714
Shi, J., Krolike, J., & Hirose, S. 2010, *ApJ* 708, 1716
Simon, J. B. & Hawley, J. F. 2009, *ApJ* 707, 833
Stone, J. M., Hawley, J. F., Gammie, C. F., & Balbus, S. A. 2010, *ApJS*, 178, 137
Stone, J. M. & Gardiner, T. A. 2010, *ApJS*, 189, 142
Yousef, T. A., *et al.* 2008, *Phs. Rev. Lett.*, 100, 4501

Advances in Plasma Astrophysics
Proceedings IAU Symposium No. 274, 2010
A. Bonanno, E. de Gouveia Dal Pino & A. G. Kosovichev, eds.

© International Astronomical Union 2011
doi:10.1017/S1743921311007411

Numerical study of jets produced by conical wire arrays on the Magpie pulsed power generator

Matteo Bocchi[1], Jerry P. Chittenden[1], Andrea Ciardi[2,3],
Francisco Suzuki-Vidal[1], Gareth N. Hall[1], Phil de Grouchy[1],
Sergei V. Lebedev[1], Simon C. Bott[4]

[1] The Blackett Laboratory, Imperial College London, SW7 2BW London, UK
email: m.bocchi@imperial.ac.uk

[2] LERMA, Université Pierre et Marie Curie, Observatoire de Paris, Meudon, France

[3] École Normale Supérieure, Paris, France. UMR 8112 CNRS

[4] Center for Energy Research, University of California, San Diego, CA, USA

Abstract. With the aim to model jets produced by conical wire arrays on the MAGPIE generator, and to strengthen the link between laboratory and astrophysical jets, we performed three-dimensional magneto-hydro-dynamic numerical simulations using the code GORGON and successfully reproduced the experiments. We found that a minimum resolution of ~ 100 μm is required to retrieve the unstable character of the jet. Moreover, arrays with less wires produce more unstable jets with stronger magnetic fields around them.

Keywords. ISM: jets and outflows, hydrodynamics, plasmas, methods: laboratory, methods: numerical

1. Introduction

Important improvements have been achieved in the comprehension of astrophysical jets. However, several questions remain open, including jet formation, propagation in an external medium and survival to potentially disruptive instabilities (Bellan *et al.*, 2009; Hardee, 2004). Jets produced by conical wire arrays are characterised by dimensionless parameters in a similar range to Young Stellar Object (YSO) jets, and are especially suitable to study the interaction of the jet with an ambient medium (Lebedev *et al.*, 2005; Ciardi *et al.*, 2008). We produced laboratory jets on the MAGPIE pulsed power generator at Imperial College, London (Mitchell *et al.*, 1996) using different setups: conical and radial wire arrays (Lebedev *et al.*, 2005) and radial foils (Ciardi *et al.*, 2009). New physical regimes for jets are possible after the recent upgrade of MAGPIE and the "Z" pulsed power facility in Sandia, USA. Our aim is to model the conical wire array experiments on MAGPIE to understand the physics and help in the design of new diagnostics and experimental setups. The three-dimensional (3D) resistive magneto-hydro-dynamic (MHD) code GORGON (Chittenden *et al.*, 2004; Ciardi *et al.*, 2007) is used for the modelling.

2. Model and numerical setup

GORGON (Chittenden *et al.*, 2004; Ciardi *et al.*, 2007) is a fully resistive, 3D MHD code based on the Van Leer algorithm. Ions and electrons are treated as a single fluid. However, the energy equations are solved separately. The vector potential **A** is diffused and advected in order to follow the electromagnetic fields. The code includes optically

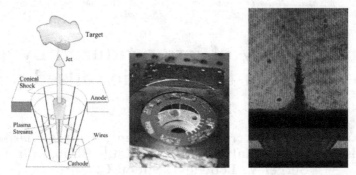

Figure 1. *Left Panel:* Scheme of a typical conical array setup. *Central Panel:* Picture of the experimental setup. *Right Panel:* Laser shadowgram of the shot *s*0301_10 taken at 355 *ns*. Array of 16 tungsten wires of 18 μm diameter.

thin radiation losses and Ohmic heating, as well as a "computational vacuum" below a density $\rho_{vac} = 10^{-4} \ Kg/m^3$, in which only the wave equation for **A** is solved.

A conical wire array is a set of wires in a conical shape, passed through by a high current pulse (see Fig. 1). The current heats the wires transforming them into plasma, and produces at the same time a global toroidal magnetic field which accelerates the plasma towards the axis of the array in a direction normal to the wires. Around the axis, a conical shock is formed. The component of the velocity parallel to the axis is conserved across the shock, so that a mainly hydrodynamic, collimated jet is formed.

We chose the numerical setup following the recent experiments carried out on the MAGPIE facility in London. In our simplified setup, two electrode plates are connected by a variable number of equally spaced wires, as seen in Fig. 1. In our reference case the cathode diameter is $d_c = 12 \ mm$, the anode diameter is $d_a = 21 \ mm$, and the array length (the distance between cathode and anode) is $l = 12 \ mm$. The inclination angle is ~ 30 degrees. We simulated arrays with 16 and 32 tungsten wires. The electric current provided by the MAGPIE generator has been modeled using the following expression:

$$I(t) = I_0 \cdot sin^2(\pi t/2\tau), \qquad (2.1)$$

where $I_0 = 1.4 \cdot 10^6$ is the peak current in Ampere and $\tau = 250 \cdot 10^{-9}$ is the peak time in seconds. The resolution employed was either 100 or 200 μm.

3. Results

Our simulations successfully reproduced the general behaviour observed in the experiments. The jet forms along the axis and displays an unstable, turbulent character in agreement with the observations (see Fig. 1, right panel). In what follows we present the differences arising when changing the number of wires and resolution, visible in Fig. 2.

First, we note that the jet appears to be turbulent only in the high resolution simulations. The non-axisimmetric features particularly visible in the case with 16 wires are not found in simulations with the low resolution of 200 μm. The profiles of the various physical quantities along the jet axis (not shown) are remarkably similar for the two different resolution employed, despite the big difference in the level of turbulence. In particular, velocity and temperature are compatible, while the density is systematically lower in the low resolution case. However, the general trend of the physical variables is preserved. We found that a minimum resolution of about 100 μm is needed to accurately reproduce the laboratory results. Indeed, the surface of the jet must be resolved numerically with enough grid points to properly follow the steep density gradients.

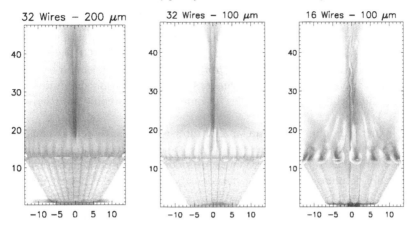

Figure 2. Emission maps of three different simulations at 400 ns. Darker shades correspond to higher emission. The axis are in *mm* units. *Left Panel:* 32 wires, 200 μm resolution. *Central Panel:* 32 wires, 100 μm resolution. *Right Panel:* 16 wires, 100 μm resolution.

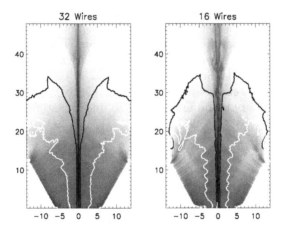

Figure 3. Density cuts in the mid-plane of two different simulations at 350 ns. Darker shades correspond to higher values. The axis are in *mm* units. *Left Panel:* 32 wires, 100 μm resolution. *Right Panel:* 16 wires, 100 μm resolution. Black contours represent the plasma beta: $\beta = 1$, higher values towards the axis. White contours represent the kinetic plasma beta (see text): $\beta_k = 10$, higher values towards the axis.

Second, the jet produced by the 16 wires array appears to be much more unstable than the one in the 32 wires array (see Fig. 3). A comparison of the magnetic field strength reveals that the field is much stronger in the 16 wires case, as it can penetrate more easily in the larger inter-wire gaps. The plasma beta (thermal over magnetic pressure), black contours in Fig. 3, seems to show that the magnetic field dominates in the region just around the jet. However, from the kinetic plasma beta (kinetic ram pressure over magnetic pressure), white contours, it is clear that the jet is kinetically dominated. The reasons for higher instability of the 16 wires simulations are not completely understood, but can be sought in the greater density modulations in the plasma flowing from the wires, visible in Fig. 3. Moreover, less wires produce less streams meeting on axis to form the jet, therefore reducing the overall symmetry of the system.

4. Conclusions

In this paper we presented the results of our numerical investigation on the jets produced by conical wire arrays on the pulsed power generator MAGPIE in London.

First, only a resolution of about $\sim 100~\mu m$ proved to be sufficient to reproduce the jet unstable character observed in the experiments. Lower resolution simulations maintained the general trend of the physical variables to some extent, but could not resolve the details in the jet.

Second, higher wire numbers mean smaller inter-wire gaps, therefore a much smoother distribution of the magnetic field and density. In the case of large inter-wire gaps however, fields can penetrate deeper in the array and are stronger around the jet. Nevertheless, the jet is kinetically dominated. Moreover, the density distribution is less regular, enhancing the unstable character of the jet.

Future work will consider, among other parameters, the effect of the inclination angle on the resulting jet. We will support the study with new sets of experiments both on MAGPIE in London and on the "Z" generator in Sandia, USA. We also plan experiments of the interaction of the jet with a gas target.

Our final goal is to connect experiments and simulations with the astrophysical scenario. YSO jets form in and propagate through molecular clouds. The subsequent interaction is mostly recognizable in the bow shock at the jet head, but is also responsible for material entrainment from the surrounding medium. The morphology and dynamics of these regions are of great interest for the study of chemical evolution and turbulent behaviour of molecular clouds. Studying such aspects in the laboratory is now possible, taking advantage of the accurate diagnostics available to plasma physicists. Therefore, laboratory jets provide very useful information to understand astrophysical jets.

Acknowledgements

Work supported by the EPSRC Grant No. EP/G001324/1 and by the NNSA under DOE Cooperative Agreements No. DE-F03-02NA00057 and No. DE-SC-0001063. Simulations were also run on the Jade supercomputer (GENCI-CINES, Paris-Montpellier, France), with support of the HPC-Europa2 project funded by the European Commission - DG Research in the Seventh Framework Programme under grant agreement No.228398.

References

Bellan, P. M., Livio, M., Kato, Y., Lebedev, S. V., Ray, T. P., Ferrari, A., Hartigan, P., Frank, A., Foster, J. M., & Nicolaï, P. 2009, *Physics of Plasmas*, 16, 041005

Hardee, P. E. 2004, *Astrophysics and Space Science*, 293, 117

Lebedev, S. V., Ciardi, A., Ampleford, D. J., Bland, S. N., Bott, S. C., Chittenden, J. P., Hall, G. N., Rapley, J., Jennings, C., Sherlock, M., Frank, A., & Blackman, E. G. 2005, *Plasma Phys. Control. Fusion*, 47, B465

Ciardi A., Ampleford, D. J., Lebedev, S. V., & Stehle, C. 2008, *ApJ*, 678, 968

Mitchell, I. H., Bayley, J. M., Chittenden, J. P., Worley, J. F., Dangor, A. E., Haines, M. G., & Choi, P. 1996, *Rev. Sci. Instrum.*, 67, 1533

Ciardi, A., Lebedev, S. V., Frank, A., Suzuki-Vidal, F., Hall, G. N., Bland, S. N., Harvey-Thompson, A., Blackman, E. G., & Camenzind, M. 2009, *ApJ*, 691, L147

Chittenden, J. P., Lebedev, S. V., Jennings, C., Bland, S. N., & Ciardi, A. 2004, *Plasma Phys. Control. Fusion*, 46, B457

Ciardi, A., Chittenden, J. P., Lebedev, S. V., Bland, S. N., Bott, S. C., Rapley, J., Hall, G. N., Suzuki-Vidal, F., Marocchino, A., Lery, T., & Stehle, C. 2007, *Physics of Plasmas*, 14, 056501

Advances in Plasma Astrophysics
Proceedings IAU Symposium No. 274, 2010   © International Astronomical Union 2011
A.Bonanno, E. de Gouveia Dal Pino & A. G. Kosovichev, eds.   doi:10.1017/S1743921311007423

Three dimensional simulations of Hall magnetohydrodynamics

Daniel O. Gómez[1,2]

[1]Instituto de Astronomía y Física del Espacio, C.C. 67 - Suc. 28,
(1428) Buenos Aires, Argentina
email: gomez@iafe.uba.ar

[2]Departamento de Física, Facultad de Ciencias Exactas y Naturales (UBA),
Ciudad Universitaria, (1428) Buenos Aires, Argentina

Abstract. Turbulent flows take place in a large variety of astrophysical objects, and often times are the source of dynamo generated magnetic fields. Much of the progress in our understanding of dynamo mechanisms, has been made within the theoretical framework of magnetohydrodynamics (MHD). However, for sufficiently diffuse media, the Hall effect eventually becomes non-negligible.

We present results from simulations of the Hall-MHD equations. The simulations are performed with a pseudospectral code to achieve exponentially fast convergence. We study the role of the Hall effect in the dynamo efficiency for different values of the Hall parameter.

Keywords. ISM: magnetic fields, MHD, turbulence.

---

1. Introduction

Numerical simulations have progressively become an important tool to study astrophysical flows. The large-scale dynamics of plasma flows can often be described within a fluidistic approximation known as one-fluid magnetohydrodynamics (MHD). Complex flows such as those corresponding to turbulent regimes are ubiquitous in astrophysics, which is consistent with their extremely large Reynolds numbers.

However, Hall currents can play a significant role in the dynamics of low density and/or low temperature astrophysical plasmas, such as dense molecular clouds (Wardle & Ng, 1999), accretion disks (Balbus & Terquem, 2001, Sano & Stone, 2002), white dwarfs and neutron stars (Yakovlev & Urpin, 1980) and in reconnection events at the Earth's magnetotail (Mozer, Bale & Phan, 2002).

In these plasmas the one-fluid MHD description needs to be extended to the so-called two-fluid MHD or Hall MHD. One of the interesting features of the Hall-MHD description, is that in the ideal limit (i.e. without resistivity) the magnetic field is stretched by the electron velocity field rather than the bulk velocity field. Since these two velocity fields can be quite different, the Hall term is expected to produce a measurable effect. For instance, the Hall term increases the reconnection rate with respect to purely MHD regimes (Ma & Bhattacharjee, 2001, Morales *et al.*, 2005). Also, the Hall term was found to enhance or suppress dynamo action depending on the relative importance of the Hall term when compared to the inductive term (Mininni, Gómez, & Mahajan, 2002). We therefore perform 3D numerical simulations of the Hall-MHD equations to explore the role of the Hall current in astrophysical scenarios.

In Section 2 we describe the so-called Hall-MHD equations, which stem from a two-fluid description of the plasma. In Section 3 we briefly describe simulations in 2.5 dimensions to study Hall magnetic reconnection. In Section 4 we present the results from 3D simulations of the Hall-MHD equations to study the efficiency of a turbulent dynamo. Finally, in Section 5 we list our conclusions.

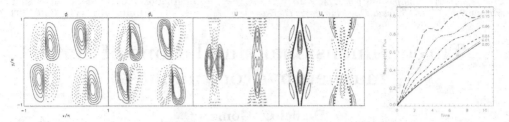

Figure 1. (a) From left to right, contour plots of the proton (ϕ) and electron (ϕ_e) stream functions, and the out of plane proton (u) and electron (u_e) velocity components. (b) Reconnected magnetic flux vs. time for different values of the Hall parameter ϵ (labelled).

2. The Hall-MHD equations

Highly conductive plasmas tend to develop thin and intense current sheets in their reconnection layers. Whenever the width of the curent sheets reaches values as low as c/w_{pi} (w_{pi} is the ion plasma frequency and c is the speed of light), it is no longer possible to neglect the Hall term in Ohm's law (Ma & Bhattacharjee, 2001). For a fully ionized plasma of protons and electrons, the generalized Ohm's law can be written as:

$$E + \frac{1}{c}v \times B = \frac{1}{\sigma}j + \frac{1}{ne}\left(\frac{1}{c}j \times B - \nabla p_e\right),\tag{2.1}$$

where n is the electron and proton density, e is the electron charge, σ is the electric conductivity, v is the plasma flow velocity, and j is the electric current density. Assuming incompressibility, the Hall-MHD equations can be cast in their dimensionless form as:

$$\partial_t v + (v \cdot \nabla)v = (\nabla \times B) \times B - \nabla p + \nu\nabla^2 v,\tag{2.2}$$

$$\partial_t B = \nabla \times [(v - \epsilon\nabla \times B) \times B] + \eta\nabla^2 B,\tag{2.3}$$

$$\nabla \cdot B = 0 = \nabla \cdot v.\tag{2.4}$$

In Eqs (2.2)–(2.4) we have normalized B and v to the Alfvén speed $v_A = B_0/\sqrt{4\pi\rho}$ (B_0: magnetic field intensity, ρ: mass density), the total gas pressure p to ρv_A^2, and longitudes and times respectively to L_0 and L_0/v_A. The dimensionless dissipation coefficients are the viscosity ν and the electric resistivity η defined as $\eta = c^2/(4\pi\sigma L_0 v_A)$. The dimensionless coefficient $\epsilon = c/(w_{pi} L_0)$ is a measure of the relative strength of the Hall effect. The dimensionless electron velocity is:

$$v_e = v - \epsilon\nabla \times B.\tag{2.5}$$

From Eqn (2.3) it is apparent that in the non-dissipative limit (i.e. $\eta \to 0$) the magnetic field remains frozen to the electron flow v_e rather than to the bulk velocity v.

3. Hall reconnection

To study the role of the Hall effect on magnetic reconnection, we performed simulations in 2.5 dimensions, i.e. assuming translational symmetry along the cartesian $\hat{z}$-direction. Under this geometrical assumption, the solenoidal fields B and v can be cast:

$$B = \nabla \times [\hat{z}a(x,y,t)] + \hat{z}b(x,y,t),\tag{3.1}$$

$$v = \nabla \times [\hat{z}\phi(x,y,t)] + \hat{z}u(x,y,t),\tag{3.2}$$

where a is the magnetic flux function, ϕ is the stream function, and b and u are the $\hat{z}$-components of these fields. The computation is carried out in a rectangular domain assuming periodic boundary conditions and the nonlinear terms are evaluated following

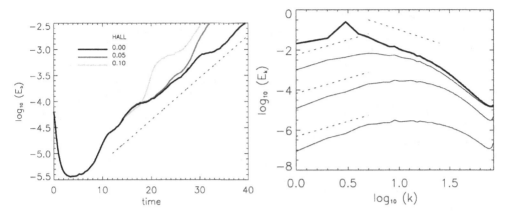

Figure 2. (a) Magnetic energy vs. time for different values of the Hall parameter (labelled). (b)Total energy spectrum (thick trace) at $t = 72$ for the case $\epsilon = 0.10$. Magnetic energy spectra at $t = 18, 36, 72$ (from bottom to top) are also shown. The Kolmogorov and Kazantsev spectra are overlaid (dotted trace) for reference.

a 2/3 dealiased pseudospectral technique. To provide a reconnection scenario, the initial condition corresponds to two oppositely oriented current sheets to satisfy periodicity (see also Morales *et al.*, 2006). We performed numerical simulations with a spatial resolution of 512×512 grid points, and different values of the Hall parameter ϵ. The dissipation coefficients are set to $\eta = \nu = 0.01$ to ensure that all the lengthscales are properly resolved. Note that pseudospectral methods conserve the energy of the system, i.e. no numerical dissipation is artificially added to the simulation.

The first two panels of Figure 1a show contour plots of the proton (ϕ) and electron $(\phi_e = \phi - \epsilon b)$ stream functions at $t = 1$. The difference between these patterns contributes to the in-plane electric current density, which in turn generates the out-of-plane magnetic field b. The two right hand panels in Figure 1a show contours of the out-of-plane velocities of protons (u) and electrons $(u_e = u - \epsilon j)$. The two species show entirely different velocity patterns. Note that in the ideal limit the magnetic field remains frozen to the electron flow, which is faster than the proton flow. The magnetic flux reconnected as a function of time at the X point can be calculated in terms of the difference of the magnetic potential in the X-point and the O-point, i.e. $a_X(t) - a_O(t)$. The effect of the Hall term on the reconnected flux is shown in Figure 1b. As ϵ is increased the reconnection process becomes more efficient, as evidenced by the total reconnected flux.

4. Hall dynamo

Haugen, Brandenburg, & Dobler (2004) have shown that not only helical flows can generate magnetic fields, but also non-helical flows are able to produce the so-called small-scale dynamos. To study the role of the Hall effect on the generation of magnetic fields by non-helical turbulent flows, we performed 3D simulations of the Hall MHD equations (i.e. Eqs (2.2)–(2.3)) with 256^3 spatial resolution. We first generate a stationary and non-helical hydrodynamic turbulence by applying an external forcing. In a second stage, a random and small magnetic field is introduced at small scales. The exponentially fast growth of magnetic energy is displayed in Figure 2a for runs with different values of the Hall parameter (labelled) (see also Gómez, Mininni, & Dmitruk, 2010). There is an initial linear stage for which the runs with $\epsilon = 0.05$ (moderate Hall) and $\epsilon = 0.10$ (large Hall) are indistinguishable from the case with $\epsilon = 0.00$ (purely MHD). The slope for these three runs during this stage is indicated by the dotted straight line in Fig. 2a, which corresponds to the growth rate of the MHD case (i.e. $\epsilon = 0.00$). Later in time the run with large Hall

($\epsilon = 0.10$) departs from this linear regime and starts growing faster. A similar behavior is observed somewhat later for the run corresponding to moderate Hall ($\epsilon = 0.05$). Therefore, in non-helical flows the Hall effect enhances the dynamo efficiency causing magnetic energy to grow super-exponentially, as has also been observed for large-scale dynamos in helical flows (Mininni, Gómez, & Mahajan, 2005, also Gómez, & Mininni, 2004, Mininni, Gómez, & Mahajan, 2003). The growth of the magnetic energy spectrum is shown in Figure 2b. The Kazantsev slope $E_k \propto k^{3/2}$ (Kazantsev, 1968) provides a reasonable approximation at small wavenumbers for all these cases, while the kinetic energy spectrum remains always close to Kolmogorov (i.e. $E_k \propto k^{-5/3}$). At saturation, the total magnetic energy reaches a sizeable fraction of the total kinetic energy (15% to 20%). Furthermore, magnetic energy remains smaller than kinetic energy at all spatial scales, which is to be expected for small-scale dynamos.

5. Conclusions

In the present paper, we call the attention on the potential relevance of the Hall effect in the dynamics of a number of astrophysical flows, specially those characterized by low electron densities. We quantitatively assess the role of the Hall effect in a number of astrophysical applications, by performing numerical integrations of the Hall MHD equations. As a first application we study magnetic reconnection in 2.5D simulations, showing the enhancement of the reconnection rate caused by the Hall effect. Gómez, Dmitruk, & Mahajan (2008) have also extended the so-called reduced MHD equations to include the Hall term (see also Martín, Dmitruk, & Gómez, 2010), which is the relevant approximation in plasma configurations with a strong external magnetic field. We also studied the role of the Hall effect in the generation of magnetic fields by turbulent non-helical flows. We show that even though the Hall term does not affect the kinematic dynamo stage during which the magnetic energy grows exponentially fast, it is responsible for a subsequent non-linear state (in between the kinematic and saturation stages) during which the magnetic energy grows super-exponentially fast.

References

Balbus, S. A. & Terquem, C. 2001, *ApJ*, 552, 235.
Gómez, D. O., Mininni, P. D., & Dmitruk, P. 2010, *Phys. Rev. E*, 82, 036406.
Gómez, D. O., Dmitruk, P., & Mahajan, S. M. 2008, *Phys. Plasmas*, 15, 102303.
Gómez, D. O. & Mininni, P. D. 2004, *Nonlin. Proc. Geophys.*, 11, 619.
Haugen, N. E. L., Brandenburg, A., & Dobler, W. 2004, *Phys. Rev. E* 70, 016308.
Kazantsev, A. P. 1968, *Sov. Phys. JETP* 26, 1031.
Ma, Z. & Bhattacharjee, A. 2001, *J. Geophys. Res.*, 106, 3773.
Martín, L., Dmitruk, P., & Gómez, D. O. 2010, *Phys. Plasmas*, 17, 112304.
Mininni, P. D., Gómez, D. O., & Mahajan, S. M. 2005, *ApJ*, 619, 1019.
Mininni, P. D., Gómez, D. O., & Mahajan, S. M. 2003, *ApJ*, 587, 472.
Mininni, P. D., Gómez, D. O., & Mahajan, S. M. 2002, *ApJ*, 567, L81.
Morales, L., Gómez, D. O., Dasso, S., & Mininni, P. D. 2006, *Adv. Space Res.*, 37, 1287.
Morales, L., Dasso, S., Gómez, D. O., & Mininni, P. D. 2005, *J. Atm. Sci. & Sol. Terr. Phys.*, 67, 1865.
Mozer, F., Bale, S., & Phan, T. D. 2002, *Phys. Rev. Lett.*, 89, 015002.
Sano, T. & Stone, J. M. 2002, *ApJ*, 570, 314.
Wardle, M. & Ng, C. 1999, *Mon. Not. R. A. S.*, 303, 239.
Yakovlev, D. G. & Urpin, V. A. 1980, *Astr. Zh.*, 57, 526.

Advances in Plasma Astrophysics
Proceedings IAU Symposium No. 274, 2010
A. Bonanno, E. de Gouveia Dal Pino & A. G. Kosovichev, eds.

© International Astronomical Union 2011
doi:10.1017/S1743921311007435

Stationary and axisymmetric magnetized equilibria of stars and winds

Shin Yoshida[1], Kotaro Fujisawa[1], Yoshiharu Eriguchi[1], Shijun Yoshida[2] and Rohta Takahashi[3]

[1] Department of Earth Science and Astronomy, Graduate School of Arts and Sciences,
University of Tokyo
Komaba, Meguro-ku, Tokyo 153-8902, Japan
email: yoshida@ea.c.u-tokyo.ac.jp

[2] Astronomical Institute, Tohoku University
6-3 Aramaki, Aoba-ku, Sendai 980-8578, Japan

[3] Cosmic Radiation Laboratory, The Institute of Physical and Chemical Research, RIKEN, 2-1
Hirosawa, Wako, Saitama 351-0198, Japan

Abstract. We present a new formulation to compute numerically stationary and axisymmetric equilibria of magnetized and self-gravitating astrophysical fluids. Under the assumption of ideal MHD, the stream function for the flow can be chosen as a basic variable with which the Euler-Maxwell equations are cast into a set of basic equations, i.e. a generalized Bernoulli equation and a Grad-Shafranov-like equation by employing various integral conditions. A novel feature of this formulation is that systems with stars, disks and winds are treated in a simple unified picture and the magnetic field structures can contain both poloidal and toroidal components.

Keywords. stars: magnetic fields, stars: rotation, stars: winds, outflows

1. Introduction

From the scale of the solar surface activities to those of galactic nuclei, magnetic field is a very important factor in astrophysical processes. Many activities of stars, such as flares, high energy photon emission or stellar winds, are thought to be driven by their magnetic field. Evolution of a molecular cloud as a birth place of a star is affected by the presence of magnetic field, while the proto-stellar activity is mediated by magnetic field. Accretion processes onto compact objects and outflows from them are thought to be driven/affected by magnetic field.

Although in reality outflows/inflows from magnetized objects are time-dependent and asymmetric, many of the objects are regarded to possess well-defined time-averaged states with axisymmetry. Therefore a theoretical modeling of stationary and axisymmetric structure in magnetized objects has been an important issue in astrophysics. The stationary solutions of magnetized objects, however, has been rather difficult to obtain, especially with the out/inflow. Numerical solutions so far obtained are that of Weber & Davis (1967), Pneuman & Kopp (1971) and Sakurai (1985, 1987), the last two of which are the only ones of self-consistent treatment. Most of the other solutions of (quasi-) stationary states are obtained by long-time numerical MHD simulations (cf. Keppens & Goedbloed, 1999). This method has an advantage of avoiding apparent singularity of equations related to critical points of flows, while it is rather difficult to study a structure of flow and magnetic field in a large scale by MHD simulations. Thus an important problem of collimation and acceleration of astrophysical winds and jets may not be fully solved by this method (at least by the present computational resources). We therefore develop a new numerical scheme that solves stationary and axisymmetric out/inflows around magnetized stars/disks. The new scheme solves integrated form of the equation of motion of gas, rather than a differential form of it. This makes us easier to spot and handle critical points of flow which are not known a priori, but obtained only after the system of equations are solved.

2. Formulation

We assume the system to be stationary and axisymmetric. Fluid in the system is perfect, isentropic. The conductivity of it is infinite and the electromagnetic field follows that of ideal MHD

$$E^a + \frac{1}{c}\epsilon^{abc}v_b B_c = 0, \tag{2.1}$$

where ϵ^{abc} is Levi-Civita tensor and the other symbols have their usual meanings. Basic equations are 1) Continuity equation, 2) four of Maxwell's equations, 3) equation of hydromagnetic momentum balance and 4) Poisson's equation for gravity. From the assumption of stationarity and axisymmetry with 1) Continuity equation, we can define a stream function Q in such a way that the meridional velocity components are expressed as

$$v^R = -\frac{1}{\rho R}\frac{\partial Q}{\partial z}, \quad v^z = \frac{1}{\rho R}\frac{\partial Q}{\partial R}, \tag{2.2}$$

in cylindrical polar coordinate (R, φ, z). The solenoidal condition of magnetic field from Maxwell's equation implies similar expression of magnetic field as a scalar magnetic flux function Ψ, which is used in an alternative formalism of magnetized flow (Lovelace *et al.* (1987); see also Fujisawa *et al.* in this volume). From the ideal MHD condition it can be shown that the magnetic flux function is a functional of Q. On the other hand, other components of the same condition imply that there is a relation

$$\frac{B_\varphi}{\rho R^2} - \frac{d\Psi}{dQ}\frac{v_\varphi}{R^2} = \sigma[Q], \tag{2.3}$$

where σ is an arbitrary functional of Q. From the toroidal component of the momentum equation we have

$$v_\varphi - \frac{1}{4\pi}\frac{d\Psi}{dQ}B_\varphi = \ell[Q], \tag{2.4}$$

where ℓ is an arbitrary functional of Q. Another arbitrary function ν of Q is shown to exist by examining a curl of momentum equation

$$v^\varphi \frac{B^\varphi}{4\pi}\frac{d^2\Psi}{dQ^2} + v^\varphi\frac{d\ell}{dQ} + \frac{1}{4\pi}\frac{d\sigma}{dQ}B_\varphi - \frac{\zeta^\varphi}{\rho} + \frac{j^\varphi}{c\rho}\frac{d\Psi}{dQ} = \nu[Q], \tag{2.5}$$

where $\vec{\zeta}$ is vorticity of the flow and $\vec{j}$ is current density.

From the meridional components of the momentum equation has a first integral

$$\int^Q \nu(q)dq - \frac{B_\varphi}{4\pi}\sigma - \int^p \frac{dp'}{\rho} - \Phi - \frac{1}{2\rho^2 R^2}|\nabla Q|^2 - \frac{1}{2}v_\varphi v^\varphi = C \text{ (const.)}, \tag{2.6}$$

where Φ is the gravitational potential satisfying Poisson's equation

$$\Delta\Phi = 4\pi G\rho. \tag{2.7}$$

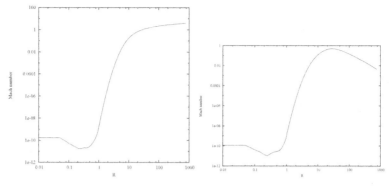

Figure 1. Profiles of Fast-wave Mach number for wind (left panel) and breeze (right panel) solutions.

From the definition of vorticity in φ, we have a Poisson-like differential equation for Q

$$\triangle \left(\frac{Q}{R} e^{i\varphi} \right) = e^{i\varphi} \left(\frac{\nabla \rho}{\rho} \cdot \frac{\nabla Q}{R} - R \zeta^\varphi \right) \tag{2.8}$$

These two Poisson PDEs are formally integrated by using proper Green's function as

$$\Phi = -G \int \frac{\rho(\vec{r'})}{|\vec{r} - \vec{r'}|} dV' \tag{2.9}$$

and

$$Q = -e^{-i\varphi} \frac{R}{4\pi} \int \frac{S_Q(\vec{r'})}{|\vec{r} - \vec{r'}|} dV' \tag{2.10}$$

where S_Q is the right hand side of Eq.(2.8).

By specifying functionals of Q (Ψ,σ,ℓ,ν), we iteratively solve Eq. (2.6), Eq. (2.9) and Eq.(2.10) .

3. Results

In the following we show some of the preliminary results obtained by our numerical code to compute fluid flow and magnetic field structure of rotating stars. The equation of state of the gas is assumed to be polytropic ones, whose indices are $N = 3$ inside the star and $N = 20$ outside, respectively. The stellar surface is defined by $\rho/\rho_c = 10^{-3}$ where ρ_c is the central density.

We show two kinds of solution here. in Fig.1, the Mach number (defined to be the ratio of poloidal velocity component to poloidal fast-wave velocity) is plotted as a function of radial distance from the center of the star. On the left panel, the Mach number increases from the surface of the star ($r = 1$) and exceeds unity at $r \sim 30$. This solution corresponds to the proper "wind" solution. On the right panel we have a solution whose Mach number always stays below unity. This solution is so-called "breeze" solution.

In Fig. 2 two typical profiles of stream lines (magnetic field lines) in the meridional section of the star are plotted. With different choices of the functionals of Q, we have complex field structure inside the star, though the flow pattern far out may be close to Parker's solution of the solar wind (Parker, 1958). In this example, the magnetic field is rather weak and the structure outside the star looks very close to the Parker's solution.

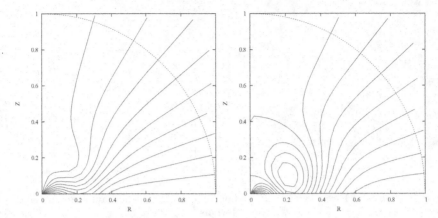

Figure 2. Examples of wind (left) and breeze (right) solutions. Stream lines (magnetic field lines) in the meridional section of a star shown as a solid lines. Blue dashed line corresponds to the surface of the star (defined as a isobaric surface of $\rho/\rho_c = 10^{-3}$).

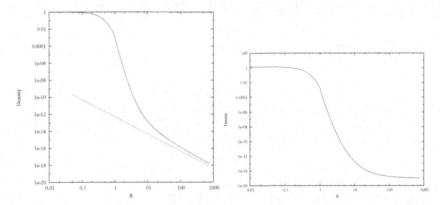

Figure 3. Density profiles of the wind (left) and the breeze (right) solutions. As the wind solution here tends asymptotically to Parker's solution far out the star, the density distributions falls as $\sim r^{-2}$, while that of the breeze solution does not.

In Fig. 3 we show density profile of wind/breeze solutions. The wind solution (left panel) tends to a Parker's solution far outside, with a density $\rho \sim r^{-2}$, while the breeze solution (left panel) tends to a constant density as expected.

References

Fujisawa, K., Yoshida, S., & Eriguchi, Y. 2010, *MNRAS, submitted*
Keppens, R. & Goedbloed, J. P. 1999, *A & A*, 343, 251
Lovelace, R. V. E., Mehanian, C., Mobarry, C. M., & Sulkanen M. E. 1986, *ApJS*, 62, 1
Parker, E. N. 1958, *ApJ*, 128, 664
Pneuman, G. W. & Kopp, R. A. 1971, *Solar Phys.*, 18, 258
Sakurai, T. 1985, *A & A*, 152, 121
Sakurai, T. 1987, *PASJ*, 39, 821
Weber, E. J. & Davis, L. 1967, *ApJ*, 148, 217

Advances in Plasma Astrophysics
Proceedings IAU Symposium No. 274, 2010
A. Bonanno, E. de Gouveia Dal Pino & A. G. Kosovichev, eds.

© International Astronomical Union 2011
doi:10.1017/S1743921311007447

Shock refraction from classical gas to relativistic plasma environments

Rony Keppens, Peter Delmont and Zakaria Meliani

Centre for Plasma Astrophysics, K.U.Leuven,
Celestijnenlaan 200B, 3001 Heverlee, Belgium
email: Rony.Keppens@wis.kuleuven.be

Abstract. The interaction of (strong) shock waves with localized density changes is of particular relevance to laboratory as well as astrophysical research. Shock tubes have been intensively studied in the lab for decades and much has been learned about shocks impinging on sudden density contrasts. In astrophysics, modern observations vividly demonstrate how (even relativistic) winds or jets show complex refraction patterns as they encounter denser interstellar material.

In this contribution, we highlight recent insights into shock refraction patterns, starting from classical up to relativistic hydro and extended to magnetohydrodynamic scenarios. Combining analytical predictions for shock refraction patterns exploiting Riemann solver methodologies, we confront numerical, analytical and (historic) laboratory insights. Using parallel, grid-adaptive simulations, we demonstrate the fate of Richtmyer-Meshkov instabilities when going from gaseous to magnetized plasma scenarios. The simulations invoke idealized configurations closely resembling lab analogues, while extending them to relativistic flow regimes.

Keywords. MHD, shock waves, instabilities

1. Introduction

Classical shock tube experiments have been conducted extensively, and in their most basic setup, they realize the laboratory counterpart of the theorist's Riemann problem, where two (static) gases are first separated by a membrane which is then suddenly removed. By controlling the pressure and density conditions on either side of the membrane, such in essence 1D problems have a known analytic solution. In case a high pressure gas is left to expand into a low pressure part, three signals arise, consisting typically of a rarefaction wave, contact discontinuity (CD), and shock wave as the high pressure gas sweeps up shocked matter in its sudden invasion of the low pressure medium. In fact, the Riemann problem solution is known under the more general situation where two states of arbitrary velocity, pressure and density are in contact, and its solution forms a major component of many contemporary shock-capturing numerical discretizations.

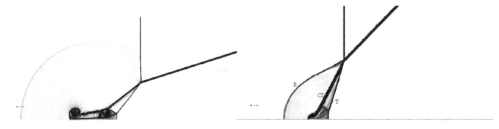

Figure 1. Irregular (left) versus regular (right) refraction pattern. A Schlieren plot of the density (i.e. an exponentially stretched quantification of the density gradient magnitude) is shown. The lower boundary is reflective.

In a more general refraction setup, one considers a (rectangular) tube with a membrane at arbitrary inclination angle α to the horizontal, seperating two static gases which differ in density. This density contrast is quantified as η, with $\eta < 1$ indicating a density decrease, or so-called slow-fast contact discontinuity, while $\eta > 1$ is referred to as a fast-slow case. This nomenclature relates to the corresponding change in sound speed. A refraction problem is then created by using a piston to generate a shock of given strength (i.e. Mach number M) at the left hand end of the tube, which traverses the tube and interacts, or refracts, with this initial contact interface. From extensive experimental campaigns, a whole zoology of shock refraction patterns is known (Abd-El-Fattah & Henderson, 1978). A recurring aspect, irrespective of whether the impinging shock is (very) weak to strong, is the transition from an irregular to a regular refraction pattern, when the angle α is varied from shallow to large values (up to 90° inclination angles, at which point the problem is a 1D shock-CD interaction problem). Fig. 1 illustrates both an irregular (left) and a regular (right) refraction, in a 2D hydrodynamic simulation where $M = 2$, $\eta = 3$ is a fast-slow transition. The figure shows a Schlieren plot of the density, and the only parameter varied is the inclination angle α. In a regular refraction case, the reflected (R), shocked contact (CD) and transmitted signal (T) meet in a single triple point at the point where the shock (vertical signal) passes over the initial contact (rightmost signal).

2. From Newtonian to relativistic hydro refractions

The specific case of regular refractions still collects several refraction patterns, since the reflected signal can be discontinuous (shock) or be of the continuous expansion fan variety. In the 1D case when $\alpha = 90°$, this is still understood from the known solution to the standard Riemann problem, as usually described in $x - t$ phase space. Indeed, as the shock then meets the (then aligned) initial contact discontinuity, the postshock state comes into contact with the state behind this CD, in turn realizing a standard Riemann problem. The (a) slow-fast versus (b) fast-slow transitions then coincide with the distinction where (a) the reflected signal is an expansion fan, while the transmitted signal moves ahead of the shock; versus (b) a case where the reflected signal is also of shock-type, and the transmitted signal lags the initial shock speed. Even when the setup is truly 2D (with $0° < \alpha < 90°$), the regular refraction is amenable to exact analytic solution, as one realizes that in the frame comoving with the triple point, one has a time-independent Riemann problem in the $x - y$ plane to solve. This 2D planar hydrodynamics refraction, and the complete Riemann-solver based solution strategy to predict critical angles for irregular to regular refractions, as well as full predictions for the 2D density, pressure and velocity variations, is described in detail in Delmont et al. (2009). Using the predictions from this solution strategy, one can quantify exactly the numerical accuracy for simulated refraction patterns in this regular refraction phase. This also allows to quantify the amount of vorticity deposited on the shocked contact discontinuity, which is the ultimate regulating factor in the further non-linear evolution where the Richtmyer-Meshkov instability ensues. Indeed, as a jump in tangential velocity is created across the shocked contact, slight perturbations of this interface will be Kelvin-Helmholtz unstable, causing a complex rippling of the shocked CD. This instability can also occur when the speeds involved are tending towards astrophysically relevant, relativistic speeds. In Fig. 2, a snapshot of a 2D relativistic hydro case is visualized, where a Mach $M = 100$ shock impinged on a fast-slow $\eta = 1000$ transition, while the postshock velocity corresponded to half the speed of light. The angle $\alpha = 45°$ is no longer obviously detectable at this time, but one can see how the darker colored regions demark the region between the (fully Richtmyer-Meshkov deformed) CD and the transmitted shock front. The intricate wave and shock patterns behind these regions correspond to the at this point multiply reflected waves, complicated by both top and bottom solid wall interactions. They each induce novel (triple) point interactions, with slip lines becoming unstable to local Kelvin-Helmholtz activity. The amount of detail visible here is due to the grid-adaptive capabilities of the `MPI-AMRVAC` software (van der Holst & Keppens, 2007) used in this study, where for this particular relativistic hydro run, 7 refinement levels realize an effective resolution of 1536×7680. Despite this high resolution, the full simulation completes within day(s) on a local desktop PC.

Figure 2. The late evolutionary stage of a relativistic hydro refraction problem. After a regular refraction phase, the shocked contact has become Richtmyer-Meshkov unstable. The reflected signal has by now had multiple interactions with the top and bottom reflective boundaries, as highlighted by this Schlieren density plot.

3. Planar MHD regular refraction

When the same setup is realized in a plasma environment, the macroscopic dynamics is governed by the magnetohydrodynamic (MHD) conservation laws. If a magnetic field purely perpendicular to the simulated plane (hence parallel to the shock front) is considered, it is possible to show that similar to the pure hydrodynamic case, three signals arise at refraction, and the ultimate instability to Richtmyer-Meshkov development is virtually identical for varying plasma beta conditions, as long as the Atwood number of the impinging shock (i.e. its density contrast) is kept fixed (Delmont *et al.*, 2009).

The MHD case under a general magnetic field configuration is more complex due to the presence of multiple reflected, as well as transmitted, wave signals. If we restrict the situation to planar MHD, where the magnetic field is initially purely normal to the shock front and uniform throughout, the shock-CD interaction will give rise to 5 signals, and a quintuple point appears in a regular refraction situation. Once more, this case is amenable to analytic treatment, where the exact variation of all state variables (density, pressure, velocity and magnetic field) about the quintuple point is predicted. One can obtain self-similar solutions where in essence all quantities vary only with polar angle ϕ about the quintuple point, and across shocks standard MHD Rankine-Hugoniot relations apply. The latter relations immediately indicate an important difference between the (planar) MHD versus the hydro case: no vorticity jump is allowed across the shocked contact, and the perturbed contact discontinuity is no longer Richtmyer-Meshkov unstable. Still, depending on governing parameters, the 2D refraction problem can show intricate transitions in its precise shock refraction dynamics. This relates to the fact that MHD shock types come in several varieties, depending on wether one crosses slow, Alfvén or fast characteristic speeds. An example transition is shown in Fig. 3, where an $M = 2$, $\eta = 3$ fast-slow case (as in Fig. 1) is compared for plasma beta $\beta = 0.5$ (left) to $\beta = 4$

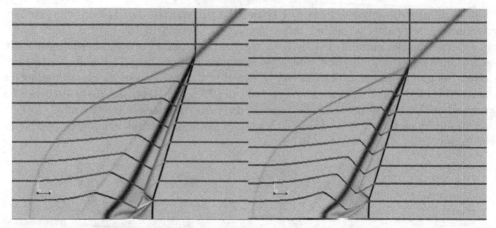

Figure 3. Planar MHD shock refraction leads to 5 signals in the regular refraction case. Left: for strong magnetic field ($\beta = 0.5$). Right: for weaker field ($\beta = 4$). The transmitted and reflected slow signals then change from slow shocks (left) to intermediate shocks (right).

(right) scenarios. This change relates to an overall decrease in the strength of the field, and the fieldlines together with the Schlieren density plots are shown. By quantifying the change in angle of the fieldlines with respect to the local shock normal, one concludes that the higher field case contains a fast shock, slow shock, CD, slow shock, fast shock refraction pattern. The lower field case at right rather has a fast shock, intermediate shock (crossing the Alfvén speed), CD, intermediate and fast shock pattern. It is to be stressed that these, and other subtle changes in the refraction pattern behavior can be quantified and predicted exactly using the Riemann solver based methodology, and that the numerical experiments as shown in Fig. 3, confirm these predictions precisely (Delmont & Keppens, 2010).

4. Outlook

Shock refractions in both gas and plasma dynamical setups provide a stringent test for shock-instability dominated dynamics. Using grid-adaptive simulations, pattern transitions and the further nonlinear developments can be studied in all details. Analytic knowledge of the solution verifies the role of magnetic fields, and one can explore the transition to speeds where relativistic effects become essential. The information gathered from laboratory experiments helps to validate code efforts, and it is easily recognized how refractions (i.e. shock front interactions with pre-existing density changes) form an essential part in many astrophysically motivated problems where jets, winds or more violent explosion fronts encounter denser or more rarified molecular cloud regions.

References

Abd-El-Fattah, A. M. & Henderson, L. F. 1978, *JFM* 86, 15
Delmont, P., Keppens, R. & van der Holst, B. 2009, *JFM*, 627, 33
Delmont, P. & Keppens, R. 2010, *J. Phys. Conf Ser.*, 216, 012007
van der Holst, B. & Keppens, R. 2007, *JCP*, 226, 925

Advances in Plasma Astrophysics
Proceedings IAU Symposium No. 274, 2010
A. Bonanno, E. de Gouveia Dal Pino & A. G. Kosovichev, eds.

© International Astronomical Union 2011
doi:10.1017/S1743921311007459

Magnetic field amplification by relativistic shocks in a turbulent medium

Yosuke Mizuno[1], Martin Pohl[2], Jacek Niemiec[3], Bing Zhang[4],
Ken-Ichi Nishikawa[1] and Philip E. Hardee[5]

[1] CSPAR, The University of Alabama in Huntsville,
320 Sparkman Drive, NSSTC, Huntsville, AL 35805, USA
email: mizuno@cspar.uah.edu

[2] Institute of Physics and Astronomy,
Universität Potsdam, 14476 Potsdam-Golm, Germany

[3] Institute of Nuclear Physics PAN,
Kraków, Poland

[4] Dept. of Physics and Astronomy, University of Nevada,
Las Vegas, NV 89154, USA

[5] Dept. Physics and Astronomy, University of Alabama,
Tuscaloosa, AL 35487, USA

Abstract. We perform two-dimensional relativistic magnetohydrodynamic simulations of a mildly relativistic shock propagating through an inhomogeneous medium. We show that the postshock region becomes turbulent owing to preshock density inhomogeneities, and the magnetic field is strongly amplified due to the stretching and folding of field lines in the turbulent velocity field. The amplified magnetic field evolves into a filamentary structure in our two-dimensional simulations. The magnetic energy spectrum is flatter than Kolmogorov and indicates that a so-called small-scale dynamo is operating in the postshock region. We also find that the amount of magnetic-field amplification depends on the direction of the mean preshock magnetic field.

Keywords. turbulence, relativistic shock, MHD

1. Introduction

In the standard GRB afterglow model (e.g., Piran 2005; Mészáros 2006), the radiation is produced in a relativistic blastwave shell propagating into a weakly magnetized plasma. Detailed studies of GRB spectra and light curves have shown that the magnetic energy density in the emitting region is a small fraction $\epsilon_B \sim 10^{-3} - 10^{-2}$ of the internal energy density (e.g., Panaitescu & Kumar 2002). However simple compressional amplification of the weak pre-existing microgauss magnetic field of the circumburst medium can not achieve this magnetization (e.g., Gruzinov 2001). The leading hypothesis for field amplification in GRB afterglows is the relativistic Weibel instability which produces filamentary currents aligned with the shock normal. These currents are responsible for the creation of transverse magnetic fields (e.g, Medvedev & Loeb 1999). However, the size of the simulated regions is orders of magnitude smaller than the GRB emission region. It remains unclear whether magnetic fields generated on scales of tens of plasma skin depths will persist at sufficient strength in the entire emission region. On the other hand, in magnetohydrodynamic (MHD) processes, if the density of the preshock medium is strongly inhomogeneous, significant vorticity is produced in the shock transition. This vorticity stretches and deforms magnetic field lines and leads to amplification (e.g, Sironi & Goodman 2007). For the long duration GRBs, i.e., associated with iron core collapse of mass-losing very massive stars, density fluctuations in the interstellar medium may arise through several processes (e.g., Sironi & Goodman 2007).

Recently, Giacalone & Jokipii (2007) have performed non-relativistic MHD shock simulations that included density fluctuations with a Kolmogorov power spectrum in the preshock medium. They observed a strong magnetic-field amplification caused by turbulence in the postshock medium; the final rms magnetic-field strength is reportedly a hundred times larger than the preshock field strength. Here we investigate the magnetic-field amplification by turbulence in two-dimensional relativistic MHD simulations of a mildly relativistic shock wave propagating through an inhomogeneous medium.

2. Numerical Setup

In order to study the propagation of a mildly relativistic shock in an inhomogeneous medium, we use the 3D GRMHD code "RAISHIN" in two-dimensional Cartesian geometry ($x - y$ plane). A detailed description of the code and its verification can be found in Mizuno *et al.* (2006).

At the beginning of the simulations, an inhomogeneous plasma with mean rest-mass density $\rho_0 = 1.0$ and containing fluctuations $\delta\rho$ uniformly flows in the positive x-direction with speed $v_0 = 0.4c$ across the whole simulation region. The density fluctuations are generated so that they have a two-dimensional Kolmogorov-like power-law spectrum of the form $P_k \propto 1/[1 + (kL)^{8/3}]$, where k is the wavenumber and L is the turbulence coherence length. The turbulence is obtained by summing over a large number of discrete wave modes. A detailed description of the method used to generate a fluctuation with a Kolmogorov-like turbulence spectrum can be found in Giacalone & Jokipii (2007). We choose the total number of modes to be $N_m = 50$, the maximum and minimum wavelength to be $\lambda_{max} = 0.5L$ and $\lambda_{min} = 0.025L$, and with a fluctuation variance of $\sqrt{<\delta\rho^2>} = 0.012\rho_0$.

We consider a low gas-pressure medium with constant $p = 0.01\rho_0 c^2$, where c is the speed of light. The equation of state is that of an ideal gas with $p = (\Gamma - 1)\rho e$, where e is the specific internal energy density and the adiabatic index $\Gamma = 5/3$. The specific enthalpy is $h \equiv 1 + e/c^2 + p/\rho c^2$.

The preshock plasma carries a weak constant magnetic field ($B_0 = 4.5\times10^{-3}(4\pi\rho_0 c^2)^{-1/2}$). We investigate the effect of the magnetic field direction with respect to the shock propagation direction, by choosing fields parallel (B^x) or perpendicular (B^y) to the shock normal. The computational domain is $(x, y) = (2L, L)$ with $N/L = 256$ grid resolution. We impose periodic boundary conditions in the y-direction. In order to create a shock wave, a rigid reflecting boundary is placed at $x = x_{max}$. The fluid, which is moving initially in the positive x-direction, is stopped at this boundary, where the velocity v^x is set to zero. As the density builds up at $x = x_{max}$, a shock forms and propagates in the $-x$-direction. New fluid continuously flows in from the inner boundary ($x = 0$) and density fluctuations are advected with the flow speed. A detailed description of initial set-up for the simulation can be found in Mizuno *et al.* (2010).

3. Simulation Results

Figure 1a shows a 2D image of the total magnetic field strength at $t = 10.0L/c$ for the parallel field (B^x) shock case. When the preshock inhomogeneous plasma encounters the shock, the shock front is rippled, leading to significant, random transverse flow behind the shock. Since the preexsiting magnetic field is much weaker than the postshock turbulence, the turbulent velocity field can easily stretch and deform the magnetic field lines. This creates regions with larger magnetic field intensity. In the region near the shock front, the vorticity scale size is small but in the region far away from the shock front,

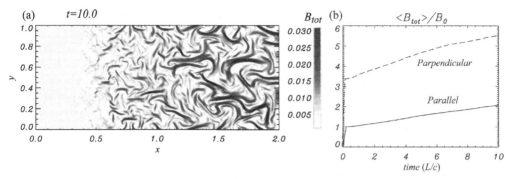

Figure 1. (*a*) Two-dimensional image of the total magnetic field strength at $t = 10.0L/c$ for the case of parallel field (B^x). (*b*) Time evolution of the volume-averaged total magnetic field strength in the postshock region normalized by the initial magnetic field strength.

the vorticity scale size becomes larger and the magnetic field is strongly amplified. The amplified magnetic field evolves in a filamentary structure. The average turbulent velocity in the postshock region is $\sim 0.02c$ at $t_s = 10$. The average sound speed and Alfvén velocity in the postshock region at $t_s = 10$ are $\sim 0.32c$ and $\sim 0.012c$. Thus the turbulent velocity is subsonic and super-Alfvénic in most of the postshock region.

Figure 1b shows the time evolution of the magnetic field indicated by the volume-averaged total magnetic field strength in the postshock region (whose size increases with time as the shock propagates away from the wall) normalized by the initial magnetic field strength. Note that the mean magnetic field strength is still increasing when the simulation was stopped. Thus, the mean magnetic field takes some time to saturate. The mean postshock magnetic field is stronger for the perpendicular field (B^y) (more than a factor 5 increase from the initial value) compared to the parallel field (B^x) (about a factor 2 increase from the initial value). This is because the magnetic field in the perpendicular field case is compressed initially by the shock. The magnetic field amplification from turbulent motion after magnetic field compression is almost same in both cases. The local maximum magnetic field strength is much larger than the mean magnetic field, about a factor of 13 and 26 times larger than initial magnetic field in the parallel and perpendicular cases respectively.

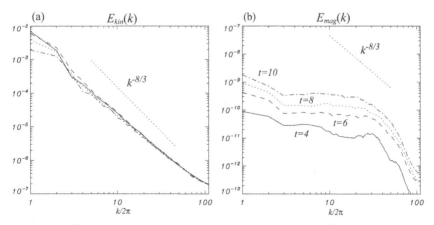

Figure 2. Spherically-integrated spectra of (*a*) the kinetic energy, (*b*) the electromagnetic energy. Different lines are for different times: $t_s = 4$ (*solid*), 6 (*dashed*), 8 (*dotted*), and 10 (*dash-dotted*). A dotted line indicates the power law $E(k) \propto k^{-8/3}$.

In order to understand the statistical properties of the turbulent fluctuations in the postshock region, it is helpful to observe their spectra.

Figure 2 shows spherically-integrated spectra of the kinetic and electromagnetic energy. The kinetic-energy spectra almost follow a Kolmogorov spectrum in all cases, $E_{kin}(k) \propto k^{-8/3}$, in two-dimensional systems. Initially the density in the preshock region is inhomogeneous with a Kolmogorov-like power spectrum, and this density power spectrum still exists in the postshock region. The kinetic-energy spectra do not change with time. The electromagnetic-energy increases over time, implying that the magnetic-field is not yet saturated. The magnetic energy power spectra are almost flat and strongly deviate from a Kolmogorov spectrum.

Spectra flatter than Kolmogorov are typical of the small-scale dynamo (e.g., Childress & Gilbert 1995; Balsara *et al.* 2004; Brandenburg & Subramanian 2005). In a small-scale dynamo, a forward cascade of magnetic energy from large scales to intermediate scales and an inverse cascade from small scales to intermediate scales are introduced. Therefore, the electromagnetic-field spectrum is flatter than Kolmogorov. A flat magnetic-energy spectrum is generally seen in turbulent-dynamo simulations (e.g., Balsara & Kim 2004).

The present simulation suggests a scenario whereby preexisting large-scale preshock density inhomogenities lead to strong magnetic field amplification in the postshock region. This process will be important in GRB and in AGN jet shocks.

Acknowledgements

This work is supported by NSF awards AST-098010 and AST-098040, NASA awards NNX08AG83G, MNiSW research project N N203 393034, and The Foundation for Polish Science through the HOMING program. The simulations were performed on the Columbia Supercomuter at NASA Ames Research Center, the SGI Altix (cobalt) at the NCSA in the TeraGrid project, and the Altix3700 BX2 at YITP in Kyoto University.

References

Balsara, D. S., Kim, J. S., Mac Low, M., & Mathews, G. J. 2004 *ApJ* 617, 339
Brandenburg, A. & Subramanian, K. 2005 *Phys. Rep.* 417, 1
Childress, S. & Gilbert, A. 1995 *Stretch, Twist, Fold: The Fast Dynamo* (Berlin: Springer)
Giacalone, J. & Jokipii, R. 2007 *ApJ* 663, L41
Medvedev, M. V. & Loeb, A. 1999 *ApJ* 526, 697
Mészáros, P. 2006 *Rep. Prog. Phys.*, 69, 2259
Panaitescu, A. & Kumar, P. 2002 *ApJ* 571, 779
Piran, T. 2005 *Rev. of Mod. Phys.* 76, 1143
Sironi, L. & Goodman, J. 2007 *ApJ* 671, 1858
Spitkovsky, A. 2008 *ApJ* 673, L39
Mizuno, Y., Nishikawa, K.-I., Koide, S., Hardee, P., & Fishman, G. J. 2006 *ArXiv Astrophysics e-prints* 0609004
Mizuno, Y., Pohl, M., Niemiec, J., Zhang, B., Nishikawa, K.-I., & Hardee, P. E. 2010 *ApJ* submitted

Advances in Plasma Astrophysics
Proceedings IAU Symposium No. 274, 2010
A. Bonanno, E. de Gouveia Dal Pino, & A. G. Kosovichev eds.

© International Astronomical Union 2011
doi:10.1017/S1743921311007460

Saturation of MRI via parasitic modes

Martin E. Pessah[1,2]

[1] Niels Bohr International Academy
Niels Bohr Institute, University of Copenhagen
Blegdamsvej 17, 2100, Copenhagen, Denmark

[2] Institute for Advanced Study
Einstein Drive, Princeton, NJ, 08540, USA
email: mpessah@nbi.dk

Abstract. Understanding the physical mechanisms that play a role in the saturation of the magnetorotational instability (MRI) has been an outstanding problem in accretion physics since the early 90's. Here, we present the summary of a study of the parasitic modes that feed off exact viscous, resistive MRI modes. We focus on the situation in which the amplitude of the magnetic field produced by the MRI is such that the instantaneous growth rate of the fastest parasitic mode matches that of the fastest MRI mode. We argue that this "saturation" amplitude provides an estimate of the magnetic field that can be generated by the MRI before the secondary instabilities suppress its growth significantly. We show that there exist two regimes, delimited by a critical Elsasser number of order unity, in which saturation is achieved via secondary instabilities that correspond to either Kelvin-Helmholtz or tearing modes.

Keywords. accretion, accretion disks, magnetohydrodynamics, instabilities, turbulence

1. Introduction

Let us consider a homogeneous, incompressible plasma in differential rotation according to $\boldsymbol{\Omega} = \Omega(r)\check{z}$ and threaded by a vertical magnetic field $\bar{\boldsymbol{B}} = \bar{B}_z\check{z}$. The equations governing the local dynamics of this MHD fluid in the shearing box approximation are

$$\frac{\partial \boldsymbol{v}}{\partial t} + (\boldsymbol{v} \cdot \boldsymbol{\nabla})\,\boldsymbol{v} = -2\boldsymbol{\Omega}_0 \times \boldsymbol{v} + q\Omega_0^2 \boldsymbol{\nabla}(r - r_0)^2 - \frac{1}{\rho}\boldsymbol{\nabla}\left(P + \frac{B^2}{8\pi}\right) + \frac{(\boldsymbol{B} \cdot \boldsymbol{\nabla})\boldsymbol{B}}{4\pi\rho} + \nu\boldsymbol{\nabla}^2\boldsymbol{v},$$

(1.1)

$$\frac{\partial \boldsymbol{B}}{\partial t} + (\boldsymbol{v} \cdot \boldsymbol{\nabla})\,\boldsymbol{B} = (\boldsymbol{B} \cdot \boldsymbol{\nabla})\,\boldsymbol{v} + \eta\boldsymbol{\nabla}^2\boldsymbol{B}, \quad \text{with} \quad \boldsymbol{\nabla} \cdot \boldsymbol{B} = 0 \quad \text{and} \quad \boldsymbol{\nabla} \cdot \boldsymbol{v} = 0.$$

(1.2)

Here, P is the pressure, ρ is the density, and the factor $q \equiv -d\ln\Omega/d\ln r$ parametrizes the magnitude of the local shear; $q = 3/2$ in the Keplerian case. Non-ideal effects due to a constant kinematic viscosity and resistivity are included in the terms proportional to ν and η. We work with dimensionless variables defined in terms of the background Alfvén speed and the local angular frequency and define the numbers $\Lambda_\nu \equiv \bar{v}_{Az}^2/\nu\Omega_0$ and $\Lambda_\eta \equiv \bar{v}_{Az}^2/\eta\Omega_0$, whose ratio is the magnetic Prandtl number, $\mathrm{Pm} \equiv \nu/\eta \equiv \Lambda_\eta/\Lambda_\nu$. The quantity Λ_η is known as the Elsasser number, while Λ_ν stands for its viscous counterpart.

The exact equations for the evolution of the secondary instabilities $\delta\boldsymbol{v}(\boldsymbol{x}, t)$ and $\delta\boldsymbol{B}(\boldsymbol{x}, t)$ affecting an MRI mode are obtained by substituting in Equations (1.1) and (1.2) the ansatz $\boldsymbol{v} = -q\Omega_0(r - r_0)\check{\phi} + \Delta\boldsymbol{v}\,e^{\Gamma t} + \delta\boldsymbol{v}$ and $\boldsymbol{B} = \bar{B}_z\check{z} + \Delta\boldsymbol{B}\,e^{\Gamma t} + \delta\boldsymbol{B}$. The first term in each of these equations accounts for the background Keplerian velocity and magnetic field. The terms proportional to $\Delta\boldsymbol{v} \equiv \boldsymbol{V}_0 \sin(Kz)$ and $\Delta\boldsymbol{B} \equiv \boldsymbol{B}_0 \cos(Kz)$ correspond to the exact, exponential fluctuations due to the MRI (Balbus & Hawley 1991), and $\Gamma(\nu, \eta, K)$ is the growth rate of the unstable MRI mode with wavelength $\lambda = 2\pi/K$ (see

Pessah & Chan 2008 for details). This substitution leads to partial differential equations for $\delta\boldsymbol{v}(\boldsymbol{x},t)$ and $\delta\boldsymbol{B}(\boldsymbol{x},t)$. However, we can gain insight into the growth rates and physical properties of the secondary instabilities by assuming that the exact (primary) MRI modes can be considered as a time-independent background from which the (secondary) parasitic modes feed off (see, e.g., Goodman & Xu 1994, Pessah & Goodman 2009, Latter *et al.* 2009, and Pessah 2010 for more details about this approach). In this framework, the dynamics of the parasitic modes is determined by

$$[(s + \nu(k_{\rm h}^2 - \partial_z^2)](k_{\rm h}^2 - \partial_z^2)\delta v_z - i(\boldsymbol{k}_{\rm h} \cdot \Delta\boldsymbol{v})(k_{\rm h}^2 - \partial_z^2 - K^2)\delta v_z$$
$$+i(\boldsymbol{k}_{\rm h} \cdot \Delta\boldsymbol{B})(k_{\rm h}^2 - \partial_z^2 - K^2)\delta B_z = 0, \qquad (1.3)$$

$$[s + \eta(k_{\rm h}^2 - \partial_z^2)]\delta B_z + i(\boldsymbol{k}_{\rm h} \cdot \Delta\boldsymbol{B})\delta v_z - i(\boldsymbol{k}_{\rm h} \cdot \Delta\boldsymbol{v})\delta B_z = 0. \qquad (1.4)$$

Here, as in Goodman & Xu (1994), we have further neglected the influence of the weak vertical background field, the Coriolis force, and the background shear flow on the dynamics of the secondary modes. The wavenumber $k_{\rm h}$ is the modulus of the horizontal wavevector $\boldsymbol{k}_{\rm h} \equiv k_x \tilde{x} + k_y \tilde{y} \equiv k_{\rm h}(\cos\theta\,\tilde{x} + \sin\theta\,\tilde{y})$ associated with the parasites.

2. Parasitic Modes

Let us focus our attention on the stability of the fastest growing MRI modes, with $K = K_{\max}(\nu,\eta)$ and $\Gamma = \Gamma_{\max}(\nu,\eta)$, and let us further consider their fastest growing parasites. It is then possible to estimate the amplitude $B_0^{\rm sat}$ such that the fastest parasitic mode, for given values of Λ_ν and Λ_η, grows as fast as the primary mode upon which it feeds. The motivation to calculate this "saturation" amplitude is that the parasite will be able to drain an amount of energy of order $(B_0^{\rm sat})^2$ from the primary mode shortly after their growth rates are comparable (see Pessah 2010 for more details).

The left panel of Fig. 1 shows the MRI saturation amplitude as a function of magnetic Prandtl number and the viscosity, while the right panel shows the dimensionless stress $\alpha_{\rm sat}\beta_{\rm sat} \equiv \bar{T}_{r\phi}/(B_0^2/8\pi)$. For $\Lambda_\nu \gtrsim 10$, the magnetic energy density presents two asymptotic regimes that correspond to Λ_η larger or smaller than unity. The associated modes correspond to Kelvin-Helmholtz and tearing modes respectively (see Fig. 2 and the discussion below). Note that in the limit Λ_ν, Pm $\gg 1$, $\alpha_{\rm sat}\beta_{\rm sat} \to 0.4$, while in the inviscid, resistive limit, i.e., $\Lambda_\nu \gg 1$ and Pm $\ll 1$, $\alpha_{\rm sat}\beta_{\rm sat} \to 0.5\,\Lambda_\eta$. Thus, despite the

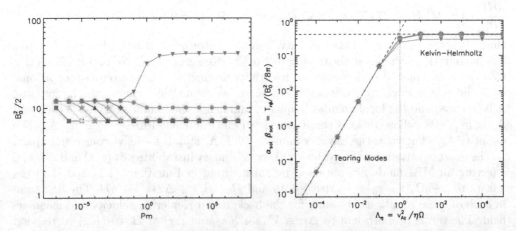

Figure 1. Predicted magnetic energy density (left) and dimensionless stress (right) for the fastest MRI mode if saturation occurs when the fastest parasitic mode matches its growth rate.

fact that the magnetic field at saturation asymptotes to a constant value, the dimensionless stress decreases linearly with Λ_η for $\Lambda_\eta \lesssim 1$. This is in qualitative agreement with the simulations in Sano & Stone (2002) (c.f., Pessah 2010; Longaretti & Lesur 2010).

In order to understand the nature of the fastest secondary modes it is useful to analyze their structure along the directions associated with their fastest growth, i.e., $\theta \equiv \theta_{\max}$. For fixed values of the dissipation coefficients, the growth rates of the secondary instabilities peak around directions which are almost aligned with either the velocity or magnetic fields of the primary MRI mode, i.e., $\theta_{\max} \simeq \theta_{\mathrm{V}}$ for $\Lambda_\eta \gg 1$ and $\theta_{\max} \simeq \theta_{\mathrm{B}}$ for $\Lambda_\eta \ll 1$. Fig. 2 shows the physical structure of the fastest parasitic modes, including the velocity and magnetic fields of the primary MRI modes, for $\Lambda_\eta = \{0.1, 1, 10\}$, from left to right, with $\Lambda_\nu \gg 1$. The arrows in the upper and lower panels correspond to the projections of the total (primary plus secondary) velocity and magnetic fields onto the plane defined by the z-axis and the direction $\theta_{\max}$. The color contours correspond to the total vorticity and current density projected onto the direction perpendicular to $\theta_{\max}$.

Tearing Modes — For the Elsasser number $\Lambda_\eta = 0.1$, the versor characterizing the direction of fastest growth, $\check{k}_{\mathrm{h}}$, points in the direction $\theta_{\max} \simeq \theta_{\mathrm{B}}$. This mode, shown in the leftmost (upper and lower) panels of Fig. 2, feeds off the current density of the primary MRI mode. The current density of the secondary modes presents maxima and minima along the planes $z = \pm n\pi/2$ where the magnetic field of the primary mode, $\Delta B = B_0 \cos(Kz)$, reverses sign. Thus, the fluctuations induced by these fastest resistive secondary modes tend to promote reconnection of the MRI field. The observed mode structure is qualitatively insensitive to the value of the Elsasser number as long as $\Lambda_\eta < 1$ and $\Lambda_\nu \gg 1$. We thus conclude that the fastest parasitic modes correspond to tearing

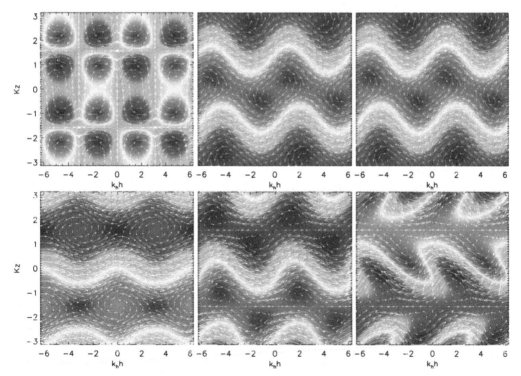

Figure 2. Physical structure of the fastest parasitic modes, including the velocity and magnetic fields of the primary modes, for $\Lambda_\eta = \{0.1, 1, 10\}$, from left to right, with $\Lambda_\nu \gg 1$.

modes for $\Lambda_\eta < 1$. These parasitic modes are enabled by non-zero resistivity and are thus absent in the ideal MHD regime studied by Goodman & Xu (1994).

Kelvin-Helmholtz Modes — The two rightmost sets of panels in Fig. 2 show the fastest secondary modes for the Elsasser numbers $\Lambda_\eta = \{1, 10\}$. The versors $\check{\mathbf{k}}_\mathrm{h}$ characterizing the direction of fastest growth point in the direction $\theta_\mathrm{max} \simeq \theta_\mathrm{V}$. These modes feed off the shear in the velocity field of the primary MRI modes. The velocity and vorticity fields show a periodic structure similar to what is obtained in the stability analysis of a periodic set of equidistant vortex sheets distributed along the $\check{z}$ direction and alternating sense. The structure of these modes is quantitatively insensitive to the value of the Elsasser number as long as $\Lambda_\eta \geqslant 1$ and $\Lambda_\nu \gg 1$. We thus conclude that the fastest parasitic modes correspond to Kelvin-Helmholtz modes for $\Lambda_\eta > 1$. In the limit $\Lambda_\eta \gg 1$, these correspond to the Kelvin-Helmholtz modes alluded to in Goodman & Xu (1994).

3. Discussion

In order to solve for the dynamics of the parasitic modes we have made a number of assumptions which might affect the value of the saturation amplitudes presented here. Despite these limitations, the properties of the parasitic modes that we described provide valuable analytical guidance and a basic framework to design and interpret tailored numerical experiments of the nonlinear saturation of the MRI. The following is a summary of our findings†. When the magnetic fields involved are weak enough so that the incompressible limit holds, the parameter driving the behavior of the growth rates of the MRI and its parasites, and thus the magnetic energy density and stresses at saturation, is the Elsasser number Λ_η. In particular, we found that, as long as viscous dissipation is small, i.e., $\Lambda_\nu \gtrsim 10$, then there exists two regimes: (i) quasi-ideal MHD, where the physical properties of the MRI and its parasitic instabilities are insensitive to dissipation. This holds as long as $\Lambda_\eta > 1$, which is applicable to the fully ionized regions of accretion disks around compact objects. (ii) inviscid, resistive MHD, where all the relevant dependencies on Λ_ν and Pm are only through the product $\mathrm{Pm}\,\Lambda_\nu$, i.e., the Elsasser number Λ_η. This regime corresponds to $\Lambda_\eta < 1$, and characterizes poorly ionized regions of protoplanetary disks. The Elsasser number for current Taylor-Couette MRI experiments is close to unity and thus both types of modes present similar growth rates in this regime.

References

Balbus, S. A. & Hawley, J. F. 1991, *ApJ*, 376, 214
Goodman, J. & Xu, G. 1994, *ApJ*, 432, 213
Jamroz, B., Julien, K., & Knobloch, E. 2008, *PhST*, 132, 4027
Latter, H. N., Lesaffre, P., & Balbus, S. A. 2009, *MNRAS*, 394, 715
Longaretti, P. Y. & Lesur, G. 2010, *A&A*, 516, 51
Pessah, M. E. & Chan, C. K. 2008, *ApJ*, 684, 498
Pessah, M. E. & Goodman, J. 2009, *ApJ*, 698, L72
Pessah, M. E. 2010, *ApJ*, 716, 1012
Sano, T., Inutsuka, S. I., & Miyama, S. M. 1998, *ApJ*, 506, L57
Sano, T. & Stone, J. M. 2002, *ApJ*, 577, 534
Umurhan, O. M., Regev, O., & Menou, K. 2007, *Phys. Rev. E*, 76, 6310
Vishniac, E. 2009, *ApJ*, 696, 1021

† For reasons of space, it is not possible to make justice to a number of different approaches addressing the saturation of the MRI. We refer the reader to the works by, e.g., Sano *et al.* (1998), Umurhan *et al.* (2007), Jamroz *et al.* (2008), Vishniac (2009), Latter *et al.* (2009), Longaretti & Lesur (2010), and the relevant references therein, for other perspectives into this problem.

Advances in Plasma Astrophysics
Proceedings IAU Symposium No. 274, 2010
A. Bonanno, E. de Gouveia Dal Pino & A. G. Kosovichev, eds.

© International Astronomical Union 2011
doi:10.1017/S1743921311007472

Formation of electron clouds during particle acceleration in a 3D current sheet

Valentina V. Zharkova[1] and Taras Siversky[1]

[1]Department of Mathematics, University of Bradford,
Bradford BD7 1DP, UK
email: v.v.zharkova@brad.ac.uk

Abstract. Acceleration of protons and electrons in a reconnecting current sheet (RCS) is investigated with the test particle and particle-in-cell (PIC) approaches in the 3D magnetic configuration including the guiding field. PIC simulations confirm a spatial separation of electrons and protons towards the midplane and reveal that this separation occur as long as protons are getting accelerated. During this time electrons are ejected into their semispace of the current sheet moving away from the midplane to distances up to a factor of $10^3 - 10^4$ of the RCS thickness and returning back to the RCS. This process of electron circulation around the current sheet midplane creates a cloud of high energy electrons around the current sheet which exists as long as protons are accelerated. Only after protons gain sufficient energy to break from the magnetic field of the RCS, they are ejected to the opposite semispace dragging accelerated electrons with them. These clouds can be the reason of hard X-ray emission in coronal sources observed by RHESSI.

Keywords. Sun: flares, plasmas, acceleration of particles, magnetic field

1. Introduction

Simulations of particle trajectories by using a test particle approach in a 3D magnetic configuration with the guiding field emulating a simple case of magnetic reconnection were carried for a constant reconnection electric field (Zharkova & Gordovskyy, 2004) or for those enhanced near the X-nullpoint due to anomalous resistivity (Wood & Neukirch, 2005). Zharkova & Gordovskyy (2004) showed that the trajectories of particles with the opposite charges (electrons or protons) can be either fully symmetric or strongly asymmetric towards the midplane of the RCS depending on the ratio between the magnetic field components.

As a result, some fraction of the released magnetic energy is transformed into kinetic energy of accelerated particles. The energy spectra of these particles also depend on a magnetic field topology, electric field strength and the dependence of transverse magnetic field on a distance from the X-nullpoint. Accelerated particles gain energies up to 100 keV for the electrons and up to 1 MeV for the protons (Zharkova & Gordovskyy, 2005a; Wood & Neukirch, 2005; Zharkova & Agapitov, 2009). Spectral indices of energy spectra gained at acceleration vary from 2 for electrons and 1.5 for protons if the transverse magnetic field linearly increases with the distance. Spectral indices for both electrons and protons become much higher if the transverse magnetic field increases exponentially with a distance from the X-nullpoint (Zharkova & Gordovskyy, 2005b).

More realistic approach considering plasma feedback to accelerated particles is archived with PIC simulation allowing to reproduce electro-magnetic fields generated by accelerated particles (Birn *et al.*, 2001; Tsiklauri & Haruki, 2007; Siversky & Zharkova, 2009) for reduced proton-to-mass ratios. These studies reproduced well magnetic reconnection rates (Birn *et al.*, 2001) and some key features of accelerated particles; power law energy spectrum and energy gains up to 100 keV.

Recent study of 3D magnetic reconnection with PIC by Drake *et al.* (2006) carried out for reduced proton skin depths related to a reduced magnitude for the speed of light, $c = 20V_A \approx 6 \cdot 10^6$ m/s, where V_A is the Alfvén velocity, revealed formation of two magnetic islands with X null points from both sides caused by tearing mode instability. Particles (electrons) then are supposed to be accelerated in stochastic second order Fermi acceleration by moving between these magnetic islands; although this mechanism does not seem to work for protons. This approach used a reduce simulation region size by reducing the skin depths for electron and protons, making all the reconnection features rather microscopic while in real flare events the current sheet thickness can achieve thousands km see, for example,]sui03.

2. Description of models

In this paper we apply iterations of test particle and PIC approaches for the simulation of particle acceleration in a 3D reconnecting current sheet with simple magnetic field topology. We assume that the current sheet plasma has typical coronal density of 10^{16} m^{-3}, which is in equilibrium with the background magnetic field. However, in order to avoid the problem with the small Debye length in the PIC approach, only a small fraction of the plasma particles (with density of 10^{10} m^{-3}) is included in the PIC simulation. This makes the ratio δ_i/λ_D to be of the order of 10. Electric and magnetic fields generated by particles in the PIC technic were then used in TP approach in order to investigate particle trajectories and densities.

Magnetic field topology

Since the electron or proton acceleration time is much shorter than the time of a reconnecting magnetic field variation as shown by Priest & Forbes (2000), one can assume the background magnetic field to be stationary. Also, from the TP simulations one finds that travel distances of accelerating particles along the RCS are of the order of 10 km at most (for the protons) (Zharkova & Agapitov, 2009) and this length is much shorter than the length scale of magnetic field variations along the current sheet. In addition, we assume that the magnetic field variations across the current sheet have much shorter length scale than along the current sheet, i.e. $L_x \ll L_z, L_y$.

The simulation domain is a small part of the RCS (see model in Siversky & Zharkova, 2009), but still large enough to contain the full trajectories of accelerated particles. The main component B_z depends on x as follows: $B_z(x) = -B_{z0} \tanh\left(\frac{x}{L_x}\right)$. Similar to Zharkova & Gordovskyy, 2004, the B_x component is assumed to be constant inside the simulation domain, i.e. $B_x = -B_{x0}$. The guiding (out-of-plane) magnetic field B_y is maximal in the midplane and vanishes outside the RCS: $B_y(x) = B_{y0}\text{sech}\left(\frac{x}{L_x}\right)$. Contrary to the TP simulations, plasma particles in the PIC simulations are considered to generate their own electric and magnetic fields, which is now self-consistently taken into account as described by Siversky & Zharkova (2009).

Reconnection electric field

In order to provide the inflow of plasma in our simulation domain we set up a background electric field, as those drifted in with velocity V_{in} by a magnetic diffusion process as shown by Priest & Forbes (2000): $E_{y0} = V_{in} \times B_{z0} + \frac{1}{\sigma\mu}\frac{\partial B_z}{\partial x}$, where V_{in} is the inflow velocity, σ is the ambient plasma conductivity, μ is magnetic permeability. On the boundaries of an RCS we ignore the gradient of the magnetic component B_z over x-coordinate by putting the second term to zero.

Simulations are carried out for the following current sheet parameters: the main component of the magnetic field $B_{z0} = 10^{-3}$ T, the current sheet half-thickness $L_x = 1$ m, the drifted electric field $E_{y0} = 250$ V/m and the guiding, B_{y0}, and transverse, B_{x0}, components of the magnetic field are selected to range from $(0.1 - 10) \times 10^{-4}$ T to cover acceleration at the various parts of RCS.

3. Results

The PIC simulations shown that the induced magnetic field $\tilde{\mathbf{B}}$ is much smaller than the background $\mathbf{B_z}$. On the other hand, the electric field $\tilde{\mathbf{E}}_\mathbf{x}$, called a polarization field, induced by the separation of electrons from protons is much larger than the reconnection (drifted) field E_y as pointed by Zharkova & Agapitov (2009) and Siversky & Zharkova (2009).

Polarization electric field induced by accelerated particles

In this subsection we consider this polarization electric field $\tilde{E}_x$ which is perpendicular to the current sheet and plotted in Fig. 3 as a function of x averaged over the z coordinate for various magnetic field parameters of B_{x0} and B_{y0}. The appearance of polarization electric field leads to a local non-neutrality of the plasma which becomes stronger if B_{x0} decreases or B_{y0} increases.

Particle trajectories in the presence of polarization electric field

In order to reconstruct the particle trajectories, we use the TP code, where the induced electric field $\tilde{E}_x$ obtained from PIC is added to the background electro-magnetic configuration described above. The trajectories of the two protons in the x-V_z phase plane entering the current sheet from the opposite directions are shown in Fig. 3. These trajectories are rather similar to those obtained in the TP simulations without $\tilde{E}_x$ where the "bounced" proton during the acceleration phase has a wider orbit while the "transit" proton has a narrower one producing energy distributions with two peaks.

Electrons entering from the $x < 0$ semispace have a dynamics similar to those of the "transit" protons, e.g. the electrons drift towards the midplane, become accelerated and ejected to the $x > 0$ semispace. However, the polarization field $\tilde{E}_x(x)$, which extends beyond the current sheet and has a component parallel to the magnetic field B_z, decelerates the ejected electrons. For chosen magnitudes of B_x and B_y, the majority of the electrons are unable to escape away from the RCS, instead they are dragged back to the current sheet and become indistinguishable from other electrons entered from the $x > 0$ semispace.

The electrons entering from the positive x side which are "bounced" from RCS in the absence of $\tilde{E}_x$, are able to reach its midplane if polarization field is present. In the vicinity of the midplane, the electrons become unmagnetized and oscillate with the gyrofrequency determined by B_y until they gain sufficient energy to break from the magnetic field. Although, if the electron initial velocity is small, it can be quasi-trapped inside the RCS (Fig. 3, bottom left plot). Such electrons are accelerated at the midplane, ejected from it, then decelerated outside the RCS and returned back to it because of the polarization field appeared due to protons still being accelerated in the RCS. This cycle is repeated for many times, forming electron clouds around the current sheet (Fig. 3, bottom right plot) which can be observed as a coronal source in solar flares. These clouds exist until the protons gain sufficient energy to leave the current sheet and, since the magnitude of the polarisation field $\tilde{E}_x(x)$ is smaller in the current model at $x < 0$ than at $x > 0$ (see Fig. 3, bottom plots), for electrons it is easier to escape to the $x < 0$ semispace, where the protons are ejected.

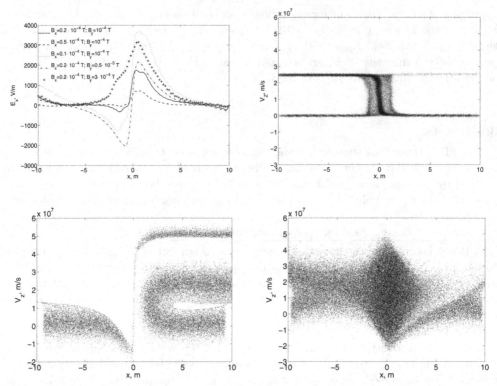

Figure 1. Top left plot: Electric field $\tilde{E}_x$ induced by particles in the PIC simulations for different values of B_{x0} and B_{y0}, for $B_{z0} = 10^{-3}$ T. Top right plot: Trajectories of protons in the V_z-x phase plane with the magnetic field magnitude of $B_{z0} = 10^{-4}$ T for the guiding magnetic field magnitude of 10^{-5} T (left bottom plot). Bottom plots: trajectories of electrons simulated for the guiding field of 10^{-4} T, $B_{z0} = 10^{-2}$ T and $B_{x0} = 10^{-4}$ T (left plot), $B_{z0} = 10^{-3}$ T and $B_{x0} = 0.5 \times 10^{-5}$ T (right plot).

4. Conclusions

The PIC simulation has shown that accelerated particles produce a strong polarisation field, $\tilde{E}_x$, caused by the charge separation of accelerated particles across the current sheet. This polarisation field is shown to be very high and leads to essential modification of the trajectories of electrons, and, to some extent, those of protons. It affects differently the "transit" particles (those which enter from and are ejected to the opposite semispaces towards the midplane) and "bounced" particles (those which enter from and are ejected to the same semispace towards the midplane) with the "bounced" protons having wider orbits than the "transit" ones. As result, in the model with polarization field the protons are ejected less asymmetrically with respect to the midplane compared to TP simulations.

The electrons, which are "bounced" in the TP approach, in the PIC simulations are dragged by the polarisation field $\tilde{E}_x$ back towards the midplane. At the same time, the "transit" electrons, which are ejected to the opposite semispace from protons in the TP approach travel from the current sheet to a distance up to 10 times the RCS thickness and then are dragged back to the RCS by the polarisation field $\tilde{E}_x$. This process results in the formation of electron clouds around current sheets which exist as long as protons are being accelerated. The clouds disappear only when protons gain sufficient energy to leave the RCS, dragging electrons with them. However, more research is required to investigate particle trajectories and energy spectra from the present models for different magnetic field topologies.

References

Birn, J., Drake, J. F., Shay, M. A., *et al.* 2001, *Journal of Geophysics Research*, 106, 3715

Drake, J. F., Swisdak, M., Che, H., & Shay, M. A. 2006, *Nature*, 443, 553

Priest, E. & Forbes, T. 2000, Magnetic Reconnection (Magnetic Reconnection, by Eric Priest and Terry Forbes, pp. 612. ISBN 0521481791. Cambridge, UK: Cambridge University Press, June 2000.)

Siversky, T. V. & Zharkova, V. V. 2009, Journal of Plasma Physics, 75, 619

Sui, L. & Holman, G. D. 2003, *Astrophysical Journal, Letters*, 596, L251

Tsiklauri, D. & Haruki, T. 2007, Physics of Plasmas, 14, 2905

Wood, P. & Neukirch, T. 2005, *Solar Physics*, 226, 73

Zharkova, V. V. & Agapitov, O. V. 2009, Journal of Plasma Physics, 75, 159

Zharkova, V. V. & Gordovskyy, M. 2004, *Astrophysical Journal*, 604, 884

Zharkova, V. V. & Gordovskyy, M. 2005a, Monthly Notices of the Royal Astronomical Society, 356, 1107

Zharkova, V. V. & Gordovskyy, M. 2005b, Space Science Reviews, 121, 165

Advances in Plasma Astrophysics
Proceedings IAU Symposium No. 274, 2010
A. Bonanno, E. de Gouveia Dal Pino & A. G. Kosovichev, eds.
© International Astronomical Union 2011
doi:10.1017/S1743921311007484

3D turbulent reconnection driven current-sheet dynamics: solar applications

Lapo Bettarini and Giovanni Lapenta

Centrum voor Plasma-Astrofysica, Departement Wiskunde, Katholieke Universiteit Leuven,
Celestijnenlaan 200B, 3001 Leuven, Belgium
email: lapo.bettarini@wis.kuleuven.be

Abstract. We provide a complete three-dimensional picture of the reconnecting dynamics of a current-sheet. Recently, a two-dimensional non-steady reconnection dynamics has been proved to occur without the presence of any anomalous effect (Lapenta, 2008, Skender & Lapenta, 2010, Bettarini & Lapenta, 2010) but such a picture must be confirmed in a full three-dimensional configuration wherein all instability modes are allowed to drive the evolution of the system, i.e. to sustain a reconnection dynamics or to push the system along a different instability path. Here we propose a full-space analysis allowing us to determine the longitudinal and, possibly, the transversal modes driving the different current-sheet disruption regimes, the corresponding characteristic time-scales and to study system's instability space- parameter (plasma beta, Lundquist and Reynolds numbers, system's aspect ratio). The conditions leading to an explosive evolution rather then to a diffusive dynamics as well as the details of the reconnection inflow/outflow regime at the disruption phase are determined. Such system embedded in a solar-like environment and undergoing a non-steady reconnection evolution may determine the formation both of jets and waves influencing the dynamics and energetic of the upper layers and of characteristic down-flows as observed in the low solar atmosphere.

Keywords. Instabilities, (magnetohydrodynamics:) MHD, plasmas, Sun: magnetic fields

1. Introduction

The power of magnetic fields to affect plasma dynamics on large scales is revealed dramatically by space-borne and ground-base missions (Cirtain *et al.* 2007, Shibata *et al.* 2007). The challenge is to capture the dynamics of magnetized plasmas in our simulations, and thus to understand them. Numerical experiments and analytical studies suggested that within the pure resistive MHD framework it is not possible to have a magnetic field-line reconnecting dynamics that spontaneously evolves from a slow, resistive reconnection regime to a fast, high-power phase. We show how the conversion of magnetic field energy via magnetic reconnection can progress in a fast, fully three-dimensional, volume-filling regime characterized by a chaotic evolution of the system. The process does not require any pre-existing turbulence seed which often is not observed in the host systems prior to the onset of energy conversion. Here, we show the preliminary results for two different magnetic field configurations in a harris-sheet equilibrium (with and without a guide field) which is a well-known system whose (reconnecting) instability dynamics is critical to model plenty of disrupting and explosive phenomena taking place both with the solar-terrestrial environment and, in general, in astrophysics.

2. Numerical settings

We numerically solve the viscous-(purely) resistive compressible 3D one-fluid MHD equations parametrized by the global Lundquist and kinetic Reynolds number, S and

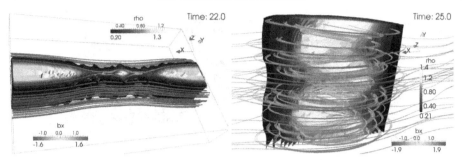

Figure 1. case **A**. On the right panel, the disruption of the initial laminar current-sheet in a multi-islands configuration (at $t = 22$, dimensionless time). Here it is shown the contour of the density and the magnetic field lines whose colors are related to the intensity of the X component of the magnetic field. On the left panel, the disruption in multiple pieces of the current-sheet is followed by a coalescence process (at about $t = 25$, dimensionless time) leading to the formation of an unique plasmoid at the centre of the numerical domain box. Afterwards the system is driven by a kinking instability.

R_M, set to 10^4. We define a stream-wise direction (X), wherein periodic boundary conditions are set at $X = 0$ and L_X, and a cross-stream and span-wise direction $(Z$ and Y respectively) where we set reflecting boundary conditions. We consider an ideal equation of state with polytropic index γ such that $p = \rho\,(\gamma - 1)\,I$ where p, ρ and I are respectively the kinetic pressure, the density and the enthalpy of the plasma. We use the code FLIP3D-MHD applying a (implicit) particle-in-cell method (FLIP algorithm, (Brackbill, 1991) that allows us (a) to solve an Eulerian-Lagrangian formulation of MHD equations describing the dynamics of our system and (b) to reduce drastically numerical effects. We consider 360 (in X), 60 (in Y), 240 (in Z) Lagrangian markers arrayed initially in a 33×3 uniform formation in each of the $120 \times 20 \times 80$ cells of our numerical grid. As already pointed out in the introduction, our system consists in a harris-sheet current-sheet with no guide field (case **A**) and with a guide field in the span-wise direction case **B**).

3. Results

We provide an exhaustive three-dimensional picture of the reconnecting dynamics of a current-sheet by means of low-diffusive numerical simulations. In both explored cases (A and B) we observe that the initial laminar state spontaneously evolve to a turbulent configuration through a two-stage process. First, the current-sheet is tearing unstable and it breaks to reach a multiple-island configuration (as shown in the left panel of Fig. 1 for case A) undergoing to a coalescence instability that lead to the formation of three-island system with a unique plasmoid at the center of the numerical domain. The presence of the guide field in case B acts to slow down the overall process that anyhow evolves according the same instability path. Secondly, according to the initial configuration we observe a different evolution because the non-steady dynamics critically depends on the interplay of perturbations developing along the magnetic field lines and across them (and so this process is possible only in three-dimensions).

case **A**: In the right panel of Fig. 1, which is a detailed view of the central plasmoid, it is evident the sinuous mode of the kinking dynamics of the structure and the complex closed loops of the magnetic field following the instability dynamics of the island. This secondary instability destroys the system and drives the high-density fragments which accelerate towards the sides of the box and drive a Rayleigh-Taylor instability of the high-density region there located.

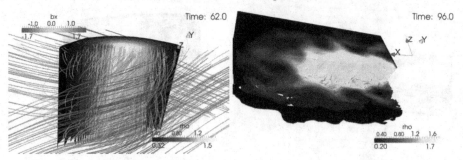

Figure 2. Case B. On the left panel, as in the case A we have a multiple breakdown of the current sheet and the several plasmoids coalesce to form a unique island that is not driven by any three-dimensional instability because of the presence of the guide field: it is shown the contour of the density and the magnetic field lines whose colors are related to the intensity of the X component of the magnetic field ($t = 62$, dimensionless time). On the right panel, clip of the density at the end of the simulation (at about $t = 96$, dimensionless time): the disruption of the central island and the related acceleration field trigger a Rayleigh-Taylor instability of the high-density regions located at the sides of the numerical domain.

case B: In the left panel of Fig. 2 the guide field stabilizes the central plasmoid against any kinking evolution and the whole structure is preserved. However the island undergoes the acceleration field resulting from the initial tearing evolution of the current-sheet (and consequent coalescence dynamic): this determines the instability evolution of the high-density regions at the sides of our box as shown in the right panel of Fig. 2. Here we show that current-sheet system can spontaneously evolve from an initial three- dimensional laminar state to a chaotic configuration via a non-steady self-feeding reconnection dynamics, thus confirming recent previous two-dimensional simulation studies (Lapenta, 2008, Skender & Lapenta, 2010). Yet the details of the magnetic configuration are key to determine the final state of the instability dynamics. The understanding of mechanisms connecting the micro-scales where reconnection takes place to the evolution of a macroscopic current-sheet configuration is the fundamental step to produce realistic models of all those phenomena requiring fast (and high power) triggering events such solar (stellar) flares and coronal mass ejections (Bettarini & Lapenta, 2010).

Acknowledgements

The research leading to these results has received funding from the European Commissions Seventh Framework Programme (FP7/2007 ? 2013) under the grant agreement N. 218816 (SOTERIA project: http://www.soteria-space.eu).

References

Bettarini, L. & Lapenta, G. 2010, *A&A*, 518, A57
Brackbill, J. U. 1991, *J. Comp. Phys.*, 96, 163
Cirtain, J. W., Golub, L., Lundquist, L., van Ballegooijen, A., Savcheva, A., Shimojo, M., DeLuca, E., Tsuneta, S., Sakao, T., Reeves, K., Weber, M., Kano, R., Narukage, N., & Shibasaki, K. 2007, *Science*, 318, 1580
Lapenta, G. 2008, *Phys. Rev. Lett.*, 100, 235001
Shibata, K., Nakamura, T., Matsumoto, T., Otsuji, K., Okamoto, T. J., Nishizuka, N., Kawate, T., Watanabe, H., Nagata, S., UeNo, S., Kitai, R., Nozawa, S., Tsuneta, S., Suematsu, Y., Ichimoto, K., Shimizu, T., Katsukawa, Y., Tarbell, T. D., Berger, T. E., Lites, B. W., Shine, R. A., & Title, A. M. 2007, *Science*, 318, 1591
Skender, M. & Lapenta, G. 2010, *Phys. Plasmas*, 17, 022905

Advances in Plasma Astrophysics
Proceedings IAU Symposium No. 274, 2010
A. Bonanno, E. de Gouveia Dal Pino & A. G. Kosovichev, eds.
© International Astronomical Union 2011
doi:10.1017/S1743921311007496

Decay of trefoil and other magnetic knots

Simon Candelaresi, Fabio Del Sordo and Axel Brandenburg

NORDITA, AlbaNova University Center, Roslagstullsbacken 23, SE-10691 Stockholm, Sweden
and
Department of Astronomy, Stockholm University, SE 10691 Stockholm, Sweden

Abstract. Two setups with interlocked magnetic flux tubes are used to study the evolution of magnetic energy and helicity on magnetohydrodynamical (MHD) systems like plasmas. In one setup the initial helicity is zero while in the other it is finite. To see if it is the actual linking or merely the helicity content that influences the dynamics of the system we also consider a setup with unlinked field lines as well as a field configuration in the shape of a trefoil knot. For helical systems the decay of magnetic energy is slowed down by the helicity which decays slowly. It turns out that it is the helicity content, rather than the actual linking, that is significant for the dynamics.

Keywords. Sun: magnetic fields

Magnetic helicity has been shown to play an important role in the dynamo process Brandenburg & Subramanian, 2005. For periodic systems where helicity is conserved simulations have shown that with increasing magnetic Reynolds number $\mathrm{Re_M}$ the saturation magnetic field strength decreases like $\mathrm{Re_M}^{-1/2}$ (Brandenburg & Dobler, 2001). This is problematic for astrophysical bodies since for the Sun $\mathrm{Re_M} = 10^9$ and galaxies $\mathrm{Re_M} = 10^{14}$. In order to alleviate this quenching the magnetic helicity of the small scale fields needs to be shed (Brandenburg *et al.*, 2009).

In the active regions of the Sun twisted magnetic field lines have been observed (Pevstov *et al.*, 1995). Later it was shown (Leka *et al.*, 1996) that the magnetic field in sunspots gets twisted before it emerges out of the surface. Manoharan *et al.*(1996) and Canfield *et al.*, (1999) demonstrated that helical structures are more likely to erupt into coronal mass ejections. This suggests that the Sun sheds helicity.

The magnetic helicity is related to the mutual linking for two non-intersecting flux tubes via (Moffatt, 1969)

$$H = \int_V \boldsymbol{A} \cdot \boldsymbol{B} \ \mathrm{d}V = 2n\phi_1\phi_2,$$

where H is the magnetic helicity, $\boldsymbol{B} = \nabla \times \boldsymbol{A}$ is the magnetic field in terms of the vector potential $\boldsymbol{A}$, ϕ_1 and ϕ_2 are the magnetic fluxes through the tubes and n is the linking number. The flux tubes may not have internal twist. In the limit of large $\mathrm{Re_M}$ H is a conserved quantity as well as the linking number.

In presence of magnetic helicity the magnetic energy decay is constrained via the realizability condition (Moffatt, 1969) which gives a lower bound for the spectral magnetic energy

$$M(k) \geqslant k|H(k)|/2\mu_0 \quad \text{with} \quad \int M(k) \ \mathrm{d}k = \langle \boldsymbol{B}^2 \rangle/2\mu_0, \quad \int H(k) \ \mathrm{d}k = \langle \boldsymbol{A} \cdot \boldsymbol{B} \rangle,$$

the magnetic permeability μ_0, where $\langle . \rangle$ denotes volume integrals.

In this work we extend earlier work (Del Sordo *et al.*, 2010) where the dynamics of interlocked flux rings, with and without helicity, was studied as well as a non-interlocked configuration. Here we also study a self-interlocked flux tube in the form of a trefoil knot.

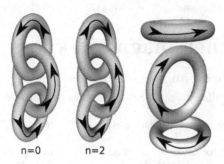

Figure 1. The three triple ring configurations for the initial time. From left to right: interlocked rings with no helicity, interlocked rings with finite helicity and non-interlocked rings without helicity. The arrows indicate the direction of the magnetic field. Adapted from Del Sordo *et al.*, 2010.

Figure 2. The initial magnetic field configuration for the trefoil knot.

The three-rings setups consist of three magnetic flux tubes. In two configurations they are interlocked where in one the helicity is zero and in the other one it has a finite value, as shown in Fig. 1. In the third setup we instead consider unlocked rings. Since the rings do not have internal twist, the helicity of this last configuration is zero. We also study the evolution of a self-interlocked flux tube having the form of a trefoil knot, with finite helicity (Fig. 2). In this case we have $H = 3\phi^2$, so the linking number is $n = 3/2$. All of these setups evolve according to the full resistive equations of MHD for an isothermal compressible medium. The Alfvén time is used as time unit.

As a consequence of the realizability condition the magnetic energy cannot decay faster than the helicity. The setups with finite H show a slower decay than the setups with no helicity (Fig. 3). The decay of the trefoil knot follows approximately the same decay law as the other configuration consisting of three rings with finite H. Within the simulation time H decays only to about one half of the initial value conserving then the topology. During later times field lines reconnect and the helicity seems to go into internal twist, which is topologically equivalent to linking; see Fig. 4.

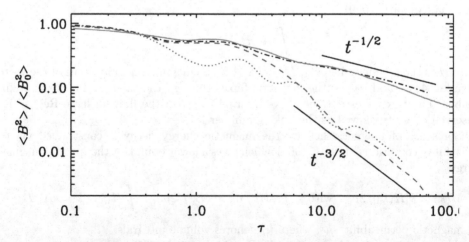

Figure 3. Evolution of the normalized magnetic energy for the trefoil know (solid/red line) compared with various three-ring configurations with $n = 2$ (dash-dotted line), $n = 0$ (dashed/blue line), and the non-interlocked case (dotted/blue line).

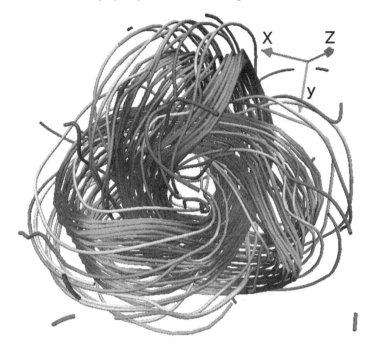

Figure 4. Magnetic field lines at 5 Alfvén times for the trefoil knot. The colors represent the magnitude of the magnetic field. Note that internal twist generation is weak.

The slow decay of H conserves the topology of the system. The linking is then eventually transformed into internal twisting during magnetic reconnection. Since both non-helical setups evolve similarly we conclude that it is mainly the magnetic helicity and not the actual linking which influences the dynamics. The helical trefoil knot evolves in a similar manner. This confirms the hypothesis that the decay of interlinked flux structures is governed by magnetic helicity and that higher-order invariants, advocated for example by Yeates *et al.*, 2010, may not be essential for describing this process.

In conclusion, we can say that magnetic helicity is decisive in controlling the decay of interlocked magnetic flux structures. If the magnetic helicity is zero, resistive decay will be fast while with finite magnetic helicity the decay will be slow and the speed of decay of magnetic energy depends on the speed at which magnetic helicity decays. This is likely an important aspect also in magnetic reconnection problems that has not yet received sufficient attention.

References

Brandenburg, A. & Subramanian, K. 2005, *Phys. Rep.* 417, 1
Brandenburg, A. & Dobler, W. 2001, *Astron. Astrophys.* 369, 329
Brandenburg, A., Candelaresi, S., & Chatterjee, P. 2009, *MNRAS* 398, 1414
Canfield, R. C., Hudson, H. S., & McKenzie, D. E. 1999, *Geophys. Res. Lett.* 26, 627
Moffatt H. K. 1969, *J. Fluid Mech.* 35, 117
Del Sordo, F., Candelaresi, S., & Brandenburg, A. 2010, *Phys. Rev. E* 81, 036401
Gibson, S. E., Fletcher, L., Del Zanna, G. *et al.* 2002 *ApJ* 574, 1021
Leka, K. D., Canfield, R. C., McClymont A. N., & van Driel-Gesztelyi, L. 1996, *ApJ* 462, 547
Manoharan, P. K., van Driel-Gesztelyi, L., Pick, M., & Demoulin P. 1996, *ApJ* 468, L73
Pevtsov, A. A., Canfield, R. C., & Metcalf, T. R. 1995, *ApJ* 440, L109
Yeates, A. R., Hornig, G., & Wilmot-Smith, A. L. 2010, *Phys. Rev. Lett.*, 105, 085002

Advances in Plasma Astrophysics
Proceedings IAU Symposium No. 274, 2010
A. Bonanno, E. de Gouveia Dal Pino & A. G. Kosovichev, eds.

© International Astronomical Union 2011
doi:10.1017/S1743921311007502

Magnetic helicity fluxes in $\alpha\Omega$ dynamos

Simon Candelaresi and Axel Brandenburg

NORDITA, AlbaNova University Center, Roslagstullsbacken 23, SE-10691 Stockholm, Sweden
Department of Astronomy, Stockholm University, SE 10691 Stockholm, Sweden

Abstract. In turbulent dynamos the production of large-scale magnetic fields is accompanied by a separation of magnetic helicity in scale. The large- and small-scale parts increase in magnitude. The small-scale part can eventually work against the dynamo and quench it, especially at high magnetic Reynolds numbers. A one-dimensional mean-field model of a dynamo is presented where diffusive magnetic helicity fluxes within the domain are important. It turns out that this effect helps to alleviate the quenching. Here we show that internal magnetic helicity fluxes, even within one hemisphere, can be important for alleviating catastrophic quenching.

Keywords. Sun: magnetic fields

The magnetic fields of astrophysical bodies like stars and galaxies show strengths which are close to equipartition. The scale of the magnetic field is larger then the dissipation length and reaches the order of the size of the object. The mechanism which creates those fields is believed to be a dynamo. Large- and small-scale magnetic helicity with opposite signs are created. For high magnetic Reynolds numbers, $\mathrm{Re_M}$, this makes the dynamo saturate only on a resistive time scale and reduces the saturation field strength much below equipartition (Brandenburg & Subramanian, 2005). This effect is called catastrophic quenching and increases with increasing Reynolds number, because for the Sun $\mathrm{Re_M} = 10^9$ and for galaxies $\mathrm{Re_M} = 10^{14}$. This suggests that helicity has to be shed. Observations have shown (Manoharan et al., 1996, Canfield et al., 1999) that helical structures on the Sun's surface are more likely to erupt into coronal mass ejections (CMEs). This suggests that the Sun sheds magnetic helicity by itself.

In our earlier work (Brandenburg et al., 2009) we have considered a one-dimensional mean-field model in the z-direction of a dynamo with wind-driven magnetic helicity flux where the wind increases with distance from the midplane. Magnetic helicity evolution is taken into account by using what is known as the "dynamical quenching" formalism that is described in our earlier paper and in references therein. We augment these studies by imposing a constant shear throughout the domain which facilitates the growth of the magnetic energy. We perform simulations in one hemisphere where we set the magnetic field in the z-direction to be symmetric (S) or antisymmetric (A) at the midplane. The outer boundaries are set to either vertical field (VF) or perfect conductor (PC). The free parameters are the dynamo numbers C_α and C_S. By varying both numbers we find the critical values for dynamo action (Fig. 1 and Fig. 2). The critical values for C_α decrease when the shear increases. This is expected, since larger shear leads to stronger toroidal field which enhances the dynamo effect.

In the rest of this paper we study in more detail the case $C_S = -10$ and consider positive values of C_α. In Figure 3 we compare the dynamical α quenching model (using a magnetic Reynolds number of $\mathrm{Re_M} = 10^5$) with the standard (non-catastrophic) α quenching where $\alpha \propto 1/(1 + \overline{\boldsymbol{B}}^2/B_{\mathrm{eq}}^2)$ with $\overline{\boldsymbol{B}}$ being the mean field and B_{eq} the equipartition value. Note that in the former case, the energies cross. Nevertheless, the A solution is stable in both cases – at least for $C_\alpha \leqslant 10$. This is demonstrated in Figure 4, where we show that after about 40 diffusive times, $\eta_t k_1 t = 40$, where η_t is the turbulent magnetic

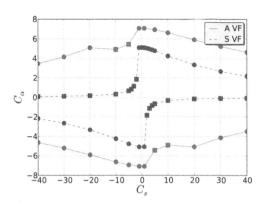

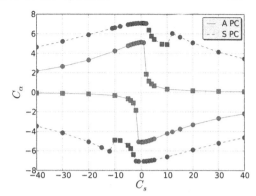

Figure 1. Critical values for the strength of the forcing C_α and the shear C_S for which dynamo action occurs for the cases of vertical field boundary conditions and antisymmetric (solid, red) and symmetric (dashed, blue) equator. The circles and squares represent oscillating and stationary solutions respectively.

Figure 2. Critical values for the strength of the forcing C_α and the shear C_S for which dynamo action occurs for the cases of perfect conductor boundary conditions and antisymmetric (solid, red) and symmetric (dashed, blue) equator. The circles and squares represent oscillating and stationary solutions respectively.

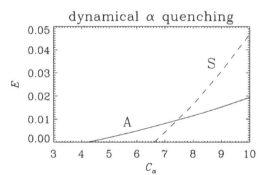

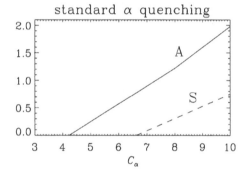

Figure 3. Comparison of the bifurcation diagrams for dynamical and standard α quenching.

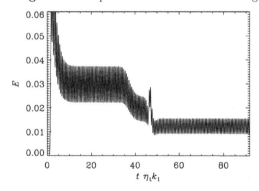

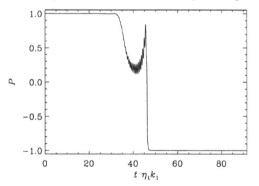

Figure 4. Evolution of the magnetic energy (left) and parity (right) for a solution that was initially even about the midplane (S or quadrupolar solution), but this solution is unstable and developed an odd parity (A or dipolar solution).

diffusivity and k_1 the basic wavenumber, the magnetic energy E decreases and the parity P swaps from $+1$ to -1; see Brandenburg *et al.*, 1989 for details on similar studies.

The crossing of the energies in the dynamical quenching model is somewhat surprising. In order to understand this behavior, we need to look at the profiles of the α effect; see

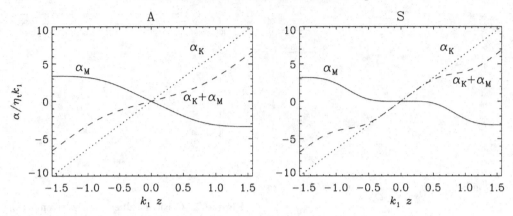

Figure 5. Profiles of α_K, α_M, and their sum for the A and S solutions at $C_\alpha = 10$.

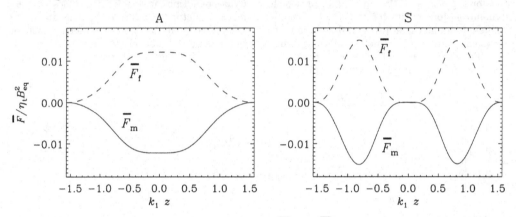

Figure 6. Time-averaged magnetic helicity fluxes, $\overline{F}_f$ and $\overline{F}_m$, of fluctuating and mean fields, for the A and S solutions, respectively, at $C_\alpha = 10$. Note that $\overline{F}_f + \overline{F}_m \approx 0$.

Figure 5. In this model, α is composed of a kinetic part, α_K, and a magnetic part, α_M, which has typically the opposite sign, which leads to a reduction of $\alpha = \alpha_K + \alpha_M$. The quenching can be alleviated by reducing α_M, for example when the divergence of the magnetic helicity flux of the small-scale field, $\overline{F}_f$, becomes important.

Naively, we would have expected that the A solution should have a larger energy, because only this solution allows a magnetic helicity flux through the equator; see Figure 6. This is however not the case, which may have several reasons. Even though the magnetic helicity flux flux small at the equator ($z = 0$), there can be significant contributions from within each hemisphere which contributes to alleviating the catastrophic quenching. The details of this will be address in more detail elsewhere.

References

Brandenburg, A. & Subramanian, K. 2005, *AN*, 326, 400
Brandenburg, A., Krause, F., Meinel, R., et al. 1989, *A&A*, 213, 411
Brandenburg, A., Candelaresi, S., & Chatterjee, P. 2009, *MNRAS* 398, 1414
Manoharan, P. K., van Driel-Gesztelyi, L., Pick, M., & Demoulin, P. 1996, *ApJ* 468, L73
Canfield, R. C., Hudson, H. S., & McKenzie, D. E. 1999, *Geophys. Res. Lett.* 26, 627

Advances in Plasma Astrophysics
Proceedings IAU Symposium No. 274, 2010
A.Bonanno, E. de Gouveia Dal Pino & A. G. Kosovichev, eds.

© International Astronomical Union 2011
doi:10.1017/S1743921311007514

A first model of stable magnetic configuration in stellar radiation zones

Vincent Duez[1], Jonathan Braithwaite[1] and Stéphane Mathis[2]

[1] Argelander Institut für Astronomie, Universität Bonn, Auf dem Hügel 71, D-53111 Bonn,
Germany, email: vduez@astro.uni-bonn.de, jonathan@astro.uni-bonn.de

[2] Laboratoire AIM, CEA/DSM-CNRS-Université Paris Diderot, IRFU/SAp Centre de Saclay,
F-91191 Gif-sur-Yvette, France, email: stephane.mathis@cea.fr

Abstract. We test the stability of a magnetic equilibrium configuration using numerical simulations and semi-analytical tools. The tested configuration is, as described by Duez & Mathis (2010), the lowest energy state for a given helicity in a stellar radiation zone. We show using 3D magneto-hydrodynamic (MHD) simulations that the present configuration is stable with respect to all submitted perturbations, that would lead to the development of kink-type instabilities in the case of purely poloidal or toroidal fields, both well known to be unstable. We also discuss, using semi-analytic work, the stabilizing influence of one component on the other and show that the found configuration actually lies in the stability domain predicted by a linear analysis of resonant modes.

Keywords. stars: magnetic fields, MHD

1. Introduction

The large-scale, ordered nature of magnetic fields detected at the surface of some Ap, O and B type stars and the scaling of their strengths (according to the flux conservation scenario) favour a fossil hypothesis, whose origin is not yet elucidated. To have survived since the stellar formation, a field must be stable on a dynamic (Alfvén) timescale. It was suggested by Prendergast (1956) that a stellar magnetic field in stable axisymmetric equilibrium must contain both poloidal (meridional) and toroidal (azimuthal) components, since both are unstable on their own (Tayler 1973; Wright 1973). This was confirmed recently by numerical simulations (Braithwaite & Spruit 2004; Braithwaite & Nordlund 2006) showing that an arbitrary initial field evolves on an Alfvén timescale into a stable configuration; axisymmetric mixed poloidal-toroidal fields were found. On the other hand, magnetic equilibria models displaying similar properties have been re-examined analytically by Duez & Mathis (2010). We here address the question of their stability usng both numerical and semi-analytical tools as recently reported in Duez *et al.* (2010).

2. The model

We deal with non force-free magnetic configurations (*i.e.* with a non-zero Lorentz force) in equilibrium inside a conductive fluid in absence of convection. Several reasons inclined us to focus on non force-free equilibria. First, Reisenegger (2009) reminds us that no configuration can be force-free everywhere. Although there do exist "force-free" configurations, they must be confined by some region or boundary layer with non-zero or singular Lorentz force. Second, non force-free equilibria have been identified in plasma physics as the result of relaxation (self-organization process involving magnetic reconnections in resistive MHD), *e.g.* by Montgomery & Phillips (1988). Third, as shown by Duez & Mathis (2010), this family of equilibria is a generalization of Taylor states (force-free relaxed equilibria; see Taylor 1974) in a stellar context, where the stratification of the medium plays a crucial role. Finally, we know (Chandrasekhar 1958) that in the ideal MHD limit the mass encompassed in magnetic flux surfaces is conserved in the axisymmetric case. We hence assume here that it is roughly conserved during the non-ideal relaxation phase, which leads automatically to non-force-free states (Woltjer 1959).

The equilibrium obtained is described in detail in Duez & Mathis (2010) as the lowest energy configuration conserving the invariants of the problem (during the relaxation phase) which are the magnetic helicity (preventing the rapid energy decay) and the mass enclosed in magnetic poloidal flux surfaces (to account for the non force-free property) which is due to the stable stratification.

3. Stability numerical analysis

We use the Stagger code (Nordlund & Galsgaard 1995). We model the star as a self-gravitating ball of ideal gas ($\gamma = 5/3$) with radial density and pressure profiles initially obeying the polytropic relation $P \propto \rho^{1+(1/n)}$, with index $n = 3$, therefore stably stratified. More details on the numerical model setup can be found in Braithwaite & Nordlund 2006. The configuration is then submitted to white perturbations (1% in density). The dynamical evolution of the mixed configuration is compared to its purely poloidal and its purely toroidal components, whose behaviour are well known to be unstable due to kink-type instabilities. The magnetic and velocity amplitudes are plotted on Fig. 1. As

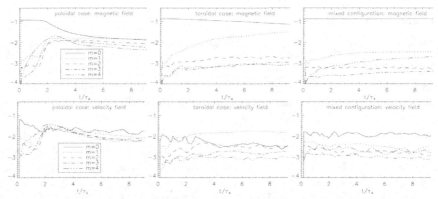

Figure 1. Time evolution of the (log) amplitudes in azimuthal modes $m = 0$ to 4 averaged over the stellar volume of the magnetic field (top row) and the velocity field (bottom row) in the simulations with the purely poloidal field (left), purely toroidal field (middle) and the mixed field (right). Initially, all the magnetic energy is in the $m = 0$ mode.

we can see, in contrast to these unstable configurations, the mixed poloidal-toroidal one does not exhibit any sign of instability, even for high azimuthal wavenumbers (up to about 40). The magnetic and velocity amplitudes are plotted on the right of fig. 1, where we see an absence of growing modes. The kinetic energy present results simply from the initial perturbation and the oscillations and waves it sets up.

To better examine the potentially unstable regions, we use Tayler's stability criteria (Tayler, 1973) *for purely toroidal fields* and estimate the stabilisation from the poloidal component, following Braithwaite (2009). In fig. 2 we plot Tayler's criteria for modes $m = 0$ and $m = 1$ – the $m = 0$ mode is unstable almost everywhere and the $m = 1$ mode is unstable in a large cone around the poles; however the poloidal field stabilises these modes in most of the meridional plane except near the equatorial plane where it merely stabilises all wavelengths small enough to fit into the available space. We can examine closely the behaviour of the field in the vicinity of the magnetic axis, where it can be

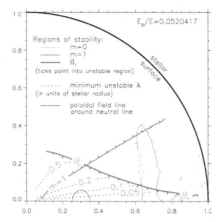

Figure 2. Half of the meridional plane, showing the regions unstable against the $m = 0$ and 1 Tayler modes in the absence of the poloidal component.

approximated as the addition of an axial and a toroidal field (cylindrical geometry). Bonanno (2008a) outlined that in this case magnetic configurations can be subject to non-axisymmetric resonant instability. They determined the dependency of the Tayler instability maximum growth rate as a function of the azimuthal wave-number m and of the ratio ε of the axial field to the toroidal one. In our case, close to the center the flux function exhibits a behaviour in $\Psi \propto r^2$, so the azimuthal field is proportional to $s = r \sin \theta$ corresponding to the Bonanno *et al.*'s parameter $\alpha = 1$. As underlined by the authors, in that case the maximum growth rate changes remarkably slowly with m for all modes with $m \geqslant 2$ and the instability is weakly non-anisotropic. If we take as a value for s_1 the radius of the neutral line or the one where the azimuthal field is strongest, we obtain respectively $\varepsilon = 0.64$ or $\varepsilon = 0.79$. According to their study (see Bonanno, 2008a; fig. 7), we fulfill the stability criterion for the modes $m = 0, 1$ and 2. Our results are therefore in agreement with their linear analysis. In the simulations we run, the mixed configuration has a poloidal energy fraction $E_{\rm p}/E = 0.052$. The magnetic-to-thermal energy ratio $E/U \approx 1/400$, which should mean that for stability we require $E_{\rm p}/E \gtrsim 0.04$ (Braithwaite, 2009). We see then that this value of E/U is near the upper limit for stability – in other words, we are near the boundary of validity of the weak-field approximation, reinforcing the result.

References

Bonanno, A. & Urpin, V. 2008, *A&A*, 488, 1
Bonanno, A. & Urpin, V. 2008, *A&A*, 477, 35
Braithwaite, J. & Nordlund, Å 2006, *A&A*, 450,1077
Braithwaite, J. & Spruit, H. 2004, *Nature*, 431,819
Chandrasekhar, S. 1958, *Proc. Nat. Acad. Sci.*, 44, 842
Duez, V. & Mathis, S. 2010, *A&A*, 517, A58
Duez, V., Braithwaite, J., & Mathis, S. 2010, *ApJL*, 524, L34
Montgomery, D. & Philips, L. 1988, *Phys. Rev. A*, 38, 2953
Nordlund, Å. & Galsgaard, K. 1995, http://www.astro.ku.dk/ aake/papers/95.ps.gz
Prendergast, K. H. 1956, *ApJ*, 123, 498
Reisenegger, A. 2009, *A&A*, 499, 557
Tayler, R. J. 1973, *MNRAS*, 161, 365
Tayler, J. B. 1974, *Phys. Rev. Lett.*, 33, 1139
Woltjer, L. 1959, *ApJ*, 130, 405
Wright, G. A. E. 1973, *MNRAS*, 162, 339

Advances in Plasma Astrophysics
Proceedings IAU Symposium No. 274, 2010
A. Bonanno, E. de Gouveia Dal Pino & A.G. Kosovichev, eds.

© International Astronomical Union 2011
doi:10.1017/S1743921311007526

Kinetic Simulations of Type II Radio Burst Emission Processes

Urs Ganse[1], Felix Spanier[1] and Rami Vainio[2]

[1] Lehrstuhl für Astronomie, Universität Würzburg,
Am Hubland, 97074 Würzburg, Germany

email: uganse@astro.uni-wuerzburg.de
email: fspanier@astro.uni-wuerzburg.de

[2] Department of Astronomy, University of Helsinki
email: rami.vainio@helsinki.fi

Abstract. Using a fully relativistic, 3D particle in cell code we have studied Langmuir- and electromagnetic wave processes in a CME foreshock plasma with counterstreaming electron beams. Langmuir wave excitation in resonance with the plasma frequency is observed, with timescales in accordance with theoretical predictions. However, no three wave interaction leading to emission of electromagnetic waves were detectable within the timeframe of our simulations.

Keywords. methods: numerical, radiation mechanisms: nonthermal, shock waves, Sun: radio radiation, waves

1. Introduction

Since the beginning of solar radio observations in the 1950ies, transient radio phenomena, so called solar radio bursts, have been observed. While consistent theories for the emission mechanism of most radio burst types have been found in the past, the so called type II radio bursts (correlated with coronal mass ejections) emission processes are still insufficiently explained.

Since CMEs are very extended structures whose shock fronts span areas from the low corona out towards interplanetary space, the large range of plasma densities should lead to a correspondingly large variation in plasma frequency, and hence, broadband emission. Observations however show a narrowband structure of two emission peaks, one approximately at the plasma frequency near the tip of the CME shock front (the so called *fundamental* emission), and one at its first harmonic.

2. The Model

In a model suggested by Cairns *et al.* (2003) emission is limited to a small spatial area in the foreshock of the CME, where electrons accelerated by shock drift acceleration at a quasi-perpendicular point along the curved shock front form an electron beam in the otherwise quiescent, thermal foreshock solar wind.

By a beam-driven instability, the electrons excite Langmuir waves, which can then undergo three-wave interaction processes similar to the creation of type III bursts (Melrose, 1986).

In addition to analytical calculation (Li *et al.* 2004), there are multiple in-situ spacecraft measurements of foreshock regions for interplanetary shocks (Pulupa and Bale, 2008) which confirm the existence of electron beams and Langmuir wave creation.

Using fundamental plasma simulations, we are trying to confirm this picture of the type II radio burst emission process. Similar simulations with reduced complexity have been

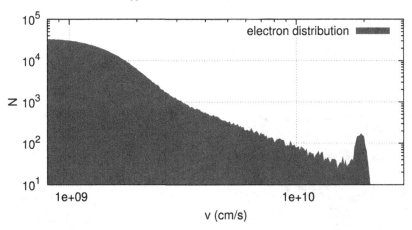

Figure 1. Model velocity distribution of electrons in the CME foreshock region: in addition to the thermal bulk, a superthermal electron beam component at 5 thermal velocities is present.

performed in 1D (Karlický and Vandas, 2007), but no complete 3-dimensional simulations exist to this date.

3. PiC simulations using ACRONYM

The numerical simulations for this research were conducted using ACRONYM, the particle in cell code developed at the university of Würzburg during the last 3 years. It is a fully-relativistic, second order 3D PiC-code written in C++, with MPI parallelization and excellent scaling behaviour even on large supercomputers.

Since the PiC method requires the plasma Debye length to be resolved, simulations encompassing the complete CME are unfeasible. Instead, the simulation scenario is focussed on the microphysics within the foreshock region. A homogeneous solar wind plasma fills the box, with two counterstreaming electron beams added to create an approximation of the electron distribution function (see fig. 1).

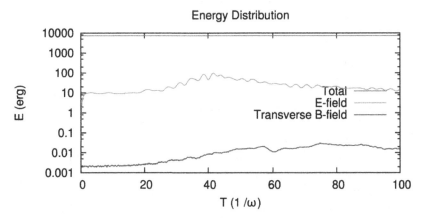

Figure 2. Evolution of energy distribution in the simulation. Kinetic energy of the electron beams is converted into electric fields, when Langmuir waves are excited. The decay of these waves deposits energy into the transverse B-field (B_x and B_y component).

Input parameters for the simulation were a plasma frequency of 200 MHz, a temperature of 10^6 K, background magnetic field of $|\mathbf{B}| = 0.1$ G parallel to the electron beam direction, the correct physical mass ratio of $m_p/m_e = 1836$ and density of $2.5 \cdot 10^7 \mathrm{cm}^{-3}$.

Simulation sizes varied from $256 \times 256 \times 512$ cells (elongated in streaming direction) to 768^3 cells, with 20 particles per cell. With a Debye length of $\lambda_D = 1.35$ cm, this leads to physical extents of up to $10 \times 10 \times 10$ m.

4. Results and Conclusion

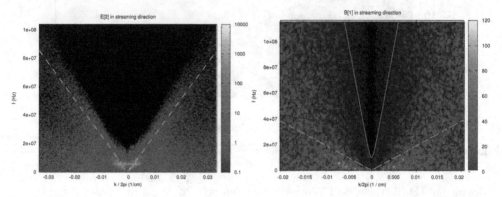

Figure 3. Dispersion plots (created using temporal and spatial fourier transform) of the longitudinal component of the electric field (**left**) and a transverse component of the magnetic field (**right**) along the streaming direction. A strong signal of Langmuir waves is visible near resonance with the plasma frequency. No electromagnetic emissions of large intensity are detectable. The green line represents the beam velocity of the electrons, the red line indicates the theoretical dispersion relation of the electromagnetic mode.

In the energy distribution (see fig. 2), a conversion of particle kinetic energy into electric field energy is observed. The dispersion plot of the longitudinal electric field along the streaming direction (see the left side of fig. 3) confirms a strong signal of Langmuir waves near resonance with the plasma frequency.

Creation of electromagnetic waves within the simulation timeframe is not visible. Instead, rise in magnetic energy (in the components perpendicular to the background field and streaming velocity) appears in figure 2. Both dispersion and magnetization of this wave indicate that this is a beam driven mode (Wiles & Cairns, 2000).

The quick decay of the Langmuir wave excitation, visible in the energy plot, suggests that other interaction processes than the one leading to radio emission may be dominant in our simulations, hence making radio emissions invisible. A possible reason is the high numerical noise in the PiC-code, leading to overly strong density fluctuations. Work is now ongoing to reduce these fluctuations (by increasing particle number and improved smoothing) and verify their effect on Langmuir wave stability.

5. Acknowledgements

We thank the Jülich supercomputing centre for the grant of computing time on Jugene. UG acknowledges support by the Elite Network of Bavaria. FS acknowledges support by the German Science Foundation, Grant SP1124/3.

References

Cairns, I., Knock, Robinson, P., & Kuncic *Space Sci. Revs.* , Vol. 107, April (2003), p. 27–34.
M. Karlický & M. Vandas Planet. and Space Sci., Vol. 55, December (2007), p. 2336-2339.
Li, B., Willes, A., Robinson, P., & Cairns, I. *APS Meeting Abstracts*, Nov (2004), p. 1035P-+.
Melrose, D. B. *Instabilities in Space and Laboratory Plasmas*, Cambridge Univ. Press, Aug (1986)
Pulupa & Bale 2008, *ApJ*, Vol. 676 , April (2008), p. 1330-1337.
Willes, A. & Cairns, I. *Physics of Plasmas, Vol. 7, Aug (2000), p. 3167-3180.*

Advances in Plasma Astrophysics
Proceedings IAU Symposium No. 274, 2010
A. Bonanno, E. de Gouveia Dal Pino & A. G. Kosovichev, eds.

© International Astronomical Union 2011
doi:10.1017/S1743921311007538

Turbulent magnetic pressure instability in stratified turbulence

K. Kemel[1,2], A. Brandenburg[1,2], N. Kleeorin[3], and I. Rogachevskii[3]

[1]NORDITA, AlbaNova University Center, Roslagstullsbacken 23, SE-10691 Stockholm, Sweden
[2]Department of Astronomy, Stockholm University, SE–10691 Stockholm, Sweden
[3]Department of Mechanical Engineering, Ben-Gurion University of the Negev,
POB 653, Beer-Sheva 84105, Israel

Abstract. A reduction of total mean turbulent pressure due to the presence of magnetic fields was previously shown to be a measurable effect in direct numerical simulations. However, in the studied parameter regime the formation of large-scale structures, as anticipated from earlier mean-field simulations, was not found. An analysis of the relevant mean-field parameter dependency and the parameter domain of interest is conducted in order to clarify this apparent discrepancy.

Keywords. turbulence, MHD, sunspots

Strong magnetic fields at the solar surface are generally thought to originate from the coherent rise of magnetic flux tubes from the tachocline through the solar convection zone. While the idea is elegant, the question remains whether it can be considered as more than a toy model, as the physics of the creation and rise of these flux tubes is not sufficiently understood Parker, 2009. Thus far any 'successful' numerical simulation of this process had to rely on strong assumptions, be it in the initial conditions or in simplified equations such as the thin flux tube approximation Spruit, 1981. As such it makes sense to explore alternative mechanisms of magnetic structure formation. Several models have been proposed where large-scale magnetic field concentrations are created through instabilities at the solar surface (Kitchatinov & Mazur, 2000, Brandenburg *et al.* 2010a, hereafter BKR).

Turbulence is generally associated with enhanced transport effects. However, it can also generate structures on much larger scales than its driving scales; see as an example the inverse cascade in 2D hydrodynamic turbulence (Kraichnan, 1967) or 3D MHD turbulence with magnetic helicity (Frisch *et al.* 1975). We study here the interaction of the turbulence with a background magnetic field. From the approximate conservation of total turbulent energy $E_{\rm tot}$, BKR find a reversed feedback from the magnetic fluctuations on the turbulent pressure (Rogachevskii & Kleeorin, 2007, hereafter RK):

$$P_{\rm turb} = -\tfrac{1}{6}\overline{b^2}/\mu_0 + \tfrac{2}{3}E_{\rm tot}.$$

It can be seen that the effective mean magnetic pressure force is reduced and can be reversed in a certain parameter range. It was suggested by RK that this positive feedback could lead to an instability, resulting in the concentration of magnetic flux.

BKR confirmed the validity of approximate turbulent energy conservation using direct numerical simulations (DNS) of homogeneous isothermal turbulence and they also demonstrated the basic phenomenon of magnetic flux concentration through the interaction between turbulence and the mean Lorentz force in mean-field MHD simulations which led to a linear instability for sufficiently strong stratification. These results were recently corroborated using DNS of inhomogeneous (stratified) isothermal turbulence of

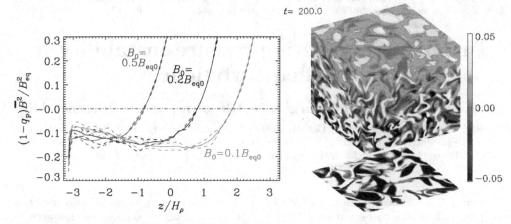

Figure 1. *Left*: Normalized effective mean magnetic pressure as a function of depth for $B_0 = 0.1B_{eq0}$, $B_0 = 0.2B_{eq0}$, and $B_0 = 0.5B_{eq0}$ using $Re = 120$ and $Pm = 1$. Adapted from Brandenburg *et al.* 2010b. *Right*: Visualization of $B_y - B_0$ on the periphery of the computational domain $B_0 = 0.1B_{eq0}$, and $Pm = 2$. Adapted from Kemel *et al.* 2010.

a cubic computational domain. An example is shown in Figure 1 where we show that the normalized effective mean magnetic pressure does indeed become negative in a large part of the domain for different imposed field strengths. However, the formation of large-scale structures is not observed, as can be seen from the right-hand panel of Figure 1.

It is important to understand whether there is really a conflict between DNS and mean-field results, or if these two studies simply apply to different parameter regimes. In order to have a chance to resolve this question, we conduct a systematic parameter survey of the instability in the mean-field model. More specifically, we determine the functional dependence of the growth rate on the input variables in order to find the relevant parameter space for the instability to develop.

In an isothermal stratified box we solve the equations in two dimensions:

$$\frac{\partial \overline{U}}{\partial t} = -\overline{U} \cdot \nabla \overline{U} - c_s^2 \nabla \ln \overline{\rho} + g + \overline{\mathcal{F}}_{M} + \overline{\mathcal{F}}_{K,tot}, \tag{0.1}$$

where

$$\overline{\rho}\,\overline{\mathcal{F}}_{M} = -\tfrac{1}{2}\nabla[(1 - q_p)\overline{B}^2] + \overline{B} \cdot \nabla\left[(1 - q_s)\overline{B}\right] \tag{0.2}$$

is the mean-field Lorentz force and

$$\overline{\mathcal{F}}_{K,tot} = (\nu_t + \nu)\left(\nabla^2 \overline{U} + \nabla\nabla \cdot \overline{U} + 2\overline{S}\nabla \ln \overline{\rho}\right) \tag{0.3}$$

is the total (turbulent and microscopic) viscous force with **S** being the viscous stress tensor. In addition, we solve the continuity and uncurled induction equations,

$$\frac{\partial \ln \overline{\rho}}{\partial t} = -\overline{U} \cdot \nabla \ln \overline{\rho} - \nabla \cdot \overline{U}, \quad \frac{\partial \overline{A}}{\partial t} = \overline{U} \times \overline{B} - (\eta_t + \eta)\overline{J}. \tag{0.4}$$

We adopt a Cartesian coordinate system, (x, y, z). The mean field is given by $\overline{B} = (0, B_0, 0) + \nabla \times \overline{A}$, and the vertical gravitational acceleration is $g = (0, 0, -g)$. The other input parameters of the simulations are the sound speed c_s, the density at the top of the box ρ_{top}, the pressure coefficient q_p (RK), the magnetic Prandtl number Pm and the molecular diffusivity η.

We start by isolating the dependence of the growth rate on the turbulent diffusivity and find approximately linear behaviour. From a data survey we observe, depending on

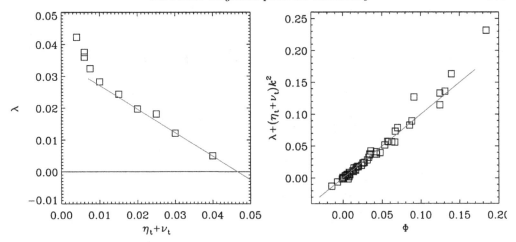

Figure 2. *Left*: dependence of growth rate on turbulent diffusivity and viscosity. *Right*: modified growth rate as a function of Φ.

the applied field, different diffusivity values above which the instability does not develop, which would imply

$$\lambda = \Phi(B) - (\eta_t + \nu_t)k^2 \tag{0.5}$$

where $1/k$ is a length scale introduced for dimensional reasons. We find that Φ is proportional to the Alfvén speed at the top of the box, where the instability initiates. Eventually we arrive at the fit formula

$$\lambda + \left(1 + \mathrm{Pr}_M^{\mathrm{turb}}\right)\eta_t k^2 = v_A k_x \left(1 + q_p/q_p^*\right)\exp\left(\ell_z/H_\rho\right), \tag{0.6}$$

as shown in Figure 2. Here, $\mathrm{Pr}_M^{\mathrm{turb}} = \nu_t/\eta_t$, ℓ_z is some typical vertical length scale, and $H_\rho = c_s^2/g$ is the density scale height. The instability resulting from this feedback effect was verified but not observed to generate large-scale structures in DNS with the current scale separation and parameter range studied so far.

An extension of this work will be the inclusion of more physics (e.g. radiative cooling) in the DNS and the comparison with mean-field models in a 3D setup. As was already pointed out in BKR, the mean-field model predicts additional short-wavelength perturbutions along the direction of the mean magnetic field, that were not included in the present study.

References

Brandenburg, A. & Dobler, W. 2002, *Comp. Phys. Comm.*, *147*, 471

Brandenburg, A., Kleeorin, N., & Rogachevskii, I. 2010a, *Astron. Nachr.*, *331*, 5 (BKR)

Brandenburg, A., Kemel, K., Kleeorin, N., & Rogachevskii, I. 2010b, arXiv:1005.5700

Frisch, U., Pouquet, A., Léorat, J., & Mazure, A. 1975, *J. Fluid Mech.*, *68*, 769

Kemel, K., Brandenburg, A., Kleeorin, N., & Rogachevskii, I. 2010, *Physics of Sun and Star Spots*, IAU Symp. 273, eds. D.P. Choudhary & K.G. Strassmeier, arXiv:1010.1659

Kraichnan, R. H. 1967, *Phys. Fluids*, *10*, 1417

Kitchatinov, L. L. & Mazur, M. V. 2000, *Solar Phys.*, *191*, 325

Parker, E. N. 2009, *Spa. Sci. Rev.*, 144, 15

Rogachevskii, I. & Kleeorin, N. 2007, *Phys. Rev. E, 76*, 056307 (RK)

Spruit, H. C. 1981, *A&A, 102*, 129

Advances in Plasma Astrophysics
Proceedings IAU Symposium No. 274, 2010
A. Bonanno, E. de Gouveia Dal Pino & A. G. Kosovichev, eds.

© International Astronomical Union 2011
doi:10.1017/S174392131100754X

Current-Driven Kink Instability in Relativistic Jets

Yosuke Mizuno[1], Philip E. Hardee[2], Yuri Lyubarsky[3], and Ken-Ici Nishikawa[1]

[1] CSPAR, The University of Alabama in Huntsville,
320 Sparkman Drive, NSSTC, Huntsville, AL 35805, USA
email: mizuno@cspar.uah.edu

[2] Dept. Physics and Astronomy, University of Alabama, Tuscaloosa, AL 35487, USA

[3] Physics Department, Ben-Gurion University, Beer-Sheva 84105, Israel

Abstract. We have investigated the development of current-driven (CD) kink instability in relativistic jets via 3D RMHD simulations. In this investigation a static force-free equilibrium helical magnetic field configuration is considered in order to study the influence of the initial configuration on the linear and nonlinear evolution of the instability. We found that the initial configuration is strongly distorted but not disrupted by the CD kink instability. The linear growth and nonlinear evolution of the CD kink instability depends moderately on the radial density profile and strongly on the magnetic pitch profile. Kink amplitude growth in the nonlinear regime for decreasing magnetic pitch leads to a slender helically twisted column wrapped by magnetic field. On the other hand, kink amplitude growth in the nonlinear regime nearly ceases for increasing magnetic pitch.

Keywords. relativistic jets, instabilities, MHD

1. Introduction

Relativistic jets occur in active galactic nuclei (AGN), occur in microquasars (μQSOs), and are thought responsible for the gamma-ray bursts (GRB). The most promising mechanism for producing relativistic jets involves magnetohydrodynamic acceleration from an accretion disk around a black hole (Blandford & Payne 1982), and/or involves the extraction of energy from a rotating black hole (Blandford & Znajek 1977).

GRMHD simulations with a spinning black hole indicate jet production with a magnetically dominated high Lorentz factor spine with $v \sim c$, and a matter dominated sheath with $v \gtrsim c/2$ possibly embedded in a lower speed, $v \ll c$, disk/coronal wind involving helically twisted magnetic fields (e.g., Hawley & Krolik 2006; McKinney 2006; Hardee *et al.* 2007).

In configurations with a strong toroidal magnetic field the current driven (CD) kink mode is unstable. This instability excites large-scale helical motions that can strongly distort or even disrupt the system. For static cylindrical force-free equilibria, the well-known Kruskal-Shafranov criterion states that the instability develops if the length of the column, ℓ, is long enough for the field lines to go around the cylinder at least once (Bateman 1978): $|B_p/B_\phi| < \ell/2\pi R$. However, rotation and shear motions could significantly affect the instability criterion. For relativistic force-free configurations, the linear instability criteria have been studied by a number of researchers (e.g., Begelman 1998; Lyubarskii 1999; Narayan *et al.* 2009).

The linear mode analysis provides conditions for the instability but says little about the impact the instability has on the system. To evaluate the impact of the potentially disruptive kink mode found from a linear analysis the instability must be followed into the non-linear regime.

In this work, we study the kink instability in relativistic systems. By relativistic we mean not only relativistically moving systems but any with magnetic energy density comparable to or greater than the plasma energy density, including the rest mass energy. In particular, we present 3D results of the CD kink instability of a static plasma column (Mizuno *et al.* 2009).

2. Numerical Setup and Results

In order to study time evolution of the CD kink instability in the relativistic MHD (RMHD) regime, we use the 3D GRMHD code "RAISHIN" in Cartesian coordinates (Mizuno *et al.* 2006). For our simulations we will choose a force-free helical magnetic field as the initial configuration. A pitch profile parameter α determines the radial profile of the magnetic pitch $P = RB_z/B_\phi$, and provides a measure of the twist of the magnetic field lines. If the pitch profile parameter $0.5 < \alpha < 1$, the magnetic pitch increases with radius. If $\alpha > 1$, the magnetic pitch decreases. When $\alpha = 1$, the magnetic pitch is constant. This configuration is the same as that used in previous non-relativistic work (Appl *et al.* 2000; Baty 2005).

In our simulations we consider a low gas pressure medium with constant $p = p_0 = 0.02\rho_0 c^2$ for the equilibrium state, and a non-uniform density profile decreasing proportional to the magnetic field strength, $\rho = \rho_1 B^2$ with $\rho_1 = 10.0\rho_0$. The magnetic field amplitude is $B_0 = 0.4\sqrt{4\pi\rho_0 c^2}$ and leads to a low plasma-β near the axis. In order to investigate the effect of different radial pitch profiles, we perform simulations with: constant pitch, $\alpha = 1$, increasing pitch, $\alpha = 0.75$, and decreasing pitch $\alpha = 2.0$.

The simulation grid is periodic along the axial direction. The grid is a Cartesian (x, y, z) box of size $4L \times 4L \times 2L$ with grid resolution of $\Delta L = L/40$, where L is a simulation scale unit. We impose outflow boundary conditions on the transverse boundaries at $x = y = \pm 2L$.

The initial MHD equilibrium configuration is perturbed by a small radial velocity component with profile given by $v_R = \delta v \exp(-R/R_a) \cos(m\theta) \sin(2\pi n z/L_z)$ with $\delta v = 0.01c$ and $R_a = 0.5L$. We choose $m = 1$ and $n = 1$ in the above formula. This is identical to imposing $(m, n) = (-1, -1)$, because of the symmetry between (m, n) and $(-m, -n)$ pairs. A detailed description of numerical set-up can be found in Mizuno *et al.* (2009).

As an indicator of the growth of the CD kink instability we use the volume-averaged kinetic energy transverse to the z-axis ($E_{kin,xy}$). The quantity $E_{kin,xy}$ allows determination of different evolutionary stages, e.g., initial exponential (linear growth phase) growth and subsequent non-linear evolution. In all cases, the initial growth regime is characterized by an exponential increase in $E_{kin,xy}$ by several orders of magnitude to a maximum amplitude followed by a slow decline in the non-linear regime. The increasing pitch case ($\alpha = 0.75$) grows more slowly and reaches a maximum at a later time with a smaller value for $E_{kin,xy}$ than the constant pitch case ($\alpha = 1.0$). On the other hand, the decreasing pitch case ($\alpha = 2.0$) grows more rapidly and reaches a maximum at an earlier time with a larger value for $E_{kin,xy}$ than the constant pitch case. Although the transition time from linear to non-linear evolution is different for each pitch case, the maximum radial velocity is almost the same at transition. The different growth rates as a function of the radial pitch profile are consistent with a non-relativistic linear analysis (Appl *et al.* 2000).

Figure 1 shows density isosurfaces with magnetic field lines for the constant pitch, increasing pitch and decreasing pitch cases for the non-uniform density profile at time $t_A = 96(L/8v_{A0})$. Displacement of the initial force-free helical magnetic field leads to a helically twisted magnetic filament winded around the density isosurface. In the constant pitch case the radial displacement of the high density region slows significantly at longer times. Continuing outwards radial motion is confined to a lower density sheath around the high density core.

(a) Constant Pitch   (b) Increasing Pitch   (c) Decreasing Pitch

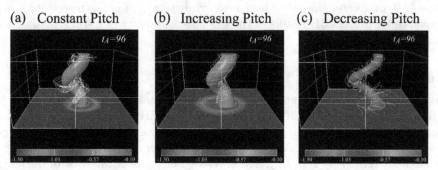

Figure 1. 3D density isosurface for (a) constant pitch, (b) increasing pitch, and (c) decreasing pitch cases. Colors show the logarithm of the density with (white) magnetic field lines.

The 3D density structure of the increasing pitch case looks similar to that of the constant pitch case. The CD kink instability initially grows exponentially. A transverse density slice shows little outwards motion of the high density region in the non-linear stage similar to the constant pitch case; however, there is little outwards motion in the low density sheath surrounding the high density region. This result is somewhat different from the constant pitch case and suggests a significant reduction in kink amplitude growth. Results from the decreasing pitch case are very different. Figure 2c shows a more slender helical density structure wrapped by the magnetic field. While the density cross section is similar to that of the constant pitch case, radial motion continues in the non-linear stage to the end of the simulation.

Acknowledgements

This work is supported in part by NSF awards AST-0506719, AST-0506666, AST-0908010, AST-0908040, NASA awards NNG05GK73G, NNX07AJ88G, NNX08AG83G, and US-Israeli BSF 2006170. The simulations were performed on the NAS Division Columbia Supercomputer at the NASA Ames Research Center and the Altix3700 BX2 at YITP in Kyoto University.

References

Appl, S., Lery, T., & Baty, H. 2000 *A&A* 355, 818
Bateman, G. 1978, *MHD Instabilities* (Cambridge, MA: MIT Press), p. 270
Baty, H. 2005 *A&A* 430, 9
Begelman, M. C. 1998 *ApJ* 493, 291
Blandford, R. D. & Payne, D. G. 1982, *MNRAS* 199, 883
Blandford, R. D. & Znajek, R. L. 1977, *MNRAS* 179, 433
Hardee, P., Mizuno, Y., & Nishikawa, K.-I. 2007 *Ap&SS* 311, 283
Hawley, J. F. & Krolik, J. H. 2006, *ApJ* 641, 103
Lyubarskii, Y. E., 1999 *MNRAS* 208, 1006
McKinney, J. C. 2006, *MNRAS* 368, 1561.
Mizuno, Y., Lyubarsky, Y., Nishikawa, K.-I., & Hardee, P. E. 2009 *ApJ* 700, 684
Mizuno, Y., Nishikawa, K.-I., Koide, S., Hardee, P., & Fishman, G. J. 2006 *ArXiv Astrophysics e-prints* 0609004
Narayan, R., Li, J., & Tchekhovskoy, A. 2009, *ApJ* 697, 1681

Advances in Plasma Astrophysics
Proceedings IAU Symposium No. 274, 2010
A.Bonanno, E. de Gouveia Dal Pino & A. Kosovichev, eds.

© International Astronomical Union 2011
doi:10.1017/S1743921311007551

High-order methods for the simulation of hydromagnetic instabilities in core-collapse supernovae

T. Rembiasz[*1], **M. Obergaulinger**[1], **M. Angel Aloy**[2], **P. Cerdá-Durán**[1] and **E. Müller**[1]

[1] Max-Planck-Institut für Astrophysik, 85748 Garching, Germany
[*] email: `rembiasz@mpa-garching.mpg.de`

[2] Universitat de València, 46100 Burjassot (Valencia), Spain

Abstract. We present an assessment of the accuracy of a recently developed MHD code used to study hydromagnetic flows in supernovae and related events. The code, based on the constrained transport formulation, incorporates unprecedented ultra-high-order methods (up to 9th order) for the reconstruction and the most accurate approximate Riemann solvers. We estimate the numerical resistivity of these schemes in tearing instability simulations.

Keywords. reconstruction schemes, Riemann solvers, hydromagnetic instabilities

1. Introduction

If the weak magnetic fields of progenitor stars are amplified efficiently in the post-collapse phase, they may affect the dynamics of core-collapse supernovae significantly. Prime candidates for this amplification are hydromagnetic instabilities in the stellar core, e.g., magneto-convection and the magneto-rotational instability (e.g., Obergaulinger *et al.* 2009). Leading to small-scale turbulence, these instabilities demand the use of highly accurate methods in supernova simulations, while the wide range of physics involved and the presence of supersonic flows and strong shock waves call for flexibility and stability.

The properties of turbulence developing due to MHD instabilities can depend crucially on transport coefficients such as the resistivity, η. In any numerical simulation, numerical diffusion and dissipation, described approximately by a numerical resistivity, η_{num}, add to the effects of the physical (microscopic) resistivity. A clear distinction between both effects is important for a correct interpretation of the simulations.

Therefore, we try to quantify the numerical resistivity of our MHD code, an implementation of a finite-volume constraint-transport scheme. Here we show a comparison of the numerical resistivity for different resolutions, Riemann solvers (Lax-Friedrichs (LF) (e.g., Toro 2009), Harten, Lax, van Leer (HLL) (e.g., Toro 2009), the approximate six-stage MHD solver (HLLD) (Miyoshi & Kusano 2005)) and reconstruction schemes (piecewise linear (PL) (e.g., Toro 2009), monotonicity preserving (MP) (Suresh & Huynh 1997) of 5th, 7th and 9th order).

2. Simulations

We performed 2D simulations of the tearing instability in a current sheet in force-free magneto-hydrostatic equilibrium with constant gas pressure and density using similar

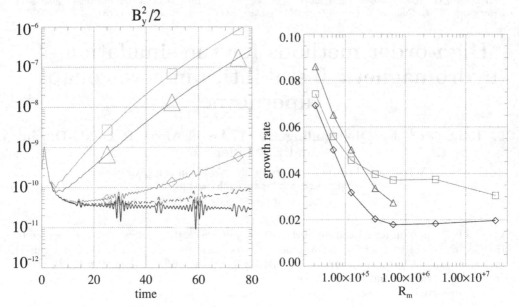

Figure 1. *Left*: Time evolution of $B_y^2/2$ for different magnetic Reynolds numbers R_m. HLL Riemann solver, MP 5th order and grid of 512×1024 zones were used. $R_m = \pi \times 10^4$ (orange squares), $R_m = \pi \times 10^5$ (red triangles), $R_m = 2\pi \times 10^5$ (green diamonds), $R_m = 2\pi \times 10^7$ (dashed blue), $R_m = \pi \times 10^7$ (solid black).
Right: Growth rates for different Riemann solvers: LF (green squares), HLL (black diamonds), HLLD (blue triangles). The simulations were performed with 7th order MP reconstruction on grid of 128×256 zones.

physical parameters as in Landi *et al.* (2008) for different resistivities. The initial config-uration is perturbed with transverse velocity fluctuations, which are much smaller than both the Alfvén and the sound speed. Due to non-zero resistivity the tearing instability sets in. This can bee seen in Fig. 1, depicting the time evolution of the average of the transverse (y) component of the magnetic energy. It grows exponentially during the first phase of evolution and its growth rate is proportional to resistivity. (Note that the on-set of the instability is delayed for higher Reynolds numbers, defined as $R_m = c_A/L_y\eta$, where L_y is the transverse box size.) However, this proportionality holds only above a critical resistivity η^* being approximately equal to the numerical resistivity η_{num}. This sets the maximum Reynolds number R_m^* which can be resolved (for a given scheme and resolution).

Resolution studies are presented in the left panel of Fig. 2. With the coarsest grid we are not able to resolve Reynolds numbers higher than $R_m^* \approx 10^5$, because the numerical resistivity dominates over the physical one. Doubling the number of grid zones reduces the numerical resistivity approximately fourfold for the used scheme. In the right panel of Fig. 2 we compare different reconstruction methods. All simulations were performed with a fixed number of the grid zones. 9th order MP is approximately twice less resistive than 7th order MP and four times less resistive than 5th order MP. The PL scheme seems to be less reliable than MP schemes for these simulations. A comparison of three different Riemann solvers is shown in the right panel of Fig. 1.

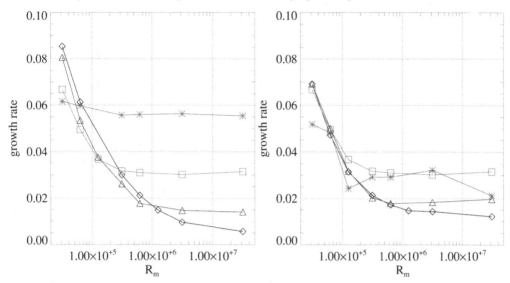

Figure 2. *Left*: Resolution studies with HLL Riemann solver and MP of 5th order. Grids of: 64 × 128 (red asterisks), 128 × 256 (green squares), 256 × 512 (blue triangles), 512 × 1024 zones (black diamonds) were used.

Right: Growth rates for different reconstruction schemes: PL with a combination of minmod and MC slope limiter (red asterisks), MP of 5th order (green squares), MP of 7th order (blue triangles), MP of 9th order (black diamonds). In the simulations HLL Riemann solver and grid of 128 × 256 zones were used.

3. Conclusions

We conclude that with the aforementioned simulations we are able to measure the numerical resistivity of our code. This allows us to choose a proper grid resolution and reconstruction schemes such that the numerical resistivity is lower than the microscopic resistivity with which we want to simulate the physical phenomena of interest.

In the performed tests we reduced by approximately the same factor the numerical resistivity of the simulations with 5th order MP, either by choosing 9th order MP, or by doubling the resolution. The latter is computationally much more expensive. Therefore, in order to resolve higher Reynolds numbers, one should start by choosing higher order reconstruction methods, rather than by increasing resolution, if the computational time is a matter of concern.

References

S. Landi, P. Londrillo, M. Velli, & L. Bettarini *Physics of Plasmas* **15** (2008) 012302

T. Miyoshi & K. Kusano, *J. Comput. Phys.* **208** (2005) 315-344

M. Obergaulinger, P. Cerdá-Durán, M. Angel Aloy, & E. Müller, *A&A* **1323** (2009)

A. Suresh & H. T. Huynh, *J. Comput. Phys.* **136** (1997) 83

E. Toro, *Riemann Solvers and Numerical Methods for Fluid Dynamics*, 3rd edition, Springer-Verlag Berling Heidelberg 2009

Advances in Plasma Astrophysics
Proceedings IAU Symposium No. 274, 2010
A. Bonanno, E. de Gouveia Dal Pino, & A. G.Kosovichev, eds.

© International Astronomical Union 2011
doi:10.1017/S1743921311007563

Dynamo in the Intra-Cluster Medium: Simulation of CGL-MHD Turbulent Dynamo

R. Santos-Lima[1], E. M. de Gouveia Dal Pino[1], A. Lazarian[2], G. Kowal[1], and D. Falceta-Gonçalves[3]

[1] IAG, Universidade de São Paulo, Rua do Matão 1226, São Paulo 05508-090, Brazil
email: `rlima@astro.iag.usp.br`

[2] Astronomy Department, University of Wisconsin, Madison, WI, USA

[3] NAC, Universidade Cruzeiro do Sul, Rua Galvão Bueno 868, São Paulo 01506-000, Brazil

Abstract. The standard magnetohydrodynamic (MHD) description of the plasma in the hot, magnetized gas of the intra-cluster (ICM) medium is not adequate because it is weakly collisional. In such collisionless magnetized gas, the microscopic velocity distribution of the particles is not isotropic, giving rise to kinetic effects on the dynamical scales. These kinetic effects could be important in understanding the turbulence as well as the amplification and maintenance of the magnetic fields in the ICM. It is possible to formulate fluid models for collisonless or weakly collisional gas by introducing modifications in the MHD equations. These models are often referred as kinetic MHD (KMHD). Using a KMHD model based on the CGL-closure, which allows the adiabatic evolution of the two components of the pressure tensor (the parallel and perpendicular components with respect to the local magnetic field), we performed 3D numerical simulations of forced turbulence in order to study the amplification of an initially weak seed magnetic field. We found that the growth rate of the magnetic energy is comparable to that of the ordinary MHD turbulent dynamo, but the magnetic energy saturates in a level smaller than that of the MHD case. We also found that a necessary condition for the dynamo to operate is to impose constraints on the anisotropy of the pressure.

Keywords. magnetic fields, intergalactic medium, plasmas, MHD

---

1. Kinetic MHD description of a weakly collisional plasma

The hypotheses underlying the MHD description of a plasma are not justified when the level of *collisionality* of the gas is low. Hence, a pure MHD description of the hot, magnetized gas of the ICM medium is not adequate because it is weakly collisional. There, the typical ion Larmour radius ρ_i is much smaller that the mean free path λ_i. For instance, in the Hydra A cluster, $\rho_i \sim 10^5$ km and $\lambda_i \sim 10^{15}$ km (based on data presented in Enßlin & Vogt 2006).

Nevertheless, it is still possible to formulate fluid models for collisionless or weakly collisional magnetized gas by introducing modifications in the MHD equations. The basic equations of these models are the usual ideal MHD equations for conservation of mass, momentum and the induction equation, with a pressure tensor allowing for distinct parallel and perpendicular components to the magnetic field:

$$P_{ij} = p_\perp \delta_{ij} + (p_\parallel - p_\perp) b_i b_j \tag{1.1}$$

where the b_i are the components of the unitary vector parallel to the magnetic field. These models are often referred as kinetic MHD (KMHD).

The lowest order closure (no heating conduction) to the set of macroscopic equations is given by (Kulsrud 1983):

$$\frac{d}{dt}\left(\frac{p_\perp}{\rho B}\right) = 0, \qquad \frac{d}{dt}\left(\frac{p_\parallel B^2}{\rho^3}\right) = 0 \qquad (1.2)$$

The resulting model is called CGL-MHD approximation (Chew *et al.* 1956). These equations of state ensure the conservation of the magnetic momentum of the particles and conservation of entropy of the gas.

The kinetic MHD models reveal linear instabilities originating from the destabilization of the MHD-analogous waves, when the difference between the pressures components reaches some level. The growth rate of these instabilities, in the linear regime, are proportional to the wave number of the perturbation.

We observe the spontaneous development of pressure anisotropy when we perform simulations of forced turbulence using the CGL-MHD model. The instabilities arising from these anisotropies give rise to small-scale structures and therefore, a considerable amount of kinetic and magnetic energy accumulates in the smallest scales of the simulation (limited by the numerical dissipation), changing the usual Kolmogorov power law of the turbulent spectrum in the inertial range. In the absence of numerical dissipation, the CGL-MHD approximation would reveal a serious problem: the growth rate of the highest wavenumbers increases without limits.

2. Anisotropy limits

When the magnetic field changes in a frequency higher than the Larmor frequency of the ions, the assumption of conservation of angular momentum is not reasonable anymore, as pitch angle scattering should occur. Therefore, there is a limit to the maximum growth rate of the kinetic instabilities above which the anisotropy of the pressure is reduced as a result of this pitch angle scattering.

Based on kinetic considerations (see Sharma *et al.* 2006 and references therein), we impose thresholds for the anisotropy of the pressure:

$$1 - \frac{p_\perp}{p_\parallel} - \frac{2}{\beta_\parallel} \lesssim \zeta, \qquad \frac{p_\perp}{p_\parallel} - 1 \lesssim \frac{2\xi}{\beta_\perp} \qquad (2.1)$$

where $\beta_\parallel = p_\parallel/(B^2/8\pi)$ and $\beta_\perp = p_\perp/(B^2/8\pi)$. Following Sharma *et al.* 2006, we use $\zeta = 0.5$ and $\xi = 3.5$.

When the anisotropy overcomes any of these thresholds, the pitch angle scattering acts to reduce the anisotropy back to the threshold. It imposes some limits to the growth rate of the kinetic instabilities.

3. Numerical simulation of the CGL-MHD turbulent dynamo

Using a Godunov-MHD code modified to evolve the CGL-MHD equations (see Kowal *et al.* 2010), we performed 3D numerical simulations of forced turbulence, starting with a weak seed magnetic field. The turbulence is forced in a periodic box by a random, non-helical, solenoidal force acting around a scale 2.5 times smaller than the side of the cubic computational domain. We work with dimensionless variables defined in terms of the background sound speed and the dimensions of the box. In code units, the random velocity is kept close to unity, hence one turn-over time of the turbulence is ≈ 0.4. The initial fields are uniform and have the following values: $\rho = 1$ (density), $p_\parallel = p_\perp = 1$, and $B = 10^{-4}$ (magnetic field). The resolution employed is 64^3.

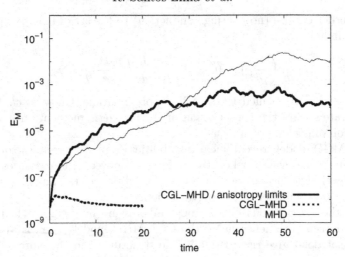

Figure 1. Evolution of the magnetic energy for the CGL-MHD turbulent dynamo. Here we compare two models: with and without the anisotropy limits (see equations 2.1). The ordinary MHD case is showed for comparison. Numerical resolution is 64^3.

Figure 1 shows the evolution of the magnetic energy for two models: one without any constraint on the anisotropy, and another imposing the thresholds of equations (2.1). For comparison, the curve for an ordinary MHD model, employing similar parameters (with an isothermal equation of state) is also shown. For our employed parameters, the growth rate of the magnetic field for the CGL-MHD model with limited anisotropy is similar to the MHD turbulent dynamo. However, the magnetic energy of the CGL-MHD model seems to saturate in a value about two orders of magnitude smaller than in the MHD case. For the CGL-MHD model without limited anisotropy, the dynamo fails and the magnetic energy does not grow at all. We also observe failure of the dynamo in simulations employing a KMHD model with isothermal equations of state for the pressure components, whenever $p_\perp > p_\parallel$. A suppression of the growth of the magnetic energy by the mirror instability was also detected by Sharma *et al.* (2006), in the context of magneto-rotational instability.

4. Conclusions and perspectives

We have shown that the turbulent CGL-MHD dynamo can amplify the magnetic energy at a rate similar to the MHD turbulent dynamo – provided that we constrain the pressure anisotropy – although the saturation of the magnetic field is much smaller. We still have to assess the influence of the numerical resolution on the results above as well as the influence of the threshold values for the anisotropy. Additionally, the structure of the magnetic field in the saturated state for the CGL-MHD model has to be studied and carefully compared with observations of the ICM.

References

Chew, G. F., Goldberger, M. L., & Low, F. E. 1956, Royal Society of London Proceedings Series A, 236, 112

Enßlin, T. A. & Vogt, C. 2006, *A&A*, 453, 447

Kowal, G., Falceta-Goncalves, D. A., & Lazarian, A. 2010, arXiv:1012.5125

Kulsrud, R. M. 1983, Basic Plasma Physics: Selected Chapters, Handbook of Plasma Physics, Volume 1, 1

Sharma, P., Hammett, G. W., Quataert, E., & Stone, J. M. 2006, *ApJ*, 637, 952

Author Index

Object Index

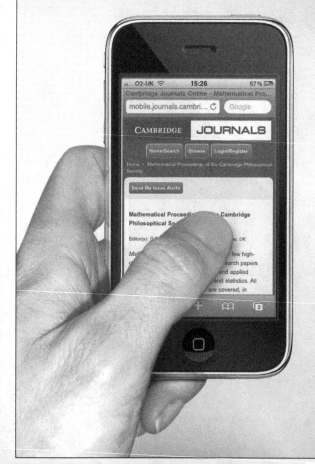

CAMBRIDGE JOURNALS

International Journal of Astrobiology

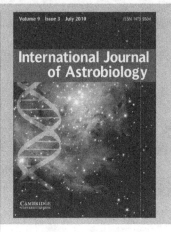

Volume 9 Issue 3 July 2010 ISSN 1473 5504

International Journal of Astrobiology

CAMBRIDGE

Managing Editor
Simon Mitton, University of Cambridge, UK

International Journal of Astrobiology is the peer-reviewed forum for practitioners in this exciting interdisciplinary field. Coverage includes cosmic prebiotic chemistry, planetary evolution, the search for planetary systems and habitable zones, extremophile biology and experimental simulation of extraterrestrial environments, Mars as an abode of life, life detection in our solar system and beyond, the search for extraterrestrial intelligence, the history of the science of astrobiology, as well as societal and educational aspects of astrobiology. Occasionally an issue of the journal is devoted to the keynote plenary research papers from an international meeting. A notable feature of the journal is the global distribution of its authors.

International Journal of Astrobiology
is available online at:
http://journals.cambridge.org/ija

**To subscribe contact
Customer Services**

in Cambridge:
Phone +44 (0)1223 326070
Fax +44 (0)1223 325150
Email journals@cambridge.org

in New York:
Phone +1 (845) 353 7500
Fax +1 (845) 353 4141
Email
subscriptions_newyork@cambridge.org

Price information
is available at: **http://journals.cambridge.org/ija**

Free email alerts
Keep up-to-date with new material – sign up at
http://journals.cambridge.org/ija-alerts

For free online content visit:
http://journals.cambridge.org/ija

CAMBRIDGE
UNIVERSITY PRESS

Printed in the United States
by Baker & Taylor Publisher Services